Abstract Algebra

A STUDENT-FRIENDLY APPROACH

Order	Commutative	Noncommutative
1	$C_1 \cong \mathbb{Z}_1 \cong S_1 \cong A_1 \cong A_2$	
2	$C_2 \cong \mathbb{Z}_2 \cong S_2$	
3	$C_3 \cong \mathbb{Z}_3 \cong A_3$	
4	$C_4 \cong \mathbb{Z}_4 \cong U_4 \cong Dic_4, C_2 \times C_2 \cong V_4$	
5	$C_5 \cong \mathbb{Z}_5$	
6	$C_6 \cong \mathbb{Z}_6 \cong C_2 \times C_3$	$D_3 \cong S_3$
7	$C_7 \cong \mathbb{Z}_7$	
8	$C_8 \cong \mathbb{Z}_8, C_2 \times C_2 \times C_2, C_2 \times C_4$	D_4 (octic group), $Q_8 \cong Dic_8$
9	$C_9 \cong \mathbb{Z}_9, C_3 \times C_3$	
10	$C_{10} \cong \mathbb{Z}_{10} \cong C_2 \times C_5$	D_5
11	$C_{11} \cong \mathbb{Z}_{11}$	
12	$C_{12} \cong \mathbb{Z}_{12} \cong C_4 \times C_3, C_2 \times C_2 \times C_3$	$D_6, A_4 \cong$ tetrahedral group, Dic_{12}

Laura L. Dos Reis
Anthony J. Dos Reis

Abstract Algebra: A Student-Friendly Approach edition 1
Copyright © 2017 by Laura L. Dos Reis, all rights reserved.

ISBN-13: 978-1539436072

ISBN-10: 1539436071

To Mom

Table of Contents (with page numbers)

Preface, *x*
Notation, *xi*
How to Use This Book, *xii*

1   **Sets**, 1
       What is a Set, 1
       Specifying Sets, 2
       Subsets and Supersets, 5
       Power Set, 6
       Review Questions, 8
       Answers to the Review Questions, 8
       Homework Questions, 8

2   **Mathematical Statements**, 9
       Definition of a Mathematical Statement, 9
       Negating Statements, 9
       Predicates, 10
       Negating Predicates, 10
       Universal and Existential Quantifiers, 11
       Negating Quantified Statements, 13
       Multiple Quantifiers, 15
       Review Questions, 17
       Answers to the Review Questions, 17
       Homework Questions, 17

3   **Compound Statements**, 18
       and Logical Operator, 18
       or Logical Operators: Inclusive and Exclusive, 18
       Deriving a Statement from a Truth Table, 19
       Implication, 23
       Biconditional, 24
       Identical Truth Tables, 25
       Definitions that Use Implication, 25
       Negating Compound Statements, 26
       Tautologies and Absurdities, 28
       Converse and Contrapositive, 28
       Review Questions, 29
       Answers to the Review Questions, 30
       Homework Questions, 30

4   **Proof Techniques**, 31
       Direct Proof, 31
       Proof by Contradiction, 34
       Contrapositive Proof, 35
       Proving a Biconditional, 35
       Proving a Disjunction, 36
       Using the Well-Ordering Principle, 37
       Proof by Induction, 39
       Proof by Cases, 41
       Without Loss of Generality, 42
       Pigeonhole Principle, 42
       Review Questions, 43
       Answers to the Review Questions, 43
       Homework Questions, 44

5   **Operations on Sets**, 45
       Intersection, Union, and Complement, 45

Venn Diagrams, 45
Pairwise Disjoint Sets, 46
Proving the Subset Relation, 47
Proving Two Sets are Equal, 48
More Laws on Sets, 49
Principle of Inclusion-Exclusion, 50
Review Questions, 50
Answers to the Review Questions, 51
Homework Questions, 51

6 Relations and Functions, 52
Relations, 52
Functions, 55
Special Categories of Functions, 57
Proving a Function is Onto, 59
Proving a Function is One-to-One, 60
Function Composition, 60
Inverse of a Function, 62
Representing Functions with Tables, 65
Bijections in Table Form, 66
Cardinality, 68
Review Questions, 71
Answers to the Review Questions, 71
Homework Questions, 71

7 Binary Operations, 73
Definition of a Binary Operation, 73
Properties of Binary Operations, 75
Identity, 76
Inverses, 77
What a Cayley Table Can Tell Us, 79
Review Questions, 82
Answers to the Review Questions, 82
Homework Questions, 83

8 Introduction to Groups, 84
Definition of a Group, 84
Multiplicative and Additive Notation, 85
Uniqueness of the Identity and Inverses in a Group, 85
Cancellation Law for Groups, 87
Solving Equations, 89
Properties of Cayley Tables for Groups, 90
Other Properties of a Group, 91
Review Questions, 92
Answers to the Review Questions, 92
Homework Questions, 93

9 Symmetric Group of Degree n, 94
An Unusual Group, 94
Cycle Representation of a Permutation, 95
Decomposing Permutations into Transpositions, 100
Order of an Element in a Group, 104
Review Questions, 106
Answers to the Review Questions, 106
Homework Questions, 106

10 Divisibility Properties of the Integers, 107
Division Algorithm, 107
Some Divisibility Proofs, 109
Greatest Common Divisor, 111

Prime Numbers, 113
Euclid's Lemma, 114
Fundamental Theorem of Arithmetic, 115
Irrationality of $\sqrt{2}$, 116
Factorization of Squares into Primes, 117
Trailing Zeros in $n!$, 118
Relatively Prime Numbers, 119
$U(n)$, 119
Least Common Multiple, 120
Determining gcd(m, n) and lcm(m, n), 121
Finding Multipliers, 122
Review Questions, 123
Answers to the Review Questions, 123
Homework Questions, 124

11 Equivalence Relations, 125
Review of Relations, 125
Special Types of Relation, 125
Partitions, 129
Quotient Structures, 130
Review Questions, 131
Answers to the Review Questions, 131
Homework Questions, 132

12 Congruence, 133
Basics, 133
Thinking about Congruence Modulo n, 134
Congruence Classes, 136
Replacement Rules, 137
Cancellation Laws, 138
Quotient Structure under Congruence Modulo n, 139
Quotient Group of Congruence Classes, 140
Modular Arithmetic, 142
Group of Integers Modulo n, 144
Isomorphic Groups, 144
Review Questions, 145
Answers to the Review Questions, 146
Homework Questions, 146

13 Symmetries of a Regular Polygon, 147
Symmetries of an Equilateral Triangle, 147
Groups of Order 3, 2, and 1, 151
Generating Sets, 153
Constructing the Graph of a Group, 153
Getting the Cayley Table from the Graph, 154
Dihedral Groups, 156
D_3 and S_3, 161
Groups of Orders 1 to 7, 162
Review Questions, 163
Answers to the Review Questions, 163
Homework Questions, 163

14 Subgroups, 165
Basics, 165
CI Subgroup Test, 167
Subgroup Generated by a Subset X, 168
CF Subgroup Test, 169
One-Step Subgroup Test, 170
Review Questions, 171
Answers to the Review Questions, 171

Homework Questions, 172

15 Isomorphic Groups and Cayley's Theorem, 173
    How to Show Two Groups Are Isomorphic, 173
    Cayley's Theorem, 175
    Review Questions, 176
    Answers to the Review Questions, 177
    Homework Questions, 177

16 Cyclic Groups, 178
    Definition of a Cyclic Group, 178
    Characterizing Cyclic Groups, 178
    Cyclic Subgroups of a Group, 181
    Isomorphism of Cyclic Groups of the Same Order, 181
    Order of r^k in C_n, 182
    Subgroups of Cyclic Groups, 184
    Order of a Subgroup of a Cyclic Group, 186
    Order of a Subgroup of a Finite Group, 187
    Review Questions, 187
    Answers to the Review Questions, 188
    Homework Questions, 188

17 Coset Decomposition and Lagrange's Theorem, 189
    Properties of Cosets, 189
    Normal Subgroups, 194
    Interesting Consequence of Lagrange's Theorem, 196
    Review Questions, 197
    Answers to the Review Questions, 197
    Homework Questions, 198

18 Quotient Groups, 199
    Quasi-Commutative Law for Normal Subgroups, 199
    Operation on Cosets, 200
    Graph of a Quotient Group, 203
    Review Questions, 204
    Answers to the Review Questions, 204
    Homework Questions, 205

19 Group Homomorphisms, 206
    Definition of a Group Homomorphism, 206
    Properties of Group Homomorphisms, 207
    Using Logarithms to Compute a Product, 209
    Fundamental Theorem of Group Homomorphisms, 210
    Test Your Mettle, 214
    Review Questions, 215
    Answers to the Review Questions, 215
    Homework Questions, 216

20 Some More Groups, 217
    $<\mathbb{Z}_n-\{0\}, \cdot_n>$ for Prime n, 217
    Direct Products, 219
    Order of an Element in a Direct Product, 220
    Fundamental Theorem of Finite Abelian Groups, 222
    Subgroups of Direct Products, 223
    Graphs of Direct Products of Cyclic Groups, 224
    Symmetries of Regular Polyhedra, 225
    Quaternion Group, 230
    Generators and Relations, 231
    Dicyclic Groups, 232
    Groups of Order 1 to 12, 234

Review Questions, 235
Answers to the Review Questions, 235
Homework Questions, 235

21 Introduction to Rings, 236
Semigroups, 236
Definition of a Ring, 237
Basic Properties of a Ring, 238
Generalized Distributive Laws, 240
Characteristic of a Ring, 241
Multiplying a Sum of 1's by a Sum of 1's, 243
Trivial Ring, 244
Ring Isomorphisms, 245
Divisors of Zero, 246
Units Cannot Be Divisors of Zero, 247
Cancellation Law for Multiplication in a Ring, 247
Direct Sum of Rings, 249
An Interesting Family of Rings, 250
Caveat, 253
Some Specializations of Rings, 254
Review Questions, 254
Answers to the Review Questions, 254
Homework Questions, 255

22 Subrings and Quotient Rings, 256
Definition of a Subring, 256
Subring Test, 256
Properties of Subrings, 257
Coset Decomposition of a Ring, 258
Absorption Rule for Cosets, 259
Ideals, 260
Principal Ideals, 262
Quotient Rings, 263
Review Questions, 265
Answers to the Review Questions, 266
Homework Questions, 266

23 Ring Homomorphisms, 267
Definition of a Ring Homomorphism, 267
Kernel of a Ring Homomorphism, 267
Fundamental Theorem of Ring Homomorphisms, 268
Review Questions, 270
Answers to the Review Questions, 270
Homework Questions, 270

24 Integral Domains and Fields, 271
Definition of an Integral Domain, 271
Cancellation Law for Integral Domains, 273
Characteristic of an Integral Domain, 273
Fields, 274
A Finite Integral Domain is a Field, 275
Units in Ideals, 276
Prime Ideals, 276
Maximal Ideals, 279
Field of Quotients of an Integral Domain, 281
Ring Extensions, 284
Review Questions, 284
Answers to the Review Questions, 285
Homework Questions, 285

25 Polynomials Part 1, 286
  A Different Way of Viewing Polynomials, 286
  Addition and Multiplication of Polynomials, 287
  Determining the Degree of a Product and of a Sum, 289
  Properties of R that Carry Over to $R[x]$, 290
  Powers-of-x Representation, 291
  Adjoining x to a Ring, 293
  Division of Polynomials, 294
  Roots of a Polynomial, 298
  Review Questions, 300
  Answers to the Review Questions, 300
  Homework Questions, 301

26 Polynomials Part 2, 302
  Principal Ideals of $F[x]$, 302
  Ideal Generated by a Nonzero Constant Polynomial, 304
  $F[x]$ is a Principal Ideal Domain, 304
  Cosets of $<p(x)>$, 305
  Irreducible Polynomials, 307
  Monic Polynomials, 309
  Enumerating Irreducible Monic Polynomials over $\mathbb{Z}_p$, 310
  Irreducible Polynomials Are Maximal Ideals, 312
  Constructing Fields Using Irreducible Polynomials, 313
  Polynomials in Place of Cosets as Field Elements, 313
  Fundamental Theorem of Field Theory, 317
  Review Questions, 318
  Answers to the Review Questions, 319
  Homework Questions, 319

27 Vector Spaces, 320
  Definition of a Vector Space, 320
  Examples of Vector Spaces, 322
  Field Over a Subfield, 323
  Vector Space of Polynomials, 324
  Subspaces, 324
  Linear Combination of Vectors, 325
  Linear Independence, 325
  Basis of a Vector Space, 326
  Review Questions, 327
  Answers to the Review Questions, 327
  Homework Questions, 328

28 Partial Orders, Lattices, and Boolean Algebra, 329
  Partial Orders, 329
  Lattices, 330
  Boolean Algebra, 336
  Review Questions, 338
  Answers to the Review Questions, 338
  Homework Questions, 338

Final Exam, 339
Answers to the Final Exam, 343
Index, 348

Preface

As a kid, I would occasionally read a comic strip that depicted a winter scene. Invariably, snow would be falling from the sky, with each snowflake in the shape of an intricate geometric pattern. Of course, real snowflakes do not look like this, or so I thought. When I saw real snow falling from the sky, it appeared to consist of amorphous clumps of white stuff. However, one day I examined one of these clumps closely. To my astonishment, each clump consisted of hundreds of tiny flakes, each with its own unique, intricate, and beautiful pattern. What appeared amorphous had, on closer examination, an incredible structure. As you read this book, you will have a similar experience with the algebraic structures you will study. At first, you will see them as having only a few gross features. But with further examination, you will find a truly amazing structure underlying those gross features.

Learning abstract algebra is not hard. It is like getting to know the deep forest—its trails, streams, lakes, flora, and fauna. It takes time, effort, and a willingness to venture into new territory. It is a task that cannot be done overnight. But with a good guide (this book!), it should be an exciting excursion with, perhaps, only a few bumps along the way.

Students—even students who have done very well in calculus—often have trouble with abstract algebra. Our objective in writing this book is to make abstract algebra as accessible as elementary calculus and, we hope, a real joy to study. Our textbook has three advantages over the standard abstract algebra textbook. First, it covers all the foundational concepts needed for abstract algebra: sets, mathematical statements, relations, functions, permutations, logic, proof techniques, binary operations, equivalence relations, divisibility properties of the integers, and congruence. The only prerequisite for the book is high school algebra. Second, it is easier to read and understand (so it is ideal for self-learners). Third, it gets the reader to think mathematically and to do mathematics—to experiment, make conjectures, and prove theorems—while reading the book. The result is not only a better learning experience but also a more enjoyable one. *Abtract Algebra: A Student-Friendly Approach* is definitely the go-to book if you are having any trouble understanding your professor or your required textbook.

This book can be used as the principal textbook in an abstract algebra course, or as a supplement for such a course. It covers all the standard topics—groups, rings, integral domains, and fields—in an introductory abstract algebra course, as well as vector spaces, lattices, and Boolean algebra. An answer key to all the homework problems is available to instructors using this textbook in their classes. We welcome any feedback you might have. Please send your comments and suggestions to the email address dosreist@newpaltz.edu.

I would like to express my gratitude to Thom Achatz and Douglas Brozovic for their helpful reviews and to Professor Mihai Stoiciu at Williams College, who taught me how to think like a mathematician.

Laura L. Dos Reis
Boston University School of Medicine

When I was in seventh grade at Albert Leonard Junior High in New Rochelle, New York, my math teacher, Mr. Grossman, occasionally made comments that I did not understand but piqued my interest. For example, he asked, "How do we know that the solution to this equation is unique?" He never expected the answer from us, nor did he tell us the answer. The underlying math was too advanced for us seventh graders. But he made me think about math. Seven years later, I am an electrical engineering student at Cornell University, where I have learned a great deal of math. But it is all manipulative math: no-proof calculus, matrix algebra, and differential equations. Wanting a broader exposure to mathematics, I borrowed a delightful book on group theory titled *Groups and Their Graphs* coauthored by none other than Mr. Grossman, my math teacher in seventh grade. In it, Mr. Grossman answers all those intriguing questions he asked when I was in his class. For introducing me to the mystery and beauty of abstract mathematics, I would especially like to thank Mr. Grossman. I would also like to thank David Hobby and Donald Silberger at SUNY New Paltz, whose insights on algebra have been invaluable to me during the preparation of this book, and Oday Sawaqed for proofreading the manuscript. Last but not least, I would like to thank my coauthor, Laura, who conceived of and initiated this project and invited me to be her coauthor.

Anthony J. Dos Reis
SUNY New Paltz

Notation (with page numbers)

$a \in A$	a is an element of the set A, 1	$[k]$	equivalence class containing k, 128
$\emptyset$	empty set, 2	$a \equiv b \pmod{n}$	a is congruent to b modulo n, 133
$\mathbb{Z}$	set of integers, 3	$+_n$	addition modulo n, 142
$\mathbb{Q}$	set of rational numbers, 3	$\mathbb{Z}_n$	set of integers modulo n: $\{0, 1, ..., n-1\}$, 144
$\mathbb{R}$	set of real numbers, 3	$\cdot_n$	multiplication modulo n, 145
$\mathbb{Z}^+$	set of positive integers, 3	$G_1 \cong G_2$	G_1 is isomorphic to G_2, 145
$\mathbb{Q}^+$	set of positive rationals, 4	U_4	fourth roots of unity, 145
$\mathbb{R}^+$	set of positive reals, 4	r	rotation symmetry, 148
$\mathbb{Z}^-$	set of negative integers, 4	C_n	cyclic rotation group of order n, 150
$\mathbb{Q}^-$	set of negative rationals, 4	f	flip symmetry, 156
$\mathbb{R}^-$	set of negative reals, 4	D_n	dihedral group of order $2n$, 160
$\mathcal{U}$	universal set, 4	$<a>$	cyclic group generated by a, 168
$A \subseteq B$	A is a subset of B, 5	$Z(G)$	center of G, 171
$A \subset B$	A is a proper subset of B, 6	$C(g)$	centralizer of g, 172
$\cdot$	Multiplication on $\mathbb{Z}, \mathbb{Q}, \mathbb{R}$, 6	G/H	set of cosets of H of a group G, 202
$\mathcal{P}(A)$	power set of A, 7	$\ker \varphi$	kernel of the homomorphism φ, 210
$\sim p$	negation of the statement p, 9	FTGH	fundamental thm of grp homomorphisms, 214
$\forall$	"for all" quantifier, 12	Q_8	quaternion group, 231
$\exists$	"there exists" quantifier, 12	Dic_{4n}	dicyclic group of order $4n$, 232
$\|x\|$	absolute value of x, 13	$<R, +, \cdot>$	ring under the operations $+$ and $\cdot$, 237
$\wedge$	"and" logical operator, 19	$R_1 \oplus R_2$	direct sum of rings R_1 and R_1, 249
$\vee$	"or" logical operator, 19	$<a>$	ideal generated by a, 262
$\underline{\vee}$	exclusive "or" logical operator, 19	$GF(p^n)$	Galois field of order p^n, 274
$\Rightarrow$	"implies" logical operator, 23	$\deg p(x)$	degree of polynomial $p(x)$, 281
$\Leftrightarrow$	"if and only if" logical operator, 24	$R[x]$	polynomials over R with indeterminate x, 291
∎	end-of-proof marker, 31	$+$	join, 332
$a \mid b$	a divides b, 33	$\cdot$	meet, 332
$\cup$	set union, 45		
$\cap$	set intersection, 45		
A^c	complement of the set A, 45		
$\|A\|$	number of elements in the set A, 45		
$A - B$	$\{x : (x \in A) \wedge (x \notin B)\}$, 46		
$A \triangle B$	symmetric difference: $(A \cup B) - (A \cap B)$, 50		
$A \times B$	Cartesian product: $\{(x, y) : x \in A \text{ and } y \in B\}$, 52		
R^{-1}	inverse of the relation R, 54		
$\circ$	function composition operator, 61		
$*$	generic binary operation symbol, 73		
$<G, *>$	the group G under the operation $*$, 84		
$\|G\|$	order of a group, 85		
a^{-1}	multiplicative inverse of the a, 85		
$-a$	additive inverse of a, 85		
V_4	Klein four group, 90		
S_n	symmetric group of degree n, 94		
A_n	alternating group of degree n, 102		
$\|g\|$	order of the element g, 104		
$a \bmod b$	remainder produced when a is divided by b, 109		
$\gcd(a, b)$	greatest common divisor of a and b, 111		
FTOA	fundamental theorem of arithmetic, 115		
$U(n)$	set of elements in $\{0, ..., n-1\}$ relatively prime to n, 119		
$\text{lcm}(a, b)$	least common positive multiple of a and b, 120		

How to Use This Book

Each chapter of the book is divided into numbered frames. Each frame consists of some information followed by a question. The answer to each question appears at the beginning of the frame that follows the question. An example follows:

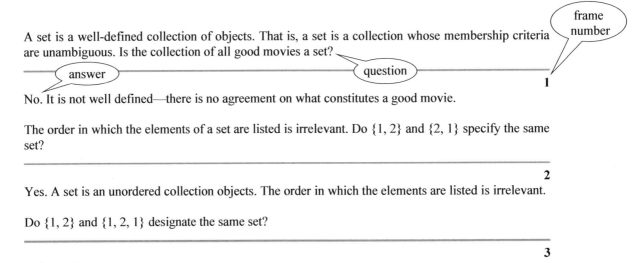

A set is a well-defined collection of objects. That is, a set is a collection whose membership criteria are unambiguous. Is the collection of all good movies a set?

1

No. It is not well defined—there is no agreement on what constitutes a good movie.

The order in which the elements of a set are listed is irrelevant. Do {1, 2} and {2, 1} specify the same set?

2

Yes. A set is an unordered collection objects. The order in which the elements are listed is irrelevant.

Do {1, 2} and {1, 2, 1} designate the same set?

3

References within the chapters use frame numbers rather than page numbers. The summary of notation on the preceding page, the table of contents, and the index use page numbers.

Be sure to think about each question and at least attempt an answer before looking at the answer that is provided. George Bernard Shaw claimed that most people never think, and he had become famous by thinking only once a week. It is not essential that you get the correct answer or even that you get some answer. But *it is essential* that you think about each question and attempt to answer it before looking at the answer provided.

Studies have shown that students who learn new material will forget most of it if they do not review it within 24 hours. Frequent review of new material is an essential part of the learning process. Accordingly, we have included a set of review questions at the end of each chapter. Be sure to do these questions, and check your answers with the answers provided. If you perform poorly on the review questions for some chapter of the book, review that chapter before continuing on to the next chapter.

1 Sets

What is a Set?

1

A *set* is a well-defined collection of zero or more objects. The objects in a set are called the ***elements*** or the ***members*** of the set. The objects in a set can be anything—for example, numbers, planets in the solar system, or students in a class. One way to specify a set is to list its elements and enclose the list with braces. For example, we can specify the set with elements 1, 2, and 3 with {1, 2, 3}.

We will often denote a set with a capital letter. For example, by writing $A = \{1, 2, 3\}$, we are indicating that A denotes the set {1, 2, 3}. To indicate that an object is in a set, we use the symbol $\in$. For example, to indicate that 2 is in the set A, we write $2 \in A$. We can list multiple elements to the left of $\in$ to indicate that they are all in the specified set. For example, to indicate that both 1 and 2 are in A, we write $1, 2 \in A$. We use $\notin$ ($\in$ with a superimposed slash) to indicate that an object is not an element of a set. For example, $4 \notin A$ indicates that 4 is not an element of A.

A set is a well-defined collection of objects. That is, a set is a collection whose membership criteria are unambiguous. Is the collection of all good movies a set?

2

No. It is not well-defined—there is no agreement on what constitutes a good movie.

The order in which the elements of a set are listed is irrelevant. Do {1, 2} and {2, 1} specify the same set?

3

Yes. A set is an unordered collection of objects. The order in which the elements are listed is irrelevant.

Do {1, 2} and {1, 2, 1} designate the same set?

4

Yes. By listing 1 within the braces, we indicate that 1 is in the set. By listing 1 a second time, we are simply indicating a second time that 1 is in the set. Thus, in both cases, 1 and 2 and only 1 and 2 are the elements of the set. Of course, we normally do not list elements more than once when specifying a set.

Does $3 = \{3\}$?

5

No. 3 and {3} are different objects; 3 is an integer, whereas {3} is a set whose one element is the integer 3.

We can think of a set as a box that contains elements. Accordingly, {3} is a box that contains 3; 3 by itself and a box that contains 3 are, of course, different objects. Thus, $3 \neq \{3\}$. Is $3 \in \{3\}$?

6

Yes

Sets are equal if and only if they have the same elements. Let $A = \{1, 2, 3\}$ and $B = \{1, 2, \{3\}\}$. Does $A = B$?

7

No. *A* has three elements: 1, 2, and 3. *B* also has three elements: 1, 2, {3}. But 3 is not the same element as {3}. The former is the integer 3; the latter is the set whose only element is the integer 3.

How many elements are in the set {1, {2, 3, 4, 5}}? List them.

8

Two: 1 and {2, 3, 4, 5}. The second element is itself a set.

Using the box metaphor, describe the set {1, {2, 3, 4, 5}}.

9

It is a box with two elements. One element is 1; the other element is another box that contains 2, 3, 4, and 5.

The set { } is the set with no elements. Using the box metaphor, how would you describe the set { }?

10

It is an empty box.

The set { } is called the ***empty set.*** We denote it with either { } or the symbol ∅. Does { } = ∅?

11

Yes.

Does { } = {0}?

12

No; { } has no elements, whereas {0} has one element (the integer 0).

Does { } = {{ }}?

13

No; { } has no elements, whereas {{ }} has one element (the empty set).

The set { } is like an empty box, but {{ }} is like an empty box within a box. Thus, { } ≠ {{ }}. Rewrite { } and {{ }} using ∅.

14

∅, {∅}

Is ∅ ∈ ∅?

15

No; ∅ has no elements. Thus, $x \notin \emptyset$, no matter what x is.

Is ∅ ∈ {∅}?

16

Yes. {∅} is a set with ∅ as an element. Thus, ∅ ∈ {∅}.

Specifying Sets

17

It becomes difficult to list all the elements of a set if the number of elements is large. Of course, it is impossible to list them all if the set is infinite. However, if the list forms an obvious pattern, we can specify a set without listing all the elements. We list just enough elements to make the pattern obvious, replacing the omitted sequence of elements with an *ellipsis* ("..."). For example, the set of integers from 1 to 1000 is specified by {1, 2, 3, ..., 1000}.

Using an ellipsis, specify the set of all positive integers.

18

{1, 2, 3, ...}

What is wrong with specifying the set of positive even integers with {2, 4, ...}?

19

From just the first two numbers, the pattern is not clear. It could be 2, 4, 6, 8, ..., but it also could be 2, 4, 8, 16, ...

Using two ellipses, specify the set of all integers.

20

{..., −3, −2, −1, 0, 1, 2, 3, ...}

We will designate this set—the set of all integers—with $\mathbb{Z}$ ($\mathbb{Z}$ is from *Zahlen,* the German word for "number"). $\mathbb{Z}$ includes all the *negative integers* (i.e., integers less than zero), zero, and all the *positive integers* (i.e., all the integers greater than zero).

Is zero a positive number, a negative number, or neither?

21

Neither.

The following are some other important sets and their symbolic designations:

- $\mathbb{Z}^+ = \{1, 2, 3, ...\}$, the set of all positive integers. We also call this set the set of *natural numbers.*
- $\mathbb{Q}$ is the set of all rational numbers. A *rational number* is a number that can be represented as the ratio of two integers with the bottom integer nonzero. For example, $\frac{-1}{3}$, $\frac{5}{7}$, $\frac{100}{3}$, and $\frac{5}{1}$ are all in $\mathbb{Q}$. Every integer k can be represented as a ratio with k on top and 1 on the bottom. For example, the integer 3 can be represented as $\frac{3}{1}$. Thus, every integer is in $\mathbb{Q}$.
- $\mathbb{R}$ is the set of all real numbers. A *real number* is any number that corresponds to some point on the number line that extends infinitely in both directions. For example, 3 corresponds to the point on the number line that is three units to the right of the zero point:

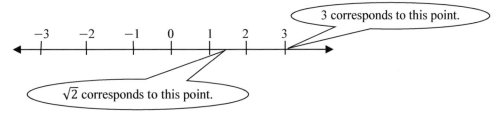

Thus, $3 \in \mathbb{R}$. The number $\sqrt{2}$, the positive square root of 2, corresponds to a point on the number line somewhere between 1 and 2 (it is approximately equal to 1.414). Thus, it, too, is in $\mathbb{R}$. However, $\sqrt{2}$ cannot be expressed as the ratio of two integers We prove this in frame **586**). Thus, $\sqrt{2} \notin \mathbb{Q}$. We call a number, like $\sqrt{2}$, that is in $\mathbb{R}$ but not in $\mathbb{Q}$ an *irrational number.*

Another way to specify a set is to use *set-builder notation.* With set-builder notation, we specify the property required for membership. Every object that has the specified property, and only such objects, are in the set. For example, consider the set

$$\underbrace{\{x : x \in \mathbb{Z} \text{ and } 1 \leq x \leq 100\}}_{\text{required property}}$$

The required property follows the colon. Read the initial segment "$\{x$" as "the set of all x." Read the colon as "such that." Thus, this is the set of all x such that $x \in \mathbb{Z}$ and $1 \leq x \leq 100$. The value 2 satisfies the required property. That is, if we substitute 2 for x in the property, we get the true statement

$$2 \in \mathbb{Z} \text{ and } 1 \leq 2 \leq 100.$$

Thus, 2 is in the set. Similarly, all the integers between 1 and 100 (including 1 and 100) satisfy the property and, therefore, are in the set. However, 2.5 does not satisfy the property (2.5 is between 1 and 100, but it is not in $\mathbb{Z}$). Thus, 2.5 is not in the set.

Is 200 in the set?

22

No; 200 is not less than or equal to 100.

Give a word description of the set $\{x : x \in \mathbb{Z} \text{ and } x < 0\}$.

23

The set of all negative integers. We denote this set with $\mathbb{Z}^-$. $\mathbb{Z}$ is the set of all integers; $\mathbb{Z}^-$ is the set of all negative integers; $\mathbb{Z}^+$ is the set of all positive integers.

What sets are denoted with $\mathbb{Q}^+$, $\mathbb{Q}^-$, $\mathbb{R}^+$, and $\mathbb{R}^-$?

24

$\mathbb{Q}^+$: the positive rationals; $\mathbb{Q}^-$: the negative rationals; $\mathbb{R}^+$: the positive reals; $\mathbb{R}^-$: the negative reals.

If in some discussion we are dealing exclusively with sets whose elements are all drawn from some set $\mathcal{U}$, then we call $\mathcal{U}$ the **universe** or the **universal set**. We often specify the universe when using set builder notation. For example, in the set builder specification of $\mathbb{Z}^-$ in frame **22**, $x \in \mathbb{Z}$ indicates that all the elements in $\mathbb{Z}^-$ come from $\mathbb{Z}$. $\mathbb{Z}$ here is the universal set.

When the specification of a set includes the specification of a universe, the universe can be given either before the colon or after. For example, we can specify $\mathbb{Z}^-$ with either $\{x \in \mathbb{Z} : x < 0\}$ or $\{x : x \in \mathbb{Z} \text{ and } x < 0\}$.

How would you read this set?

25

The set of all x in $\mathbb{Z}$ such that x is less than zero.

Rewrite the following set so that the specification of the universe precedes the colon. Also write the set using the list approach: $\{x : x \in \mathbb{Z} \text{ and } 1 < x \leq 5\}$

26

$\{x \in \mathbb{Z} : 1 < x \leq 5\}$, $\{2, 3, 4, 5\}$

Give a word description of the following set: $\{k \in \mathbb{Z} : k > 0\}$.

27

The set of all positive integers.

We can define this set using x rather than k like so: $\{x \in \mathbb{Z} : x > 0\}$. In these two specifications of $\mathbb{Z}^+$, x and k are simply placeholders that indicate where we plug in an object to determine if that object is in the set. Thus, the name of the placeholder does not affect the set defined.

Let's specify the set E of all even integers using set-builder notation: $E = \{x : x \text{ is an even integer}\}$. This specification relies on the reader knowing what an even integer is. We can get a more useful specification if we specify a property that can be used to distinguish even integers from other objects. An even integer is an integer into which 2 divides evenly. Equivalently, it is a number that can be factored into integers, one of which is 2. For example, $10 = 2 \cdot 5$. Thus, 10 is even. Similarly, $0 = 2 \cdot 0$. Thus, 0 is also even. Even numbers are precisely the integers that are equal to 2 times some integer. Using this property, we can specify the set of even numbers with

$$E = \{x : x = 2k \text{ and } k \in \mathbb{Z}\}$$

We can also specify this set without using the placeholder x, as follows:

$$E = \{2k : k \in \mathbb{Z}\}$$

Here we are saying that E is the set of all integers whose value is given by $2k$, where $k \in \mathbb{Z}$.

The notation $2\mathbb{Z}$ is commonly used to denote the set obtained from $\mathbb{Z}$ by multiplying each element of $\mathbb{Z}$ by 2. That is, $2\mathbb{Z} = \{2k : k \in \mathbb{Z}\}$. Thus, $2\mathbb{Z}$ is the set of _____ integers.

28

even

What set is denoted by $3\mathbb{Z}$?

29

$3\mathbb{Z}$ is the set of integers that are a multiple of 3.

Using set-builder notation, specify the set O of all odd integers. *Hint*: Each odd integer is one more than some even integer.

30

$O = \{x : x = 2k + 1 \text{ and } k \in \mathbb{Z}\}$ or $O = \{2k + 1 : k \in \mathbb{Z}\}$

Subsets and Supersets

31

Let A and B be sets. We say that A is a **subset** of B (denoted by $A \subseteq B$) if and only if every element in A is also in B. For example, suppose $A = \{2, 3\}$ and $B = \{1, 2, 3\}$. Then A is a subset of B. We know this by observing that every element of A (2 and 3) is in B. However, B is not a subset of A (denoted by $B \nsubseteq A$). We know this by observing that 1 is in B, but 1 is not in A. Thus, every element in B is not in A.

We say that B is a **superset** of A (denoted by $B \supseteq A$) if and only if every element in A is in B. B is a superset of A if and only if A is a subset of B. In other words, saying that A is a subset of B is equivalent to saying that B is a superset of A.

Is $\emptyset \subseteq A$, where $A = \{2, 3\}$?

32

Yes. This statement is false only if there is an element in the empty set $\emptyset$ that is not in A. But there are no elements in $\emptyset$. Thus, the statement $\emptyset \subseteq A$ must be true.

Note that our argument that $\emptyset$ is a subset of A does not depend on A. Thus, $\emptyset$ is the subset of *every* set. We can state this in the following way: $\emptyset \subseteq A$ for every set A. Here A represents any set—not some specific set.

Is $A \subseteq A$ for every set A?

33

Yes. Every element in some set A obviously is an element in A. Thus, every set A is a subset of itself.

Is $\{1\} \subseteq \{\{1\}\}$?

34

No. If this were true, then 1 (the element in $\{1\}$) would be an element of $\{\{1\}\}$. But the only element in $\{\{1\}\}$ is $\{1\}$. 1 and $\{1\}$ are different objects.

Is $\{1\} \in \{\{1\}\}$?

Yes

Let $A = \{1, \{1\}\}$. Which of the following statements are true: $\emptyset \in A$, $1 \in A$, $\{1\} \in A$, $\emptyset \subseteq A$, $1 \subseteq A$, $\{1\} \subseteq A$, $\{\{1\}\} \subseteq A$ $\{1, \{1\}\} \subseteq A$

$\emptyset \in A$ is false, $1 \in A$ is true, $\{1\} \in A$ is true, $\emptyset \subseteq A$ is true, $1 \subseteq A$ is false, $\{1\} \subseteq A$ is true, $\{\{1\}\} \subseteq A$ is true, $\{1, \{1\}\} \subseteq A$ is true.

A is a ***proper subset*** of B (denoted by $A \subset B$) if $A \subseteq B$ and $A \neq B$, in which case there is at least one element in B that is not in A. For example, if $A = \{1, 2\}$ and $B = \{1, 2, 3\}$, then A is a proper subset of B.

Give an example of sets A and B such that both $A \subseteq B$ and $A = B$.

$A = \{1\}, B = \{1\}$

Give an example of sets A and B such that both $A \subseteq B$ and $A \neq B$.

$A = \{1\}, B = \{1, 2\}$

Is $\emptyset \subset B$, where $B = \{1, 2\}$?

Yes. $\emptyset$ is a subset of B. Moreover, 1 is an element in B but not an element in $\emptyset$. Thus, $\emptyset$ is a proper subset of B.

Is $\emptyset$ a proper subset of A for every set A?

No; $\emptyset$ is a proper subset of A for any set A, except for $\emptyset$. No set is a proper subset of itself.

Power Set

List all the subsets of $\{1, 2\}$.

$\emptyset, \{1\}, \{2\}, \{1, 2\}$. There are four subsets.

List all the subsets of $\{1, 2, 3\}$.

$\emptyset, \{1\}, \{2\}, \{3\}, \{1, 2\}, \{1, 3\}, \{2, 3\}, \{1, 2, 3\}$. There are eight subsets.

One way to determine the number of subsets a finite set can have is to determine the number of ways a subset can be constructed from the elements of the given set. For example, let's construct a subset from the set $\{1, 2, 3\}$. We can put 1 in the subset we are building or not put it in. So we have two possibilities for the element 1, namely, in or not in. Similarly, we have two possibilities for the element 2 (in or not in) and two possibilities for the element 3 (in or not in). Thus, there are $2 \cdot 2 \cdot 2 = 2^3 = 8$ ways we can construct a subset from $\{1, 2, 3\}$, each yielding a distinct subset (note that we use "·" to represent multiplication). Every subset of $\{1, 2, 3\}$ can be constructed in this way. Thus, there are exactly eight (i.e., eight and only eight) subsets of $\{1, 2, 3\}$.

How many subsets does $\{1, 2, 3, 4\}$ have?

44

$2 \cdot 2 \cdot 2 \cdot 2 = 2^4 = 16$

Suppose we have a set with *n* elements. Each element can either be in or not in a subset. How many subsets does the set have?

45

2^n

Let's denote the number of elements in a set A with $|A|$. Using $|A|$, express the number of subsets of a finite set

46

$2^{|A|}$

We call the set of all the subsets of a set A the ***power set*** of A. For example, suppose $A = \{1, 2\}$. A has the following subsets: $\emptyset, \{1\}, \{2\}, \{1, 2\}$. The power set of A is the set of all these subsets. That is, it is the set $\{\emptyset, \{1\}, \{2\}, \{1, 2\}\}$.

List all the subsets of $\{1\}$.

47

$\emptyset, \{1\}$

The set of all these sets is the power set of $\{1\}$. What is the power set of $\{1\}$?

48

$\{\emptyset, \{1\}\}$

List all the subsets of $\emptyset$.

49

$\emptyset$

The set of all these sets is the power set of $\emptyset$. What is the power set of $\emptyset$?

50

$\{\emptyset\}$

The power set of A is denoted by $\mathcal{P}(A)$. For example, if $A = \{1, 2\}$, then $\mathcal{P}(A) = \{\emptyset, \{1\}, \{2\}, \{1, 2\}\}$. $\mathcal{P}(A)$, the power set of A, is defined as $\{X : X \subseteq A\}$. That is, $\mathcal{P}(A)$ is the set of all subsets X of A.

What is $\mathcal{P}(A)$ if $A = \{1, \{2\}\}$? *Hint*: A has two elements so there are 2^2 subsets. List all the subsets of A, separating each from the next with a comma. Then enclose your list with braces. Your list should include $\emptyset$ and A itself (both are subsets of A). 1 is an element of A so $\{1\}$ is a subset of A, and, therefore, should be on your list. Similarly, $\{2\}$ is an element of A so $\{\{2\}\}$ is a subset of A that should be on your list.

51

$\{\emptyset, \{1\}, \{\{2\}\}, \{1, \{2\}\}\}$

Suppose A is finite. Using $|A|$, give an expression for $|\mathcal{P}(A)|$.

52

$|\mathcal{P}(A)|$ is the number of subsets of A. From frame **46**, we know that the number of subsets of a finite set A is $2^{|A|}$; thus, $|\mathcal{P}(A)| = 2^{|A|}$.

Review Questions

1. How many elements are in the following sets: $\emptyset$, $\{\emptyset\}$, $\{\emptyset, \{\emptyset\}\}$, $\{\emptyset, \{\emptyset, \{\emptyset\}\}\}$?
2. Specify the set of all nonzero integers using the set-builder notation.
3. How many subsets does $\{a\}$ have?
4. What is the power set of $\{a\}$?
5. What is the power set of $\{\emptyset, \{\emptyset\}\}$?
6. Specify the set of integers that are both a multiple of 2 and a multiple of 3.
7. List all proper subsets of $\{a, b, c\}$.
8. Which of the following statements are true: $a \in \{a\}$, $a \subseteq \{a\}$, $\{a\} \subseteq \{a, b\}$, $\{a\} \in \{a, b\}$?
9. Specify the set that consists of all real numbers, except for the real numbers between 2 and 3 inclusive.
10. Describe $\{\{\{\{5\}\}\}\}$ using the box analogy.

Answers to the Review Questions

1. $\emptyset$: 0, $\{\emptyset\}$: 1, $\{\emptyset, \{\emptyset\}\}$: 2, $\{\emptyset, \{\emptyset, \{\emptyset\}\}\}$: 2
2. $\{x \in \mathbb{Z} : x \neq 0\}$
3. 2
4. $\{\emptyset, \{a\}\}$
5. $\{\emptyset, \{\emptyset\}, \{\{\emptyset\}\}, \{\emptyset, \{\emptyset\}\}\}$
6. $\{x \in \mathbb{Z} : x = 2k \text{ and } x = 3j \text{ for some } k, j \in \mathbb{Z}\}$
7. $\emptyset, \{a\}, \{b\}, \{c\}, \{a, b\}, \{a, c\}, \{b, c\}$
8. $a \in \{a\}$ is true, $a \subseteq \{a\}$ is false, $\{a\} \subseteq \{a, b\}$ is true, $\{a\} \in \{a, b\}$ is false.
9. $\{x \in \mathbb{R} : x < 2 \text{ or } x > 3\}$
10. 5 in a box in a box in a box in a box

Homework Questions

1. What is the power set of $\{a, \{b, \{c\}\}\}$?
2. Using the list approach, specify the set of all odd integers.
3. What is the power set of $\{\ \}$?
4. Does the set $A = \{1, 2\}$ have itself as an element?
5. Let $X = \{S : S \text{ is a set and } S \notin S\}$. Is $\{1, 2\} \in X$?
6. Is $X \in X$, where X is as defined in homework question 5?
7. Is $5.1 \in \mathbb{Z}$? Is $5.1 \in \mathbb{Q}$? Is $5.1 \in \mathbb{R}$?
8. Suppose $A = \{1, 2\}$ and $B = \{2, 3, 4\}$. Specify the following sets using the list approach: $\{x : x \in A \text{ and } x \in B\}$, $\{x : x \in A \text{ and } x \notin B\}$.
9. List the elements in this set: $\{n \in \mathbb{Z} : n^2 = 9\}$.
10. Does $\{\emptyset, 1\} = \{1\}$?

2 Mathematical Statements

Definition of a Mathematical Statement

53

A *statement* or *proposition* in the mathematical sense is an assertion that is either true or false. For example, "2 + 2 = 4" is a true statement; "2 + 2 = 5" is a false statement.

Which of the following are statements?

1. 2 is an even integer.
2. 2 > 5
3. Study your math.
4. What does 2 + 2 equal?

54

Choices 1 and 2 are statements (1 is true, 2 is false). Choice 3 is a command. Choice 4 is a question. Neither choice 3 nor choice 4 has a true or false value. Thus, they are not statements.

Negating Statements

55

Mathematical reasoning often requires the negation of statements. For example, a proof by contradiction requires the negation of a statement (we discuss proof by contradiction in chapter 4). So it is important to be able to negate statements.

The *complement* of a truth value is the "opposite" truth value. Thus, the complement of true is false; the complement of false is true. Let p be a statement. Then a *negation* of p (denoted by $\sim p$) is a modification of p that results in a new statement whose truth value is the complement of the truth value of p. Thus, if p is true, then $\sim p$ is false; if p is false, then $\sim p$ is true. For simple statements, the negation is obtained by the insertion or removal of the word "not" or by an equivalent action.

What is the negation of "2 + 2 is equal to 4"?

56

2 + 2 is not equal to 4.

What is the negation of "2 + 2 is not equal to 4"?

57

2 + 2 is equal to 4.

Statements that use a mathematical symbol are often negated simply by slashing the symbol, or by removing the slash if one is present. What is the negation of $2 \in A$?

58

$2 \notin A$

Predicates

59

Is the following a statement: 2 is less than 5?

60

Yes. It has a true or false value (its value is true).

Now consider the following sentence:

$$x \text{ is less than } 5.$$

Because x here does not have a specific value, this statement is neither true nor false. Thus, it is not a statement. However, if we replace x with an integer, then we get a statement. For example, if we replace x with 2, we get the true statement

$$2 \text{ is less than } 5.$$

If, alternatively, we replace x with 10, we get the false statement

$$10 \text{ is less than } 5.$$

Because x can be replaced with any one of many values, we call it a ***variable.*** These statement-like assertions that contain one or more variables are called ***predicates.***

If the same variable appears more than once in a predicate, then whatever value replaces one of the instances of the variable must also replace all the other instances of that variable. For example, in the predicate

$$x + x = 10,$$

if we replace the first x with 3, then we must also replace the second x with the same value, to get the statement

$$3 + 3 = 10.$$

We cannot replace the first x with 3 and the second x with some other value, such as 5. However, if a predicate has distinct variables, then distinct variables can be replaced independently. For example, in the predicate

$$x + y = 10,$$

we can replace x with 3 and y with 5, to get the statement

$$3 + 5 = 10.$$

Is $\{2\} \subseteq \mathbb{Z}$ a statement or a predicate?

61

It is a statement, because $\mathbb{Z}$ is not a variable. $\mathbb{Z}$ denotes a specific set (the set of all integers).

Is $X \subseteq \mathbb{Z}$ a statement or a predicate, where X does not designate a specific set?

62

It is a predicate (X here is a variable—it does not designate a specific set).

Negating Predicates

63

We can negate predicates as well as statements. The negation of a given predicate is the predicate that has the opposite truth value as the given predicate for all possible values the variables can assume. For example, the negation of the predicate is

x is equal to 5.

x is not equal to 5.

No matter what we replace x with, the resulting statements have complementary truth values. Thus, each predicate is the negation of the other.

Predicates that use a mathematical symbol are often negated simply by slashing the symbol, or by removing the slash if one is present. What is the negation of $x = 5$?

64

$x \neq 5$

We simply replace "=" with "≠". The symbol "=" means "is equal to." The symbol "≠" means "is not equal to." This modification, in effect, inserts "not."

Negate $A \subseteq B$.

65

$A \nsubseteq B$

What is the negation of $x < y$?

66

$x \not< y$, or, equivalently, $x \geq y$.

Universal and Existential Quantifiers

67

One way to convert a predicate into a statement is to replace each variable with an element from some set. For example, $x < 5$ is a predicate. By replacing x with an element from $\mathbb{Z}$, we get a statement. For example, if we replace x with 2, we get the statement $2 < 5$.

Another way of converting a predicate into a statement is by *quantifying* the variable (or variables, if there is more than one). We quantify a variable using either the phrase "for all" or the phrase "there exists," or their equivalents. For example, let's quantify the variable x in the predicate $x < 5$ with "for all." Let's also indicate that the replacements of x are restricted to the integers by writing $x \in \mathbb{Z}$. We get the statement

For all $x \in \mathbb{Z}$, $x < 5$.

Read this statement as "for all x in $\mathbb{Z}$, x is less than 5." This statement is true if every allowable replacement for x converts the predicate, $x < 5$, into a true statement; otherwise, the statement is false.

Is the quantified statement above true?

68

No. Replacing x with any integer less than 5 converts the predicate into a true statement. However, replacing x with any integer greater than or equal to 5 converts the predicate into a false statement. Because the predicate is not converted into a true statement for *all* allowable replacements of x, it is false.

What does $x \in \mathbb{Z}$ indicate in the statement in the preceding frame?

69

It indicates that x can be replaced by any element of $\mathbb{Z}$.

We call the set of allowable replacements in a quantified statement the *universe* or the *domain of quantification.* Thus, in the quantified statement in frame **67**, $\mathbb{Z}$ is the universe. If the universe is clear from context, we can omit its specification when quantifying variables. For example, if it is clear that the universe is $\mathbb{Z}$, we can omit $x \in \mathbb{Z}$ in the quantified statement in frame **67** to get

$$\text{For all } x, x < 5.$$

Let's assume in the examples that follow that the universe is $\mathbb{Z}$, unless a different universe is explicitly specified.

Is the following statement true: For all x, $x < x + 1$.

70

Yes, because every allowable replacement of x yields a true statement.

The "for all" quantifier is called the ***universal quantifier.*** The symbol $\forall$ (an upside-down capital letter A) is used to represent the universal quantifier. For example, we can write the statement

$$\text{For all } x, x < x + 1.$$

as

$$(\forall x)(x < x + 1)$$

We use parentheses to separate the quantifier part $(\forall x)$ from the predicate $(x < x + 1)$. If we explicitly specify the universe, the statement becomes $(\forall x \in \mathbb{Z})(x < x + 1)$.

Now let's turn our attention to the other quantifier, "there exists." A statement that uses this quantifier is true if *at least one* replacement of the variable converts the predicate into a true statement. Otherwise, it is false. For example, consider the statement

$$\text{There exists an } x \text{ such that } x < 5.$$

Recall that we are assuming that our universe is $\mathbb{Z}$. Thus, x can be replaced by any integer. Is this quantified statement true?

71

Yes, because at least one replacement for x (2, for example) converts the predicate $x < 5$ into a true statement.

Is the following statement true: There exists an x such that $x^2 = 5$.

72

No. No integer squared equals 5. Remember, we are assuming that our universe is $\mathbb{Z}$. If, however, the universe is the set of real numbers, then the statement is true. What replacements for x convert the predicate $x^2 = 5$ into a true statement?

73

$\sqrt{5}$ and $-\sqrt{5}$

The "there exists" quantifier is called the ***existential quantifier.*** The symbol $\exists$ (a backward capital letter E) is used to represent the existential quantifier. When we use $\exists$, we omit the phrase "such that." For example, we write the statement

$$\text{There exists an } x \text{ such that } x^2 = 5.$$

as

$$(\exists x)(x^2 = 5)$$

For a statement that uses the universal quantifier to be true, _____ allowable replacements of the variables must convert the predicate into a true statement. For an existentially quantified statement to be true, _____ of the allowable replacements must convert the predicate into a true statement.

74

every; at least one

A variety of words and phrases can designate the universal quantifier. In addition to "for all," we can use "all," "every," "any," "each," and "if-then." For example, the following statements all use the universal quantifier and are equivalent:

- $(\forall x \in A)(x \in B)$
- For all $x \in A$, $x \in B$
- All elements in A are in B.
- Every element in A is an element in B.
- Any element in A is an element in B.
- Each element in A is an element in B.
- If $x \in A$, then $x \in B$.

For the existential quantifier, we can use "there exists," "there is," "at least one," or "for some." All of the following statements use existential quantifiers and are equivalent:

- $(\exists x \in \mathbb{Z})(x < 5)$
- There exists an integer x such that $x < 5$.
- There is an integer x less than 5.
- At least one integer is less than 5.
- $x < 5$ for some integer x.

Rewrite the following statements using $\forall$ or $\exists$. Explicitly specify the universe. $|x|$ is the absolute value of x.

For every integer x, $|x| \geq x$.
$x + 6 = 20$ for some integer x.
If x is an integer, then $x^2 \geq 0$.

75

$(\forall x \in \mathbb{Z})(|x| \geq x)$
$(\exists x \in \mathbb{Z})(x + 6 = 20)$
$(\forall x \in \mathbb{Z})(x^2 \geq 0)$

Using the quantifier "for all", rewrite the following statement: 1 is the smallest positive integer. Explicitly specify the universe $\mathbb{Z}^+$ (the positive integers): *Hint*: This statement asserts that 1 is less than or equal to *every* positive integer.

76

For all $x \in \mathbb{Z}^+$, $1 \leq x$.

In the given statement "1 is the smallest positive integer," none of the words that commonly denote the universal quantifier are used. However, the suffix "-est" in the word "smallest" indicates that 1 has some kind of relationship with *all* the elements in $\mathbb{Z}^+$. Thus, universal quantification is implicitly indicated.

Negating Quantified Statements

77

Now let's consider how to negate statements that use quantifiers. What is the negation of the following?

All elements of A are in B.

This is a tricky question because of the use of the universal quantifier "all." Simply changing "are" to "are not" to get

All elements of A are not in B.

does *not* yield the negation of the given statement. To see this, note that if the given statement is false, the statement with "not" is not necessarily true. For example, suppose some of the elements of A are in B, and some are not. Then both the given statement and the statement with "not" are false. Thus, the statement with "not" is not the negation of the given statement. However, suppose we put "not" at the front of the given statement to get

Not all elements of A are in B.

Chapter 2: Mathematical Statements

The "not" at the beginning of the statement negates the entire given statement, giving us the negation of the given statement. Unfortunately, it uses the nonstandard quantifier "not all."

Convert the "not all" statement to an equivalent statement that uses the standard existential quantifier. *Hint*: If not all of the elements in A have a certain property, then there must be at least one element in A that does not have the property, and vice versa.

78

There is at least one element of A not in B.

We can replace "is at least one" with "exists an" without changing the meaning of the statement. We get the following:

There exists an element in A not in B.

In this example, two changes are required to negate the given universally quantified statement:

- "Not" is inserted into the predicate.
- "All" is changed to "there exists."

Now let's consider the negation of an existentially quantified statement such as the following:

There exists an element in A that is in B.

What is its negation? *Hint*: Start the statement with "No elements in A are …"

79

No elements in A are in B.

This is the negation of the given statement. There is an equivalent negation starting with "All elements in A are …" What is this statement?

80

All elements in A are not in B.

To convert the existentially qualified statement

There exists an element in A that is in B.

to its negation

All elements in A are not in B.

two changes are required:

- "Not" is inserted into the predicate.
- "There exists" is changed to "all."

On the basis of the preceding examples, we can formulate the following rule for negating quantified statements:

When negating a statement with the word "all," change "all" to "there exists." When negating a statement with "there exists," change "there exists" to "all." In either case, negate the predicate. That is, insert or remove "not," or perform an equivalent modification.

When written with the quantifiers $\forall$ and $\exists$, statements are easy to negate: simply change $\exists$ to $\forall$ or $\forall$ to $\exists$ and negate the predicate within the second pair of parentheses.

Negate $(\forall x \in A)(x \in B)$.

$(\exists x \in A)(x \notin B)$

Negate $(\exists x \in \mathbb{Z})(x^2 = 4)$

82

$(\forall x \in \mathbb{Z})(x^2 \neq 4)$

Let $P(x)$ represent some predicate that uses the variable x. Negate $(\forall x)(P(x))$ (represent the negation of $P(x)$ with $\sim P(x)$).

83

$(\exists x)(\sim P(x))$

Negate $(\exists x)(P(x))$

84

$(\forall x)(\sim P(x))$

Using $P(x)$ to represent an arbitrary predicate that uses x, let's summarize the rules for negating quantified statements:

- $\sim[(\forall x)(P(x))]$ is $(\exists x)(\sim P(x))$
- $\sim[(\exists x)(P(x))]$ is $(\forall x)(\sim P(x))$

We simply "flip" the quantifier (i.e., change $\forall$ to $\exists$, or change $\exists$ to $\forall$) and negate the predicate. We can think of the negation process as moving the $\sim$ symbol to the right until it reaches the predicate. As $\sim$ passes over the quantifier, it flips the quantifier.

Is $x^2 > 4$ a statement or a predicate?

85

Predicate. Its truth value depends on what we plug in for the variable x.

We call x a ***free variable*** in $x^2 > 4$ because it is not under the control of a quantifier. But now suppose in the universe of integers we quantify x with the universal quantifier. We get the statement $(\forall x \in \mathbb{Z})(x^2 > 4)$. Similarly, if we quantify the predicate with the existential quantifier, we get $(\exists x \in \mathbb{Z})(x^2 > 4)$. In both cases, x is no longer a free variable because it is controlled by a quantifier. We refer to variables under the control of a quantifier as ***bound variables.*** For a predicate to become a statement, all its variables must be bound. For example, $(\exists y)(x + y = 0)$ is a predicate rather than a statement because not all of its variables are bound (x is free).

What are the true/false values of $(\forall x \in \mathbb{Z})(x^2 > 4)$ and $(\exists x \in \mathbb{Z})(x^2 > 4)$.

86

False (use 1 for x) and true (use 3 for x), respectively.

Multiple Quantifiers

87

How many quantifiers does the following statement have: "For all x, there exists a y such that $x + y = 0$"?

88

Two: a universal quantifier ("for all") and an existential quantifier ("there exists").

Rewrite the statement in the preceding frame using mathematical notation.

89

$(\forall x)(\exists y)(x + y = 0)$

Is this statement true if the universe for the two quantifiers is $\mathbb{Z}$?

90

Yes. For every x, the required y exists (it is $-x$). For example, if $x = 5$, then the required y is -5. If $x = 0$, what is the required y?

91

0

The statement $(\forall x)(\exists y)(x + y = 0)$ is a statement rather than a predicate because all its variables are bound, but $x + y = 0$ by itself is a predicate because it has free variables (both x and y are free). Similarly, $(\exists y)(x + y = 0)$ by itself is a predicate rather than a statement because it has a free variable, namely, x. We can view the statement $(\forall x)(\exists y)(x + y = 0)$ as consisting of a *single* quantifier $(\forall x)$ followed by the predicate $(\exists y)(x + y = 0)$. In other words, we can think of

$$(\forall x)(\exists y)(x + y = 0)$$

as equal to

$$(\forall x)(P(x))$$

where $P(x)$ is the predicate $(\exists y)(x + y = 0)$.

Thus, $\sim[(\forall x)(\exists y)(x + y = 0)] = \sim[(\forall x)(P(x))]$, which, by the first negation rule in frame **84**, is $(\exists y)(\sim P(x))$. Substituting $(\exists y)(x + y = 0)$ for $P(x)$, we get $(\exists y)(\sim[(\exists y)(x + y = 0)])$. Applying the second negation rule in frame **84**, we get _____.

92

$(\exists x)(\forall y)(x + y \neq 0)$

Thus, the negation of $(\forall x)(\exists y)(x + y = 0)$ is $(\exists x)(\forall y)(x + y \neq 0)$. To get the negation of the given statement, we "flip" each quantifier and negate the innermost predicate.

What is the negation of $(\forall x)(\exists y)(P(x, y))$, where $P(x, y)$ is an arbitrary predicate that contains the variables x and

---

93

$(\exists x)(\forall y)(\sim P(x, y))$

Here are the rules for negating statements with two quantifiers. There are four cases:

- $\sim[(\forall x)(\forall y)(P(x, y))]$ is $(\exists x)(\exists y)(\sim P(x, y))$
- $\sim[(\forall x)(\exists y)(P(x, y))]$ is $(\exists x)(\forall y)(\sim P(x, y))$
- $\sim[(\exists x)(\forall y)(P(x, y))]$ is $(\forall x)(\exists y)(\sim P(x, y))$
- $\sim[(\exists x)(\exists y)(P(x, y))]$ is $(\forall x)(\forall y)(\sim P(x, y))$

For each case, to negate, we "flip" the quantifiers and negate the innermost predicate.

Write the following statement in mathematical notation: "For every x and z, there is a y such that $x + y = z$."

94

$(\forall x)(\forall z)(\exists y)(x + y = z)$

Is this statement true if the universe for all three quantifiers is $\mathbb{Z}$? If it is $\mathbb{Z}^+$?

95

True for $\mathbb{Z}$ (let $y = z - x$). Not true for $\mathbb{Z}^+$. To show this, take $x = 5$ and $z = 1$. Then y has to be equal to -4. But $-4 \notin \mathbb{Z}^+$.

What is the negation of the statement in the preceding frame?

$(\exists x)(\exists z)(\forall y)(x + y \neq z)$

Review Questions

1. Is "$x < 2$" a statement?
2. Negate: $x < 2$.
3. Negate: $(\forall x)(x \geq 7)$.
4. Convert the following statement to an equivalent statement that uses the universal quantifier: -1 is the largest negative integer.
5. Negate the following statement: $(\forall x)(\forall z)(\exists y)(x + y = z)$.
6. Identify the free variables and the bound variables in $(\forall z)(\exists y)(x + y = z)$. Is it a predicate?
7. Express $A \subseteq B$ as a universally quantified statement.
8. Convert the following statement to an equivalent statement that uses $\exists$: "Some months are cold." In your answer, use M to represent the set of months in a year.
9. Change the statement "If x is a positive integer, then $x + 1$ is a positive integer" to a quantified statement that uses $\forall$.
10. What is the negation of "Some mathematicians are not smart"?

Answers to the Review Questions

1. No. It is a predicate.
2. $x \geq 2$
3. $(\exists x)(x < 7)$
4. For all negative integers x, $x \leq -1$.
5. $(\exists x)(\exists z)(\forall y)(x + y \neq z)$
6. x is free; y and z are bound. It is a predicate because it has at least one free variable.
7. For all $x \in A$, $x \in B$.
8. $(\exists m \in M)(m \text{ is cold})$
9. $(\forall x \in \mathbb{Z}^+)(x + 1 \in \mathbb{Z}^+)$
10. All mathematicians are smart.

Homework Questions

1. Negate: "Not all mathematicians know Latin."
2. Convert to a "there exists" statement: "Not all mathematicians know Latin."
3. Is the following statement true (assume the universe is $\mathbb{Z}$): $(\forall x)(\forall y)(\exists z)(xz = y)$?
4. Is the following statement true (assume the universe is $\mathbb{Z}$): $(\exists x)(\forall y)(\forall z)(xz = y)$?
5. Negate: $(\exists x)(\forall y)(\forall z)(xz = y)$.
6. Express $A \subset B$ as a quantified statement.
7. Convert the following statement to an equivalent statement that uses the universal quantifier: "The square root of 2 is not an integer."
8. Convert to an equivalent statement that uses $\forall$: "6 is the smallest positive integer that is divisible by both 2 and 3."
9. Convert the following statement to a statement that uses the universal quantifier: "There is no largest integer."
10. Does $(\forall x)(\forall z)(\exists y)(x + y = z)$ have the same truth value as $(\exists y)(\forall x)(\forall z)(x + y = z)$? Assume the universe is $\mathbb{Z}$. Justify your answer.

3 Compound Statements

and Logical Operator

Using the word "and," we can create *compound statements,* that is, statements that contain subparts that are themselves statements. We can similarly create compound predicates. When used in this way, "and" is called a *logical operator* or a *logical connective.* For example, here are two statements:

$$2 + 2 = 4$$
$$2 \geq 3.$$

Combining them with "and," we get the compound statement

$$2 + 2 = 4 \text{ and } 2 \geq 3.$$

A compound statement formed with "and" is true if and only if its two substatements are *both* true. Thus, the preceding compound statement is false (because $2 \geq 3$ is false). A compound statement consisting of two substatements joined by "and" is called a *conjunction.*

Suppose p and q are statements. A convenient way to represent the values of the compound statement "p and q" as a function of the value of p and the value of q is with the following *truth table*:

p	q	p and q	
F	F	F	first row
F	T	F	second row
T	F	F	third row
T	T	T	fourth row

The first two elements in each row correspond to a particular assignment of truth values to p and q ("F" represents false, and "T" represents true). The third element in these rows shows the value of "p and q" for that assignment. For example, the truth values in the third row indicate that if p is true and q is false, then the compound statement "p and q" is false. Note that "p and q" is true only in the fourth row, where both p is true and q is true.

or Logical Operators: Inclusive and Exclusive

We can also create compound statements using the word "or." The compound statement of the form

$$p \text{ or } q$$

is true if either p is true or q is true or if both p and q are true. A compound statement consisting of two substatements joined by "or" is called a *disjunction.* Here is the truth table for a disjunction:

p	q	p or q	
F	F	F	first row
F	T	T	second row
T	F	T	third row
T	T	T	fourth row

The statement "p or q" is true if either p alone is true (third row) or q alone is true (second row) or if both p and q are true (fourth row). The interpretation of "or" as described by the preceding truth table is often at odds with the meaning of "or" in everyday English. For example, consider the meaning of the following sentences:

- Today, Bert will work, or he will go fishing.
- Jane will go to Paris, or she will go to London.
- The student will study, or the student will flunk.

All three suggest that the subject of the sentence will do one of two alternatives, but not both. The word "or" used in this way—with the "both" case excluded—is called the **exclusive or.** However, the "or" used in mathematics, unless indicated otherwise, is the inclusive "or"—the "or" that includes the "both" case. For example, when we write

$$x = 1 \text{ or } y = 2$$

we mean either $x = 1$ or $y = 2$ or both $x = 1$ and $y = 2$.

Because "and" and "or" are words commonly used in English, it is sometimes confusing to use "and" or "or" when creating compound statements in mathematics. To avoid this sort of ambiguity, mathematicians often use the symbols ∧, ∨, and ⊻ to denote the logical operators "and," "or," and exclusive "or," respectively, when writing compound statements. For example, we can write "p and q," "p or q," and "p exclusive or q" as $p \land q$, $p \lor q$, and $p \veebar q$, respectively. It is easy to remember that ∧ is the "and" operator—it is the letter "A" in "And" with the horizontal bar missing.

Give the truth table for $p \veebar q$ (the exclusive "or").

99

p	q	$p \veebar q$
F	F	F
F	T	T
T	F	T
T	T	F

← $p \veebar q$ and $p \lor q$ differ in this row only.

It is identical to the truth table for $p \lor q$ (the inclusive "or"), except for the "both" case—the case in which both p and q are true.

Deriving a Statement from a Truth Table

100

Recall that a conjunction is a compound statement consisting of two subparts joined by ∧ or "and." The truth table of a conjunction has exactly one (i.e., one and only one) row in which the value of the conjunction is T. Let's call this row the ***true row.*** For example, consider the conjunction $\sim p \land q$. Because negation has higher precedence than conjunction, in $\sim p \land q$ we apply $\sim$ to p and then do the conjunction. That is, we interpret $\sim p \land q$ as $(\sim p) \land q$. Here is the truth table for $\sim p \land q$:

p	q	$\sim p \land q$
F	F	F
F	T	T
T	F	F
T	T	F

← $\sim p \land q$ is T only in this row.

For the conjunction $\sim p \land q$ to be true, both $\sim p$ and q have to be true, or, equivalently, p must be false and q true. Thus, this conjunction is true in only one row of its truth table—the row in which p is false and q is true.

~p ∧ ~q is true in which row of its truth table?

101

The row in which both *p* and *q* are false.

Now let's start with a truth table in which only one row is T and then determine the corresponding conjunction. What conjunction corresponds to the following truth table (it is true in only the third row)?

p	q	? ∧ ?	
F	F	F	first row
F	T	F	second row
T	F	T	third row
T	T	F	fourth row

102

p ∧ ~*q*

We can determine the required conjunction using the following procedure. First, look for the row in the truth table in which the conjunction is true. In that row, if the value of a variable is T, then that variable appears in the conjunction without ~. If the value of the variable is F, then the variable appears in the conjunction with ~. In the preceding example, *p* in the third row is T, so it appears in the conjunction without ~; *q* in the third row is F, so it appears in the conjunction with ~. Thus, the conjunction is *p* ∧ ~*q*.

What conjunction corresponds to the following truth table?

p	q	? ∧ ?
F	F	T
F	T	F
T	F	F
T	T	F

103

~*p* ∧ ~*q*

What conjunction corresponds to the following truth table?

p	q	? ∧ ?
F	F	F
F	T	F
T	F	F
T	T	T

104

p ∧ *q*

Key idea: Every truth table with exactly one true row corresponds to a conjunction.

But what about a truth table in which more than one row is a true row? For example, consider

p	q	?
F	F	T
F	T	F
T	F	F
T	T	T

We know that ~*p* ∧ ~*q* is true in the first row only and that *p* ∧ *q* is true in the fourth row only. What then can we say about (~*p* ∧ ~*q*) ∨ (*p* ∧ *q*)?

105

In the first row, ~p ∨ ~q is true, which makes the disjunction (~p ∧ ~q) ∨ (p ∧ q) true for this row. Similarly, in the fourth row, p ∧ q is true, making the disjunction true in this row also. This disjunction is true in precisely the first and fourth rows. Thus, it corresponds to the truth table in the preceding frame.

Key idea: Every truth table with more than one true row corresponds to a disjunction of conjunctions. We need one conjunction for each true row of the truth table.

We have been analyzing conjunctions and their relationship to rows in a truth table that have the value T. We can proceed similarly for disjunctions. However, for disjunctions, we focus on *false rows* in a truth table—rows in which the value of the disjunction is F rather than T.

Consider the truth table for the conjunction p ∨ q. How many rows are false rows? *Hint*: For a disjunction to be false, *both* sides of the disjunction must be false.

106

One (only in the row in which both p and q are F).

How many rows are false rows in the truth table for ~p ∨ q?

107

One. For ~p ∨ q to be false, both ~p and q must be false. Thus, p must be true and q must be false.

A disjunction of two variables is false in only one row of its truth table—the row in which both sides of the disjunction are false.

What is the disjunction that corresponds to the following truth table?

p	q	? ∨ ?	
F	F	T	first row
F	T	F	second row
T	F	T	third row
T	T	T	fourth row

108

p ∨ ~q. Here we want a disjunction that is false in the second row of its truth table, that is, when p is false and q is true. The statement p ∨ ~q is false when p is false and q is true.

We can determine the required disjunction using the following procedure. First, look for the row in the truth table in which the disjunction is false. In that row, if the value of a variable is T, then that variable appears in the disjunction with ~. If the value of the variable is F, then the variable appears in the disjunction without ~. In the preceding example, p in the second row is F, so it appears in the disjunction without ~; q in the second row is T, so it appears in the disjunction with ~. Thus, the disjunction is p ∨ ~q.

What is the disjunction that corresponds to the following truth table?

p	q	? ∨ ?
F	F	T
F	T	T
T	F	F
T	T	T

109

~p ∨ q

Key idea: Every truth table with exactly one false row corresponds to a disjunction.

What about a truth table in which more than one row has the value F? For example, consider

p	q	?	
F	F	T	first row
F	T	F	second row
T	F	F	third row
T	T	T	fourth row

We know that $p \lor \sim q$ is false in the second row only and that $\sim p \lor q$ is false in the third row only. What then can we say about $(p \lor \sim q) \land (\sim p \lor q)$?

110

In the second row, $p \lor \sim q$ is false, which makes the conjunction $(p \lor \sim q) \land (\sim p \lor q)$ false for this row. Similarly, in the third row, $\sim p \lor q$ is false, making the conjunction false in this row also. This conjunction is false in precisely the second and third rows. Thus, it corresponds to the truth table in the preceding frame.

What is the conjunction of disjunctions that corresponds to the following truth table?

p	q	?
F	F	F
F	T	T
T	F	T
T	T	F

Hint: Determine a disjunction that is false in the first row. Also determine a disjunction that is false in the fourth row. Then form the conjunction of these two disjunctions. That conjunction will be false in precisely the first and fourth rows.

111

$(p \lor q) \land (\sim p \lor \sim q)$

Key idea: Every truth table with more than one false row corresponds to a conjunction of disjunctions. We need one disjunction for each false row.

Give both a disjunction of conjunctions (derived from the true rows) and a conjunction of disjunctions (derived from the false rows) that correspond to the following table:

p	q	?
F	F	F
F	T	T
T	F	F
T	T	T

112

$(\sim p \land q) \lor (p \land q)$ (derived from rows whose value is T)
$(p \lor q) \land (\sim p \lor q)$ (derived from rows whose value is F)

Give two distinct compound statements that correspond to the following truth table:

p	q	?
F	F	F
F	T	F
T	F	F
T	T	T

Hint: One is a single conjunction; the other expression is a conjunction of three disjunctions.

113

$p \land q$ (from the row whose value is T)

$(p \vee q) \wedge (p \vee {\sim} q) \wedge ({\sim} p \vee q)$ (from the three rows whose value is F)

Implication

114

We can also create a compound statement using the words "if" and "then." We call such a compound statement an ***implication***. For example, a statement of the form "if p then q" is an implication. We call p and q the ***antecedent*** and ***consequent***, respectively. The symbol for implication is "$\Rightarrow$." For example, we can write "if p then q" as "$p \Rightarrow q$."

Under what circumstances is "if p then q" true? Under what circumstances is it false? Suppose I say, "If I win the lottery, then I will buy a car." What circumstances would make this statement a lie? Clearly the statement is a lie if I win the lottery but do not buy a car. If, alternatively, I win the lottery and buy the car, the statement is true. But what if I do not win the lottery? Then whether or not I buy a car is irrelevant. The statement is certainly not a lie in either case. Thus, if the antecedent p is false, the implication is true *regardless* of the truth value of the consequent q.

Construct the truth table for "if p then q."

115

p	q	if p then q
F	F	T
F	T	T
T	F	F
T	T	T

Key idea: "If p then q" is false if and only if p is true and q is false.

In the implication,

> if 1 equals 2, then the moon is made of cheese,

the antecedent, "1 equals 2," is unrelated to the consequent, "the moon is made of cheese." Mathematicians generally do not use implications of this sort; rather, they use implications in which the consequent is not only related to the antecedent but also logically follows from the antecedent. For example, here is a typical implication a mathematician would use:

> If n is even, then n^2 is even.

Both the antecedent and the consequent involve n and the concept of evenness. Clearly, in this example, the antecedent and the consequent are related. Moreover, there is a causal relationship between them. Specifically, the antecedent logically follows from the antecedent: to show the consequent, we have to use the antecedent.

Is the statement "if 1 equals 2, then the moon is made of cheese" true?

116

Yes, because the antecedent is false. This example illustrates that the consequent in a true implication can be false.

An implication can appear in a variety of forms. In addition to "if p then q," we can use any of the following forms:

1. $p \Rightarrow q$
2. p implies q
3. q follows from p
4. q if p
5. p is sufficient for q
6. q is necessary for p
7. p only if q

Versions 2, 5, 6, and 7 require some explanation. In version 2, we are using the word "implies." In everyday English, *imply* means "to suggest" or "to hint at." But in mathematics, "implies" means to logically follow. That is, "p implies q" means q is true whenever p is true. This statement is not a hint or a suggestion; it is an unambiguous assertion. In version 5, we are saying that establishing that p is true is sufficient to establish that q is true. That is, p implies q. In version 6, we are saying that if p is true, then it is necessary that q is true. That is, p implies q. Version 7 is difficult to comprehend. Suppose I say, "Bert will move only if he gets a dog." If Bert subsequently moves, what can we conclude? Answer: Bert got a dog. But this means that Bert's moving implies that Bert got a dog.

Key idea: "p only if q" means that p implies q.

Construct the truth table for $p \Rightarrow q$ and $\sim(p \Rightarrow q)$.

117

p	q	$p \Rightarrow q$	$\sim(p \Rightarrow q)$
F	F	T	F
F	T	T	F
T	F	F	T
T	T	T	F

Because the truth table for implication has exactly one row with the value F, there is a single disjunction that has the same truth table. What is this disjunction? *Hint*: See frame **108**.

118

$\sim p \vee q$

Give a conjunction that has the same truth table as $\sim(p \Rightarrow q)$. *Hint*: Its truth table has only one true row (see the truth table in the preceding frame).

119

$p \wedge \sim q$

Key idea: $\sim p \vee q$ is equivalent to $p \Rightarrow q$ (i.e., they have identical truth tables); $p \wedge \sim q$ is equivalent to the negation of $p \Rightarrow q$.

These equivalences are not hard to remember: $\sim p \vee q$ captures precisely those cases in which $p \Rightarrow q$ is true (p is false or q is true). Thus, it is equivalent to the given implication. Similarly, $p \wedge \sim q$ captures the one case in which the implication is false (p is true and q is false). Thus, it is the negation of the given implication.

Biconditional

120

So far, we have studied the logical operators $\wedge$ (and), $\vee$ (or), $\veebar$ (exclusive or), and $\Rightarrow$ (implication). We have one more operator to study: the biconditional, denoted by the symbol $\Leftrightarrow$. A ***biconditional*** is a two-way implication: $p \Leftrightarrow q$ means both $p \Rightarrow q$ and $q \Rightarrow p$. Each side of the biconditional implies the other side.

A biconditional can appear in a variety of forms. In addition to $p \Leftrightarrow q$, we can use any of the following forms:

1. p implies q and q implies p
2. $p \Rightarrow q$ and $q \Rightarrow p$
3. p if and only if q
4. p iff q
5. p is necessary and sufficient for q
6. p is true if and only if q is true
7. p is false if and only if q is false

In version 3, the "if" part of "if and only if" indicates that q implies p. The "only if" part indicates that p implies q (see frame **116** for a discussion of "only if"). In version 4, we are using "iff," the abbreviation for "if and only if." In version 5, we are using "necessary" to indicate $q \Rightarrow p$ and "sufficient" to indicate $p \Rightarrow q$, so "necessary and sufficient" indicates a two-way implication. In versions 6 and 7, we are describing the truth table for a biconditional.

Here is the truth table for a biconditional $p \Leftrightarrow q$. It is true if and only if p and q are either both true or both false:

p	q	$p \Leftrightarrow q$
F	F	T
F	T	F
T	F	F
T	T	T

Identical Truth Tables

121

Construct the truth tables for $p \Rightarrow q$ and for $\sim p \vee q$.

122

p	q	$p \Rightarrow q$	$\sim p \vee q$
F	F	T	T
F	T	T	T
T	F	F	F
T	T	T	T

Columns are identical.

$p \Rightarrow q$ and $\sim p \vee q$ are **equivalent**. That is, they have identical truth tables. We denote equivalence with the symbol $\equiv$. For example, we write $p \Rightarrow q \equiv \sim p \vee q$.

Suppose a given expression contains as a subpart $p \Rightarrow q$. If we replace $p \Rightarrow q$ with $\sim p \vee q$, what effect does the change have on the truth table for the expression?

123

None. Because $p \Rightarrow q$ and $\sim p \vee q$ have the same truth table, they have the same truth value for every possible assignment of T and F to p and q. Thus, replacing $p \Rightarrow q$ with $\sim p \vee q$ in a complex expression does not change the expression's value for any assignment of T or F to p and q. But this means that the truth table for the modified expression is the same as the original.

Construct the truth tables for p and $\sim\sim p$.

124

p	$\sim\sim p$
F	F
T	T

Their truth tables are identical. Thus, substituting p for $\sim\sim p$ in an expression does not change the truth table for that expression. For example, $\sim\sim p \vee \sim\sim q$ simplifies to $p \vee q$. Without using a truth table, give an argument that $(p \vee q) \wedge \sim(p \wedge q)$ is equivalent to $p \veebar q$. *Hint*: What is the effect of $\sim(p \wedge q)$ on the truth table for the given expression?

125

$\sim(p \wedge q)$ forces the expression to be false only in the case when both p and q are true. Thus, the expression is $p \vee q$ (the inclusive "or"), with the "both" case forced to false. The result is the exclusive "or."

Definitions That Use Implication

126

Definitions in mathematics textbooks are often stated in the form of an implication. However, they are meant to be interpreted as biconditionals. For example, consider the following definition of a rational number:

26 Chapter 3: Compound Statements

A number is ***rational*** if it can be represented in the form $\frac{x}{y}$, where x and y are integers and $y \neq 0$.

Note that this definition is an implication in the form of "q if p." Read strictly as an implication, it tells us that a particular type of number (i.e., a number that can be expressed as the ratio of two integers) is rational. However, it does not rule out the possibility that other types of numbers are also rational. For example, it does not rule out $\sqrt{2}$ as a rational number. But $\sqrt{2}$ is *not* a rational number. Thus, *as an implication,* our definition does not fully capture the concept of a rational number.

But now let's suppose the definition is a biconditional, that is, written in the form "q if and only if p." Then our definition tells us not only that a number is rational *if* it can be expressed as a ratio of two integers but also that a number is rational *only if* it can be expressed as a ratio of two integers. The "only if" part of the biconditional rules out the possibility that a rational number can be a number other than one that can be expressed as a ratio of two integers. Interpreted as a biconditional, our definition fully captures the concept of a rational number.

Key idea: In mathematics, interpret definitions of the form of an implication as if they were biconditionals." This interpretation applies *only* to definitions. Theorems, on the other hand, should be interpreted exactly as they are written.

If definitions should be interpreted as biconditionals, then why not write them as biconditionals? Many textbook authors, in fact, do write definitions as biconditionals. For those who do not, here is their justification: definitions are necessarily biconditionals; otherwise, they would not fully capture the concept they are defining. For example, suppose the rational numbers did include $\sqrt{2}$ in addition to all the numbers that can be expressed as a ratio of two integers. Then the definition of a rational number would obviously *have to state this.* Because the definition does not state this, we can assume that numbers that cannot be expressed as a ratio of integers are not rational.

Negating Compound Statements

127

The truth tables for a statement and its negation have complementary values. For example, the truth table for $p \wedge q$ is

p	q	$p \wedge q$
F	F	F
F	T	F
T	F	F
T	T	T

By "flipping" the truth values in the third column of this truth table, we get the truth table for $\sim(p \wedge q)$:

p	q	$\sim(p \wedge q)$
F	F	T
F	T	T
T	F	T
T	T	F

Using the technique in frame **108**, determine the disjunction that has the same truth table as this new table.

128

$\sim p \vee \sim q$

To indicate that $\sim(p \wedge q)$ and $\sim p \vee \sim q$ have identical truth tables, we write $\sim(p \wedge q) \equiv \sim p \vee \sim q$. This one of the two **DeMorgan's laws** (we will see the other DeMorgan's law shortly). It tells us that we negate $p \wedge q$ by negating the subparts of the conjunction individually, and, in addition, changing $\wedge$ to $\vee$. The result is $\sim p \vee \sim q$.

Consider the statement "it's raining cats, and it's raining dogs." Let's determine its negation using DeMorgan's law. We change "and" to "or" (the inclusive or) and negate the two substatements. We get "it is not raining cats, or it is not raining dogs."

Using DeMorgan's law, negate $\sim p \wedge \sim q$ and simplify.

129

$\sim\sim p \lor \sim\sim q$, which simplifies to $p \lor q$.

Here is the truth table for the negation of $p \lor q$:

p	q	$\sim(p \lor q)$
F	F	T
F	T	F
T	F	F
T	T	F

Derive a conjunction from the truth table using the technique in frame **102**.

130

$\sim p \land \sim q$

This is the negation of $p \lor q$. Thus, $\sim(p \lor q) \equiv \sim p \land \sim q$. This equivalence is the other DeMorgan's law (we saw the first DeMorgan's law in frame **128**). Here are both DeMorgan's laws:

- $\sim(p \land q) \equiv \sim p \lor \sim q$
- $\sim(p \lor q) \equiv \sim p \land \sim q$

Notice the similarity of the two laws: in both cases, the subparts of the compound statement are individually negated. In addition, $\lor$ gets changed to $\land$, and $\land$ gets changed to $\lor$.

Using DeMorgan's laws, negate $p \lor \sim q$ and simplify.

131

$\sim p \land \sim\sim q$, which simplifies to $\sim p \land q$

Using DeMorgan's laws, negate $p \land (q \lor r)$. *Hint*: Use DeMorgan's laws twice.

132

$\sim p \lor \sim(q \lor r)$. Using DeMorgan's laws a second time, we get $\sim p \lor (\sim q \land \sim r)$.

Using DeMorgan's laws, negate $(p \land \sim q) \lor r$.

133

$\sim(p \land \sim q) \land \sim r$, which is equivalent to $(\sim p \lor \sim\sim q) \land \sim r$, which simplifies to $(\sim p \lor q) \land \sim r$.

Consider the statement "I will eat ice cream or I will eat pie," where "or" here is the inclusive or. Thus, I will eat one or both of the two choices. Use DeMorgan's law to get its negation.

134

I will not eat ice cream, and I will not eat pie.

This answer makes sense: because the given statement asserts that I will eat one or both choices, then its negation must assert that I will eat neither choice. That is, I will not eat ice cream *and* I will not eat pie.

We determined DeMorgan's laws mechanically from truth tables in frames **127** and **129**. Alternatively, we can determine them with simple reasoning about negation. For example, to determine the negation of $p \land q$, consider what must be true if $p \land q$ is false: p or q or both must be false, or, equivalently, $\sim p$ or $\sim q$ or both must be true. The statement that is true when $\sim p$ or $\sim q$ or both are true is $\sim p \lor \sim q$.

In the same way, argue that the negation of $p \lor q$ is $\sim p \land \sim q$.

135

Chapter 3: Compound Statements

The statement $p \vee q$ is true when p is true or q is true or both p and q are true. Thus, its negation is true when both p and q are false. The statement that is true when both p and q are false is $\sim p \wedge \sim q$.

Tautologies and Absurdities

136

Give the truth table for $p \vee \sim p$.

137

p	$p \vee \sim p$
F	T
T	T

Regardless of the value of p, $p \vee \sim p$ is true. We call these "always true" statements ***tautologies.***

Give the truth table for $p \wedge \sim p$.

138

p	$p \wedge \sim p$
F	F
T	F

Regardless of the value of p, $p \vee \sim p$ is false. We call these "always false" statements ***contradictions*** or ***absurdities.***

Converse and Contrapositive

139

We call $q \Rightarrow p$ the ***converse*** of $p \Rightarrow q$. Here are the truth tables for $p \Rightarrow q$ and its converse:

p	q	$p \Rightarrow q$	$q \Rightarrow p$	
F	F	T	T	first row
F	T	T	F	second row
T	F	F	T	third row
T	T	T	T	fourth row

Note that $p \Rightarrow q$ and $q \Rightarrow p$ are not equivalent. That is, they do not have identical truth tables. From their truth tables, we can see that if $p \Rightarrow q$ is true, then $q \Rightarrow p$ is not necessarily true (see the second row of the truth table), and vice versa (see the third row of the truth table).

Is the implication "if $x = 3$, then $x^2 = 9$" true?

140

Yes

Give its converse.

141

If $x^2 = 9$, then $x = 3$.

Is the converse true?

142

No; x could be -3.

This example illustrates that the converse of a true statement is not necessarily true.

What is the converse of the statement "if $x + 1 = 4$, then $x = 3$"?

143

If $x = 3$, then $x + 1 = 4$.

Are the given statement and its converse both true?

144

Yes

From these examples, we can see that the converse of a true statement may or may not be true. For some statements, the converse is true. For other statements, the converse is not true. Because the converse of a true statement may not be true, a proof of a statement cannot be used as a proof of its converse.

Construct the truth tables for $p \Rightarrow q$ and for $\sim q \Rightarrow \sim p$.

145

p	q	$p \Rightarrow q$	$\sim q \Rightarrow \sim p$
F	F	T	T
F	T	T	T
T	F	F	F
T	T	T	T

We call $\sim q \Rightarrow \sim p$ the **contrapositive** of $p \Rightarrow q$. We obtain the contrapositive of $p \Rightarrow q$ by both switching and negating p and q. From the preceding truth table, we can see that $p \Rightarrow q \equiv \sim q \Rightarrow \sim p$ (i.e., they have identical truth tables). Thus, when an implication is true, its contrapositive must also be true. Conversely, when the contrapositive is true, the implication must also be true. Thus, a proof for one is a proof the other. For example, to prove that n^2 is even $\Rightarrow n$ is even, we can prove its contrapositive instead: n is not even $\Rightarrow n^2$ is not even.

In frame **120**, we learned that a biconditional is a two-way implication. That is, $p \Leftrightarrow q$ is equivalent to $(p \Rightarrow q) \wedge (q \Rightarrow p)$. If we take the contrapositive of the two implications, we get the equivalent statement $(\sim q \Rightarrow \sim p) \wedge (\sim p \Rightarrow \sim q)$, which, in turn, is equivalent to $\sim p \Leftrightarrow \sim q$. Thus, $p \Leftrightarrow q$ is equivalent to $\sim p \Leftrightarrow \sim q$.

Key idea: $p \Leftrightarrow q$ is true if and only if $\sim p \Leftrightarrow \sim q$ is true.

Review Questions

1. Determine the conjunction that corresponds to

p	q	?
F	F	T
F	T	F
T	F	F
T	T	F

2. Determine disjunction that corresponds to

p	q	?
F	F	F
F	T	T
T	F	T
T	F	T

3. Determine the disjunction of conjunctions that corresponds to

p	q	?
F	F	F
F	T	T
T	F	F
T	T	T

4. Determine the conjunction of disjunctions that corresponds to the truth table in review question 3.
5. Negate and simplify $p \land \sim q$.
6. Negate and simplify $p \lor \sim q$.
7. Negate and simplify $p \land (\sim q \lor r)$.
8. Negate and simplify $(p \land r) \lor (\sim q \lor r)$.
9. What is the converse of "if x is odd, then x^2 is odd"?
10. What is the contrapositive of "if x is odd, then x^2 is odd"?

Answers to the Review Questions

1. $\sim p \land \sim q$
2. $p \lor q$
3. $(\sim p \land q) \lor (p \land q)$
4. $(p \lor q) \land (\sim p \lor q)$
5. $\sim p \lor q$
6. $\sim p \land q$
7. $\sim p \lor (q \land \sim r)$
8. $(\sim p \lor \sim r) \land (q \land \sim r)$
9. If x^2 is odd, then x is odd.
10. If x^2 is not odd, then x is not odd.

Homework Questions

1. Determine the disjunction of conjunctions that corresponds to

p	q	r	?
F	F	F	F
F	F	T	T
F	T	F	F
F	T	T	T
T	F	F	T
T	F	T	T
T	T	F	F
T	T	T	T

2. Determine the conjunction of disjunctions that corresponds to the truth table in homework question 1.
3. Because $\lor$ and $\land$ are not on a standard keyboard, we often use concatenation (i.e., placing the two subexpressions next to each other) in place of $\land$ and + in place of $\lor$. For example, we can write $(p \land q) \lor (r \land s)$ as $(pq) + (rs)$. Moreover, if we assume that conjunction has higher precedence than disjunction, we can omit the parentheses around the conjunctions to get $pq + rs$. We use this notation in this problem and in homework problems 4, 5, 6, 7, and 8. Rewrite $(p \lor q) \land (r \lor s)$ using this new notation.
4. Does the "and" operation left distribute over the "or" operation? That is, is $p(r + s)$ equivalent to $pr + ps$ (see homework question 3)? Justify your answer.
5. Does the "or" operation left distribute over the "and" operation? That is, is $p + rs$ equivalent to $(p + r)(p + s)$ (see homework question 3)?
6. Construct the truth table for $p + pq$ (see homework question 3).
7. Is the truth value of $p + pq$ dependent on q?
8. Give the simplest possible expression for $p + pq$? *Hint*: See homework question 7.
9. What is the converse of the contrapositive of $p \Rightarrow q$?
10. Give a conjunction that is equivalent to the negation of $p \Rightarrow q$.

4 Proof Techniques

Direct Proof

146

Suppose that we are given the statement p, and we want to prove that the statement q logically follows from p. In a ***direct proof***, we create a sequence of statements, starting with the given statement p. Each statement in the sequence after the given statement p should be either implied by one or more of the statements that precede it in the proof, or already known to be true, or a definition. For example, the proof of q given p could take the following form:

1. p Given
2. statement   Definition
3. statement   2
4. statement   1, 3
5. statement   Definition
6. statement   Known to be true
7. statement   4, 5, 6
8. q 7 ∎

Each statement is numbered. The justification for each statement is provided to its right, unless it is obvious to the intended reader. In our example, statement 4 is implied by statements 1 and 3. Thus, we write "1, 3" to the right of statement 4 to indicate that statements 1 and 3 imply statement 4. The sequence ends with q, establishing that q logically follows from p. Proofs are usually terminated with a special mark, such as ∎ or QED (Latin for "which was to be demonstrated").

In practice, proofs are rarely written in the numbered tabular form illustrated by the preceding example. Most often, they are in the form of a sequence of sentences that use words like "thus," "hence," "then," "therefore," and "whence" to show the logical sequence and words like "since," "because," "by definition," and "we are given that" to provide the justifications for the steps of the proof.

Sometimes a direct proof consists of a sequence of statements in which each statement implies the next one. We can then simply write out the sequence of statements, separating each from the next with "⇒" to indicate that each statement implies the next one in the sequence. An n-statement direct proof would then have the following form:

$p \Rightarrow \text{statement}_2 \Rightarrow \text{statement}_3 \Rightarrow \ldots \Rightarrow \text{statement}_{n-1} \Rightarrow q$ ∎

In a direct proof, does every step have to be justified?

147

Not if the justification should be obvious to the intended reader.

Let's now do a simple example of a direct proof, first in tabular form, then with a sequence of sentences. In this proof, we use the definitions of even and odd integers:

An integer n is ***even*** if and only if $n = 2i$ for some $i \in \mathbb{Z}$. An integer m is ***odd*** if and only if $m = 2j + 1$ for some $j \in \mathbb{Z}$.

If an integer is not even, it is odd; if it is not odd, it is even (we prove this in frame **165**). Let's prove the statement "if n and m are odd, then $n + m$ is even." Here is the proof:

1. Let *n, m* be two odd integers.   Given
2. $n = 2i + 1$ and $m = 2j + 1$ for some $i, j \in \mathbb{Z}$.   1 and the definition of an odd integer
3. $n + m = (2i + 1) + (2j + 1) = 2(i + j) + 2 =$ 2
   $2(i + j + 1) = 2k$ where k is the integer $i + j + 1$.
4. $n + m$ is an even integer.   3 and the definition of an even integer ∎

Here is a shorter version—a version more likely to be in a mathematics textbook:

Let *n, m* be odd. Then, $n = 2i + 1$ and $m = 2j + 1$ for some $i, j \in \mathbb{Z}$. Then $n + m = (2i + 1) + (2j + 1) = 2(i + j + 1)$. Thus, $n + m$ is even. ∎

Finally, here is a third version that uses ⇒ to show that each statement implies the next:

Let *n, m* be two odd integers
⇒ $n = 2i + 1$ and $m = 2j + 1$ for some $i, j \in \mathbb{Z}$
⇒ $n + m = (2i + 1) + (2j + 1) = 2(i + j + 1)$
⇒ $n + m$ is an even number. ∎

Once we prove a statement, we call it a ***theorem.*** The theorem that we just proved makes an assertion about *all* pairs of odd integers. Thus, it is a _____ quantified statement.

148

universally

Restate the theorem using the universal quantifier explicitly.

149

For all odd integers *m* and *n*, $m + n$ is even.

The universe for the universal quantifier in this statement is the set of odd numbers. That is, *m* can be any odd number and *n* can be any odd number. In our proof, we made no assumptions about *m* and *n* other than that they are odd. Thus, our argument that $m + n$ is even applies to *any* pair of odd numbers.

    The preceding proof illustrates how we prove a universally quantified statement: we advance an argument that makes no assumptions about the elements selected from the universe (other than that they are from the universe). Thus, our argument applies to all the elements in the universe.

    To prove a statement that uses the existential quantifier, we need to find only one selection of elements from the universe that makes the predicate true. For example, to prove $(\exists x \in \mathbb{Z})(x < x + 1)$, we simply observe that $x < x + 1$ if $x = 0$. However, this approach is incorrect for a universally qualified statement. If the predicate in a universally quantified statement is true for one value of *x*, it is not necessarily true for all values of *x* in the universe. For example, observing that $x < x + 1$ if $x = 0$ does *not* prove $(\forall x \in \mathbb{Z})(x < x + 1)$.

Given that $0 < 1$ is a true statement, prove that $(\forall x \in \mathbb{Z})(x < x + 1)$.

150

Let *x* be an arbitrary element of $\mathbb{Z}$. We know that $0 < 1$. Adding *x* to both sides of this equality, we get $x < x + 1$. Because we are making no assumptions about *x* other than it is in $\mathbb{Z}$, our argument that $x < x + 1$ works for all $x \in \mathbb{Z}$. Thus, $(\forall x \in \mathbb{Z})(x < x + 1)$. ∎

Our proof in frame **147** that the sum of two odd integers is even is not difficult to construct. We do not have to be particularly creative or ingenious. We simply express the oddness of *m* and *n* mathematically, add *n* and *m,* and show that the result has 2 as a factor.

    Let's now try another direct proof. This one is less obvious. Before we get to the proof, let's introduce some terminology. Suppose *b* is an integer that can be factored into two integers, *a* and *k*, where $a \neq 0$ (*k* can be zero or nonzero). That is, there are integers *a* and *k* with $a \neq 0$ such that $b = ak$. To describe this situation, we can use any of the following statements:

- *a* is a factor of *b*.
- *a* divides *b*.
- *a* is a divisor of *b*.
- *b* is divisible by *a*.
- *b* is a multiple of *a*.

For example, 64 can be factored into $2 \cdot 32$. Thus, 2 is a factor of 64, 2 divides 64, 2 is a divisor of 64, 64 is divisible by 2, and 64 is a multiple of 2. To indicate that *a* divides *b*, we write $a \mid b$. For example, $2 \mid 12$ means that 2 divides 12. That is, there is some integer k such that $12 = 2k$. Recall that we defined an even integer as any integer n such that $n = 2k$ for some integer k. We can equivalently define an even integer as any integer n such that $2 \mid n$.

Show that 16 divides 64.

151

$16 \cdot 4 = 64$. This equation shows that 16 and 4 both divide 64

From $10 \cdot 6.4 = 64$, we cannot conclude that 10 divides 64 (it does not) because this factorization of 64 is not all integers.

List all the divisors of 4. Be sure to include all the negative divisors.

152

$\pm 1, \pm 2, \pm 4$

Let's prove that if 3 divides the sum of the digits in a positive two-digit decimal number, then 3 divides the number. For example, the sum of the digits in $72 = 7 + 2 = 9$. Because 3 divides 9 (the sum of the digits), according to the statement we are about to prove, 3 should also divide 72. It does, as demonstrated by $72 = 3 \cdot 24$. For this proof, we need to represent the sum of the two digits of the given number. To do this, we simply represent the two-digit number as $d_1 d_0$, where d_1 is the digit in the tens position and d_0 is the digit in the units position. Then the sum of the two digits is $d_1 + d_0$. We are given that 3 divides the sum of the digits in a two-digit number. That is, $3 \mid (d_1 + d_0)$. What does this imply about $d_1 + d_0$?

153

$d_1 + d_0 = 3k$ for some $k \in \mathbb{Z}$

Consider the two-digit decimal number 72; 7 is in the tens position, and 2 is in the units position. Express the value of 72 in terms of 7 and 2.

154

$10 \cdot 7 + 2$

Similarly, express the value of the two-digit decimal $d_1 d_0$.

155

$10 d_1 + d_0$

We have to show that 3 divides $10d_1 + d_0$ (the given number) if 3 divides $d_1 + d_0$ (the sum of the individual digits). Do this. *Hint*: Show that $10d_1 + d_0$ has a 3 factor if $d_1 + d_0$ has a 3 factor by solving for d_0 in $d_1 + d_0 = 3k$ and then substituting the result for d_0 in $10d_1 + d_0$.

156

$d_1 + d_0 = 3k \Rightarrow d_0 = 3k - d_1$. Substituting $3k - d_1$ for d_0 in $10d_1 + d_0$, we get $10d_1 + 3k - d_1 = 9d_1 + 3k = 3(3d_1 + k) \Rightarrow 3$ divides the given number. ∎

Constructing the preceding proof is not a trivial task. However, we are provided with substantial guidance by two observations:

1. An arbitrary two-digit number is not necessarily divisible by 3. Thus, in our proof, we obviously have to use the condition $d_1 + d_0 = 3k$ for some integer k.

2. We have to show $10d_1 + d_0 = 3j$ for some $j \in \mathbb{Z}$.

So how do we accomplish the objective given in 2 above using the condition given in 1? Substitution is one way of transforming a mathematical expression. So we try it to see if it works. In this case, it does.

In a direct proof of q given p, you list statements starting with p, working forward until you get q. If you get stuck working forward, try working backward from q. That is, determine a statement—let's call it s—that implies q. Place s in the proof just before q. Having s and q (the last two statements in the proof) may give you a clue how to continue working forward in the proof. If not, then continue working backward. Working forward and backward often makes a proof easier to complete than working forward exclusively.

Proof by Contradiction

157

Let's consider another proof technique: proof by contradiction. In a ***proof by contradiction***, we prove a statement p by assuming its negation, $\sim p$, and then deriving a contradiction through a sequence of logically sound steps. A contradiction occurs when a statement appears in the sequence that is obviously false, is conflicting with some previous statement in the sequence, or is conflicting with a statement already known to be true. A logically sound sequence of statements can produce a contradiction only if the starting statement is false. Thus, if a logically sound sequence of statements that starts with $\sim p$ produces a contradiction, $\sim p$ must be false, or equivalently, p must be true.

Let's prove by contradiction that there is no largest integer. This statement says something about all integers. Thus, it is universally quantified. Restate this statement, explicitly specifying the universal quantifier.

158

For all $x \in \mathbb{Z}$, x is not the largest integer.

What is the negation of this statement (remember that "all" changes to "there exists" and that the predicate is negated)?

159

There exists an $x \in \mathbb{Z}$ such that x is the largest integer. This statement—the negation of the statement we want to prove—starts our proof by contradiction. We prefix it with the word "assume" or "suppose" to indicate that we are simply assuming it to be true. Here is the proof:

Assume there exists an $x \in \mathbb{Z}$ such that x is the largest integer. Note that $x + 1$ is an integer, and $x + 1 > x$. But then x is not the largest integer. We now have our contradiction: we have both that x is the largest integer and that x is not the largest integer. From this contradiction, we can conclude that our initial assumption—that there is a largest integer—is false. ∎

Let's prove by contradiction that $\emptyset \subseteq A$ for all sets A. Negate this statement.

160

$\emptyset \not\subseteq A$ for some set A.

The given statement is universally quantified. Thus, to negate it, we change "for all" to "for some" and $\subseteq$ to $\not\subseteq$. This statement, prefixed with "assume," starts our proof by contradiction. Do the proof.

161

Assume $\emptyset \not\subseteq A$ for some set A. Then there exists an $x \in \emptyset$ such that $x \notin A$. But there are no elements in $\emptyset$. From this contradiction, we can conclude that our initial assumption—that $\emptyset \not\subseteq A$ for some set A—is false. Thus, its negation, $\emptyset \subseteq A$ for all sets A, is true. ∎

Some theorems have the form "if p then q," which is equivalent to "q given p," "q if p," and "$p \Rightarrow q$." What is the negation of $p \Rightarrow q$? *Hint*: $p \Rightarrow q$ is not true if and only if p is true and q is false.

162

$p \wedge \sim q$

Accordingly, a proof by contradiction of a theorem in the form $p \Rightarrow q$ starts with the statement "suppose p and $\sim q$." Let's do an example.

Prove that if n^2 is even, then n is even. Negate this statement.

163

n^2 is even and n is not even.

Thus, our proof starts with the following statement: "suppose n^2 is even and n is not even." Do the proof.

164

Suppose n^2 is even and n is not even. Because n is not even, it must be odd, which implies that $n = 2k + 1$ for some $k \in \mathbb{Z}$. Thus, $n^2 = (2k + 1)^2 = 4k^2 + 4k + 1 = 2(2k^2 + 2) + 1 = 2j + 1$, where $j = 2k^2 + 2$. Thus, n^2 is odd. But n^2 is even. ∎

Note that we ended the proof as soon as the contradiction appeared. This should not cause any confusion as long as it is clear that the proof technique used is proof by contradiction. To make a proof perfectly clear, it is a good idea to indicate the proof technique used right at the beginning of the proof. For example, a writer might start a proof by contradiction with the words "Proof by contradiction."

Prove by contradiction the statement "an integer cannot be both even and odd." *Hint*: Start by letting x be both an even integer and an odd integer (which is the negation of the statement we want to prove). Then derive a contradiction.

165

Proof by contradiction: Let x be an even integer and an odd integer $\Rightarrow x = 2i$ and $x = 2j + 1$ for some $i, j \in \mathbb{Z} \Rightarrow 2i = 2j + 1 \Rightarrow 2i - 2j = 1 \Rightarrow 2(i - j) = 1 \Rightarrow i - j = \frac{1}{2}$. But $i - j$ is an integer. ∎

Contrapositive Proof

166

Let's investigate another important proof technique: the ***contrapositive proof.*** In a contrapositive proof, we prove $p \Rightarrow q$ by proving its contrapositive $\sim q \Rightarrow \sim p$. The implications $p \Rightarrow q$ and $\sim q \Rightarrow \sim p$ are equivalent. Thus, we can prove either one by proving the other one.

Let's do a contrapositive proof of the following statement: If $3n$ is odd, then n is odd. This statement is difficult to prove directly (i.e., with a direct proof), but its contrapositive has a simple, direct proof.

What is the contrapositive of the given statement?

167

If n is not odd, then $3n$ is not odd.

Here is the proof:

n is not odd $\Rightarrow n$ is even $\Rightarrow n = 2k$ for some $k \in \mathbb{Z} \Rightarrow 3n = 2(3k) \Rightarrow 3n$ is even $\Rightarrow 3n$ is not odd. ∎

Good advice: If you are having trouble proving $p \Rightarrow q$, try proving $\sim q \Rightarrow \sim p$ instead. As in the preceding example, proving the contrapositive is sometimes easier than proving the given statement.

Proving a Biconditional

168

Suppose we want to prove a biconditional—that is, a statement that has any of the following forms (they are all equivalent):

Chapter 4: Proof Techniques

- $p \Leftrightarrow q$
- p if and only if q
- $p \Rightarrow q$ and $q \Rightarrow p$
- p is necessary and sufficient for q

One approach is to directly prove $p \Rightarrow q$ and directly prove $q \Rightarrow p$. Alternatively, we can use a contrapositive proof for either or both of these implications. For example, instead of proving $q \Rightarrow p$, we can prove the equivalent $\sim p \Rightarrow \sim q$. There are, in fact, four variations for a proof of the biconditional $p \Leftrightarrow q$:

1. Prove $p \Rightarrow q$, and prove $q \Rightarrow p$.
2. Prove $p \Rightarrow q$, and prove $\sim p \Rightarrow \sim q$.
3. Prove $\sim q \Rightarrow \sim p$, and prove $q \Rightarrow p$.
4. Prove $\sim q \Rightarrow \sim p$, and prove $\sim p \Rightarrow \sim q$.

Let's prove the following biconditional

m is odd $\Leftrightarrow m^2$ is odd

using variation 2 above. We have to prove both of the following statements:

- m is odd $\Rightarrow m^2$ is odd.
- m is not odd $\Rightarrow$ if m^2 is not odd.

Here is the proof of "m is odd $\Rightarrow m^2$ is odd."

m is odd
$\Rightarrow m = 2k + 1$ for some $k \in \mathbb{Z}$
$\Rightarrow m^2 = 4k^2 + 2k + 1 = 2(2k^2 + k) + 1 = 2j + 1$ where $j = 2k^2 + k$
$\Rightarrow m^2$ is odd.

Complete the proof by proving "m is not odd $\Rightarrow$ if m^2 is not odd."

169

m is not odd
$\Rightarrow m$ is even
$\Rightarrow m = 2k$ for some $k \in \mathbb{Z}$
$\Rightarrow m^2 = 4k^2 = 2(2k^2) = 2j$ where $j = 2k^2$
$\Rightarrow m^2$ is even
$\Rightarrow m^2$ is not odd. ∎

Key idea: A common way to prove $p \Leftrightarrow q$ is to prove $p \Rightarrow q$ and $\sim p \Rightarrow \sim q$.

Proving a Disjunction

170

The truth table for $p \Rightarrow (q \vee r)$ has exactly one false row. What are the values of p, q, and r for this row?

171

For an implication to be false, the antecedent must be true and the consequent must be false. Thus, for $p \Rightarrow (q \vee r)$ to be false, p must be true and q and r must be false (both q and r must be false for $q \vee r$ to be false).

Now consider $(p \wedge \sim q) \Rightarrow r$. Its truth table has exactly one false row. What are the values of p, q, and r for this row?

172

p must be true and q and r must be false—the same values as for $(p \wedge \sim q) \Rightarrow r$. Thus, they have identical truth tables.

Because their tables are identical, to prove $p \Rightarrow (q \vee r)$, we can instead prove $(p \wedge \sim q) \Rightarrow r$. For example, let's prove that if $ab = 0$, then $a = 0$ or $b = 0$, where a and b are real numbers. This statement is of the form $p \Rightarrow (q \vee r)$, where p is "$ab = 0$," q is "$a = 0$," and r is "$b = 0$." We will prove it by proving the equivalent statement $(p \wedge \sim q) \Rightarrow r$:

If $ab = 0$ and $a \neq 0$, then $b = 0$.

Here is the proof:

Assume $a \neq 0$. Then $1/a$ exists (if $a = 0$, then $1/a$ is undefined). Multiplying both sides of $ab = 0$ by $1/a$, we get $(1/a)ab = (1/a)0$, which simplifies to $b = 0$. ∎

Using the Well-Ordering Principle

173

Suppose that, on a line that extends infinitely in both directions, we mark off the integers at equal intervals:

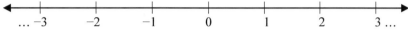

Then each point on this line corresponds to a number. For example, the midpoint between 0 and 1 corresponds to the number 0.5. The set consisting of every number that corresponds to some point on the number line above is called the set of **real numbers,** denoted by $\mathbb{R}$. Every integer is a real number. Thus, $\mathbb{Z}$ is a _____ of $\mathbb{R}$.

174

subset

Consider the set $S = \{x \in \mathbb{R} : 0 \leq x \leq 1\}$. What is the smallest element in this set?

175

0

Now consider the set $M = \{x \in \mathbb{R} : 0 < x < 1\}$. Is 0 its smallest element in M?

176

No; 0 is not an element of M, because all the elements of M are greater than 0.

So what is the smallest element of M? It is not 0.1, because there are positive real numbers smaller than 0.1. Similarly, it is not 0.0000000001, because there are positive real numbers smaller than 0.0000000001. Given any real number greater than 0, we can always find a smaller real number greater than 0. For example, we can always divide by 10 to get a smaller number greater than 0. We conclude that *M does not have a smallest number.*

Does $\mathbb{Z}$ have a smallest element?

177

No.

Prove this with a proof by contradiction. *Hint*: The proof is similar to our proof in frame **159** that there is no largest integer.

178

Assume that $\mathbb{Z}$ has a smallest element. Let k be this integer. $k - 1 \in \mathbb{Z}$ because subtracting integers always yields an integer. $k - 1 < k$. But then k is not the smallest element in $\mathbb{Z}$. We have a contradiction. We resolve this contradiction by concluding that our initial assumption—that $\mathbb{Z}$ has a smallest element—is false. ∎

Does the set $\{2, 5, 11\}$ have a smallest element?

179

Yes. It is 2.

Does the set {5, 7, 9, 11, ...} have a smallest element?

180

Yes. It is 5.

Does every nonempty set of *positive* integers have a smallest element?

181

Yes

That every nonempty set of positive integers has a smallest element is called the ***well-ordering principle.*** The well-ordering principle is often used in contradiction proofs to show that a subset *S* of the set of positive integers is empty: we start by assuming that *S* is nonempty and therefore, by the well-ordering principle, has a smallest element. We then show that *S* has an element smaller than the smallest element—an obvious contradiction. We can then conclude that our original assumption—that *S* is nonempty—is false.

Let's do an example. Before we start, let's review some properties of inequalities. Consider the inequality 2 < 5. This inequality is true. Multiply both components of this inequality by 10 to get a new inequality.

182

20 < 50. Is this inequality also true?

183

Yes

Now multiply the original inequality by −10.

184

−10 < −50. Is this inequality true?

185

No

However, if we reverse the direction of the inequality (i.e., change "<" to ">"), then we get the true inequality −10 > −50. Here is the general rule:

Suppose $x < y$. If k is positive, then $kx < ky$. If k is negative, then $kx > ky$.

What if $k = 0$?

186

Then $kx = ky = 0$.

We have a different rule for addition: If an integer k is added to both sides of an inequality, the direction of the inequality does not change for any value of k:

Suppose $x < y$. Then $x + k < y + k$ for all k.

Now let's get back to our proof by contradiction in which we use the well-ordering principle. Let's show that the smallest positive integer is 1. We do this by showing that the set *S* of positive integers less than 1 is empty.

Assume that *S* is nonempty. What do we know about *S*? *Hint*: Use the well-ordering principle.

187

It has a smallest element.

Let m be this smallest element. What do we know about m? Because it is in S, it must be positive and less than 1. Express this using mathematical symbols.

188

$0 < m < 1$

Now multiply all three components of the preceding inequality by m. What is the new inequality that results?

189

$0 < m^2 < m$

The direction of the inequalities remains the same because we are multiplying by a positive number. From the preceding frame, we have that $m < 1$. Thus, $0 < m^2 < m < 1$, which implies $m^2 \in S$. But then m is not the smallest element in S. We have our contradiction. We conclude that our initial assumption—that S is nonempty—is false. hat is, there is no positive integer less than 1. ∎

Proof by Induction

190

If $x = 5$, does $x + 1 = 6$? We can equivalently ask, is it true that $(x = 5) \Rightarrow (x + 1 = 6)$?

191

Yes. Just add 1 to both sides of $x = 5$ to get $x + 1 = 6$. Thus, $x + 1 = 6$ logically follows from $x = 5$. But we cannot conclude that $x + 1 = 6$. Why not?

192

The antecedent in the implication $(x = 5) \Rightarrow (x + 1 = 6)$ may not be true.

Suppose, for example, $x = 3$. Then, $x + 1 = 4$. If, however, x is indeed equal to 5, then, of course, $x + 1 = 6$. To summarize, if it is true that $(x = 5) \Rightarrow (x + 1 = 6)$, we cannot conclude that $x + 1 = 6$. However, if it is true that $x = 5$ *and* $(x = 5) \Rightarrow (x + 1 = 6)$, then we can correctly conclude that $x + 1 = 6$.

In general, if $p \Rightarrow q$ is true, then it is not necessarily the case that q is true (see the first row of the truth table below). However, if both p and $p \Rightarrow q$ are true, then q must also be true (see the fourth row of the truth table). That q follows from p and $p \Rightarrow q$ is a law of logic called ***modus ponens.***

p	q	$p \Rightarrow q$	
F	F	T	first row
F	T	T	second row
T	F	F	third row
T	T	T	fourth row

Suppose that we have an infinite sequence of statements, $s_1, s_2, s_3, \ldots$, in which each statement implies the next. That is,

$$s_1 \Rightarrow s_2 \Rightarrow s_3 \Rightarrow \ldots$$

Can we conclude that all the statements are true?

193

No. s_1 could be false, in which case s_2 could also be false, in which case s_3 could also be false, and so on. But suppose we also know that s_1 is true. Can we then conclude that all the statements are true?

194

Yes. Because s_1 and $s_1 \Rightarrow s_2$ are true, s_2 is true by modus ponens. Then, because s_2 and $s_2 \Rightarrow s_3$ are true, s_3 is true. Each time we show that a statement in the sequence is true, we then use it and modus ponens to show that the next statement is also true. By

Chapter 4: Proof Techniques

repeatedly applying modus ponens, we can show that all the statements in the sequence are true. This line of logical reasoning is called the ***principle of mathematical induction.***

Suppose we are given an infinite sequence of statements: $s_1, s_2, s_3, \ldots$. If we can show that

1. s_1 is true, and
2. $s_1 \Rightarrow s_2 \Rightarrow s_3 \Rightarrow \ldots$ is true,

then by the principle of mathematical induction, every statement in our sequence is true. That is, s_n is true for all $n \geq 1$. But how do we do step 2 above? It consists of an infinite number of implications. We obviously cannot prove each one individually. We simply prove just one implication, $s_n \Rightarrow s_{n+1}$, where we make *no* assumptions about n except that it is a positive integer. Thus, proving that $s_n \Rightarrow s_{n+1}$ proves that every statement in our sequence implies the next—that is $s_1 \Rightarrow s_2 \Rightarrow s_3, \Rightarrow \ldots$.

Using mathematical induction, let's prove a theorem that gives us a formula for determining the sum of the consecutive positive integers starting from 1 up to n:

$$1 + 2 + \cdots + n = n(n + 1)/2 \text{ for } n \geq 1$$

This theorem makes a statement for each value of n. Thus, it represents a sequence of statements $s_1, s_2, s_3, \ldots$, corresponding to $n = 1, 2, 3, \ldots$, where

s_1: $1 = 1(1 + 1)/2$
s_2: $1 + 2 = 2(2 + 1)/2$
s_3: $1 + 2 + 3 = 3(3 + 1)/2$
$\ \vdots$
s_n: $1 + 2 + \cdots + n = n(n + 1)/2$
s_{n+1}: $1 + 2 + \cdots + n+1 = [n+1]([n+1] + 1)/2 = (n+1)(n+2)/2$
$\phantom{s_{n+1}:}\ \vdots$

To prove by induction that each statement is true (and therefore, our theorem is true), we perform two steps:

1. show that s_1 is true, and
2. show that $s_n \Rightarrow s_{n+1}$, where we make no assumptions about n other than it is a positive integer.

Step 1 is called the ***basis step***; step 2 is called the ***induction step.*** In the induction step, we assume s_n and then prove s_{n+1}. The assumption we make in the induction step—that s_n is true—is called the ***inductive hypothesis.***

What is s_1?

195

$1 = 1(1 + 1)/2$

Is s_1 true?

196

Yes. Both sides of the equation in the preceding frame equal 1.

What is s_n?

197

$1 + \cdots + n = n(n + 1)/2$

What is s_{n+1}?

198

$1 + 2 + \cdots + n + 1 = [n+1]([n+1] + 1)/2 = (n+1)(n+2)/2$

In the induction step, we have to show that $s_n \Rightarrow s_{n+1}$. To do this is easy. Simply start with the equation for s_n and add $n + 1$ to both sides. Do this.

199

$1 + \cdots + n + (n+1) = n(n+1)/2 + (n+1)$

The left side of this new equality is the left side of s_{n+1}. If we then show that its right side equals the right side of s_{n+1}, then we have shown that our new equation—which we derived from s_n—is s_{n+1}. Thus, we have shown that $s_n \Rightarrow s_{n+1}$.

Let's show that the right side of the equation above is the the right side of s_{n+1}. To do this, combine the terms on right side of the preceding equality (put everything over 2).

200

$1 + \cdots + n + (n+1) = [n(n+1) + 2(n+1)]/2$

Now, factor out $n+1$ in the numerator on the right side.

201

$1 + \cdots + n + (n+1) = (n+1)(n+2)/2$

This equation is s_{n+1}. So, from s_n, we can get s_{n+1}. Our reasoning did not assume any particular value of n. Thus, $s_n \Rightarrow s_{n+1}$ for *all* positive integers n. In other words, we have proven $s_1 \Rightarrow s_2 \Rightarrow s_3 \Rightarrow \ldots$. Because we have also shown that s_1 is true (frame **195**), by the principle of mathematical induction, we can conclude that s_n is true for all positive integers n. ∎

Let s_n be the statement $n^2 > n$. Suppose we want to prove that s_n is true for all integers $n \geq 2$. That is, we want to prove that all the statements in the sequence $s_2, s_3, s_4, \ldots$ are true. The first statement in the sequence is s_2, not s_1 (s_1, in fact, is not true). We can still use induction. We simply show in the basis step that s_2—not s_1—is true. Then, in the induction step, we show that s_n implies s_{n+1} for all $n \geq 2$. In this proof, we need to use the ***transitivity law of inequalities***:

If $a < b$ and $b < c$, then $a < c$.

Do the proof. *Hint*: In the basis step, start with the inequality $2 > 1$. In the induction step, add $2n + 1$ to both sides of s_n.

202

Basis: $2 > 1$. Multiply both sides of this inequality by 2. We get $2^2 > 2$, which is s_2.

Induction: Assume s_n is true. That is, $n^2 > n$. Add $2n + 1$ to both sides of this inequality. We get $n^2 + 2n + 1 > n + 2n + 1$.

Finish the proof. *Hint*: Factor $n^2 + 2n + 1$.

203

$n^2 + 2n + 1 = (n+1)^2$. Thus, $(n+1)^2 > n + 2n + 1$. We know that $n + 2n + 1 > n + 1$, because $2n$ is positive. Then, by the transitivity law of inequalities, $(n+1)^2 > n + 1$, which is s_{n+1}. ∎

In this proof, how did we know to add $2n + 1$ to both sides of s_n? The left side of s_{n+1} is $(n+1)^2 = n^2 + 2n + 1$. The left side of s_n is n^2. Thus, to get the left side of s_{n+1} from s_n, we simply add $2n + 1$ to the left side of s_n (and to the right side of s_n to maintain the equality). We then check to see if the right side of s_n to which we added $2n + 1$ equals the right side of s_{n+1}.

Proof by Cases

204

Sometimes a proof can be simplified if it is broken into cases, and a proof is provided for each case. Let's use this approach to prove that $n^2 - n$ is even for all integers n. We need proofs for two cases: for n even and for n odd.

Case 1: n is even.
Then $n = 2k$ for some integer k, and $n^2 - n = (2k)^2 - 2k = 2(2k^2 - k)$. Thus, $n^2 - n$ is even.
Case 2: n is odd.

Do case 2.

205

Then $n = 2k + 1$ for some integer k. $n^2 - n = (2k+1)^2 - (2k+1) = 4k^2 + 4k + 1 - (2k+1) = 4k^2 + 2k = 2(2k^2 + k)$. Thus, $n^2 - n$ is even. ∎

Without Loss of Generality

206

Suppose that we want to prove that the sum of two integers, one even and one odd, is odd, and to do this, we assume that $x = 2$ and $y = 3$ are our two integers. We then show that $x + y$ is odd. Such a "proof," of course, is incorrect. It shows only that for the particular pair of even and odd integers 2 and 3, their sum is odd. Because we make the assumption that $x = 2$ and $y = 3$, our proof loses its generality. It does not show that the sum of an even and odd integer is odd for *every* pair of even and odd integers. Our assumption that $x = 2$ and $y = 3$ simplifies our proof (which is desirable), but it makes the proof incorrect (which is unacceptable). However, sometimes we can make assumptions in a proof that simplify the proof but do not result in a loss of generality. When we do so, we typically indicate that there is no loss of generality by stating "without loss of generality" (commonly abbreviated "wlog"). For example, a proof that the sum of two integers, one even and one odd, is odd might proceed as follows:

Let $x, y \in \mathbb{Z}$. Without loss of generality, assume that x is even and that y is odd. Then $x = 2i$ and $y = 2j + 1$ for some $i, j \in \mathbb{Z}$. Adding x and y, we get $x + y = 2i + 2j + 1 = 2(i + j) + 1 \Rightarrow x + y$ is odd. ∎

The case in which x is even and y is odd is really no different from the case in which y is even and x is odd. In both cases, one number drawn from $\mathbb{Z}$ is even and the other is odd. Thus, we do not lose generality by assuming that x is even and y is odd. Moreover, this assumption simplifies the proof—we do not have to show that $x + y$ is odd for the y is even and x is odd case.

Pigeonhole Principle

207

Suppose we have five pigeonholes and four pigeons. Can we place all the pigeons into the pigeonholes so that no pigeonhole has more than one pigeon?

208

Yes. Simply place each pigeon in a pigeonhole that is not already occupied by a pigeon. Then four of the five pigeonholes will have exactly one pigeon, and the remaining pigeonhole will be empty.

We can also avoid dual occupancy of a pigeonhole if we have five pigeons. But if we have more than five pigeons, then at least one pigeonhole will have at least two pigeons. In general, if we have more pigeons that pigeonholes, then at least one pigeonhole will have at least two pigeons. We call this observation the ***pigeonhole principle***. This principle seems obvious, but let's prove it with a proof by contradiction: Suppose we have n pigeonholes and p pigeons, with $p > n$. Suppose we can place the p pigeons so that each pigeonhole has at most one pigeon. Then, because we have n pigeonholes, there can be at most n pigeons. But we have $p > n$ pigeons. Thus, our initial assumption that each pigeonhole has at most one pigeon is incorrect. At least one pigeonhole has more than one pigeon. ∎

Restate the pigeonhole principle in terms of a function from A to B, where A and B are finite sets with $|A| > |B|$. *Hint*: The elements of A are the pigeons; the elements of B are the pigeonholes.

209

If $|A| > |B|$, then there is no one-to-one function from A to B.

Suppose we have ten pigeonholes. How many pigeons do we need to guarantee that at least one pigeonhole has at least two pigeons?

210

At least 11. We need one more pigeon that we have pigeonholes.

Suppose we have ten pigeonholes. How many pigeons do we need to guarantee that at least one pigeonhole has at least *three* pigeons?

211

At least 21. We can accommodate 20 pigeons without exceeding two pigeons per pigeonhole by placing a pair of pigeons in each pigeonhole. But then the 21st pigeon would have to occupy a pigeonhole that already had two pigeons.

Suppose students take an exam that is graded from 1 to 10. How many students would have to take the exam to guarantee that at least three students received the same grade?

212

At least 21. In this problem, each possible grade is a "pigeonhole," and each student is a "pigeon." We can accommodate 20 students without exceeding two students per grade by pairing the students and giving each pair a distinct grade. But then the 21st student would have to get a grade that a pair of students already has.

Review Questions

1. Prove: If $A \subseteq B$ and $B \subseteq C$ then $A \subseteq C$.
2. Prove: If the sum of the digits in a three-digit integer is divisible by 3, then the given integer is divisible by 3.
3. Prove: The product of two rational numbers is rational.
4. Prove: If $3n + 2$ is even, then n is even.
5. Prove: The product of two odd numbers is odd.
6. How many people are required to guarantee that three are born on the same day of the year. *Hint*: There are 366 possible birthdays.
7. Prove by induction: $n < 2^n$ for all positive integers n.
8. Prove: There exists a positive integer n equal to the sum of all the positive integers less than n.
9. Prove: The sum of an irrational number and a rational number is an irrational number.
10. Prove by cases: $9n^2 + 3n - 2$ is even for all $n \in \mathbb{Z}$.

Answers to the Review Questions

1. Let $x \in A$. Because $x \in A$ and $A \subseteq B$, $x \in B$. Because $x \in B$ and $B \subseteq C$, $x \in C$. Thus, $A \subseteq C$. ∎
2. Suppose that $n = d_2 d_1 d_0$ and $3|(d_0 + d_1 + d_2)$. Then $d_0 + d_1 + d_2 = 3k$ for some $k \in \mathbb{Z}$, and $n = 100d_2 + 10d_1 + d_0$. Then $d_0 = 3k - d_2 - d_1$. Substituting in $n = 100d_2 + 10d_1 + d_0$, we get $n = 100d_2 + 10d_1 + 3k - d_2 - d_1 = 99d_2 + 9d_1 + 3k = 3(33d_2 + 3d_1 + k) \Rightarrow 3 \mid n$. ∎
3. Let n and m be rational numbers. Then $n = \frac{a}{b}$ and $m = \frac{c}{d}$, where $a, b, c, d \in \mathbb{Z}$, $b \neq 0$, and $d \neq 0$. $nm = \left(\frac{a}{b}\right)\left(\frac{c}{d}\right) = \frac{ac}{bd}$. $ac, bd \in \mathbb{Z}$. Moreover, $bd \neq 0$, because $b \neq 0$ and $d \neq 0$. Thus, nm is rational. ∎
4. *Contrapositive proof*: Suppose n is not even. Then $n = 2k + 1$ for some $k \in \mathbb{Z}$. Substituting in $3n + 2$, we get $3n + 2 = 3(2k + 1) + 2 = 6k + 4 + 1 = 2(3k + 2) + 1 \Rightarrow 3n + 2$ is not even. ∎
5. Suppose n and m are odd. Then $n = 2k + 1$ and $m = 2j + 1$, where $k, j \in \mathbb{Z}$. $nm = (2k + 1)(2j + 1) = 4kj + 2j + 2k + 1 = 2(2kj + j + k) + 1$. Thus, nm is odd. ∎
6. $2 \cdot 366 + 1 = 733$
7. *Basis step* $n = 1$: $n = 1 < 2 = 2^1 = 2^n$.
   Induction step: Assume $n < 2^n$. Then $n + 1 < 2^n + 1 < 2^n + 2^n = 2^{n+1}$. ∎
8. $3 = 1 + 2$ ∎
9. *Proof by contradiction*: Let x be rational, y irrational, and $z = x + y$. Suppose z is rational. Solving for y, we get $y = z - x$. $z - x$, the difference of two rational numbers, is rational. Thus, y is rational. But y is irrational. ∎
10. *Case 1*: n is even. Then $9n^2$, $3n$, and 2 are all even. Thus, $9n^2 + 3n - 2$ is even because the sum of evens is even.
    Case 2: n is odd. Then $9n^2$ and $3n$ are odd. Then $9n^2 + 3n$ is even because the sum of two odds is even. Thus, $(9n^2 + 3n) - 2$ is even because the sum of evens is even. ∎

Homework Questions

1. Prove: $1 + 2 + 3 + \cdots + 2^n = 2^{n+1} - 1$.
2. Prove: The sum of any two consecutive integers is odd.
3. Prove: $(1 - \frac{1}{2})(1 - \frac{1}{3})(1 - \frac{1}{4}) \cdots (1 - \frac{1}{n}) = \frac{1}{n}$ for all $n \geq 2$.
4. Prove: $5 \mid (n^5 - n)$ for all $n \geq 0$.
5. Prove: $1^2 + 2^2 + 3^2 + \cdots + n^2 = n(n + 1)(2n + 1)/6$.
6. Prove: If nm is odd then n is odd.
7. Prove: $2^{n+1} \leq 3^n$ for all $n \geq 2$.
8. Prove without using induction: The number of the first n positive integers is $n(n + 1)/2$. *Hint*: Consider the sum of $1 + 2 + \cdots + n$ and $n + (n - 1) + \cdots + 1$.
9. Prove: n is even if and only if $7n + 4$ is even.
10. Prove: The product of two consecutive even integers plus 1 is equal to the square of the intervening odd number, for example, $2 \cdot 4 + 1 = 3^2$.

5 Operations on Sets

Intersection, Union, and Complement

213

We can perform operations on sets to create new sets. For example, the **intersection** of the sets A and B (denoted by $A \cap B$) is the set of elements that are in both A and B. That is, $A \cap B = \{x : x \in A \wedge x \in B\}$. The union of the sets A and B (denoted by $A \cup B$) is the set that contains all the elements in A or B or both. That is, $A \cup B = \{x : x \in A \vee x \in B\}$. For example, if $A = \{1, 2\}$ and $B = \{2, 3\}$, then $A \cap B = \{2\}$ and $A \cup B = \{1, 2, 3\}$.

Suppose we restrict our attention to the subsets of some set $\mathcal{U}$. That is, the elements in all our sets are drawn from $\mathcal{U}$. We then say that $\mathcal{U}$ is the **universal set.** The **complement of the set A** (denoted by A^c) with respect to $\mathcal{U}$ is the set of elements in $\mathcal{U}$ that are not in A. That is, $A^c = \{x : x \in \mathcal{U} \wedge x \notin A\}$.

The elements in A^c depend not only on what is in A but also on what is in the universal set. For example, suppose the universal set is $\{1, 2, 3, 4, 5\}$. Thus, all our sets will have elements exclusively from this set. With this universal set, the complement of $\{1, 2\}$ is $\{3, 4, 5\}$. If, on the other hand, our universal set is $\mathbb{Z}$, then the complement of $\{1, 2\}$ is the set of all integers, except for 1 and 2.

If $x \in A \cup B$, is $x \in A$?

214

Not necessarily: x might be in B only. Then it would be in $A \cup B$ but not in A.

If $x \in A \cap B$, is $x \in A$?

215

Yes. If x is in $A \cap B$, it must be in both A and B.

Venn Diagrams

216

We can graphically represent set operations using Venn diagrams. In a **Venn diagram,** we place circles on a rectangle. The rectangle represents the universal set; the circles represent sets whose elements are from the universal set. For example, we can represent the intersection of the sets A and B with the diagram that follows. To facilitate referring to specific regions of this diagram, we have labeled each region with an identifying number.

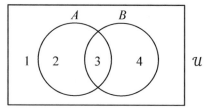

Regions 1, 2, 3, and 4 together (i.e., the entire rectangle) represent the universal set; the left circle (i.e., regions 2 and 3) represents A; the right circle (i.e., regions 3 and 4) represents B. The overlap of the A and B circles (i.e., region 3) represents the intersection of A and B.

Which regions in the preceding Venn diagram represent $A \cup B$?

The A and B circles together (i.e., regions 2, 3, and 4) represent $A \cup B$.

Which regions represent A^c?

218

Regions 1 and 4 (i.e., everything outside the A circle).

The ***set difference of A and B*** (denoted by $A - B$) is the set of all the elements in A that are not in B. That is, $A - B = \{x : x \in A \wedge x \notin B\}$. For example, if $A = \{2, 3\}$ and $B = \{3, 4\}$, then $A - B = \{2\}$.

Which regions in the Venn diagram in frame **216** represent $A - B$?

219

Region 2

Which regions in the Venn diagram in frame **216** represent $A \cap B^c$? These are all the elements inside the A circle that are also outside the B circle.

220

Region 2

Compare your answers for $A \cap B^c$ and for $A - B$. They are the same. What does this suggest?

221

$A - B = A \cap B^c$

This equality is easy to see: Because B^c is the set of all elements in the universal set not in B, $A \cap B^c$ contains all the elements that are both in A and not in B. Thus, $A \cap B^c$ contains precisely the elements in $A - B$.

What are the elements in $\mathbb{R} - \mathbb{Q}$?

222

$\mathbb{R} - \mathbb{Q}$ is the set of irrational numbers.

Pairwise Disjoint Sets

223

Two sets A and B are ***disjoint*** if their intersection is the empty set. In other words, A and B are disjoint if they have no elements in common.

Let $A = \{1, 2\}$, $B = \{2, 3\}$, and $C = \{4, 5\}$. Determine the set $(A \cap B) \cap C$ (i.e., intersect A with B, then intersect the result with C). Also determine the set $A \cap (B \cap C)$ (intersect A with the intersection of B with C).

224

Both result in $\emptyset$.

In the preceding example, the intersection of all three sets is the empty set. However, there is a common element between two of the sets (2 is in both A and B). Clearly, if the intersection of three sets is the empty set, we *cannot* conclude that there are no sets in the collection with common elements. This observation also applies to intersections of more than three sets. However, it does not apply to the intersection of two sets. If $A \cap B = \emptyset$, then it follows directly from the definition of intersection that A and B have no elements in common.

If each set in a collection of sets has no elements in common with any other set in the collection, then we say the sets are ***pairwise disjoint*** (*pairwise* means "every pair"). The sets A, B, and C in the preceding frame are *not* pairwise disjoint because A and B are not disjoint.

How do we test if a collection of sets is pairwise disjoint? We saw in the example in the preceding frame that we cannot simply intersect all the sets and then check if the result is ∅. To test if a collection of sets is pairwise disjoint, we have to intersect every possible pair of sets. If each of the intersections yields ∅ (indicating that the two sets in the pair have no elements in common), then we can conclude the collection of sets is pairwise disjoint.

Let's apply this procedure to $A = \{1, 2\}$, $B = \{2, 3\}$, and $C = \{4, 5\}$. Determine $A \cap B$, $A \cap C$, and $B \cap C$.

225

$A \cap B = \{2\}$, $A \cap C = \emptyset$, $B \cap C = \emptyset$. Because not all of these intersections yield ∅, A, B, and C are not pairwise disjoint.

To determine if a collection of sets is pairwise disjoint, we do not also have to test intersections in which the sets are in reverse order. For example, if we test $A \cap B$, then we do not have to test $B \cap A$, because $B \cap A = A \cap B$ for all sets A and B. The equality $B \cap A = A \cap B$ for all sets A and B is called the ***commutative law of intersection.*** We say that the sets A and B ***commute*** under intersection. That is, commuting (i.e., interchanging) A and B in the intersection $A \cap B$ does not affect the result of the intersection.

Proving the Subset Relation

226

Give an informal argument using the concept of a Venn diagram to show that if $A \cup B = A$, then $B \subseteq A$ for all sets A and B.

227

Informal argument: In a Venn diagram, the only way to get $A \cup B = A$ is to have B completely inside A.

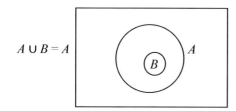

Let's now do a formal proof of this theorem. Here is our approach: we let $x \in B$. That is, we let x be an *arbitrary* element in B. We then show that x is also in A. Because x is an arbitrary element in B, our argument that x is also in A applies to *every* element in B. Thus, every element in B is in A, or, equivalently, $B \subseteq A$. Here is the proof:

Let $x \in B$. Then, by the definition of the union operator, $x \in A \cup B$. We are given that $A \cup B = A$. Thus, $x \in A$. ∎

This technique—showing that an arbitrary element in B is in A—is not the only way to prove our theorem. Let's also do a contrapositive proof.

What is the contrapositive of the given statement: if $A \cup B = A$, then $B \subseteq A$ for all sets A and B?

228

If $B \nsubseteq A$, then $A \cup B \neq A$ for all sets A and B.

Our contrapositive proof assumes $B \nsubseteq A$ and then shows $A \cup B \neq A$. What does $B \nsubseteq A$ tell us?

229

If $B \nsubseteq A$, then there exists an x such that $x \in B$ and $x \notin A$.

$x \in B$ implies what about x and $A \cup B$?

230

$x \in A \cup B$

We have shown that $x \notin A$ and $x \in A \cup B$. What can we conclude about $A \cup B$ and A?

231

They cannot be equal. That is, $A \cup B \neq A$. ∎

48 Chapter 5: Operation on Sets

Let's now prove the same theorem with a proof by contradiction. We want to prove that $A \cup B = A \Rightarrow B \subseteq A$. Our theorem is of the form $p \Rightarrow q$. Recall from frame **162**, in proof by contradiction of $p \Rightarrow q$, that we start with "suppose p and $\sim q$," from which we derive a contradiction. Accordingly, we start this proof with the following: Suppose $A \cup B = A$ and $B \not\subseteq A$.

Do the proof. *Hint*: Starting with $B \not\subseteq A$, show that $A \cup B \neq A$.

232

Suppose $A \cup B = A$ and $B \not\subseteq A$. Then there exists an x such that $x \in B$ and $x \notin A$. Because $x \in B$, $x \in A \cup B$. We also have that $x \notin A$. Thus, $A \cup B \neq A$. But $A \cup B = A$. ∎

Prove that for all sets A and B, $A \subseteq B$ if $A \cap B = A$. Use the technique that assumes that $x \in A$ and then shows that $x \in B$.

233

Let $x \in A$. We are given that $A \cap B = A$. Thus, x is also in $A \cap B$. Then, from the definition of set intersection, $x \in B$. Thus, $A \subseteq B$. ∎

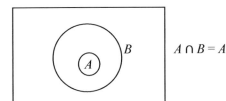

Proving Two Sets Are Equal

234

One way to prove that two sets are equal is to show that each is a subset of the other. Let's do an example. Let's prove the following: $(A \cup B)^c = A^c \cap B^c$. First, we show $(A \cup B)^c \subseteq A^c \cap B^c$:

Let $x \in (A \cup B)^c \Rightarrow x \notin A \cup B \Rightarrow x \notin A$ and $x \notin B \Rightarrow x \in A^c$ and $x \in B^c \Rightarrow x \in A^c \cap B^c$. Thus, $(A \cup B)^c \subseteq A^c \cap B^c$.

Next, we show that $A^c \cap B^c \subseteq (A \cup B)^c$. How do we start the proof?

235

Let $x \in A^c \cap B^c$.

Using the definition of intersection, what can we say about x?

236

$x \in A^c$ and $x \in B^c$.

This, in turn, implies what? *Hint*: Change $\in$ to $\notin$.

237

$x \notin A$ and $x \notin B$.

If x is in neither A nor B, what can we conclude?

238

$x \notin A \cup B$.

But this implies what? *Hint*: Change $\notin$ to $\in$.

239

$x \in (A \cup B)^c$. Thus, $A^c \cap B^c \subseteq (A \cup B)^c$. In frame **234**, we proved that $(A \cup B)^c \subseteq A^c \cap B^c$. These two results together prove that $(A \cup B)^c = A^c \cap B^c$. ∎

Let's state this equality with words: the complement of a union is the intersection of the complements. Using a similar proof, we can also show that $(A \cap B)^c = A^c \cup B^c$. State this equality with words.

240

The complement of an intersection is the union of the complements.

DeMorgan's Laws	Set Equalities
$\sim(p \lor q) \equiv \sim p \land \sim q$	$(A \cup B)^c = A^c \cap B^c$
$\sim(p \land q) \equiv \sim p \lor \sim q$	$(A \cap B)^c = A^c \cup B$

Notice how the set equalities in the preceding frame parallel DeMorgan's laws from chapter 3. In fact, these set equalities are also called DeMorgan's laws.

More Laws on Sets

241

One of the laws of sets is the ***associative law of intersection***. That is, $A \cap (B \cap C) = (A \cap B) \cap C$ for all sets A, B, and C. By this law, the order in which we perform the intersection operations (left to right or right to left) does not matter. Thus, we can dispense with the parentheses when we write a sequence of two intersections. For example, we can simply write $A \cap B \cap C$.
  Is set union commutative? That is, does $A \cup B = B \cup A$ for all sets A and B? Is it associative?

242

Set union is both commutative and associative. Thus, the order in which we perform the union operations and the order in which we write the sets do not matter.

Is addition on the integers commutative and associative?

243

Yes

Is subtraction on the integers commutative and associative?

244

No; $5 - 3 \neq 3 - 5$, so subtraction is not commutative; $(3 - 2) - 1 = 0$, but $3 - (2 - 1) = 2$, so it also is not associative.

The following table summarizes the laws on sets:

	Commutative laws	
$A \cup B = B \cup A$		$A \cap B = B \cap A$
	Associative laws	
$(A \cup B) \cup C = A \cup (B \cup C)$		$(A \cap B) \cap C = A \cap (B \cap C)$
	Distributive laws	
$A \cup (B \cap C) = (A \cup B) \cap (A \cup C)$		$A \cap (B \cup C) = (A \cap B) \cup (A \cap C)$
	Identity laws	
$A \cup \emptyset = A$		$A \cap \mathcal{U} = A$
	Complement laws	
$A \cup A^c = \mathcal{U}$	$(A^c)^c = A$	$A \cap A^c = \emptyset$
	Domination laws	
$A \cup \mathcal{U} = \mathcal{U}$		$A \cap \emptyset = \emptyset$
	Absorption laws	
$A \cup (A \cap B) = A$		$A \cap (A \cup B) = A$
	DeMorgan's laws	
$(A \cup B)^c = A^c \cap B^c$		$(A \cap B)^c = A^c \cup B^c$

All the laws with ∪ and ∩ come in pairs. Moreover, in each pair, you can obtain one from the other by switching ∩ with ∪ and switching ∅ (the empty set) with $\mathcal{U}$ (the universal set). For example, one of the identity laws is $A \cup \emptyset = A$. To get the other identity law, we replace ∪ with ∩ and ∅ with $\mathcal{U}$ to get $A \cap \mathcal{U} = A$.

When we union ∅ with any set A, we get A back. More precisely, $\emptyset \cup A = A \cup \emptyset = A$. For this reason, we call ∅ an *identity* under union. Similarly, $\mathcal{U}$ is an identity under intersection. That is, when we intersect $\mathcal{U}$ with any set A, we get A back.

In ℝ, what is the identity with respect to addition?

245

0, because $a + 0 = 0 + a = a$ for all $a \in \mathbb{R}$.

What is the identity with respect to multiplication?

246

1, because $a \cdot 1 = 1 \cdot a = a$ for all $a \in \mathbb{R}$.

Principle of Inclusion-Exclusion

247

Recall that we use the vertical bars around a set to denote the number of elements in the set. For finite sets A and B, does $|A \cup B| = |A| + |B|$? That is, is the number of elements in $A \cup B$ equal to the number of elements in A plus the number of elements in B? Let's consider an example: $A = \{1, 2\}$, $B = \{2, 3\}$, and $A \cup B = \{1, 2, 3\}$. What is $|A \cup B|$ and $|A| + |B|$?

248

$|A \cup B| = 3$, $|A| + |B| = 4$

The reason why we get different values is because $|A| + |B|$ counts the elements common to A and B twice. To compensate for this, we need to subtract out the number of elements common to A and B (that way, they are counted only once). The number of elements common to A and B is $|A \cap B|$. Thus, the correct formula for $|A \cup B|$ is

$$|A \cup B| = |A| + |B| - |A \cap B|.$$

This formula is called the ***principle of inclusion-exclusion.***

Suppose that at a college 50 students are majoring in math and 80 students are majoring in biology. By comparing the list of math majors with the list of biology majors, it is determined that 10 students are on both lists. That is, 10 students are majoring in both math and biology. What is the total number of students majoring in math and/or biology?

249

Let M be the set of math students, and let B be the set of biology students: $|M \cup B| = |M| \cup |B| - |M \cap B| = 50 + 80 - 10 = 120$.

Review Questions

1. Using DeMorgan's laws, simplify $[(A \cap B^c) \cup C]^c$.
2. Show that $A \cup B \subseteq A \cup B \cup C$.
3. $A \triangle B$, the *symmetric difference* of A and B, is the set containing those elements in A or in B, but not in both. Which regions in the Venn diagram in frame **216** correspond to $A \triangle B$?
4. If $A - B = B - A$, what can we conclude about A and B?
5. Let $A = \{1, 2, 3\}$, $B = \{2, 3, 4, 5\}$, and $C = \{1\}$. Determine the following sets: $A \cap B$, $A \cup B$, $A \cap B \cap C$, $A - B$, $B - A$, $A \triangle B$, $C \triangle \mathbb{Z}$, $A \cup (B \cap C)$. See review question 3.
6. If $A \cap B = A \cap C$, does $B = C$? In other words, can we cancel A?
7. If $A \cup B = A \cup C$, does $B = C$? In other words, can we cancel A?
8. Prove that $(A - B) \cap (B - A) = \emptyset$.

9. Does $\mathcal{P}(A) \cap A = \emptyset$ for all sets A? $\mathcal{P}(A)$ is the power set of A.
10. The math department currently has 20 students majoring in math. The physics department currently has 50 students majoring in physics. Seven students are majoring in both math and physics. How many students are majoring in math and/or physics?

Answers to the Review Questions

1. $[(A \cap B^c) \cup C]^c = (A \cap B^c)^c \cap C^c = (A^c \cup B) \cap C^c$
2. Let $x \in A \cup B \Rightarrow x \in (A \cup B) \cup C = A \cup B \cup C$. Thus, $A \cup B \subseteq A \cup B \cup C$. ∎
3. Regions 2 and 4.
4. $A = B$
5. $A \cap B = \{2, 3\}$, $A \cup B = \{1, 2, 3, 4, 5\}$, $A \cap B \cap C = \emptyset$, $A - B = \{1\}$, $B - A = \{4, 5\}$, $A \triangle B = \{1, 4, 5\}$, $C \triangle \mathbb{Z}$ = all the integers except 1, $A \cup (B \cap C) = \{1, 2, 3\}$
6. No. If $A = \emptyset$, then any B and C satisfy the given equation.
7. No. If B is any subset of A, and C is any subset of A, then $A \cup B = A \cup C$.
8. Let $x \in A - B \Rightarrow x \notin B \Rightarrow x \notin B - A \Rightarrow A - B$ and $B - A$ are disjoint. That is, $(A - B) \cap (B - A) = \emptyset$. ∎
9. No. Let $A = \{1, \{1\}\}$. $\mathcal{P}(A) = \{\emptyset, \{1\}, \{\{1\}\}, \{1, \{1\}\}\}$ and $\mathcal{P}(A) \cap A = \{\{1\}\}$.
10. $|M \cup P| = |M| + |P| - |M \cap P| = 20 + 50 - 7 = 63$

Homework Questions

1. Prove that $A \cup (B \cap C) = (A \cup B) \cap (A \cup C)$.
2. Prove that $A \cap (B \cup C) = (A \cap B) \cup (A \cap C)$.
3. Prove that $(A - B) \cup (A \cap B) \cup (B - A) = A \cup B$.
4. Prove or disprove: $\mathcal{P}(A) \subseteq \mathcal{P}(B)$ if and only if $A \subseteq B$.
5. Let X = the set of all sets. Is $X \in X$?
6. The set $\{1, 2\}$ has 1 and 2 as its only elements. The set $\{1, 2\}$ is not an element of $\{1, 2\}$. That is, $\{1, 2\} \notin \{1, 2\}$. For most sets S, $S \notin S$. Let $R = \{S : S \notin S\}$. Is $R \in R$? Is $R \notin R$?
7. Suppose that, in a room, there are 20 people who are rich or tall. If 10 people in the room are rich and 50 people are tall, how many people are both rich and tall?
8. Suppose $A \cup B = \{1, 2, 3, 4, 5, 6, 7\}$, $A \cap B = \{1, 7\}$, and $A - B = \{3, 6\}$. What are the elements in A? In B?
9. Let A be the set of even integers, B the set of integers that are multiples of 3, and C the set of integers that are multiples of 6. Determine $A \cap B$, $A \cup B$, $A \cap B \cap C$, and $A \cup B \cup C$.
10. Complete the following equation: $|A \cup B \cup C| = |A| + |B| + |C|$ ....

6 Relations and Functions

Relations

250

A set is an unordered collection of objects. Does {2, 3} = {3, 2}?

251

Yes. The order in which the elements of a set are listed is irrelevant. However, in an **ordered pair**, order matters. We specify an ordered pair by enclosing the pair with parentheses. For example, the ordered pair with 2 as the first component and 3 as the second component is (2, 3).

Does (2, 3) = (3, 2)?

252

No. Order matters in an ordered pair.

Suppose we are given that $(a, b) = (c, d)$, where a, b, c, and d are integers. What can we conclude about a, b, c, and d?

253

$a = c$ and $b = d$. That is, the first components of the two pairs must be equal, and the second components of the two pairs must be equal.

Given two sets A and B, we can create ordered pairs by taking the first component of each pair from A and the second component from B. For example, suppose $A = \{1, 2\}$ and $B = \{3\}$. List all the possible ordered pairs in which the first component is from A and the second component is from B.

254

(1, 3), (2, 3)

The set of all these order pairs is called the **Cartesian product** of A and B (denoted by $A \times B$). We define $A \times B$ as follows: $A \times B = \{(x, y) : x \in A \text{ and } y \in B\}$. In words, $A \times B$ is the set of all ordered pairs whose first component is from A and whose second component is from B. What is $A \times B$ if $A = \{1, 2\}$ and $B = \{1, 3\}$?

255

$A \times B = \{(1, 1), (1, 3), (2, 1), (2, 3)\}$

The two sets involved in a Cartesian product can be the same. If $A = \{1, 2\}$, what is $A \times A$?

256

$\{(1, 1), (1, 2), (2, 1), (2, 2)\}$

A **relation from A to B** is any subset of $A \times B$. For example, suppose $A = \{1, 2\}$ and $B = \{3, 4\}$.

Is $\{(1, 3), (2, 3)\}$ a relation from A to B?

257

Yes, because $\{(1, 3), (2, 3)\}$ is a subset of $A \times B$.

Is $A \times B$ a relation from A to B?

258

Yes, because $A \times B$ is a subset of $A \times B$.

Is $\emptyset$ a relation from A to B?

259

Yes, because $\emptyset$ is a subset of $A \times B$. *Any* subset of $A \times B$ is a relation from A to B.

Let $A = \{1, 2\}$. Is $\{(1, 1)\}$ a relation from A to A?

260

Yes, because $\{(1, 1)\}$ is a subset of $A \times A$.

A relation from A to A is any subset of $A \times A$. We generally call such a relation simply a ***relation on A.*** In a relation on A, the first and second components of each ordered pair are elements of A. Let $A = \{1, 2\}$. Is $\{(1, 1), (1, 3)\}$ a relation on A?

261

No. The given set of ordered pairs is not a subset of $A \times A$ (the ordered pair $(1, 3)$ is not in $A \times A$).

Relations from A to B can be nicely represented with an ***arrow diagram***—a diagram consisting of arrows connecting the elements of A with the elements of B. Each arrow represents an ordered pair in the relation. For example, suppose $A = \{1, 2\}$ and $B = \{3, 4\}$. One relation from A to B is $\{(1, 3), (1, 4)\}$. Here is its arrow diagram:

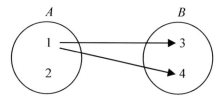

The arrow from 1 to 3 represents the ordered pair $(1, 3)$. Similarly, the arrow from 1 to 4 represents the ordered pair $(1, 4)$. An arrow corresponding to an ordered pair points from the first component of the pair to the second component.

The technical name for an arrow diagram is ***directed graph*** or ***digraph.*** However, we will use the more descriptive name "arrow diagram." Let $A = \{1, 2\}$. What is the arrow diagram of the following relation on A: $\{(1, 1), (1, 2)\}$? In your diagram, represent A twice.

262

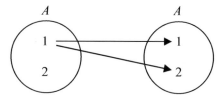

Here we have represented A twice, each with a circle. However, for relations on A, it is not necessary to represent A twice. Instead, we can represent it only once with a single circle and put all our arrows on that circle. Using this approach, represent the relation on A given by $\{(1, 1), (1, 2)\}$.

263

54 Chapter 6: Relations and Functions

We often give a name or symbol to a relation. For example, let $X = \{(1, 1), (1, 2)\}$. Would it be correct to state that $(1, 2) \in X$?

264

Yes, because the ordered pair $(1, 2)$ is one of the elements of X.

We can also indicate that $(1, 2)$ is in X by writing $1 \, X \, 2$. In other words, $1 \, X \, 2$ is shorthand for $(1, 2) \in X$. We call this alternate notation *infix notation* because the name of the relation is *fixed in* the middle of the two components of the ordered pair.

Is $1 \, X \, 1$ true for the relation X given in the preceding frame?

265

Yes, because the ordered pair $(1, 1) \in X$.

With infix notation, we can also indicate that an ordered pair is not in a relation by slashing the symbol for the relation. For example, suppose R is some relation. Then $2 \, \not R \, 1$ indicates that $(2, 1) \notin R$.

How can we tell from an arrow diagram for R if $2 \, \not R \, 1$?

266

There is no arrow from 2 to 1.

Let L be the relation on $\mathbb{Z}$ consisting of all ordered pairs in which the first component is less than the second component. Is $(1, 300) \in L$? Is $(300, 1) \in L$?

267

$(1, 300) \in L$ because 1 is less than 300.
$(300, 1) \notin L$ because 300 is not less than 1.

Rewrite $(1, 300) \in L$ and $(300, 1) \notin L$ using infix notation.

268

$1 \, L \, 300$
$300 \, \not L \, 1$

We have been using the name L to represent the relation on $\mathbb{Z}$ in which the first number in each pair is less than the second number. However, there is a well-established symbol for this relation. We should certainly use it rather than L. Can you guess what it is?

269

$<$

Thus, instead of writing $(1, 300) \in L$ or $1 \, L \, 300$, we should write $(1, 300) \in <$ or, using infix notation, $1 < 300$; $1 < 300$, of course, is the common way to indicate that $(1, 300)$ is in the relation designated by the symbol $<$. However, we could just as well use $(1, 300) \in <$. The symbol $<$, like L, represents a relation on $\mathbb{Z}$ (and on $\mathbb{Q}$ and on $\mathbb{R}$). Similarly, $>, \leq, \geq, =$, and $\neq$ also represent relations on $\mathbb{Z}$ (and on $\mathbb{Q}$ and on $\mathbb{R}$). We call these symbols *relational operators.* Each relational operator represents a set of ordered pairs.

Using the list approach, specify the relation on $\mathbb{Z}$ represented by the symbol $=$.

270

$\{\ldots, (-2, -2), (-1, -1), (0, 0), (1, 1), (2, 2), \ldots\}$

If R is a relation, then the *inverse of R* (denoted by R^{-1}) is the set of ordered pairs obtained by switching the first and second components of each ordered pair in R. For example, if $R = \{(1, 1), (1, 3), (2, 3)\}$, then $R^{-1} = \{(1, 1), (3, 1), (3, 2)\}$.

How do we obtain the arrow diagram for R^{-1} given the arrow diagram for R?

271

Reverse the direction of each arrow, which in effect switches the components of the corresponding ordered pair.

How do we then get the arrow representation of the inverse of the inverse of R?

272

Switch the arrows a second time. But then we get back the arrow diagram of R. Thus, the inverse of the inverse of a relation R is equal to _____ .

273

R. Symbolically, $(R^{-1})^{-1} = R$.

Functions

274

A *function* from a set A to a set B is a relation from A to B with the following property: in its arrow diagram, there must be exactly one arrow exiting each element of A pointing to some element of B. If this not the case, then the relation is not a function from A to B. We call the set A the **domain** of the function and the set B the **codomain** of the function. Thus, the domain is the set that has arrows exiting it, and the codomain is the set that has arrows entering it.

Consider the relations from A to B as specified in the following arrow diagrams:

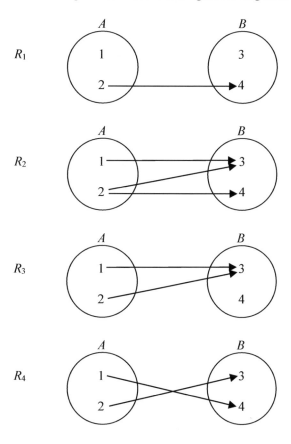

R_1 is not a function from A to B because 1 in A has no arrow exiting it. R_2 is also not a function for a different reason: 2 has more than one arrow exiting it. R_3 has exactly one arrow exiting each element of A pointing to some element in B; thus, it is a function from A to B. A is the domain of R_3; B is the codomain of R_3.

Is R_4 a function from A to B?

275

Yes. R_4 has exactly one arrow exiting each element of A pointing to some element in B.

Chapter 6: Relations and Functions

Here is the formal definition of a function:

> A **function** f from a set A to a set B is a relation from A to B with the property that, for every $a \in A$, there is exactly one $b \in B$ such that $(a, b) \in f$.

It is important that you understand this definition. You should convince yourself that it is equivalent to saying that in the arrow diagram, there is exactly one arrow exiting each element of A, and each of these arrows points to some element of B.

Is $(2, 3) \in R_3$, where R_3 is as in the preceding frame?

276

Yes

Another way to indicate that $(2, 3) \in R_3$ is with infix notation: $2\, R_3\, 3$. We can also use **standard function notation**: $R_3(2) = 3$.

Indicate that $(1, 3)$ is in the function R_3 using $\in$, infix notation, and standard function notation.

277

$(1, 3) \in R_3$
$1\, R_3\, 3$
$R_3(1) = 3$

The concept that a function is a set of ordered pairs may be new to you. Functions are typically specified with a rule rather than with a set of ordered pairs. For example, the square function on the integers is typically specified as $sq(x) = x^2$ for all $x \in \mathbb{Z}$. But this is just a way of specifying a set of ordered pairs. The rule $sq(x) = x^2$ for all $x \in \mathbb{Z}$ indicates that (x, x^2) is an ordered pair for each $x \in \mathbb{Z}$. For example, it indicates that $(3, 3^2)$—which is equal to $(3, 9)$—is an ordered pair in the function.

Specify the sq function by listing its ordered pairs.

278

$\{..., (-3, 9), (-2, 4), (-1, 1), (0, 0), (1, 1), (2, 4), (3, 9), ...\}$

Draw a graph of the function $\{(-2, -4), (-1, -2), (0, 0), (1, 2), (2, 4)\}$. Your graph should have x and y axes and a point for each ordered pair.

279

Domain = $\{-2, -1, 0, 1, 2\}$
Range = $\{-4, -2, 0, 2, 4\}$

A graph is another way to specify the set of ordered pairs of a function. List four ways to specify a function.

280

1. List its ordered pairs.
2. Draw its arrow diagram.
3. Give a rule for determining its ordered pairs.
4. Draw its graph.

A shorthand way to indicate that f is a function from A to B is to write $f: A \to B$. The arrow points from the domain to the codomain. In addition to the domain and codomain, a function has a range. The **range** of a function f from A to B (denoted by $f(A)$) is the set of the elements in B to which at least one arrow points in the arrow diagram for f. That is, the range of f is $\{b \in B : \text{there exists an } a \in A \text{ such that } f(a) = b\}$.

To indicate that $f(a) = b$, we generally say any of the following:

- f **maps** a to b.
- b is the **image** of a under f.
- a is a **preimage** of b under f.

Consider the function f_1 from A to B given by the following arrow diagram:

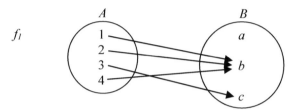

What is its domain, codomain, and range? What is the image of 3? What are the preimages of b? The function f_1 maps 3 to which element?

281

Domain: $\{1, 2, 3, 4\}$; codomain: $\{a, b, c\}$; range: $\{b, c\}$; image of 3: c; preimages of b: 1, 2, and 4; f_1 maps 3 to c.

Suppose $sq : \mathbb{Z} \to \mathbb{Z}$ is defined by $sq(x) = x^2$. What is the domain of sq? What is the range of sq?

282

The domain is $\mathbb{Z}$. The range is the set of integer squares.

What is the domain and range of the function $f(x)$ defined by $f(x) = 5$ for all $x \in \mathbb{Z}$?

283

The domain is $\mathbb{Z}$. The range is $\{5\}$, because every element in the domain is mapped to 5. Thus, in the arrow diagram for f, 5 is the only element in the codomain to which an arrow points.

Special Categories of Functions

284

Let's consider some special categories of functions. An **onto function** (also called a **surjection**) from A to B is a function whose arrow diagram has at least one arrow pointing to each element in the codomain B. For example, f_2, as follows, is an onto function. Each element in B has at least one arrow pointing to it:

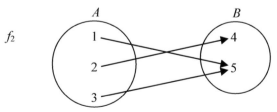

Here is the formal definition of an onto function:

A function f from a set A to a set B is **onto** if and only if for every $b \in B$, there is at least one $a \in A$ such that $f(a) = b$.

If f is an onto function from A to B, we indicate this by saying "f is a function from A **onto** B." Give a definition of an onto function that uses the concept of preimage.

58 Chapter 6: Relations and Functions

285

A function from a set *A* to a set *B* is onto if and only if every element in *B* has *at least one* preimage in *A*.

A **one-to-one function** (also called an **injection**) from a set *A* to a set *B* is a function whose arrow diagram has no element in *B* (the codomain) with more than one arrow pointing to it. In a one-to-one function, different elements in *A* map to different elements in *B*. For example, f_3, as follows, is a one-to-one function. Each element in *B* has at most one arrow pointing to it:

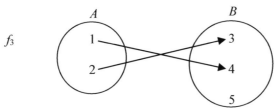

Why is f_2 in the preceding frame not a one-to-one function?

286

Because 5 in *B* has more than one arrow pointing to it.

Here is the formal definition of a one-to-one function:

A function *f* from a set *A* to a set *B* is one-to-one if and only if for all $a, b \in A$, $f(a) = f(b) \Rightarrow a = b$.

This definition is important to understand. In the definition, "$f(a) = f(b)$" tells us that *f* maps *a* and *b* to the same element. Thus, the implication in the definition tells us that if *f* maps *a* and *b* maps to the same element, then *a* and *b* must be the *same* element. Thus, different elements of the domain must map to the different elements of the codomain.

We can get an equivalent definition of a one-to-one function simply by replacing the implication in the preceding definition with its equivalent contrapositive. Do this.

287

A function *f* from set *A* to a set *B* is one-to-one if and only if, for all $a, b \in A$, $a \neq b \Rightarrow f(a) \neq f(b)$.

This form of the definition is much easier to understand: "$a \neq b$" tells us that *a* and *b* are different elements; "$f(a) \neq f(b)$" tells us that *a* and *b* map to different elements of the codomain. That is, a function is one-to-one if and only if different elements in the domain map to different elements in the codomain.
    Our first definition of a one-to-one function is the definition typically found in textbooks. It is generally easier to show that a function is one-to-one using this definition. That is why it is the more common definition.

Give a definition of a one-to-one function that uses the concept preimage.

288

A one-to-one function is a function if and only if each element of the codomain has *at most one* preimage in the domain.

If a function is both onto and one-to-one, then we call it a **bijection** or a **one-to-one correspondence.** An onto function has at least one arrow pointing to each element of its codomain. A one-to-one function has at most one arrow pointing to each element in its codomain. Thus, a function that is both onto and one-to-one (i.e., a bijection) has _____ _____ arrow pointing to each element of the codomain.

289

exactly one

Is f_4 a bijection from *A* to *B*?

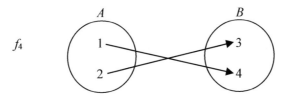

290

Yes. Each element in *B* has exactly one arrow pointing to it.

Give a definition of a bijection that uses the concept of preimage.

291

A function is a bijection if and only if each element of the codomain has *exactly one* preimage in the domain.

A ***function on A*** is a function from *A* to *A*. Thus, *A* is both the domain and the codomain. Is the relation on *A* given by the following arrow diagram a function on *A*?

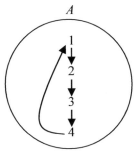

292

Yes. Every element of domain *A* has exactly one arrow exiting it pointing to an element of *A*.

Is it onto?

293

Yes. Every element of the codomain *A* has at least one arrow pointing to it.

Is it one-to-one?

294

Yes. Every element in the codomain *A* has at most one arrow pointing to it.

Is it a bijection?

295

Yes. Every element in the codomain *A* has exactly one arrow pointing to it.

Proving a Function Is Onto

296

Here is the general approach for proving that a function is onto: we show that for an arbitrary element in the codomain, there is an element in the domain that is mapped to it. Because the argument we advance is for an *arbitrary* element in the codomain, our argument applies to *every* element in the codomain. Thus, we can conclude that every element in the codomain has a preimage in the domain and, therefore, the function is onto.

Let's prove that the function $f: \mathbb{R} \to \mathbb{R}$ defined by $f(x) = 2x$ is onto. Here is the proof: Let $y \in \mathbb{R}$; y is our arbitrary element in the codomain of the function. We have to show that there is some element x *in the domain* that is mapped to it. This is easy to do: we simply note that $x = y/2$ is in the domain of the function and that $f(x) = f(y/2) = 2(y/2) = y$. ∎

Would this proof also work if the domain and codomain of $f(x)$ were $\mathbb{Z}$ instead of $\mathbb{R}$?

297

No. $f(x)$ maps the integers to the even integers (if x is an integer, then $f(x) = 2x$ is an even integer). Thus, the odd integers have no preimages under $f(x)$.

Proving a Function Is One-to-One

298

To prove a function f is one-to-one, we show that f has the property specified in the definition of a one-to-one function in frame **286**. Specifically, we show that

for all x and y in the domain, if $f(x) = f(y)$, then $x = y$.

Equivalently, we can show the contrapositive:

for all x and y in the domain, if $x \neq y$, then $f(x) \neq f(y)$.

However, the former is generally easier to show. Using the first approach, let's prove that $f: \mathbb{R} \to \mathbb{R}$ defined by $f(x) = 2x$ is one-to-one. Here is the start of the proof: suppose $f(x) = f(y)$. Do the proof.

299

Suppose $f(x) = f(y) \Rightarrow 2x = 2y \Rightarrow x = y \Rightarrow f$ is one-to-one. ∎

Prove that the log base 10 function is one-to-one. *Hint*: Use the fact that $10^{\log(x)} = x$ for any x in the domain of log

300

Suppose $\log(x) = \log(y)$. Then $10^{\log(x)} = 10^{\log(y)}$. $10^{\log(x)} = x$ and $10^{\log(y)} = y$. Thus, $x = y$. ∎

Prove that $sq(x) = x^2$ for $x \in \mathbb{Z}$ is not one-to-one.

301

$sq(3) = 3^2 = 9 = (-3)^2 = sq(-3)$. Both 3 and −3 map to the same element in the codomain. Thus, sq is not one-to-one. ∎

Function Composition

302

Suppose the functions f and g are as represented by the following arrow diagrams:

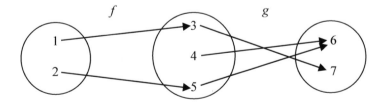

f is a function from the set A to the set B. g is a function from the set B to the set C. f and g together define a function from A to C called the **composition** of f and g (denoted by $g \circ f$).
   To determine the mapping under $g \circ f$, we simply follow the arrows from A to C. For example, f maps 1 in A to 3 in B; g maps 3 in B to 7 in C. Thus, $(g \circ f)(1) = 7$. That is, $g \circ f$ maps 1 to 7.

$g \circ f$ maps 2 to what element?

303

f maps 2 to 5; g maps 5 to 6. Thus, $(g \circ f)(2) = 6$.

To determine $(g \circ f)(2)$, we first apply f, and then we apply g. Because we apply the functions in *f*-then-*g* order, would it not make sense to denote this composition with $f \circ g$ rather than with $g \circ f$? The reason for denoting this composition with $g \circ f$ will become clear once you see the definition of function composition. Here is the definition:

Given function f from A to B and g from B to C, then $(g \circ f)(x) = g(f(x))$ for all x in A.

Using the standard notation for functions, we define the *f*-then-*g* composition by writing $g(f(x))$. Because the function names in $g(f(x))$ appear in *g*-*f* order, it in fact makes sense to denote this *f*-then-*g* composition with $g \circ f$. Unfortunately, it is easy to misinterpret $g \circ f$. We typically do things in left-to-right order. Thus, it is easy to incorrectly view $g \circ f$ as denoting a composition in which the functions are applied in *g*-then-*f* order. *Be sure to remember, in the function composition $g \circ f$, to apply the functions in right-to-left order, that is, first f, then g.*

In standard function notation, we write a function's name before the variables. However, suppose instead that we were to write a function's name after the variables. For example, suppose we were to write "$(x)f$" rather than "$f(x)$." Let's call this nonstandard function notation ***postfix notation*** because the function's name appears after the variables (*post* means "after"). With postfix notation, the *f*-then-*g* composition would be defined as $((x)f)g$. Note that the function names in the definition now appear in *f*-*g* order. Thus, with postfix notation, it would make sense to denote *f*-then-*g* composition with $f \circ g$ rather than with $g \circ f$. Some algebraists use postfix notation precisely so that an *f*-then-*g* function composition is denoted with the more natural $f \circ g$. However, we will stick with the standard function and composition notation.

Suppose that f is the function on $\mathbb{Z}$ defined by $f(x) = 2x$ and that g is the function on $\mathbb{Z}$ defined by $g(x) = x^2$. To which element in $\mathbb{Z}$ does the composition $g \circ f$ map 3?

304

f maps 3 to $2 \cdot 3 = 6$; g maps 6 to $6^2 = 36$. Thus, $g \circ f$ maps 3 to 36.

To which element in $\mathbb{Z}$ does $g \circ f$ map x? Give your answer in terms of x.

305

f maps x to $2x$; g maps $2x$ to $(2x)^2 = 4x^2$. Thus, $g \circ f$ maps x to $4x^2$. In other words, $g \circ f$ is the function $(g \circ f)(x) = 4x^2$.

Suppose $f(x) = x + 1$ and $g(x) = x^2 - 3$. What is $(g \circ f)(x)$?

306

$(g \circ f)(x) = g(f(x)) = g(x + 1) = (x + 1)^2 - 3 = x^2 + 2x - 2$

Suppose f and g are both one-to-one. Let's prove that $g \circ f$ is then also one-to-one. We start our proof by assuming $(g \circ f)(x) = (g \circ f)(y)$. We have to show that $x = y$. From the definition of $g \circ f$, what does $(g \circ f)(x) = (g \circ f)(y)$ mean?

307

$g(f(x)) = g(f(y))$

What does this equality imply given that g is one-to-one?

308

$f(x) = f(y)$

What does this equality imply given that f is one-to-one?

309

$x = y$. Thus, $g \circ f$ is one-to-one. ∎

The preceding theorem and proof are important, so be sure to understand them. Let's prove this theorem again, but this time more informally. Sometimes a less formal approach provides more insight into the subtleties of a theorem. Here is our proof:

Suppose a and b are distinct elements in the domain of f. Because f is one-to-one, a and b map to distinct elements in the domain of g. These elements, in turn, map to distinct elements in the codomain of g because g is one-to-one. Because, under $g \circ f$, distinct elements map to distinct elements, $g \circ f$ is one-to-one. ∎

Chapter 6: Relations and Functions

Suppose f is a function from A to B, and g is a function from B to C, and both f and g are onto functions. Then $g \circ f$ is onto. Prove this. *Hint*: Use the definition of "onto" twice

---

310

Let $c \in C$. Because g is onto, there is an element $b \in B$ such that $g(b) = c$. Because f is onto, there is an element $a \in A$ such that $f(a) = b$. Thus, $(g \circ f)(a) = g(f(a)) = g(b) = c$. We have shown that for any element $c \in C$, there is an element $a \in A$ such that $(g \circ f)(a) = c$. Thus, $g \circ f$ is onto. ∎

If f and g are both bijections, what can we conclude about $g \circ f$?

311

$g \circ f$ is also a bijection. From frame **309**, we know $g \circ f$ is one-to-one. From the preceding frame, we know $g \circ f$ is onto. Thus, $g \circ f$ is a bijection.

Key idea: The composition of two bijections is also a bijection.

Inverse of a Function

312

Suppose f is a function from A to B represented by the following arrow diagram:

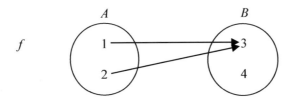

The *inverse of f* (denoted by f^{-1}) is the relation from B to A that corresponds to reversing all the arrows in the diagram for f. Redraw the diagram so that it represents f^{-1} (i.e., switch the direction of the arrows).

313

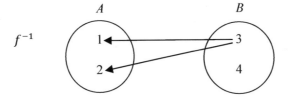

The inverse of f is the relation from B to A that corresponds to reversing all the arrows in the diagram.

Does this arrow diagram represent a function from B to A?

314

No, because 3 is mapped to more than one element in A. Moreover, 4 is not mapped to any element in A.

Now consider the function g given by the following arrow diagram:

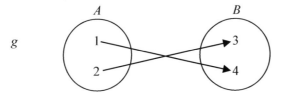

g is a function, so by definition, there must be exactly one arrow exiting from each element of A. Moreover, it is a bijection, so there is exactly one arrow pointing to each element of B.

What can we conclude about the arrow diagram for g^{-1}?

315

There is exactly one arrow exiting each element of B, and there is exactly one arrow pointing to each element of A.

Thus, g^{-1} is a _____.

316

bijection

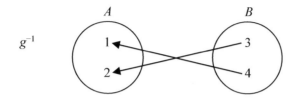

Let's summarize our observations: the inverse of a function from A to B is not necessarily a function from B to A. However, the inverse of a bijection from A to B is a function from B to A. In fact, it is a bijection from B to A.

Let's look at another example. The function f on $\mathbb{R}$ defined by $f(x) = 2x$ is a bijection. Thus, its inverse is a bijection. To determine the inverse, first replace $f(x)$ in "$f(x) = 2x$" with y.

317

$y = 2x$

Next, interchange x and y.

318

$x = 2y$

Now solve for y.

319

$y = x/2$

Finally, replace y with $f^{-1}(x)$.

320

$f^{-1}(x) = x/2$ for all $x \in \mathbb{R}$

Show that $f^{-1}(x)$ is onto and one-to-one

321

onto: Let $y \in \mathbb{R}$. We have to find an element in $\mathbb{R}$ that is mapped to y. $2y$ is the required element: $f^{-1}(2y) = 2y/2 = y$. Thus, $f^{-1}(x)$ is onto. *One-to-one*: Assume $f^{-1}(x) = f^{-1}(y) \Rightarrow x/2 = y/2 \Rightarrow x = y \Rightarrow f^{-1}(x)$ is one-to-one. ∎

Graph the function $sq: \mathbb{R} \to \mathbb{R}$ defined by $sq(x) = x^2$ for $-2 \leq x \leq 2$.

322

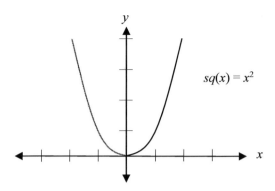

Is this function one-to-one?

323

No. Both x and $-x$ map to the same value. For example, $sq(2) = sq(-2) = 4$.

What simple test can we perform on this graph to determine if it represents a one-to-one function? *Hint*: Consider horizontal lines.

324

Check if any horizontal line intersects the graph at more than one point. Such a line means that there is a y value that has multiple x values that map to it. Thus, the function is not one-to-one.

Is $sq(x)$ a function from $\mathbb{R}$ *onto* $\mathbb{R}$?

325

No. It does not map to any negative number in $\mathbb{R}$.

What simple test can you perform on a graph to determine if it represents an onto function? *Hint*: Consider horizontal lines.

326

Check if every horizontal line intersects the graph in at least one point.

What simple test can we perform on a graph of a function from $\mathbb{R}$ to $\mathbb{R}$ to determine if it represents a bijection?

327

Check if every horizontal line intersects the graph in exactly one point.

Suppose a function f on the set A is *not* one-to-one. Then, in its arrow diagram, there must be two elements in A, say, a_1 and a_2, pointing to the same element in A, say, a_3:

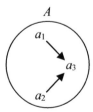

f maps both a_1 and a_2 to a_3. Thus, f^{-1} maps a_3 to both a_1 and a_2. If we start with a_1 and apply f followed by f^{-1}, we can go from a_1 to a_3 (under f), then from a_3 back to a_1 (under f^{-1}). But we can also go from a_1 to a_3 (under f) and then to a_2 (under f^{-1}). The composition $f^{-1} \circ f$ applied to a_1 has two values: a_1 and a_2. Thus, $f^{-1} \circ f$ is not a function (by definition, a function maps every element in its domain to one, and only one, element).

Now suppose that f is a bijection and that $f(a_1) = a_3$. Because f is one-to-one, we know that only a_1 is mapped to a_3 under f. Thus, f^{-1} necessarily maps a_3 back to a_1. Clearly $(f^{-1} \circ f)(a) = a$ for all $a \in A$. Using the same reasoning, we can similarly conclude that $(f \circ f^{-1})(a) = a$ for all $a \in A$. So $f^{-1} \circ f$ and $f \circ f^{-1}$ map each element of A to itself. This function is clearly one-to-one and onto and, therefore, a bijection. We call this bijection the ***identity bijection*** and denote it with e. The identity bijection maps each element in A to itself.

Representing Functions with Tables

328

Let $A = \{1, 2, 3\}$ and $B = \{4, 5, 6, 7\}$. The following arrow diagram represents one of the many possible functions from A to B:

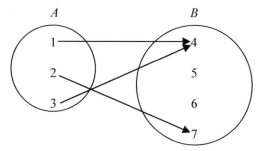

A function is a set of ordered pairs. We can represent a function with an arrow diagram in which each arrow represents an ordered pair. Alternatively, we can represent a function simply by listing the ordered pairs in table form. Here is the same function in table form: $\begin{pmatrix} 1 & 2 & 3 \\ 4 & 7 & 4 \end{pmatrix}$. Each column corresponds to an ordered pair. There are three ordered pairs in the given function. Thus, there are three columns in its table. The first column indicates that the function maps 1 to 4; the second column indicates that the function maps 2 to 7; the third column indicates that the function maps 3 to 4.

Does the following table represent a function from A to B: $\begin{pmatrix} 1 & 2 & 3 \\ 4 & 4 & 7 \end{pmatrix}$

329

Yes. The first row is a list of all the elements in A with no repeats. This property reflects the fact that in the corresponding arrow diagram, exactly one arrow exits from each element in A. The second row is a list of elements from B, where repeats are allowed (unless the function is one-to-one).

Does the following table represent a function from A to B: $\begin{pmatrix} 1 & 2 & 1 \\ 7 & 5 & 5 \end{pmatrix}$

330

No. A function maps each element in its domain to one and only one element in its range. This table indicates that 1 is mapped to both 7 and 5. Thus, it does not represent a function.

Characterize the first row of the table representation of a function from A to B.

331

The first row is a list of all the elements in A with no repeats.

Characterize the second row of the table representation of a function from A to B.

332

The second row is a list of elements from B with repeats allowed.

In table form, every function from A to B has the same first row (i.e., the list of the elements in the domain with no repeats). Thus, it is the second row that determines which function from A to B is represented by the table. Every distinct second row corresponds to a distinct function from A to B, and vice versa. Thus, the number of possible second rows is equal to the number of functions from A to B.

In our example from frame **328**, B has four elements (4, 5, 6, and 7). There are four possibilities for each position in the second row: $\begin{pmatrix} 1 & 2 & 3 \\ \square & \square & \square \end{pmatrix}$

Each can be 4, 5, 6, or 7.

Thus, the total number of possible second rows is $4 \cdot 4 \cdot 4 = 4^3 = 64$ functions from A to B. Note that in 4^3, 3 is the size of the domain A, and 4 is the size of the codomain B.

If A has two elements and B has 10 elements, how many functions are there from A to B? *Hint*: The table has two columns with 10 choices for the second component in each column.

333

$10^2 = 100$; 2 is the size of the domain, and 10 is the size of the codomain.

Suppose A and B are finite sets. Let $|A|$ represent the number of elements in A and $|B|$ the number of elements in B. How many functions are there from A to B?

334

$|B|^{|A|}$

Suppose A is finite. Let $|A|$ be the number of elements in A. How many functions are there on A?

335

$|A|^{|A|}$

Bijections in Table Form

336

In the arrow diagram of a bijection on A, every element in the codomain has exactly one arrow pointing to it. Thus, in its table representation, the second row must have each element of the codomain appearing _____ _____.

337

exactly once

For example, if $A = \{1, 2, 3\}$, one bijection on A is given by $\begin{pmatrix} 1 & 2 & 3 \\ 2 & 3 & 1 \end{pmatrix}$. In the second row, each element of A appears exactly once. Let's see how many distinct second rows we can make in which each element of $A = \{1, 2, 3\}$ appears exactly once. Each such row corresponds to a bijection on A.

In the table representation of a bijection on $\{1, 2, 3\}$, we have three choices (1, 2, or 3) for first position in the second row: $\begin{pmatrix} 1 & 2 & 3 \\ \square & \square & \square \end{pmatrix}$

three choices for this position

Because we cannot repeat an element, we have only two choices for the second position: $\begin{pmatrix} 1 & 2 & 3 \\ \square & \square & \square \end{pmatrix}$

two choices for this position

After making two choices, we have only one element in A left. Thus, this element must fill the third position: $\begin{pmatrix} 1 & 2 & 3 \\ \square & \square & \square \end{pmatrix}$

Remaining element goes here.

Thus, we can make the second row in $3 \cdot 2 \cdot 1 = 6$ ways. Each corresponds to a bijection on $A = \{1, 2, 3\}$. Thus, there are six bijections on A. The product $3 \cdot 2 \cdot 1$ is called 3-factorial (denoted by 3!).

How many bijections are on a set with four elements?

338

$4! = 4 \cdot 3 \cdot 2 \cdot 1 = 24$

How many bijections are on a set with n elements?

339

$n! = n \cdot (n-1) \cdot \cdots \cdot 2 \cdot 1$

Using tables, specify every one-to-one function on $A = \{1, 2\}$.

340

$\begin{pmatrix} 1 & 2 \\ 1 & 2 \end{pmatrix}, \begin{pmatrix} 1 & 2 \\ 2 & 1 \end{pmatrix}$

Similarly specify every onto function on A.

341

$\begin{pmatrix} 1 & 2 \\ 1 & 2 \end{pmatrix}, \begin{pmatrix} 1 & 2 \\ 2 & 1 \end{pmatrix}$

Every one-to-one function on A is an onto function on A; every onto function on A is a one-to-one function on A. Is this also true if $A = \{1, 2, 3\}$? What is the table form of each one-to-one function and each onto function on $\{1, 2, 3\}$?

342

All the one-to-one functions on $A = \{1, 2, 3\}$:

$\begin{pmatrix} 1 & 2 & 3 \\ 1 & 2 & 3 \end{pmatrix}, \quad \begin{pmatrix} 1 & 2 & 3 \\ 2 & 3 & 1 \end{pmatrix}, \quad \begin{pmatrix} 1 & 2 & 3 \\ 3 & 1 & 2 \end{pmatrix}, \quad \begin{pmatrix} 1 & 2 & 3 \\ 1 & 3 & 2 \end{pmatrix}, \quad \begin{pmatrix} 1 & 2 & 3 \\ 2 & 1 & 3 \end{pmatrix}, \quad \begin{pmatrix} 1 & 2 & 3 \\ 3 & 2 & 1 \end{pmatrix}$

These are also all the onto functions on A. Through enumeration of all the functions on $A = \{1, 2, 3\}$, we have shown that a function on $A = \{1, 2, 3\}$ is onto if and only if it is one-to-one. Let's prove the generalization of this result: a function on any finite set is one-to-one if and only if the function is onto. This statement has the form "p if and only if q." Recall that one way to prove a statement of this form is to show $p \Rightarrow q$ and $\sim p \Rightarrow \sim q$ (the latter is the contrapositive of $q \Rightarrow p$). Let's first show $p \Rightarrow q$. That is, show that if a function on a finite set is one-to-one, it is also onto.

343

Suppose a function on a finite A is one-to-one. Let $|A| = n$. In its arrow diagram, one arrow exits from each element of A, for a total of n arrows. Because the function is one-to-one, each arrow points to a different element of A. Thus, there are n distinct elements in A to which an arrow points. But there are only n elements in A. Thus, every element in A has an arrow pointing to it, which means the function is onto.

Now show $\sim p \Rightarrow \sim q$. That is, show that if a function on a finite set is not one-to-one, then it is not onto.

344

Suppose a function on a finite set A is not one-to-one. Let $|A| = n$. Because the function is not one-to-one, there must be two elements in A that the function maps to the same element. Thus, of the n arrows exiting the elements of A, two go to the same element, leaving $n - 2$ arrows, which can point to at most $n - 2$ elements. But there are $n - 1$ elements remaining in A. Thus, there must be at least one element in A to which no arrow points. This means the function is not onto. ∎

What if A is an infinite set? Is a function one-to-one if and only if it is onto?

345

No. For example, the function on $\mathbb{Z}$ defined by $f(x) = 2x$ is one-to-one but not onto (the odd integers have no preimages).

Suppose A and B are finite sets and f is a function from A to B. What can you say about $|A|$ and $|B|$ if f is one-to-one? If f is onto? If f is a bijection?

68   Chapter 6: Relations and Functions

346

If f is one-to-one, then $|A| \leq |B|$, because f maps each element in A to a distinct element of B. If f is onto, then $|A| \geq |B|$, because each element in B has at least one preimage. If f is a bijection, then f is both one-to-one and onto $\Rightarrow |A| \leq |B|$ and $|A| \geq |B| \Rightarrow |A| = |B|$.

Cardinality

347

Suppose you have a herd of sheep and a herd of goats. You want to determine if the number of sheep is the same as the number of goats. Unfortunately, your counting ability is limited: you cannot count beyond 2. So you cannot simply count the sheep, count the goats, and compare the two counts. What do you do? *Hint*: Set up a bijection.

348

Pair each sheep with a distinct goat. If you can do that with no unpaired sheep or unpaired goats left over, then you have a bijection from the sheep to the goats. Thus, by frame **346**, the number of sheep must equal the number of goats.

Suppose we pair each sheep with a goat. Such a pairing is an injection (i.e., a one-to-one function) from the sheep to the goats, in which case there must be at least as many goats as sheep. If there are no leftover, unpaired goats, then our pairing is a bijection (as well as an injection), in which case we have the same number of goats as sheep. From this example, we can see that we can use the notions of injection and bijection to compare the sizes of sets without having to count them. If A and B are sets, and an injection exists from A to B, then $|A| \leq |B|$, where $|A|$ and $|B|$ are the number of elements in A and B, respectively. If a bijection exists from A to B, then $|A| = |B|$. The beauty of this approach is that we can use it for infinite sets as well as finite sets.

The ***cardinality*** of a finite set is simply the number of elements it contains. For example, the cardinality of $\{a, b\}$ is 2. We say two sets have the ***same cardinality*** if and only if there exists a bijection between them. Saying that two finite sets have the same cardinality is equivalent to saying that they have the same number of elements. If two infinite sets have the same cardinality, each element in one set can be paired with a distinct element in the other set with no unpaired elements left over. Thus, in this sense, the two sets have the same size.

The set $\{a, b\}$ has cardinality 2; the set $\{5, 10, 17\}$ has cardinality 3. The set $\{5, 10, 17\}$ is bigger than the set $\{a, b\}$. Obviously, finite sets can have different sizes. Can infinite sets also have different sizes? That is, are there infinite sets A and B between which *no* bijection exists (in which case they do not have the same cardinality)? Let's investigate.

Every positive integer is a positive rational number. But not every positive rational number is a positive integer. For example, the positive integer 2 equals $\frac{2}{1}$, a rational number. But the rational number $\frac{1}{2}$ does not equal any positive integer. Thus, it is reasonable to expect that $\mathbb{Z}^+$ and $\mathbb{Q}^+$ do not have the same cardinality. But, in fact, they do. We prove this by constructing in table form a bijection from $\mathbb{Z}^+$ to $\mathbb{Q}^+$.

The first row of the table for the bijection lists all the numbers in $\mathbb{Z}^+$ in ascending order: 1 2 3 .... The second row lists all the rational numbers. Some rational numbers can be reduced by dividing the numerator and denominator by a common divisor. For example, $\frac{6}{4}$ can be reduced to $\frac{3}{2}$; $\frac{6}{4}$ and $\frac{3}{2}$ are two representations of the same rational number. To avoid repeats in the second row, we will not list any rational numbers that are not in reduced form. For the same reason, we will also not list any rational number whose numerator and denominator are both negative.

We should *not* start our second row this way: $\frac{1}{1}$ $\frac{1}{2}$ $\frac{1}{3}$ $\frac{1}{4}$ $\frac{1}{5}$ .... Why not?

349

An infinite number of rational numbers have 1 in the numerator. Thus, our list *will never reach* the rational numbers whose numerators are greater than 1. We need to list out the rational numbers so that every rational number will eventually appear on the list. Specifically, every rational number on our list should have a *finite number of numbers preceding it*.

Here is how we create our list: sum the numerator and denominator of each reduced rational number we are to list. Start by listing all those numbers that sum to 2 (there is only one: $\frac{1}{1}$), then those that sum to 3 ($\frac{1}{2}$ and $\frac{2}{1}$), then those that sum to 4 ($\frac{1}{3}$ and $\frac{3}{1}$), and so on. Thus, the table that represents a bijection from $\mathbb{Z}^+$ to $\mathbb{Q}^+$ looks like this:

$$\begin{pmatrix} 1 & 2 & 3 & 4 & 5 & 6 & 7 & 8 & 9 & \dots \\ \frac{1}{1} & \frac{1}{2} & \frac{2}{1} & \frac{1}{3} & \frac{3}{1} & \frac{1}{4} & \frac{2}{3} & \frac{3}{2} & \frac{4}{1} & \dots \end{pmatrix}$$

$\underbrace{}_{\text{2-group}} \underbrace{}_{\text{3-group}} \underbrace{}_{\text{4-group}} \underbrace{}_{\text{5-group}}$

Let's call the group of rational numbers in reduced form that have a numerator and denominator that sum to n the **n-group**. At most, how many numbers are in the n-group?

350

$\frac{1}{n-1}, \frac{2}{n-2}, \dots, \frac{n-1}{1}$. There are at most $n - 1$. There might be fewer than $n - 1$ because some of the numbers among the $n - 1$ may not be in reduced form. Every n-group is finite.

Let $\frac{x}{y}$ be an arbitrary positive rational number in reduced form. Show that $\frac{x}{y}$ will appear in our list after a finite number of numbers are listed. *Hint*: $\frac{x}{y}$ is in the k-group where $k = x + y$.

351

Let k represent the sum of x and y. Then, $\frac{x}{y}$ will appear in the k-group on our list. The k-group is preceded by a finite number of groups: the $(k-1)$-group, the $(k-2)$-group, and so on. Moreover, each of these groups is finite. The k-group is also finite. Thus, $\frac{x}{y}$ is preceded on the list by a finite number of numbers. We will not run into the problem we saw in frame **348**, where we never list out some of the rational numbers because they are preceded on the list by an infinite number of numbers. The table in frame **349** contains all the elements in $\mathbb{Z}^+$ with no repeats in the top row; it contains all the elements in $\mathbb{Q}^+$ with no repeats in the bottom row. Thus, the table represents a bijection from $\mathbb{Z}^+$ to $\mathbb{Q}^+$, showing that $\mathbb{Z}^+$ and $\mathbb{Q}^+$ have the same cardinality. ∎

Construct a table that shows that $\mathbb{Z}^+$ and the even positive integers have the same cardinality.

352

$$\begin{pmatrix} 1 & 2 & 3 & 4 & 5 & \dots \\ 2 & 4 & 6 & 8 & 10 & \dots \end{pmatrix}$$

Construct a table that shows that $\mathbb{Z}^+$ has the same cardinality as $\mathbb{Z}$. *Hint*: The $\mathbb{Z}$ row should list the integers in a sequence such that any integer n will appear on the list after a finite number of integers. Thus, you *cannot* list all the positive integers first or all the negative integers first.

353

$$\begin{pmatrix} 1 & 2 & 3 & 4 & 5 & \dots \\ 0 & 1 & -1 & 2 & -2 & \dots \end{pmatrix}$$

Because a bijection from $\mathbb{Z}^+$ to $\mathbb{Z}$ exists, we can conclude that $\mathbb{Z}^+$ and $\mathbb{Z}$ have the same "size" or, more precisely, the same cardinality. This conclusion is certainly counterintuitive. For every positive integer, there is a corresponding negative integer. The integers also include 0. Thus, it is reasonable (but incorrect) to think that the size of $\mathbb{Z}$ is twice the size of $\mathbb{Z}^+$ plus 1 (the plus 1 is for 0). The properties of infinite sets are strange indeed.

A ***countably infinite set*** is a set that has the same cardinality as the positive integers. A ***countable set*** is a set that is either finite or countably infinite. Thus, $\mathbb{Q}^+$ and $\mathbb{Z}$ are both countable and countably infinite. An ***uncountable set*** is a set that is not countable.

Is every infinite set countably infinite? In view of the preceding examples, this seems like a reasonable conjecture, but it is incorrect. $T = \{x \in \mathbb{R} : 0 < x \leq 1\}$, the infinite subset of real numbers between 0 and 1, is not countably infinite. Thus, the infinity of this set has a different size than the infinity of the positive integers. Like finite sets, infinite sets have different sizes.

Let's show that a set T is not countably infinite with a proof by contradiction. Assume $\mathbb{Z}^+$ and T have the same cardinality. That is, a bijection exists from $\mathbb{Z}^+$ to T. Let's represent this bijection in table form. For convenience, we will show the table as two columns rather than two rows. The table would look something like this:

70 Chapter 6: Relations and Functions

$\mathbb{Z}^+$	T
1	.234999...
2	.353333...
3	.010101...
4	.499999...
5	.111111...
⋮	⋮

The left column (call it the $\mathbb{Z}^+$ list) contains all the positive integers. The right column (call it the T list) contains all the numbers in T, with each number represented by its corresponding nonterminating form (so it appears only once). For example, 0.5 appears in the list as 0.499999.... View the list of numbers on the T list as a matrix of digits. For example, the T list shown here would correspond to the matrix

Construct a fractional real number by taking the digits from the diagonal of this matrix and replacing each 5 with 6 and any digit that is not 5 with 5. The digits in the diagonal are 2, 5, 0, 9, 1, ... Thus, our constructed number starts with 0.56555.

The first digit in our constructed number is not equal to the first digit in the diagonal. But the first digit in the diagonal is the first digit in the first number in our T list. Thus, our constructed number differs from the first number on our T list by at least the first digit. Can our constructed number equal the second number on the T list?

354

No. It differs from the second number in at least the second digit.

Can the constructed number be equal to any number of the T list?

355

No. For any $n = 1, 2, 3, ...,$ our constructed number differs from the nth number on the list in at least the nth digit. Thus, our constructed number differs from *every* number on the T list.

Is our constructed number in the set $T = \{x \in \mathbb{R} : 0 < x \leq 1\}$?

356

Yes

Is the function from $\mathbb{Z}^+$ to T represented by the table listing the elements of $\mathbb{Z}^+$ and T in frame **353** a bijection?

357

No. Our constructed number is in T but has no preimage. That is, no positive integer is mapped to it. Thus, this function is not a bijection.

We have a contradiction: we assumed a bijection exists from $\mathbb{Z}^+$ to T. But we then showed that the function is not a bijection. Our initial assumption—that a bijection from $\mathbb{Z}^+$ to T exists—must be incorrect. Thus, T is not countably infinite. ∎
    The subset of any countable set is countable. Thus, if the entire set of real numbers were countably infinite, then our set T would be countably infinite. But T is not countably infinite. Thus, $\mathbb{R}$ is also not countably infinite.

Give an informal argument that supports the assertion that any subset of a countable set is countable.

358

Suppose that $A \subseteq B$ and that B is a countable set. *Case 1*: B is finite. Then A is finite and by definition countable. *Case 2*: B is infinite and A is finite. Then then A by definition is countable. *Case 3*: A is infinite. Because B is countably infinite, there exists a list of all the elements of B. Remove from this list all the elements not in A, leaving a list of all the elements in A. Map the successive elements of this new list to 1, 2, 3, ... This mapping is a bijection from A to $\mathbb{Z}^+$. Thus, A is countably infinite and, therefore, countable. ∎

If you are uncomfortable with the conclusions about infinite sets that we have presented, or perhaps, if you disagree with them, you are in good company. When Georg Cantor first published his ideas on infinite sets, the mathematical community rejected them as absurd. The poor reception of his work made Cantor so distraught that he almost went insane. However, Cantor's work today is regarded as a major breakthrough in mathematics and is universally accepted.

Review Questions

1. What is $A \times B$, where $A = \{1, 2, 3\}$ and $B = \{3, 4\}$?
2. What is $(A \times A) \times A$, where $A = \{1\}$? Is $(A \times A) \times A = A \times (A \times A)$?
3. How many functions exist on A where $A = \{1, 2\}$? How many onto functions? How many one-to-one functions?
4. A horizontal line test can be used on a graph of a function on $\mathbb{R}$ to determine if the function is one-to-one or onto. Give a similar test that can be used to determine if a relation is a function.
5. Show informally that $\mathbb{Q}$ is countably infinite.
6. Determine the inverse of $f(x) = x^3 + 5$. Is the inverse a function?
7. What is the inverse of $\begin{pmatrix} 1 & 2 & 3 \\ 1 & 3 & 2 \end{pmatrix}$? Show the inverse in table form.
8. Let $f = \begin{pmatrix} 1 & 2 & 3 \\ 1 & 3 & 2 \end{pmatrix}$ and $g = \begin{pmatrix} 1 & 2 & 3 \\ 2 & 3 & 1 \end{pmatrix}$. Specify $f \circ g$, $g \circ f$, $f \circ f$, and $g \circ g \circ g$ in table form.
9. Let $f(x) = x + 1$ and $g(x) = x - 1$. Specify $f \circ g$ and $g \circ f$ in standard function notation.
10. What does "$f: A \to B$" mean?

Answers to the Review Questions

1. $\{(1, 3), (1, 4), (2, 3), (2, 4), (3, 3), (3, 4)\}$
2. $(A \times A) \times A = \{((1, 1), 1)\} \neq A \times (A \times A) = \{(1, (1, 1))\}$.
3. Four functions, two are onto, the same two are one-to-one.
4. Every vertical line should intersect the graph in exactly one point.
5. Create a list for $\mathbb{Q}$ as follows: insert $-\frac{a}{b}$ after each element $\frac{a}{b}$ in the list for $\mathbb{Q}^+$ in frame **349**. Insert $\frac{0}{1}$ at the beginning of the list. Then the new list contains all the elements in $\mathbb{Q}$. Moreover, any given element in $\mathbb{Q}$ has a finite number of elements preceding it on this list. Define a function that maps 1 to the first element of this list, 2 to the second element, as so on. This function is a bijection from $\mathbb{Z}^+$ to $\mathbb{Q}$. Thus, $\mathbb{Q}$ is countably infinite.
6. Switching x and y in $y = x^3 + 5$, we get $x = y^3 + 5$. Solving for y, we get $y = f^{-1}(x) = \sqrt[3]{x - 5}$, which is a function.
7. $\begin{pmatrix} 1 & 2 & 3 \\ 1 & 3 & 2 \end{pmatrix}$
8. $f \circ g = \begin{pmatrix} 1 & 2 & 3 \\ 3 & 2 & 1 \end{pmatrix}$, $g \circ f = \begin{pmatrix} 1 & 2 & 3 \\ 2 & 1 & 3 \end{pmatrix}$, $f \circ f = \begin{pmatrix} 1 & 2 & 3 \\ 1 & 2 & 3 \end{pmatrix}$, $g \circ g \circ g = \begin{pmatrix} 1 & 2 & 3 \\ 1 & 2 & 3 \end{pmatrix}$
9. $(f \circ g)(x) = x$ and $(g \circ f)(x) = x$.
10. It means "f is a function from A to B."

Homework Questions

1. Let $f(x) = x^2 + 1$ and $g(x) = x^3 - 1$. Specify $f \circ g$ and $g \circ f$ in standard function notation.
2. Determine the inverse of $f(x) = x^3 + 1$. Is the inverse a function?
3. Show that the set of functions from $\mathbb{Z}^+$ to $\{0, 1\}$ is not countably infinite. *Hint*: Use a "diagonal" argument similar to the one in frame **353**.
4. Let f be a function from $\mathbb{R}$ to $\mathbb{R}$ that is ***strictly increasing***. That is, for all $a, b \in \mathbb{R}$, $a < b \Rightarrow f(a) < f(b)$. Show that f is one-to-one.

5. Determine $\begin{pmatrix} 1 & 2 & 3 \\ 2 & 3 & 1 \end{pmatrix} \circ \begin{pmatrix} 1 & 2 & 3 \\ 2 & 3 & 1 \end{pmatrix}$.
6. Determine $\begin{pmatrix} 1 & 2 & 3 \\ 2 & 3 & 1 \end{pmatrix} \circ \begin{pmatrix} 1 & 2 & 3 \\ 2 & 3 & 1 \end{pmatrix}^{-1}$
7. Let f be a function from A to B and g be a function from B to C. Show that f is onto if $g \circ f$ is onto. Show that f is one-to-one if $g \circ f$ is one-to-one.
8. Does $f(A \cap B) = f(A) \cap f(B)$ for all functions f and all sets A and B?
9. Addition on the integers is a function. What is its domain? Its codomain?
10. Let $f = \begin{pmatrix} 1 & 2 & 3 & 4 & 5 \\ 3 & 4 & 2 & 5 & 1 \end{pmatrix}$. Let $f^2 = f \circ f$, $f^3 = f \circ f \circ f$, and so on. What is the smallest positive n such that f^n equals the identity function (i.e., the function in which each element is mapped to itself).

7 Binary Operations

Definition of a Binary Operation

359

What kind of mathematical object is subtraction on the integers? Its effect is to map each ordered pair of integers to an integer. Each pair of integers is mapped to exactly one integer. For example, subtraction maps (5, 2) to the integer $5 - 2 = 3$.

To show the action of subtraction on the ordered pair (5, 2), we can draw an arrow from the ordered pair to the integer that subtraction produces: $(5, 2) \to 3$. If we do this for every set of ordered pairs of integers, we have a complete representation of the subtraction operation. Exactly one arrow exits each ordered pair and points to some integer. Our arrow diagram has two sets—the set of ordered pairs of integers (from which arrows exit) and $\mathbb{Z}$ (to which arrows point):

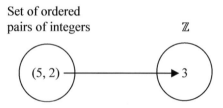

Exactly one arrow exits each ordered pair and points to some element in $\mathbb{Z}$. Such an arrow diagram represents a _____.

360

function

How do we denote the set of all ordered pairs of integers?

361

$\mathbb{Z} \times \mathbb{Z}$

Thus, subtraction maps each element in the set $\mathbb{Z} \times \mathbb{Z}$ to an element in $\mathbb{Z}$. Subtraction is a _____ from _____ to $\mathbb{Z}$.

362

function, $\mathbb{Z} \times \mathbb{Z}$

Let S be a nonempty set. We call a function from $S \times S$ to S a **binary operation on S.** For example, subtraction is a binary operation on $\mathbb{Z}$ because it is a function from $\mathbb{Z} \times \mathbb{Z}$ to $\mathbb{Z}$. Subtraction on $\mathbb{Z}^+$, on the other hand, is not a binary operation on $\mathbb{Z}^+$. It does not always produce an integer in $\mathbb{Z}^+$. For example, $2, 3 \in \mathbb{Z}^+$, but $2 - 3 = -1 \notin \mathbb{Z}^+$.

When discussing a binary operation in general, we will use the symbol $*$ to represent the binary operation. If, however, we are discussing a binary operation with a well-established symbol for that binary operation, we, of course, will use that symbol. For example, we will use "+", "−", "·", and "/" to represent, respectively, the addition, subtraction, multiplication, and division binary operations.

If an operation on elements from a set S always produces an element in S, we say that S is **closed** under that operation. For example, $\mathbb{Z}$ is closed under subtraction, but $\mathbb{Z}^+$ is not.

By definition, a binary operation on a set S always produces an element in S. Thus, a set S is closed under any binary operation on S. However, a subset of S is not necessarily closed under a binary operation on S. For example, consider the set of integers and the subset of odd integers. The integers are closed under addition. But the subset of odd integers is not closed under addition. To demonstrate this lack of closure, we need only one example that violates closure. For example, $1 + 3 = 4$; 1 and 3 are odd, but their sum is even.

Are the irrational numbers (i.e., numbers like $\sqrt{2}$ that cannot be represented as a ratio of two integers) closed under multiplication?

363

No; $\sqrt{2} \cdot \sqrt{2} = 2$. $\sqrt{2}$ is irrational, but 2 is not irrational. The product of some pairs of irrational numbers is irrational (e.g., $\sqrt{2} \cdot \sqrt{3} = \sqrt{6}$). For the closure property to hold, *every* pair of irrational numbers must produce an irrational number.

Are the integers closed under negation?

364

Yes. The negation of any integer is an integer.

Negation operates on single elements, not ordered pairs. So we call it a **unary operation** ("unary" means "one"). A binary operation operates on ordered pairs ("binary" means "two"). A **ternary operation** operates on ordered triples ("ternary" means "three"). An **n-ary operation** operates on *n*-tuples (an **n-tuple** is an ordered sequence of *n* items).

To indicate that subtraction maps the ordered pair (5, 3) to 2, we typically write $5 - 3 = 2$. However, we could instead use standard function notation. In standard function notation, the symbol representing the function precedes the *n*-tuple. Thus, in standard function notation, we indicate that (5, 3) is mapped to 2 under subtraction by writing $-(5, 3) = 2$ This notation is less convenient than the usual notation, but it does have one advantage: it emphasizes that subtraction on $\mathbb{Z}$ is a function on ordered pairs.

We can think of a binary operation as "operating on" an ordered pair, producing some result. The components of the ordered pair on which a binary operation operates are called **operands**. We call the symbol that represents the binary operation the **operation symbol** or, more simply, the **operator**. For example, in $5 - 3$, the subtraction binary operation operates on the operands 5 and 3, producing 2. The symbol "$-$" is the operation symbol for subtraction. Similarly, in $5 \cdot 3$, the multiplication binary operation operates on the operands 5 and 3, producing 15. The symbol "$\cdot$" is the operation symbol for multiplication.

When no ambiguity results, we can use **juxtaposition** to represent a binary operation. That is, we can omit the symbol for the binary operation and simply write the two operands next to each other with no intervening operation symbol. For example, instead of writing $a \cdot b$ to represent the product of *a* and *b*, we can write ab.

If there is a well-established symbol for some binary operation, we will use that symbol to represent that binary operation. For example, we will use "+" and "$-$" for addition and subtraction, respectively. If, on the other hand, a binary operation has no well-established symbol, or we are discussing binary operations in general, we will use "$*$" as the operation symbol, or we will simply use juxtaposition.

A convenient way to represent a binary operation on a finite set S is with a two-dimensional table called a **Cayley table**. For example, consider the binary operation $*$ on the set $S = \{a, b\}$ defined by the following arrow diagram:

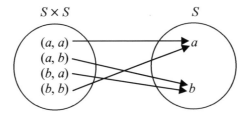

We can represent this binary operation with a table with two rows and two columns:

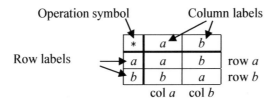

We label the rows and columns with the elements from the set S. If we have a symbol for the operation, we place it in the upper left corner. Each position in the table corresponds to an ordered pair given by the row and column labels. For example, the position at row *a* (i.e., the row labeled with *a*) and column *b* (i.e., the column labeled with *b*) corresponds to the ordered pair (a, b). Because the operation maps $a * b = b$ (i.e., the operation maps the order pair (a, b) to b), we place *b* at this position. Similarly, the position at row *b*/column *a* contains *b* to indicate that $b * a = b$. The two other positions contain *a*, indicating that the binary operation maps the corresponding ordered pairs to *a*.

In a Cayley table, the rows and the columns are usually labeled in the same order. For example, in the preceding Cayley table, the rows are labeled in *a-b* order. Thus, the columns are also labeled in *a-b* order. Be sure to be aware of the convention we are using to refer to specific rows and columns in a Cayley table: we use the label on a row to refer to that row; similarly, we use the label on a column to refer to that column. For example, in the Cayley table above, row *a* is the row whose label is *a*.

Properties of Binary Operations

365

Let's now consider some properties that a binary operation can have. Let $*$ be the operation symbol for some binary operation on the set *S*. The operator $*$ is **commutative** on a set *S* if and only if for all $a, b \in S$, $a * b = b * a$.

Addition and multiplication are commutative on $\mathbb{Z}$, $\mathbb{Q}$, and $\mathbb{R}$. For example, $2 + 3 = 3 + 2$, and $2 \cdot 3 = 3 \cdot 2$. However, subtraction is not commutative. For example, $2 - 4 \neq 4 - 2$.

A binary operation $*$ on *S* is **associative** if and only if $(a * b) * c = a * (b * c)$ for all $a, b, c \in S$. Consider the expression $a * b * c$. Here we have two operations, represented by the two occurrences of the operation symbol $*$. Keeping the operands a, b, c in that order, we have two ways of evaluating $a * b * c$: we can do the left operation $a * b$ first or the right operation $b * c$ first. If we do the left operation first, we evaluate $a * b$ and then use the result as the left operand of the right operation. We indicate this order by parenthesizing $a * b * c$ this way: $(a * b) * c$. If we do the right operation first, we evaluate $b * c$ and then use the result as the right operand of the left operation. We indicate this order by parenthesizing this way: $a * (b * c)$. If $*$ is associative, the two orders of performing the operations produce the same result. Thus, in an expression like $a * (b * c)$, we can omit the parentheses with introducing an ambiguity.

The associativity property applies to expressions with three operands and two operators, such as $a * b * c$. However, it generalizes to longer expressions, such as $a * b * c * d$. If $*$ is associative, does $(a * b) * (c * d) = a * (b * c) * d$?

366

Yes, by the generalization of the associative property.

If $*$ is associative, does $a * (b * c) = c * (a * b)$?

367

Not necessarily, because the order of *a*, *b*, and *c* on the left side differs from their order on the right side. However, if $*$ is commutative as well as associative, then $a * (b * c) = (a * b) * c = c * (a * b)$. In this sequence of equalities, the first equality holds by associativity, the second by commutativity.

Is addition on $\mathbb{Z}$ associative?

368

Yes

Evaluate $1 + 2 + 3$ doing the left operation first. Then repeat, doing the right operation first.

369

Left first: $(1 + 2) + 3 = 3 + 3 = 6$. Right first: $1 + (2 + 3) = 1 + 5 = 6$.

To show that a binary operator $*$ is not associative, we have to show that the following statement is false: for all a, b, c, $(a * b) * c = a * (b * c)$, or, equivalently, its negation is true.

What is the negation of "for all a, b, c, $(a * b) * c = a * (b * c)$"? *Hint*: Recall that to negate a universally quantified statement, change "for all" to "there exists." Also negate the predicate, which, in this example, is $(a * b) * c = a * (b * c)$.

370

There exists *a*, *b*, and *c* such that $(a * b) * c \neq a * (b * c)$.

76 Chapter 7: Binary Operations

To show that this existentially qualified statement is true, we need to find only one set of values of a, b, and c such that $(a * b) * c \neq a * (b * c)$. We call such a set of values a **counterexample**, because it "counters" (i.e., disproves) the assertion that $(a * b) * c = a * (b * c)$ for all a, b, and c.

Show that subtraction on $\mathbb{Z}$ is not associative by providing a counterexample; specifically, provide a set of integer values for a, b, and c such that $(a - b) - c \neq a - (b - c)$.

371

$(3 - 2) - 1 = 1 - 1 = 0$, but $3 - (2 - 1) = 3 - 1 = 2$.

Identity

372

Let $*$ be a binary operation on a set S. Suppose $e \in S$ and $a * e = e * a = a$ for all $a \in S$. We then call e an **identity element** or, more simply, an **identity**. For example, adding zero to an integer on either side has no effect, in the sense that the result is just the integer to which zero is added. That is, $a + 0 = 0 + a = a$ for all $a \in \mathbb{Z}$. Thus, 0 is an identity in $\mathbb{Z}$ under addition.

What is the identity in $\mathbb{Z}$ under multiplication?

373

1, because $1 \cdot a = a \cdot 1 = a$ for all $a \in \mathbb{Z}$

For an element e to be an identity in a set S under the operation $*$, it has to be *both* a **left identity** (i.e., $e * a = a$ for all $a \in S$) and a **right identity** (i.e., $a * e = a$ for all $a \in S$). For example, 0 is a both a left identity in $\mathbb{Z}$ under addition (because $0 + a = a$ for all $a \in \mathbb{Z}$) and a right identity (because $a + 0 = a$ for all $a \in \mathbb{Z}$). Thus, 0 is an identity.

Is 0 an identity in the integers under subtraction? Note that $a - 0 = a$ for all $a \in \mathbb{Z}$.

374

No. 0 is not a left identity. To see this, note that, $0 - 5 \neq 5$.

Under subtraction, 0 is not a left identity. Thus, it is not an identity under subtraction. $\mathbb{Z}$ under subtraction has no identity.

Suppose e_1 and e_2 are both identities in a set S under an operation $*$. What does $e_1 * e_2$ equal, given that e_1 is an identity?

375

$e_1 * e_2 = e_2$

What does $e_1 * e_2$ equal, given that e_2 is an identity?

376

$e_1 * e_2 = e_1$

What can you conclude about e_1 and e_2?

377

$e_1 = e_2$, because both e_1 and e_2 equal $e_1 * e_2$.

In other words, if e_1 and e_2 are both identities, then e_1 and e_2 are the *same* element. Thus, an identity, if one exists, is unique.

Key idea: a set with a binary operation has at most one identity.

A more concise way of showing that $e_1 = e_2$ is to write the following sequence of equalities: $e_1 = e_1 * e_2 = e_2$. The first equality holds because e_2 is an identity; the second equality holds because e_1 is an identity.

For a given binary operation on a set, an identity, if one exists, is unique. A set, however, can have multiple identities if it has more than one binary operation—one identity for each binary operation.

What is the identity in ℤ under addition?

378

0

What is the identity in ℤ under multiplication?

379

1

What is the identity in ℤ under subtraction?

380

It has no identity under subtraction.

Why is the following question ambiguous: what is the identity in ℤ?

381

The identity depends on the operation as well as the set.

Show that if a set has a right identity e_r and an identity e, then e_r and e are the same element. *Hint*: Fill in the blank with a product, $e_r =$ _____ $= e$.

382

$e_r = e * e_r = e$

The first equality holds because e is an identity; the second equality holds because e_r is a right identity. In a similar way, we can show that if a set has a left identity and an identity, then these two elements are the same element. Thus, in a set that has an identity, the set cannot have a distinct left identity or a distinct right identity. The only left and right identities in a set that has an identity is the identity itself.

We will generally use the letter e to represent the identity in a set, unless that set already has a well-established symbol for its identity.

Inverses

383

Suppose S is a set with a binary operation $*$ and elements x, y and e, where e is the identity. If $x * y = y * x = e$, then we say that y is the inverse of x, and x is the inverse of y. For example, when we add the integers 2 and -2 or -2 and 2, we get 0, the identity in ℤ under addition. That is, $2 + (-2) = (-2) + 2 = 0$. Thus, -2 is the inverse of 2, and 2 is the inverse of -2 under addition.

In ℚ (the rational numbers), what is the inverse of the integer 2 under multiplication? *Hint*: 1 is the identity in ℚ under multiplication.

384

$\frac{1}{2}$, because $\frac{1}{2} \cdot 2 = 2 \cdot \frac{1}{2} = 1$

Here is the formal definition of an inverse of an element under the binary operation $*$: if there is an element b such that $a * b = b * a = e$, where e is the identity, then we say that b is the **inverse** of a. Clearly, if b is an inverse of a, then a is an inverse of b.

What is the inverse of 0 under addition?

385

0 is its own inverse.

The inverse of 5 under addition is that number which, when added to 5, produces the identity under addition. The identity is 0. What is the inverse of 5 under addition?

386

−5, because 5 + (−5) = (−5) + 5 = 0

We say that −5 is the inverse of 5 under addition, or, more simply, that −5 is the *additive inverse* of 5. What is the additive inverse of a in $\mathbb{Z}$?

387

−a, because $a + (-a) = (-a) + a = 0$

What is the additive inverse of −5?

388

5

What is the inverse of $\frac{2}{3}$ in $\mathbb{Q}$ (the rational numbers) under addition?

389

$-\frac{2}{3}$, because $\frac{2}{3} + (-\frac{2}{3}) = (-\frac{2}{3}) + \frac{2}{3} = 0$ (0 is the identity under addition)

What is the inverse of $\frac{2}{3}$ in $\mathbb{Q}$ under multiplication?

390

$\frac{3}{2}$, because $\frac{2}{3} \cdot \frac{3}{2} = \frac{3}{2} \cdot \frac{2}{3} = 1$ (1 is the identity under multiplication). We call $\frac{3}{2}$ the *multiplicative inverse* of $\frac{2}{3}$.

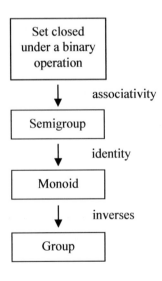

A set closed under a binary operation can have various names, depending on its additional properties.

A *semigroup* is a set closed under an associative binary operation.

A *monoid* is a semigroup with an identity.

A *group* is a monoid with the *inverses property* (i.e., every element has an inverse).

A monoid is a semigroup with an _____.

391

identity

A group is a semigroup with an _____ and the _____ property.

392

identity, inverses

What a Cayley Table Can Tell Us

393

Does the following Cayley table specify a binary operation on the set $S = \{b, c\}$?

*	b	c
b	b	d
c	c	b

394

No; d is not in S.

Recall that a binary operation on a set S is a function from $S \times S$ to S. Thus, a binary operation on S always produces an element in S. In other words, by definition, a set is closed under any binary operation *on that set*. We see from the Cayley table in the preceding frame that S is not closed under the operation represented by the table (d in row b/column c is not in S). Thus, the table does not represent a binary operation on S.

Propose a test to determine if a Cayley table represents a binary operation.

395

Each position in the table should be occupied by one of the row labels (or, equivalently, one of the column labels).

Does the following Cayley table represent a binary operation?

*	b	c
b	c	c
c	c	c

396

Yes. The set $\{b, c\}$ is closed under the operation represented by the table.

Here is a Cayley diagram for a binary operation * on the set $\{b, c\}$:

*	b	c
b	b	c
c	b	c

Is b a left identity? Is b a right identity? Is b an identity?

397

From row b, we can see that $b * b = b$ and $b * c = c$. Thus, b is a left identity. But from row c/column b, we can see that $c * b \neq c$. Thus, b is not a right identity, and therefore b is not an identity (remember that an element is an identity if and only if it is *both* a left identity and a right identity).

Show that c in the preceding frame is not a right identity.

398

From row b/column c in the Cayley diagram in frame **396**, we can see that $b * c \neq b$. Thus, c is not a right identity.

Propose a test to determine from a Cayley table if the corresponding set has a left identity. Propose a test to determine from a Cayley table if the corresponding set has a right identity.

399

If some element's row is identical to the column labels, then that element is a left identity. If some element's column is identical to the row labels, then that element is a right identity. For example, in frame **396**, row b is identical to the column labels. Row c is also identical to the column labels. Thus, both b and c are left identities.

Chapter 7: Binary Operations

Given the following Cayley table, identify the left and right identities:

*	b	c
b	b	c
c	c	c

400

Here b is a left identity and a right identity and, therefore, an identity; c is not a left identity, not a right identity and, therefore, not an identity.

Propose a test to determine from a Cayley table if the corresponding set has an identity.

401

If some element's row is identical to the column labels, and its column is identical to the row labels, then that element is the identity.

In the following Cayley diagram, is e an identity:

*	e	a	b	c	d
e	e	a	b	c	d
a	a	b	c	d	ⓔ
b	b	c	d	e	a
c	c	d	e	a	b
d	d	ⓔ	a	b	c

Mirror images with respect to the main diagonal

402

Yes. Row e is identical to the column labels; column e is identical to the row labels.

From the Cayley table in the preceding frame, determine the inverse of a.

403

From row a/column d, we see that $a * d = e$. From row d/column a, we see that $d * a = e$. Thus, the inverse of a is d.

If an identity appears on the **main diagonal** (i.e., the top-left-to-bottom-right diagonal), then the corresponding element is its own inverse. For example, consider the following Cayley table:

*	e	a	b	c
e	e	a	b	c
a	a	e	c	b
b	b	c	e	a
c	c	b	a	e

The identity e appears on the main diagonal in row a/column a. Thus, $a * a = e$, which means a is its own inverse. In this Cayley table, the main diagonal contains the identity exclusively. Thus, every element is its own inverse.

In frame **401**, which elements are their own inverses?

404

Only e.

Suppose x is an element that has an inverse. If x is the inverse of itself, then the identity is in row x/column x. If, however, x has an inverse, and it is not its own inverse, then associated with x are two occurrences of the identity in the Cayley table. These two occurrences are "mirror images" of each other with respect to the main diagonal. For example, in frame **401**, a and d are inverses of each other. That is, $a * d = e$ and $d * a = e$. The e corresponding to $a * d$ is in row a/column d. The e corresponding to $d * a$ is in row d/column a. The two e's are mirror images of each other with respect to the main diagonal of the Cayley table.

In the Cayley diagram in frame **401**, is there another pair of *e*'s that are mirror images?

405

Yes. There are two *e*'s corresponding to $b * c$ and $c * b$. Thus, *b* is the inverse of *c*, and *c* is the inverse of *b*.

Here is a test to determine if an element, say *x*, has an inverse: in row *x*, look for the identity. If there is an identity, and if it is either on the main diagonal or has a mirror image with respect to the main diagonal, then *x* has an inverse.

Using this test, determine which elements in the Cayley diagram in frame **403** have inverses.

406

All of them.

From the following Cayley table, determine the identity, the inverse of *e*, and the inverse of *b*:

*	e	b
e	e	b
b	b	e

407

e is the identity; the inverse of *e* is *e*; the inverse of *b* is *b*.

Now let's determine if the operation defined by the Cayley table in the preceding frame is associative. Unfortunately, unlike the tests for the identity and inverses, there is no simple test for associativity. One test is to simply use "brute force"—that is, test every possible triple to see if associativity holds. Using brute force to confirm associativity for the Cayley table in the preceding frame, we have to test all of the following equalities:

$$(e * e) * e = e * (e * e) \qquad (e * e) * b = e * (e * b) \qquad (e * b) * b = e * (b * b)$$
$$b * (b * b) = (b * b) * b \qquad (e * b) * e = e * (b * e) \qquad (b * e) * b = b * (e * b)$$
$$(b * e) * e = b * (e * e) \qquad (b * b) * e = b * (b * e)$$

We have one equation for each distinct triple. Each triple has three components, each of which can be either *e* or *b*. We have two choices for each component. Thus, there are $2 \cdot 2 \cdot 2 = 8$ triples and 8 corresponding equations. For larger sets, we have more triples and, therefore, more equations to test. For example, if a set has 10 elements, then there are $10 \cdot 10 \cdot 10 = 1{,}000$ triples. Clearly the brute-force approach is practical only for small sets (unless you have a computer do the testing for you).

Consider the equation $(x * y) * z = x * (y * z)$. Suppose *x* is the identity. What, then, do the left and right sides of this equation equal?

408

Left side: $(x * y) * z = y * z$. Right side: $x * (y * z) = y * z$.

Thus, the equation necessarily holds. Does the equation necessarily hold if *y* is the identity? If *z* is the identity?

409

Yes, for both cases. If *y* is the identity, then the left side is $(x * y) * z = x * z$, and the right side is $x * (y * z) = x * z$. If *z* is the identity, then the left side is $(x * y) * z = x * y$, and the right side is $x * (y * z) = x * y$. Thus, if the identity appears anywhere in a triple, we do not have to test its corresponding equation to confirm that associativity holds.

To confirm associativity for the operation specified by the Cayley table in frame **406**, which equations do we have to test?

410

Only $(b * b) * b = b * (b * b)$

Does this equality hold?

411

Yes. From the Cayley table, $b * b = e$. The left side is $(b * b) * b = e * b = b$, and the right side is $b * (b * b) = b * e = b$. Thus, the operation is associative.

Using brute force, how many equations do we have to test to confirm associativity if a set has n elements, one of which is the identity?

We have $n - 1$ choices for each component of each triplet. Thus, we have $(n - 1) \cdot (n - 1) \cdot (n - 1) = (n - 1)^3$ triples and the same number of equations to test.

Review Questions

1. Let f be the function from $\mathbb{Z} \times \mathbb{Z}$ to $\mathbb{Z}$ defined by $f(x, y) = x + y - xy$. Is the binary operation defined by f commutative? What is the identity in $\mathbb{Z}$ under f, if any?? What is the inverse of k, if any, where $k \in \mathbb{Z}$?
2. Let $B = \{0, 1\}$. Let g be the function from $B \times B$ to B defined by $g(x, y) = xy$. Construct the Cayley table for g.
3. What is the identity in the following Cayley table? What are the inverses of a, b, and c?

*	a	b	c
a	a	b	c
b	b	c	a
c	c	a	b

4. How many elements are in the domain of the binary operation given by the Cayley table in review question 3? How many elements are in the range? Is the operation one-to-one?
5. Is there an identity in the set $\{a, b\}$ under the operation given by the following Cayley table? Is there a left identity? Is there a right identity?

*	a	b
a	a	b
b	a	b

6. Is there an identity in the set $\{a, b\}$ under the operation given by the following Cayley table? Is there a left identity? Is there a right identity?

*	a	b
a	a	b
b	b	a

7. How can you determine from a Cayley table if an element is a right identity?
8. Is the operation defined by the Cayley table in review question 5 associative? Justify your answer.
9. Is the operation defined by the Cayley table in review question 6 associative? Justify your answer.
10. What characteristic does a Cayley table have if the associated operation is commutative?

Answers to the Review Questions

1. $f(x, y) = x + y - xy = y + x - yx = f(y, x) \Rightarrow f$ is commutative; $f(x, 0) = x$, $f(0, x) = x \Rightarrow 0$ is the identity. To get the inverse of k, let $f(k, y) = 0$ (i.e., $k + y - ky = 0$), and then solve for y. We get $y = k/(k - 1)$. This is the left inverse of k; it is also the right inverse of k. Thus, it is the inverse of k. $k/(k - 1) \in \mathbb{Z}$ only if k equals 0 or 2. Thus, only 0 and 2 have inverses in $\mathbb{Z}$.
2.

*	0	1
0	0	0
1	0	1

3. a is the identity; a is its own inverse. The inverse of b is c. The inverse of c is b.

4. The domain has nine elements (there are nine ordered pairs). The range has three elements (a, b, and c). The mapping is not one-to-one (three ordered pairs map to b).
5. a and b are both left identities but not right identities. Thus, neither a nor b is an identity.
6. a is the identity; a is also both a left identity and a right identity. b is not a left identity, not a right identity, and not an identity.
7. The column corresponding to that element is identical to the row labels.
8. Yes. The product of any triple is equal to the rightmost element in the triple.
9. Yes. a is the identity, so we need to check only $b*(b*b) = (b*b)*b$. Both sides equal b.
10. The table is symmetrical with respect to the main diagonal.

Homework Questions

1. Let f be the function from $\mathbb{Z} \times \mathbb{Z}$ to $\mathbb{Z}$ defined by $f(x, y) = x + y + xy$. Is the binary operation defined by f commutative? What is the identity in $\mathbb{Z}$ under f, if any? Does every element in $\mathbb{Z}$ have an inverse under f?
2. Let $B = \{0, 1\}$. Let g be the function from $B \times B$ to B defined by $g(x, y) = \max(x, y)$. Construct the Cayley table for g.
3. Construct the smallest possible Cayley table corresponding to a nonassociative operation.
4. Construct the smallest possible Cayley table corresponding to a noncommutative operation.
5. Assuming the operation $*$ is both associative and commutative, show that $(a * b) * c = c * (b * a)$.
6. Is addition a binary operation on the odd integers? Justify your answer.
7. Is multiplication a binary operation on the odd integers? Justify your answer.
8. Let f be the function from $\mathbb{Z} \times \mathbb{Z}$ to $\mathbb{Z}$ defined by $f(x, y) = x$. Is f commutative? Is f associative? Justify your answer.
9. Can a Cayley table be constructed that contains exactly one left identity and exactly one right identity, but not an identity? If your answer is yes, construct such a table. If your answer is no, justify your answer.
10. Show that if an element has an inverse, the inverse is unique if the operation on the set is associative.

Chapter 8: Introduction to Groups

Introduction to Groups

Definition of a Group

A *group* is a set together with a binary operation that has the following properties:

- Closure: The set is closed under the binary operation.
- Associativity: The binary operation is associative.
- Identity: The set has an identity.
- Inverses: Each element in the set has an inverse that is in that set.

For example, $\mathbb{Z}$ under addition is a group. It has all the required properties:

- Closure: $a + b \in \mathbb{Z}$ for all $a, b \in \mathbb{Z}$.
- Associativity: $a + (b + c) = (a + b) + c$ for all $a, b, c \in \mathbb{Z}$.
- Identity: $0 + a = a + 0 = a$ for all $a \in \mathbb{Z}$; 0 is the identity.
- Inverses: For each $a \in \mathbb{Z}$, $-a$ is its inverse, and $-a \in \mathbb{Z}$.

When discussing groups in general or a group that does not have a well-established symbol for its binary operation, we will use "$*$" to represent the operation, or simply use juxtaposition. For example, we will write $a * b$ or ab to represent the binary operation of the group applied to the elements a and b. Of course, if the binary operation of a group has a well-established symbol, we will use that symbol. For example, in the group given above—$\mathbb{Z}$ under addition—we use the well-established symbol for addition (+) to represent its binary operation.

A group is both a set and a binary operation. Thus, when referring to a group, we typically specify not only the set but also the operation. For example, to specify the group of integers under addition, we use $<\mathbb{Z}, +>$. The angle brackets indicate that $\mathbb{Z}$ and the operation represented by + together are a single algebraic structure. If the group operation is clear from context, we can omit the operator and specify only the set. For example, we might simply say "the group $\mathbb{Z}$."

Given a set and a binary operation, can we conclude the set is closed under the given binary operation? Not necessarily. For example, consider the set of odd integers and the binary operation addition. The set of *all* the integers is closed under addition, but the set of odd integers—a subset of the integers—is not closed. To see this, observe that the sum of 3 and 5, two odd integers is 8, an even integer. However, if we are given a set and a binary operation *on that set*, then by the definition of a binary operation on a set, the set is closed under the given binary operation. Suppose we modify our definition of a group by adding the phrase "on that set" and deleting the closure requirement. We get the following alternative definition: A *group* is a set together with a binary operation *on that set* that has the following properties:

- Associativity: The binary operation is associative.
- Identity: The set has an identity.
- Inverses: Each element in the set has an inverse that is in that set.

Does our new definition imply our first definition?

Yes. A binary operation *on a set*, by definition, is a function that maps each pair of elements of the set to some element in that set. Thus, a set is closed under any binary operation *on that set*. Conversely, our first definition implies our second definition. To see this, observe that closure under a binary operation means the operation maps each ordered pair of elements of the set to some element of that set. Thus, the operation is a binary operation *on that set*. Because each of our two definitions implies the other, they are equivalent.

A ***commutative*** (or ***Abelian***) ***group*** is a group whose operation is commutative. That is, a group $<G, *>$ is commutative if $a * b = b * a$ for all $a, b \in G$. Some groups are commutative; some are not. However, all groups by definition are associative.

The ***order of a group G,*** denoted by $|G|$, is the number of elements in G. If $|G| = n$ for some positive integer n, then G is called a ***finite group.*** Otherwise, G is called an ***infinite group.***

Multiplicative and Additive Notation

415

We will generally use ***multiplicative notation*** when describing groups. That is, we borrow notation and terminology from everyday multiplication. For example, we describe $a * b$ (the group operation $*$ applied to a and b) as the "product" of a and b, or as a "multiplied" by b. We use exponents as we do in high school algebra. For example, we write $a * a * a$ as a^3, the inverse of a as a^{-1}, and $a^{-1} * a^{-1} * a^{-1}$ as a^{-3}. The expression a^0 by definition is the identity. We sometimes represent the product of a and b by juxtaposing a and b—that is, by writing a and b next to each other without an intervening operation symbol.

Of course, for groups in which the operation is addition, it would be confusing to use multiplicative notation. Thus, for a group whose operation is addition, like $<\mathbb{Z}, +>$, we use ***additive notation***: we describe $a + b$ as the "sum" of and b. We write $a + a + a$ as $3a$, the inverse of a as $-a$, and $(-a) + (-a) + (-a)$ as $-3a$. The expression $0a$ by definition is the identity. When using additive notation, we never use juxtaposition to represent a sum. For example, we never write $a + b$ as ab (because juxtaposition suggests multiplication).

The following table summarizes the differences between multiplicative and additive notation:

Multiplicative notation	Additive notation
$a * b$ or ab (the product)	$a + b$ (the sum)
a^{-1} (the inverse of a)	$-a$ (the inverse of a)
$a * a * \cdots * a = a^n$ (product of n as)	$a + a + \cdots + a = na$ (sum of n as)
$a^{-1} * a^{-1} * \cdots * a^{-1} = a^{-n}$ (product of n a^{-1}'s)	$(-a) + (-a) + \cdots + (-a) = -na$ (sum of n $-a$'s)
$a^1 = a$	$1a = a$
a^0 is the identity	$0a$ is the identity

Uniqueness of the Identity and Inverses in a Group

416

In frame **377**, we showed that in a set with a binary operation, there can be *at most one* identity. By definition, a group has an identity. The phrase "has an identity" does not exclude the possibility that there is more than one identity. That is, "has an identity" means "has *at least one* identity." From frame **377**, we know that in a set with a binary operation, there can be *at most one* identity. What can we then conclude about the identity in a group?

417

It is unique. That is, there is *one and only one* identity, or, equivalently, there is *exactly one* identity.

Let's now investigate inverses in a group. Let's show that each element in a group has a unique inverse. Suppose y_1 and y_2 are both inverses of x in a group whose identity is e. We want to show that y_1 and y_2 are the same element (which implies that there can be only one inverse of x). To show that y_1 and y_2 are the same element, we will construct a sequence of equalities that has the following form: $y_1 = \cdots = y_2$.

Let start our sequence of equalities with $y_1 = y_1 * e$. We want our sequence of equalities to end with y_2. So at some point we have to eliminate y_1. How can we eliminate y_1? *Hint*: x is the inverse of y_1.

418

Multiply y_1 by its inverse, which is x. This multiplication eliminates y_1 and produces e.

What should we replace e with in $y_1 = y_1 * e$? The replacement should be a product equal to e and have on its left side x so that y_1 is eliminated when multiplied by x. *Hint*: We want to not only eliminate y_1 but also get y_2.

419

Replace e with $x * y_2$ (which equals e because y_2 is also an inverse of x). We get $y_1 = y_1 * e = y_1 * (x * y_2)$. Now apply the associative property.

420

$y_1 = y_1 * e = y_1 * (x * y_2) = (y_1 * x) * y_2$

Now simplify $y_1 * x$.

421

$y_1 = y_1 * e = y_1 * (x * y_2) = (y_1 * x) * y_2 = e * y_2$

Finally, simplify $e * y_2$.

422

$y_1 = y_1 * e = y_1 * (x * y_2) = (y_1 * x) * y_2 = e * y_2 = y_2$. ∎

A good way to write out our string of equalities that shows that y_1 is equal to y_2 is to use a separate line and provide a justification for each step. We get

y_1
$= y_1 * e$ e is the identity
$= y_1 * (x * y_2)$ y_2 is the inverse of x
$= (y_1 * x) * y_2$ associativity
$= e * y_2$ y_1 is the inverse of x
$= y_2$ e is the identity ∎

Show that in a group, an element cannot have an inverse and a distinct left inverse. *Hint*: Suppose x has an inverse y and a left inverse l. Then show $y = l$.

423

Suppose x has an inverse y and a left inverse l. Then
y
$= e * y$ e is the identity
$= (l * x) * y$ l is the left inverse of x
$= l * (x * y)$ associativity
$= l * e$ y is the inverse of x
$= l$ e is the identity ∎

We can similarly show that if x has an inverse y and a right inverse r, then y and r are the same element.

Although in a group, each element has a unique inverse, the uniqueness of inverses is *not* stated in the definition of a group (the definition states that each element has *an* inverse—not a unique inverse). Similarly, the uniqueness of the identity is not stated in the definition of a group (the definition states that a group has an identity—not a unique identity). However, the uniqueness of inverses and the identity follow from the properties that are stated in the definition. Thus, it is unnecessary and, in fact, superfluous for the definition to state the uniqueness of inverses and the identity. This is standard practice. Definitions typically provide only a minimal set of properties necessary to capture a concept. Consider the following definition of a square (assume **rectangle** has been previously defined):

A **square** is a rectangle whose four sides are of equal length and whose area is equal to the length of a side squared.
What is wrong with this definition?

424

The second part of this definition—"the area is equal to the length of a side squared"—is superfluous. It is not needed in the definition of a square.

Cancellation Law for Groups

425

In every group G, the *left cancellation law* holds:

For all $a, b, c \in G$, if $a * b = a * c$, then $b = c$.

Thus, if we are given $a * b = a * c$, we can cancel the a's on each side of the equation to get $b = c$. The left cancellation law applies only if the operand to be canceled is leftmost on each side of the equation. Thus, the left cancellation law does not allow us to cancel the a in $a * b = c * a$. Here is the proof of this important law:

$a * b = a * c$	given
$\Rightarrow a^{-1} * (a * b) = a^{-1} * (a * c)$	multiply both sides by a^{-1}
$\Rightarrow (a^{-1} * a) * b = (a^{-1} * a) * c$	associativity
$\Rightarrow e * b = e * c$	a^{-1} is the inverse of a
$\Rightarrow b = c$	e is the identity ∎

The second line of this proof requires some explanation. The first line is $a * b = a * c$. This equation means that $a * b$ and $a * c$ are the same element. To get the second line, we multiply each side of the first line with a^{-1} (the unique inverse of a). But does the equality on the second line hold? It does, but how do we know this? *Hint*: $*$ is a function.

426

We are given that $a * b = a * c$. This means that $a * b$ and $a * c$ are the same element. Thus, in the second line, $a^{-1} * (a * b) = a^{-1} * (a * c)$, the two operands on the left side (a^{-1} and $a * b$) *are the same elements in the same order* as the respective two operands on the right side (a^{-1} and $a * c$). Moreover, $*$ is a binary operation. That is, it is a function that maps each pair of elements to one and only one element in the set. Thus, $*$ necessarily maps the left pair and the right pair to the same element, guaranteeing that the equality on the second line holds.

The cancellation law we just proved is called the *left cancellation law* because it allows the cancellation of an element positioned leftmost on both sides of an equality. Similarly, the *right cancellation law* allows the cancellation of an element positioned rightmost on both sides of an equation.

Prove the right cancellation law.

427

$b * a = c * a$	given
$\Rightarrow (b * a) * a^{-1} = (c * a) * a^{-1}$	multiply both sides by a^{-1}
$\Rightarrow b * (a * a^{-1}) = c * (a * a^{-1})$	associativity
$\Rightarrow b * e = c * e$	a^{-1} is the inverse of a
$\Rightarrow b = c$	e is the identity ∎

In our proofs of the left and right cancellation laws, we assumed only the properties that every group has. Thus, these cancellation laws apply to *every* group. Suppose we are given $a * b = c * a$. The element a does not appear leftmost on both sides, nor does it appear rightmost on both sides. Thus, neither the left cancellation law nor the right cancellation law applies. We cannot cancel the a's. However, if the group is commutative, we can commute the c and a on the right side to get $a * c$ on the right side. We can then use the left cancellation law to cancel the a's to get $b = c$.

Suppose $d * a * d * e = d * a * f$ in a group. Can we cancel the a's?

428

Yes. By the left cancellation law, we can cancel the d's and then the a's.

Here is an incorrect "proof" of the statement

for all $a, b, c \in G$, if $a * b = c * a$, then $b = c$,

where G is a group. This statement is true for commutative groups but not for all groups. It says we can cancel the two a's even though one is leftmost and the other is rightmost. What is wrong with this "proof"?

88 Chapter 8: Introductions to Groups

$a * b = c * a$ given
$\Rightarrow a^{-1} * (a * b) = (c * a) * a^{-1}$ multiply both sides by a^{-1}
$\Rightarrow (a^{-1} * a) * b = c * (a * a^{-1})$ associativity
$\Rightarrow e * b = c * e$ a^{-1} is the inverse of a
$\Rightarrow b = c$ e is the identity ∎

429

The second line is incorrect. Although both sides of the equation on this line have the same pair of elements, the order of the elements differs. Thus, the equality does not necessarily hold. Remember that in a group with elements x and y, $x * y$ is not necessarily equal to $y * x$, unless the group is commutative.

Can we cancel the b's in $a * b * c = a * b * d$ in a noncommutative group?

430

Yes. Use the left cancellation law twice: first cancel the a's, then the b's, giving $c = d$.

Can we cancel the b's in $a * b * c = c * b * d$ in a noncommutative group?

431

No. The b's are neither leftmost nor rightmost. We can make one b leftmost, but not both b's. For example, if we multiply both sides by a^{-1}, we get $a^{-1} * (a * b * c) = a^{-1} * (c * b * d)$. The left side simplifies to $b * c$. Thus, the left side has b leftmost, but on the right side, $a^{-1} * c * b * d$, b is not leftmost. We also cannot make both b's rightmost.

Give an example that shows that the left cancellation law does *not* hold for $\langle \mathbb{R}, \cdot \rangle$ (the real numbers under multiplication). Recall that "·" is the symbol we use for multiplication on the real numbers (and also for multiplication on the integers and the rationals). *Hint*: Use 0 as the common factor.

432

$0 \cdot 3 = 0 \cdot 5$. Canceling 0, we get $3 = 5$. But $3 \neq 5$.

Because the left cancellation law does not hold for $\mathbb{R}$ under multiplication, $\mathbb{R}$ under multiplication *must not be a group* (recall that the left cancellation law holds for *all* groups). Which group property does $\mathbb{R}$ under multiplication *not* have? *Hint*: Consider 0.

433

The inverses property; 0 has no inverse.

Explain why 0 has no inverse in $\mathbb{R}$ under multiplication.

434

Because the identity for $\mathbb{R}$ under multiplication is 1, the inverse of 0, if one exists, is the number x such that $0x = x0 = 1$. But $0x = x0 = 0$ for all $x \in \mathbb{R}$. Thus, 0 has no inverse under multiplication. However, every other element of $\mathbb{R}$ has an inverse (the inverse of x is $\frac{1}{x}$ for all $x \neq 0$).

The proofs of the cancellation laws require the group properties of closure, associativity, and identity. However, they do not require that *every* element in the set have an inverse. Only the element we are cancelling has to have an inverse (it is the only inverse used in the proofs in frames **425** and **427**). Thus, we can apply the cancellation laws to $\mathbb{R}$ under multiplication as long as the element we cancel has an inverse (i.e., it is nonzero).

Suppose $5x = 5y$, where $x, y \in \mathbb{R}$. Can we conclude that $x = y$?

435

Yes, because 5 is nonzero and therefore has an inverse.

Suppose $0x = 0y$. Can we conclude that $x = y$?

436

No, because 0 has no inverse.

Solving Equations

437

In a group G, we can solve equations of the form $a * x = b$. Try solving for x in this equation. *Hint*: Multiply both sides on the left by a^{-1}.

438

$a * x = b$	given
$\Rightarrow a^{-1} * (a * x) = a^{-1} * b$	multiply both sides by a^{-1}
$\Rightarrow (a^{-1} * a) * x = a^{-1} * b$	associativity
$\Rightarrow e * x = a^{-1} * b$	a^{-1} is the inverse of a
$\Rightarrow x = a^{-1} * b$	e is the identity ∎

Not only can we solve for x but we know it has a unique solution (given by $a^{-1} * b$). How do we know the solution is unique?

439

Because $*$ is a function from $G \times G$ to G, it necessarily maps the pair (a^{-1}, b) to a unique element in G. We can also show uniqueness of a solution this way: Suppose r and s are both solutions to the equation $a * x = b$.

$\Rightarrow a * r = b$ and $a * s = b$	r and s are solutions to the given equation
$\Rightarrow a * r = a * s$	both sides equal b
$\Rightarrow r = s$	left cancellation law
$\Rightarrow$ the solution is unique.	the solutions r and s are the same element

Solve for x in $x * a = b$.

440

$x * a = b$	given
$\Rightarrow (x * a) * a^{-1} = b * a^{-1}$	multiply both sides by a^{-1}
$\Rightarrow x * (a * a^{-1}) = b * a^{-1}$	associativity
$\Rightarrow x * e = b * a^{-1}$	a^{-1} is the inverse of a
$\Rightarrow x = b * a^{-1}$	e is the identity ∎

To solve for x in $a * x = b$, we move a to the right side of the equation. Similarly, to solve for x in $x * a = b$, we move a to the right side. In both cases, a on the left side of the equation becomes a^{-1} on the right side of the equation. When you move an element that is *leftmost*, it becomes its inverse and is positioned *leftmost* on the other side of the equation. Similarly, when you move an element that is *rightmost*, it becomes its inverse and is positioned *rightmost* on the other side of the equation. Let's call this rule the **switching-sides rule.** Of course, if the group is commutative, then the order of the operands does not matter.

Using the switching-sides rule, can we get $x = b * a^{-1}$ from $a * x = b$?

441

No. The element a in $a * x = b$ is leftmost. Thus, a^{-1} in the new equation should be leftmost.

Move a to the right side of the equation $a = b * c$. *Hint*: Because a is alone on the left side of the equation, it is both leftmost and rightmost on the left side.

442

Because a is both leftmost and rightmost on the left side of the given equation, its inverse can end up either leftmost or rightmost on the right side. Thus, we can get either $e = bca^{-1}$ or $e = a^{-1}bc$, where e is the identity.

$\mathbb{Z}$ under addition is a group. Thus, we can use the switching-sides rule. Suppose $5 + x = 20$, where $x \in \mathbb{Z}$. Using the switching-sides rule, move 5 to the right side. *Hint*: -5 is the inverse of 5 under addition.

443

$x = -5 + 20 = 15$

As with the cancellation laws, the justification of the switching-sides rule does not require that every element has an inverse—only that the element moved has an inverse. Thus, we can apply it to $\mathbb{R}$ under multiplication with the restriction that we cannot move 0 (because 0 has no inverse in $\mathbb{R}$ under multiplication).

Suppose $5x = 15$. Move 5 to the right side.

444

5 becomes $\frac{1}{5}$ (the inverse of 5 under multiplication) on the right side. Thus, we get $x = \frac{1}{5} \cdot 15 = 3$.

In grade school, you were probably taught that to solve for x in $5x = 15$, you divide both sides by 5. Dividing by 5 is equivalent to moving 5 to the other side of the equation, where it becomes the inverse of 5 under multiplication (i.e., 5 becomes $\frac{1}{5}$). So the rule you were taught is really our switching-sides rule.

Properties of Cayley Tables for Groups

445

Does the following Cayley table correspond to a group, given that the operation represented is associative?

*	e	a	b	c
e	e	a	b	c
a	a	b	c	e
b	b	c	e	a
c	c	e	a	b

446

Yes. Each element in the table is equal to one of the row labels. Thus, the set is closed under its operation. The e row matches the column labels, and the e column matches the row labels. Thus, e is the identity. Each row and each column has an identity that is either on the main diagonal or has a mirror image with respect to the main diagonal. Thus, each element has an inverse (see frame **405**). This table has the properties that imply the closure, identity, and inverses properties. We are given that the operation is associative. Thus, this table corresponds to a group. This group is called the ***cyclic group of order 4***.

Does the following Cayley table correspond to a group, given that the operation represented is associative?

*	e	a	b	c
e	e	a	b	c
a	a	e	c	b
b	b	c	e	a
c	c	b	a	e

447

Yes. From the table, we can confirm the closure, identity, and inverses properties. We are given that the operation is associative. Thus, the table corresponds to a group. This group is called the ***Klein four group*** denoted by V_4). It is unusual in that its main diagonal contains the identity exclusively.

The diagonal in the table for V_4 implies what?

448

Each of its four elements is its own inverse.

In the Cayley tables in frames **445** and **446**, each element of the group appears exactly once in every row and exactly once in every column. Does every Cayley table for a group have this property? Let's investigate. Let G be a group with a binary operation $*$. Let's take an arbitrary row in the Cayley table for G, say row p, and an arbitrary element of G, say q. Let's then show that q appears exactly once in row p. Because we are making no assumptions about p, q, or G other than p is one of the rows in the Cayley table for an G, q is an element of G, and G is some group, we can then conclude that in the Cayley table for any group, each element of the group appears exactly once in every row.

By frame **438**, $p * x = q$ has a unique solution—let's say r. Thus, $p * r = q$. The corresponding Cayley table has the following form:

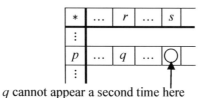

q cannot appear a second time here

The element q appears in row p/column r. Moreover, q appears only in this one position in row p. Why can a second q not appear in row p, say, in column s?

449

We would have $p * r = q$ and $p * s = q$. Then $p * x = q$ would have two solutions: r and s. But this is impossible because by frame **438**, $p * x = q$ has a unique solution. Thus, q appears exactly once in the p row. ∎

Show that in a Cayley table for a group, each element appears exactly once in every column. *Hint*: In a group, $x * p = q$ has a unique solution.

450

By frame **440**, $x * p = q$ has a unique solution, say r. Then q appears in row r/column p. Suppose q appears a second time in column p, say in row s. But then $x * p = q$ would not have a unique solution—it would have two solutions: r and s. ∎

Key idea: In a Cayley table for a group, each element appears exactly once in every row and exactly once in every column.

Other Properties of a Group

451

In a group with elements a and b, what is the inverse of $a * b$? *Hint*: Find the element x such that $a * b * x = e$, where e is the identity, using the switching-sides rule twice.

452

$a * b * x = e \Rightarrow b * x = a^{-1} * e = a^{-1} \Rightarrow x = b^{-1} * a^{-1}$

Confirm that the inverse of $a * b$ is $b^{-1} * a^{-1}$ by checking if the product of these two elements equals the identity.

453

$(a * b) * (b^{-1} * a^{-1}) = a * (b * b^{-1}) * a^{-1} = a * e * a^{-1} = a * a^{-1} = e$
$(b^{-1} * a^{-1}) * (a * b) = b^{-1} * (a^{-1} * a) * b = b^{-1} * e * b = b * b^{-1} = e$

That the inverse of $a * b$ is $b^{-1} * a^{-1}$ is sometimes referred to as the **socks-shoes property**. The order in which we take off our socks and shoes is the reverse of the order in which we put them on; analogously, the inverse of a product is the product of the inverses in reverse order. Generalizing the socks-shoes property, we get $(a_1 * a_2 * \cdots * a_n)^{-1} = a_n^{-1} * \cdots * a_2^{-1} * a_1^{-1}$.

Apply the sock-shoes property to $(a^n)^{-1}$.

92   Chapter 8: Introductions to Groups

454

By the sock-shoes property, $(a^n)^{-1} = (a^{-1})^n$, which by definition equals a^{-n}. For example, $(a^3)^{-1} = (aaa)^{-1} = a^{-1}a^{-1}a^{-1} = (a^{-1})^3 = a^{-3}$.

Let S be a set with the binary operation $*$. An element $a \in S$ is **idempotent** if and only if $a * a = a$. Prove that, in a group, an element is idempotent if and only if it is the identity. *Hint*: Use the left cancellation law.

455

$a * a = a \Rightarrow a * a = a * e$, where e is the identity. Canceling the a's on the left side, we get $a = e$. Conversely, if e is the identity, then $e * e = e$. Thus, e is idempotent. ∎

Suppose g is an element of a group G whose identity is e. If $g^2 = g^5$, what can we conclude about g^3? *Hint*: Use the left cancellation law.

456

$g^2 = g^5 \Rightarrow g * g * e = g * g * g * g * g$. Cancelling g twice, we get $e = g^3$. Another way to show that $g^3 = e$ is to rewrite $g^2 = g^5$ as $g^2 = g^2 * g^3$, which implies that g^3 is the identity.

Suppose $g^j = g^k$ with $j < k$ in a group G. What is the identity in G?

457

Canceling j times the element g, we get $e = g^{k-j}$.

Review Questions

1. Express $a^2 b^3 a^{-4}$ in additive notation.
2. Show that $5\mathbb{Z} = \{5n : n \in \mathbb{Z}\}$ under addition is a group.
3. Can a group have more than one idempotent element?
4. Suppose G is a group and $abxcd = e$, where $a, b, c, d, e, x \in G$, and e is the identity. Solve for x.
5. Suppose $H \subseteq \mathbb{Z}$ and H is a group under addition. Is H^c (i.e., the complement of H with respect to $\mathbb{Z}$) also a group under addition? Justify your answer.
6. Disprove: $a^2 = b^2 \Rightarrow a = b$, where a and b are elements of a group.
7. Are the odd integers under addition a group? Justify your answer.
8. Show that, in a group, if $a * b = b * a$, then $a^2 * b^2 = b^2 * a^2$.
9. Prove or give a counterexample: In a group, if $a^2 = e$, then $a = e$, where e is the identity.
10. Determine x^4 for each element $x \in V_4$ (see frame **446**). Similarly, determine x^4 for each element x in the cyclic group of order 4 in frame **445**.

Answers to the Review Questions

1. $2a + 3b + 4(-a)$
2. *Closure*: Let $x, y \in 5\mathbb{Z}$. Then $x = 5i$ and $y = 5j$ for some $i, j \in \mathbb{Z} \Rightarrow i + j = 5i + 5j = 5(i + j) \in 5\mathbb{Z}$. *Associativity*: Because addition is associative in $\mathbb{Z}$, it is also associative in any subset of $\mathbb{Z}$. We say that $<5\mathbb{Z}, +>$ **inherits** associativity from $<\mathbb{Z}, +>$. *Identity*: Let $x \in 5\mathbb{Z}$. $0 + x = x + 0 = x$; the element 0 is the identity. *Inverses*: Let $x \in 5\mathbb{Z}$. Then $x = 5i$ for some $i \in \mathbb{Z}$. Then $y = 5(-i)$, which is in $5\mathbb{Z}$, is the inverse of x because $5i + 5(-i) = 5(-i) + 5i = 0$.
3. No. An idempotent in a group is the identity. Because there is one and only one identity in a group, the identity is the one and only idempotent.
4. $x = b^{-1}a^{-1}d^{-1}c^{-1}$
5. No. H^c does not have the identity 0.
6. Counterexample: Suppose a and b are two distinct elements in the Klein four group (see frame **446**). Then $a^2 = b^2 = e$, but $a \neq b$.

7. No. The odd integers lack the identity property (0 is an even number) and the closure property (the sum of two odd numbers is even).
8. $a^2 * b^2 = a * a * b * b$. Because a and b commute, we can commute all the b's to the left of all the as to get $b * b * a * a = b^2 * a^2$. ∎
9. Counterexample: In the Klein four group, the square of every nonidentity element is equal to the identity. Another argument: Consider the group with exactly two elements, the identity e and a. The product $a * a$ cannot equal a (otherwise, it is idempotent and, therefore, the identity). Thus, $a * a$ must equal e.
10. For both groups, the fourth power of every element is equal to the identity. In fact, for every finite group G of order n, $g^n = e$ for all $g \in G$, where e is the identity (we prove this in frame **998**).

Homework Questions

1. Show that in a finite group, the number of elements x such that $x^2 = e$ is even.
2. Show that if $a^2 = e$, where e is the identity, for all elements a in a group, then the group is commutative. *Hint*: First show that $b * a$ is the inverse of $a * b$.
3. Suppose $H \subseteq \mathbb{Z}$ and H is a group under addition. Let $i \in H$ and $o \in \mathbb{Z} - H$. Show that is $i + o \in \mathbb{Z} - H$. In other words, show that the sum of an "inside-H" integer and an "outside-H" integer is outside H? *Hint*: Use a proof by contradiction.
4. Give an example of a finite subset of $\mathbb{Z}$ that is a group under addition.
5. Suppose $a \in G$. Let $aG = \{ag : g \in G\}$. Show that $aG = G$.
6. Show that $<\mathbb{Z}, *>$ is a group, where $*$ is defined by $a * b = a + b + 1$ for all $a, b \in \mathbb{Z}$.
7. Suppose $<G, *>$, $<H, *>$, and $<K, *>$ are groups with $H, K \subseteq G$. Show that $<H \cap K, *>$ is a group.
8. Suppose a group G is finite and $g \in G$. Then the successive powers of g cannot be all distinct (otherwise, G would be infinite). Thus, for some i, j such that $0 < i < j$, $g^i = g^j$. What can we then conclude about g^{j-i}?
9. Prove that if no element appears more than once in any row of a Cayley table for a set with a binary operation, then the left cancellation law holds. Also prove that if no element appears more than once in any column of a Cayley table for a set with a binary operation, then the right cancellation law holds.
10. Prove the converse of the following statement: If $a * x = b$ and $x * a = b$ have unique solutions for all elements a and b in a set, then each element in the set appears exactly once in each row and in each column of the Cayley table.

9 Symmetric Group of Degree *n*

An Unusual Group

458

When we think of a binary operation, we typically think of everyday addition, subtraction, multiplication, or division, all of which operate on numbers. However, binary operations are not necessarily operations on numbers. Remember that a binary operation on a set A is simply a function from $A \times A$ to A. There are no restrictions on the types of elements A can have.

In this section, we look at a binary operation that operates on a somewhat unusual set: S_n, the set of all bijections on the set $\{1, 2, ..., n\}$. As we shall see, S_n under the operation of composition is a group. We call this group the **symmetric group of degree n**. For example, S_3, the set of all bijections on $\{1, 2, 3\}$ under composition, is the symmetric group of degree 3.

Recall from frame **339** that the number of bijections on a finite set of order n is $n!$. Thus, S_n has $n!$ elements. For example, S_3 has $3! = 3 \cdot 2 \cdot 1 = 6$ bijections:

$$\begin{array}{cccccc} f_1 & f_2 & f_3 & f_4 & f_5 & f_6 \\ \begin{pmatrix} 1 & 2 & 3 \\ 1 & 2 & 3 \end{pmatrix} & \begin{pmatrix} 1 & 2 & 3 \\ 2 & 3 & 1 \end{pmatrix} & \begin{pmatrix} 1 & 2 & 3 \\ 3 & 1 & 2 \end{pmatrix} & \begin{pmatrix} 1 & 2 & 3 \\ 1 & 3 & 2 \end{pmatrix} & \begin{pmatrix} 1 & 2 & 3 \\ 2 & 1 & 3 \end{pmatrix} & \begin{pmatrix} 1 & 2 & 3 \\ 3 & 2 & 1 \end{pmatrix} \end{array}$$

A *permutation* is an ordering of objects. A bijection on a set specifies a permutation. For example, the bijection on $\{1, 2, 3\}$ given by $\begin{pmatrix} 1 & 2 & 3 \\ 1 & 3 & 2 \end{pmatrix}$ specifies the permutation 1, 3, 2. Specifically, it indicates the required replacements of the initial ordering, 1, 2, 3, needed to produce the ordering 1, 3, 2. The first column indicates that 1 remains where it is. The second column indicates that 2 is replaced by 3. The third column indicates that 3 is replaced by 2.

Conversely, a permutation of the elements of the set $\{1, 2, ..., n\}$ specifies a bijection on that set. For example, the permutation 1, 3, 2 corresponds to the bijection $\begin{pmatrix} 1 & 2 & 3 \\ 1 & 3 & 2 \end{pmatrix}$. Because a bijection on a set specifies a permutation of the elements of that set, and vice versa, a bijection is often called a permutation. In fact, it is customary to refer to the elements of S_n as permutations rather than as bijections. We will follow suit.

S_n is the set of all permutations on $\{1, 2, ..., n\}$. If you compose any two elements in S_n, do you get another permutation on $\{1, 2, ..., n\}$? In other words, is S_n closed under function composition?

459

Yes. A composition of two permutations (i.e., two bijections) on a set is also a permutation on that set (we showed this in frame **311**). Thus, composition is a function from $S_n \times S_n$ to S_n. In other words, composition on S_n is a _____ _____ on S_n.

460

binary operation

We learned in chapter 6 that the composition $(f_2 \circ f_1)(x)$ is defined as $f_2(f_1(x))$. We first apply f_1 to x and then apply f_2 to the result of applying f_1 to x. The functions are applied in right-to-left order.

Apply the definition of composition to $((p \circ q) \circ r)(x)$, where p, q, and r are permutations.

461

Applying r, we get $(p \circ q)(r(x))$. Next, we apply q to $r(x)$, and then we apply p to the result to get $p(q(r(x)))$.

Perform a similar transformation on $(p \circ (q \circ r))(x)$ using the definition of function composition. *Hint*: First apply $q \circ r$, then p.

462

$(p \circ (q \circ r))(x) = p \circ ((q \circ r)(x)) = p \circ (q(r(x))) = p(q(r(x)))$

Compositions $p \circ (q \circ r)$ and $(p \circ q) \circ r$ define the same function, namely, $p(q(r(x)))$. That is, $(p \circ q) \circ r = p \circ (q \circ r)$ for all p, q, $r \in S_n$. Composition is an _____ operation on S_n.

463

associative

The function e on $\{1, 2, ..., n\}$ defined by $e(x) = x$, $1 \leq x \leq n$, is a bijection on $\{1, 2, ..., n\}$. Thus, it is an element in S_n.

What does $(e \circ f)(x)$ equal?

464

$e(f(x))$

From the definition of e, what does $e(f(x))$ equal?

465

$f(x)$. Thus, $e \circ f = f$. Similarly, $f \circ e = f$.

Because $e \circ f = f \circ e = f$, e is the _____ in S_n under _____.

466

identity, composition

Recall that if f is a permutation, then f^{-1}, the inverse of f, is also a permutation. Moreover, the composition of f and its inverse in either order is the identity permutation e. Thus, for every f in S_n, there is a permutation f^{-1} in S_n such that $f \circ f^{-1} = f^{-1} \circ f = e$. In other words, every element in S_n has an inverse.

We have shown that S_n under composition has the closure, associativity, identity, and inverses properties. Thus, S_n under composition is a _____.

467

group

Cycle Representation of a Permutation

468

Another way to represent a permutation on a finite set is with one or more parenthesized nonrepeating lists of elements from the set. We call these lists ***cycles***. In each cycle, each element, except the last, is mapped to the element to its immediate right. The last element is mapped to the first element. Thus, a cycle is simply a circular list. For example, the cycle (1 2 3) represents the permutation in which 1 is mapped to 2, 2 is mapped to 3, and 3 is mapped to 1. In table form, this permutation is $\begin{pmatrix} 1 & 2 & 3 \\ 2 & 3 & 1 \end{pmatrix}$. We can also represent this permutation with the cycles (2 3 1) and (3 1 2). The three variations—(1 2 3), (2 3 1), and (3 1 2)—all represent the same circular list (in all three, 1 is mapped to 2, 2 is mapped to 3, and 3 is mapped to 1). Thus, they all represent the same permutation. Among the cycles that represent the same permutation, we will generally use the one that starts with the smallest element. Thus, we will use (1 2 3) rather than the equivalent (2 3 1) or (3 1 2).

We call a cycle that consists of n elements an ***n-cycle***. For example, (1 2 3) is a 3-cycle.

Do (3 1 2) and (1 3 2) represent the same permutation?

469

No. They are distinct circular lists. The first maps 3 to 1; the second maps 3 to 2.

Convert the following permutation in cycle form to table form: (1 3 5 2 4).

470

96 Chapter 9: Symmetric Group of Degree n

$\begin{pmatrix} 1 & 2 & 3 & 4 & 5 \\ 3 & 4 & 5 & 1 & 2 \end{pmatrix}$

Convert the following permutation in table form to cycle form: $\begin{pmatrix} 1 & 2 & 3 & 4 & 5 \\ 5 & 1 & 4 & 2 & 3 \end{pmatrix}$. *Hint*: This table indicates that 1 goes to 5, 5 goes to 3, and so on. Thus, the cycle starts with 1 5 3.

471

(1 5 3 4 2)

A 1-cycle indicates that its element is mapped to itself. For example, (2) indicates that 2 is mapped to 2. All the elements within a cycle must be distinct. Thus, (1 1) is not a valid cycle. To represent the mapping of 1 to 1, use the 1-cycle (1).

What is the cycle representation of $\begin{pmatrix} 1 & 2 & 3 \\ 1 & 3 & 2 \end{pmatrix}$?

472

Here 2 is mapped to 3 and 3 is mapped to 2, which gives us the cycle (2 3). In addition, 1 is mapped to 1, which gives us the cycle (1). This permutation is the composition of two cycles: (1) and (2 3). That is, it is equal to (1) ∘ (2 3).

A 1-cycle does not change the ordering of the elements, so we will generally omit them when representing permutations. For example, we will write (1) ∘ (2 3) simply as (2 3), with the understanding that the missing element is in a 1-cycle.

What is the cycle representation of $\begin{pmatrix} 1 & 2 & 3 \\ 1 & 2 & 3 \end{pmatrix}$?

473

(1) ∘ (2) ∘ (3). Here we cannot omit all the 1-cycles because the composition consists only of 1-cycles. So we omit all but one, retaining the one with the smallest element. Thus, we write this permutation as (1).

Put the following permutation in cycle form; also put its inverse in table form and in cycle form: $\gamma = \begin{pmatrix} 1 & 2 & 3 & 4 \\ 2 & 3 & 4 & 1 \end{pmatrix}$.

474

$\gamma = (1\ 2\ 3\ 4)$, $\gamma^{-1} = \begin{pmatrix} 1 & 2 & 3 & 4 \\ 4 & 1 & 2 & 3 \end{pmatrix}$, $\gamma^{-1} = (4\ 3\ 2\ 1) = (1\ 4\ 3\ 2)$

To determine the table representation of γ^{-1}, we simply view the table for γ as a mapping of the elements in the second row to the corresponding elements in the first row. For example, in the table for γ, 1 is directly above 2. Thus, γ maps 1 to 2. But then γ^{-1} maps 2 to 1. Similarly, 2 is directly above 3. Thus, γ maps 2 to 3, and γ^{-1} maps 3 to 2. We get the cycle for γ^{-1} simply by reversing the elements in the cycle for γ. For example, if $\gamma = (1\ 2\ 3\ 4)$, then $\gamma^{-1} = (4\ 3\ 2\ 1)$. Starting this cycle with the smallest element, we get (1 4 3 2). Note that in this last form, γ^{-1} has the same first element as γ, but the remaining elements are in reverse order. In general, if $\gamma = (a_1\ a_2\ \dots\ a_n)$, then $\gamma^{-1} = (a_1\ a_n\ a_{n-1}\ \dots\ a_2)$. To get the inverse, we keep the first element in the first position and reverse the remaining elements on the list.

What is the inverse of (1 5 3 2 4)?

475

(1 4 2 3 5)

Representing some permutations gives rise to more than one cycle. For example, consider $\begin{pmatrix} 1 & 2 & 3 & 4 \\ 4 & 3 & 2 & 1 \end{pmatrix}$; 1 goes to 4 and 4 goes to 1. Thus, we get the cycle (1 4). But this cycle does not specify the mapping of 2 and 3. Thus, we need additional cycles. We pick any element not in (1 4), and determine the cycle that starts with that element. Let's use 2 (which is not is (1 4)). We get the cycle (2 3). We now have cycles such that each element is in exactly one cycle. The composition of these cycles, (1 4) ∘ (2 3), represents the given permutation.

Does (1 4) ∘ (2 3) = (2 3) ∘ (1 4)?

476

Yes. Because the cycles are pairwise disjoint (i.e., no two cycles have any elements in common), the mapping in each cycle is not affected by the other cycle. For example, (2 3) maps 2 to 3 and 3 to 2. But then (1 4) has no effect on 2 or 3. Thus, it does not matter if (2 3) precedes or follows (1 4).

Put the following permutation in cycle form: $\begin{pmatrix} 1 & 2 & 3 & 4 & 5 & 6 \\ 2 & 1 & 5 & 4 & 6 & 3 \end{pmatrix}$.

477

(1 2) ∘ (3 5 6)

We omit (4) because it is a 1-cycle. We again get cycles that are disjoint. In fact, whenever we use our technique to convert the table form of a permutation to cycle form, the cycles that result, if more than one, will always be disjoint. Let's prove this. We will use a proof by contradiction.

Suppose the procedure produces cycles α and β that are not disjoint. Then there is at least one element k that is in both α and β. The elements in α are k and all its successors in the circular list defined by α. The elements of β are also k and all its successors. Thus, $\alpha = \beta$. But this is impossible. Each succeeding cycle produced by our procedure starts with an element not in any of the preceding cycles. Thus, no two cycles can be identical. We resolve this contradiction by concluding that α and β are disjoint. ∎

Convert the following permutation to cycle form: $\begin{pmatrix} 1 & 2 & 3 & 4 & 5 & 6 \\ 5 & 3 & 4 & 2 & 6 & 1 \end{pmatrix}$. Are the cycles disjoint?

478

(1 5 6) ∘ (2 3 4). The cycles are disjoint, as expected.

We can use exponent notation to represent the composition of identical cycles. For example, we can represent (1 2 3) ∘ (1 2 3) with $(1\ 2\ 3)^2$. We can also use negative exponents: α^{-k}, where k is a positive integer, is defined as $(\alpha^{-1})^k$. For example, $(1\ 2\ 3)^{-2} = [(1\ 2\ 3)^{-1}]^2$.

Convert $[(1\ 2\ 3)^{-1}]^2$ to a single cycle.

479

$[(1\ 2\ 3)^{-1}]^2 = (1\ 3\ 2)^2 = (1\ 3\ 2) \circ (1\ 3\ 2)$. The right cycle of (1 3 2) ∘ (1 3 2) maps 1 to 3. The left cycle then maps 3 to 2. Thus, the composition maps 1 to 2.

Determine how the composition maps 2 and 3.

480

2 is mapped to 3, and 3 is mapped to 1.

What, then, is the cycle that represents the same permutation as $[(1\ 2\ 3)^{-1}]^2$?

481

(1 2 3)

A permutation on $\{1, 2, ..., n\}$ is in **standard cycle form** if and only if it satisfies all of the following criteria:

- It is in the form of a single cycle or a composition of disjoint cycles.
- 1-cycles are omitted unless the permutation is the identity, in which case it is represented with (1).
- Each cycle starts with the smallest element in that cycle.
- The cycles are ordered so that their first elements are in ascending order.

(1 2) ∘ (1 3) is not in standard cycle form because the two cycles are not disjoint (1 is in both cycles). The right cycle maps 1 to 3; the left cycle has no effect on 3. Thus, in standard cycle form, this permutation starts with 1 3. Complete this cycle. The result will be the permutation in standard cycle form.

482

(1 3 2)

Put (1 2) ∘ (1 3) ∘ (1 4) ∘ (5) in standard cycle form.

483

Applying the cycles in right-to-left order, we get the following mappings:

$$\begin{aligned} 1 &\to 4 \\ 2 &\to 1 \\ 3 &\to 1 \to 2 \\ 4 &\to 1 \to 3 \\ 5 &\to 5 \end{aligned}$$

In standard cycle form, this permutation is (1 4 3 2).

Let's convert (2 3) ∘ (1 3) and (1 3) ∘ (2 3) to table form. Both permutations have the same pair of cycles, but in a different order. Let's see if order matters. Note that neither is in standard cycle form because in each permutation, the cycles are not disjoint (3 is in both cycles). Remember, we apply the functions in a composition from right to left. Thus, in (2 3) ∘ (1 3), we apply (1 3) first.

In (2 3) ∘ (1 3), (1 3) maps 1 to 3. (2 3) then maps 3 to 2. Thus, the composition maps 1 to 2. Our table representation is $\begin{pmatrix} 1 & 2 & 3 \\ 2 & ? & ? \end{pmatrix}$. Complete the table.

484

$(2\ 3) \circ (1\ 3) = \begin{pmatrix} 1 & 2 & 3 \\ 2 & 3 & 1 \end{pmatrix}$

Now convert (1 3) ∘ (2 3) to table form.

485

$(1\ 3) \circ (2\ 3) = \begin{pmatrix} 1 & 2 & 3 \\ 3 & 1 & 2 \end{pmatrix}$

We have that $(2\ 3) \circ (1\ 3) = \begin{pmatrix} 1 & 2 & 3 \\ 2 & 3 & 1 \end{pmatrix} \neq \begin{pmatrix} 1 & 2 & 3 \\ 3 & 1 & 2 \end{pmatrix} = (1\ 3) \circ (2\ 3)$.

As we observed in frame **476**, if a permutation is represented by a composition of disjoint cycles, then the order of the cycles in the composition does not matter. However, as the preceding example illustrates, *if the cycles are not disjoint, then order may matter*: changing the order may change the permutation represented, depending of the cycles.

Represent in standard cycle form the permutations f_1, f_2, f_3, f_4, f_5, and f_6 that follow (these are the elements of S_3, the symmetric group of degree 3):

$$\overset{f_1}{\begin{pmatrix} 1 & 2 & 3 \\ 1 & 2 & 3 \end{pmatrix}} \quad \overset{f_2}{\begin{pmatrix} 1 & 2 & 3 \\ 2 & 3 & 1 \end{pmatrix}} \quad \overset{f_3}{\begin{pmatrix} 1 & 2 & 3 \\ 3 & 1 & 2 \end{pmatrix}} \quad \overset{f_4}{\begin{pmatrix} 1 & 2 & 3 \\ 1 & 3 & 2 \end{pmatrix}} \quad \overset{f_5}{\begin{pmatrix} 1 & 2 & 3 \\ 2 & 1 & 3 \end{pmatrix}} \quad \overset{f_6}{\begin{pmatrix} 1 & 2 & 3 \\ 3 & 2 & 1 \end{pmatrix}}$$

486

f_1: (1), f_2: (1 2 3), f_3: (1 3 2), f_4: (2 3), f_5: (1 2), f_6: (1 3)

Let's now determine the Cayley table for $S_3 = \{f_1, f_2, f_3, f_4, f_5, f_6\}$ under composition. Let's start with $f_5 \circ f_2 = (1\ 2) \circ (1\ 2\ 3)$. Remember that in a function composition, functions are applied in right-to-left order. Thus, to determine the mapping of any element under $f_5 \circ f_2 = (1\ 2) \circ (1\ 2\ 3)$, we first apply (1 2 3) and then (1 2) (order matters here because the cycles are not disjoint). (1 2 3) maps 1 to 2. Then (1 2) maps 2 to 1. Thus, the net effect of (1 2) ∘ (1 2 3) on 1 is to map it to 1.

(1 2) ∘ (1 2 3) maps 2 to _____.

487

3, because (1 2 3) maps 2 to 3, and (1 2) has no effect on 3.

(1 2) ∘ (1 2 3) maps 3 to _____.

488

2, because (1 2 3) maps 3 to 1, and (1 2) then maps 1 to 2.

Give the table representation of (1 2) ∘ (1 2 3).

489

$$\begin{pmatrix} 1 & 2 & 3 \\ 1 & 3 & 2 \end{pmatrix}$$

Convert this permutation to cycle form.

490

(2 3)

Which element in $\{f_1, f_2, f_3, f_4, f_5, f_6\}$ is (2 3)? See frame **486**.

491

f_4. Thus, $f_5 \circ f_2 = f_4$.

Determine the table representation of (1 2 3) ∘ (1 2). This composition corresponds to $f_2 \circ f_5$.

492

$$\begin{pmatrix} 1 & 2 & 3 \\ 3 & 2 & 1 \end{pmatrix}$$

Convert this permutation to cycle form.

493

(1 3), which is f_6

So far, we have two entries for the Cayley table of S_3: $f_5 \circ f_2 = f_4$ and $f_2 \circ f_5 = f_6$. Following is the start of the Cayley table for S_3 under composition. Complete the table.

∘	f_1	f_2	f_3	f_4	f_5	f_6
f_1						
f_2					f_6	
f_3						
f_4						
f_5		f_4				
f_6						

494

∘	f_1	f_2	f_3	f_4	f_5	f_6
f_1	f_1	f_2	f_3	f_4	f_5	f_6
f_2	f_2	f_3	f_1	f_5	f_6	f_4
f_3	f_3	f_1	f_2	f_6	f_4	f_5
f_4	f_4	f_6	f_5	f_1	f_3	f_2
f_5	f_5	f_4	f_6	f_2	f_1	f_3
f_6	f_6	f_5	f_4	f_3	f_2	f_1

Row f_1 and column f_1 are easy to determine. Why?

495

f_1 is the identity element in S_3 under composition

Decomposing Permutations into Transpositions

496

A *transposition* is a 2-cycle. For example, (1 2) is a transposition. Switching the elements of a transposition does not change the permutation represented. For example, (1 2) and (2 1) are the same permutation: both map 1 to 2, and 2 to 1.

Describe the permutation given by the composition of two identical transpositions. In other words, describe the permutation whose form is $(a\ b) \circ (a\ b)$.

497

The right $(a\ b)$ maps a to b. The left $(a\ b)$ maps b to a. Thus, the net effect is to map a to a. Similarly, the net effect is to map b to b. Thus, $(a\ b) \circ (a\ b)$ is the identity permutation.

A cycle of any length can be decomposed into a transposition or a composition of transpositions. A 1-cycle can be represented by any two identical transpositions. For example, (1) = (1 2) ∘ (1 2). A 2-cycle is itself a transposition. Any cycle with more than two elements can be decomposed into a composition of transpositions using the following rule:

For $n > 2$, $(a_1\ a_2\ \ldots\ a_n) = (a_1\ a_n) \circ (a_1\ a_{n-1}) \circ \cdots \circ (a_1\ a_2)$.

For example, (1 2 3 4 5) = (1 5) ∘ (1 4) ∘ (1 3) ∘ (1 2).

Convert (1 4 2 5 3) to a composition of transpositions.

498

(1 3) ∘ (1 5) ∘ (1 2) ∘ (1 4)

The 3-cycle (1 2 3) is decomposed into two transpositions: (1 3) ∘ (1 2). How many transpositions are there in the decomposition of an *n*-cycle for $n > 2$?

499

$n - 1$. Thus, if n is even, then the number of transpositions is odd. If n is odd, then the number of transpositions is even.

Key idea: We have seen that a cycle can be decomposed into a transposition or a composition of transpositions. Moreover, because any permutation can be decomposed into cycles, which in turn can be decomposed into transpositions, any permutation can be decomposed into a transposition or a composition of transpositions.

Decompose $\begin{pmatrix} 1 & 2 & 3 & 4 & 5 & 6 & 7 \\ 2 & 1 & 4 & 5 & 6 & 3 & 7 \end{pmatrix}$ into transpositions.

500

First we decompose into cycles. We get (1 2) ∘ (3 4 5 6). We then decompose each cycle with more than two elements into transpositions. We get (1 2) ∘ (3 6) ∘ (3 5) ∘ (3 4).

Decompose (1) into transpositions.

501

(1) = (1 2) ∘ (1 2)

Decompose (1 2) into transpositions.

502

(1 2) is already a transposition.

Decompose every permutation in S_3 into a composition of transpositions. Here are the permutations in S_3: f_1: (1), f_2: (1 2 3), f_3: (1 3 2), f_4: (2 3), f_5: (1 2), f_6: (1 3).

503

f_1: (1) = (1 2) ∘ (1 2)
f_2: (1 2 3) = (1 3) ∘ (1 2)
f_3: (1 3 2) = (1 2) ∘ (1 3)
f_4: (2 3)
f_5: (1 2)
f_6: (1 3)

Decompose (1 2 3) and (2 3 1) into transpositions.

504

(1 2 3) = (1 3) ∘ (1 2)
(2 3 1) = (2 1) ∘ (2 3) = (1 2) ∘ (2 3)

Because (1 2 3) = (2 3 1), (1 3) ∘ (1 2) = (1 2) ∘ (2 3).

Is the decomposition of a permutation into a composition of transpositions unique?

505

No, as demonstrated by the example in the preceding frame. But, perhaps, all decompositions of a given permutation into transpositions have the same number of transpositions. But this is not the case.

Show this. *Hint*: $(a\ b) \circ (a\ b)$ is the identity permutation.

506

Because $(a\ b) \circ (a\ b)$ is the identity permutation, we can tack it onto either end of a composition of transpositions without changing the permutation defined.

Give a permutation equivalent to (2 3) that has three transpositions.

507

(2 3) ∘ (1 2) ∘ (1 2)

We can also replace any single transposition $(a\ b)$ with $(a\ c) \circ (b\ c) \circ (a\ c)$ in a composition of transpositions without changing the permutation. Let's do this for $(a\ b)$, where $a = 2$ and $b = 3$. Let $c = 1$. We get (2 3) = (2 1) ∘ (3 1) ∘ (2 1).

Apply the same transformation to (1 2). *Hint*: For c, use 3.

508

(1 2) = (1 3) ∘ (2 3) ∘ (1 3)

Increasing the number of transpositions using either of the approaches in the two preceding frames increases the number of transpositions by 2. Thus, they do not change the evenness or oddness of the number of transpositions. This observation leads us to the conjecture that a *permutation is either even or odd*—that is, given a permutation, all its decompositions into transpositions consist of an even number of transpositions, or they all consist of an odd number of permutations. This conjecture turns out to be true. Let's show this.

Consider the following product:

$$(1-2) \cdot (1-3) \cdot (1-4) \cdot (2-3) \cdot (2-4) \cdot (3-4)$$

It is the product of all possible differences whose operands are from {1, 2, 3, 4} in which the left operand in each difference is less than the right operand: 1 is less than 2, 3, and 4, thus, the product includes the differences $(1-2)$, $(1-3)$, and $(1-4)$; 2 is less than 3 and 4, thus, the product includes $(2-3)$ and $(2-4)$; 3 is less than 4, thus, the product includes $(3-4)$.

What is the sign of the preceding product?

509

The product contains an even number of negative factors. Thus, the product is positive.

Suppose we now apply a transposition in S_4 to the product in the preceding frame. Let's use (1 2). Then wherever 1 appears, replace 1 with 2. Wherever 2 appears, replace 2 with 1. Show the resulting product.

510

$(1-2) \cdot (1-3) \cdot (1-4) \cdot (2-3) \cdot (2-4) \cdot (3-4)$
$\quad \Downarrow \qquad \Downarrow \qquad \Downarrow \qquad \Downarrow \qquad \Downarrow \qquad \Downarrow$
$(2-1) \cdot (2-3) \cdot (2-4) \cdot (1-3) \cdot (1-4) \cdot (3-4)$

The new product is almost the same. The only difference is that $(1-2)$ is changed to $(2-1)$. All other changes occur in pairs that cancel each other. For example, $(1-3)$ becomes $(2-3)$, but $(2-3)$ becomes $(1-3)$. Thus, in the new product, there are the $(1-3)$ and $(2-3)$ factors, just as in the original product.

What is the effect of the (1 2) transposition on the sign of the product?

511

Because $(2-1)$ is the negation of $(1-2)$, the sign of the product changes. Because the original product is positive, the new product is negative. Similarly, applying *any* transposition (x, y), where $x \neq y$ and $x, y \in \{1, 2, 3, 4\}$, to the product expression in frame **508** changes the sign of the product.

What effect do three transpositions applied to the product have on the sign of the product?

512

The sign changes three times. The first two sign changes cancel each other. Thus, the net effect is a sign change.

What effect does an odd number of transpositions applied to the product have on the sign of the product?

513

The net effect is a sign change.

What effect does an even number of transpositions applied to the product have on the sign of the product?

514

No net effect.

Let's turn our attention to the permutations in S_4. Suppose β is some permutation in S_4 that can be represented with both an even number of transpositions (i.e., as the composition of an even number of transpositions) and with an odd number of transpositions (i.e., as a single transposition or as the composition of an odd number of transpositions). Because both forms represent the same permutation, they *must* have the same effect on the product in frame **508**. But, in fact, they would not have the same effect: the even one does not change the product's sign, but the odd one does. We resolve this contradiction by concluding that the representations of β as a product of transpositions are either *all even or all odd*. In the former case, we describe β as an **even permutation,** in the latter case as an **odd permutation**.

We have shown that every permutation in S_4 is either even or odd. We can use the same approach to prove the general case: for $n > 1$, every permutation in S_n is either even or odd. The one remaining symmetric group, S_1, contains only the identity permutation, which is even. Thus, every permutation in any symmetric group is either even or odd.

Key idea: If a permutation can be decomposed into an even number of transpositions, then every decomposition of that permutation into transpositions consists of an even number of transpositions. Similarly, if a permutation can be decomposed into an odd number of transpositions, then every decomposition of that permutation into transpositions consists of an odd number of transpositions.

Characterize each permutation in S_3 as even or odd (see frame **503**).

515

$f_1, f_2,$ and f_3 are even; $f_4, f_5,$ and f_6 are odd.

Let A_n be the set of all the even permutations in S_n. Prove that A_n is closed under composition.

516

All the permutations in A_n are even ⇒ the number of transpositions in a composition of any two permutations in A_n is the sum of two even numbers, which is an even number ⇒ the composition permutation is also even and, therefore, in A_n.

Prove that the identity permutation is even.

517

The identity permutation (1 2) ∘ (1 2) is even ⇒ it is in A_n. Alternatively, we can think of the identity permutation as consisting of the application of zero (an even number) of transpositions. Thus, the identity permutation is even.

Prove that every permutation in A_n has an inverse in A_n. Here is the start of a proof by contradiction: Let $\gamma \in A_n$. Thus, γ is even. Assume γ^{-1} is not in A_n. That is, assume it is odd. $\gamma \circ \gamma^{-1}$ is the identity permutation, and we know the identity permutation is even.

518

Thus, $\gamma \circ \gamma^{-1}$ is even. But $\gamma \circ \gamma^{-1}$ is odd (because γ is even and γ^{-1} is odd). We resolve this contradiction by concluding that our assumption that γ^{-1} is odd is incorrect; γ^{-1} must be even and, therefore, in A_n. ∎

In the preceding frames, we have shown that A_n under function composition has the properties of closure, identity, and inverses. In frame **462**, we showed that composition is associative. Thus, A_n itself is a group. It has all the properties required of a group: closure, associativity, identity, and inverses. A_n is a subgroup of S_n (a **subgroup** is a group within a group with the same operation as the enclosing group). A_n is called the **alternating group of degree n**.

Suppose we compose every element in S_n, where $n > 1$, with (1 2). For example, suppose γ and β are two distinct permutations in S_n. Composing each with (1 2), we get $\gamma \circ (1\ 2)$ and $\beta \circ (1\ 2)$. Is it possible that $\gamma \circ (1\ 2) = \beta \circ (1\ 2)$? *Hint*: S_n is a group, so it obeys the right cancellation law.

519

No. Canceling (1 2), we get $\gamma = \beta$. But $\gamma \neq \beta$. So it must be that $\gamma \circ (1\ 2) \neq \beta \circ (1\ 2)$.

S_n has $n!$ permutations. If we compose each one with (1 2), we get a new set of $n!$ permutations. Let's denote this new set with T_n. By the preceding frame, all the permutations in T_n are distinct. Thus, T_n, like S_n, has all the $n!$ permutations of $\{1, 2, ..., n\}$. What can we conclude about S_n and T_n?

520

$S_n = T_n$ for $n > 1$

S_1 is (1), but T_1 is (1) ∘ (1 2) = (1 2). Thus, $S_1 \neq T_1$. Why does our preceding argument that $S_n = T_n$ not not hold when $n = 1$?

521

(1 2) $\in S_n$ for $n > 1$. S_n is a group under composition. Thus, for $n > 1$, *by the closure property* of groups, composing every element of S_n with (1 2) produces an element in S_n. But (1 2) $\notin S_1$. Thus, the closure property of groups does not apply to the composition of (1) and (1 2). Composing (1), the single element in S_1, with (1 2) produces (1 2), which is not in S_1.

Let's continue our analysis of S_n, where $n > 1$, so that $S_n = T_n$. Composing every element in S_n with (1 2) changes every even permutation into an odd permutation and every odd permutation into an even permutation. Thus, the number of odd permutations in T_n equals the number of even permutations in S_n. But S_n and T_n are the same set. Thus, in S_n, the number of even permutations must equal the number of odd permutations.

How many elements are in S_n?

522

$n!$

How many elements are in A_n where $n > 1$? How many elements are in A_1?

523

Half of the permutations in S_n are even; half are odd. A_n consists of all the even permutations. Thus, $|A_n| = n!/2$ for $n > 1$. A_1 contains only one element (the identity element).

Order of an Element in a Group

524

The **order of a group G**, denoted by $|G|$, is the number of elements in G. If g is an element of a group, we can also speak of the **order of g**, denoted by $|g|$. It is defined as the smallest positive integer n such that $g^n = e$. For example, consider the following Cayley table for a group with identity e:

*	e	a	b	c
e	e	a	b	c
a	a	b	c	e
b	b	c	e	a
c	c	e	a	b

$a^1 = a$, $a^2 = b$, $a^3 = aa^2 = ab = c$, $a^4 = aa^3 = ac = e$. Thus, $|a| = 4$.

What are the orders of e, b, and c?

525

$|e| = 1$, $|b| = 2$, $|c| = 4$

Determining the order of a permutation represented by a single cycle is simple. Consider the cycle $\gamma = (a\ b\ c)$. Let's determine $\gamma^3 = (a\ b\ c) \circ (a\ b\ c) \circ (a\ b\ c)$. The right cycle maps a to b. The middle cycle then maps b to c. Finally, the left cycle maps c to a. Thus, γ^3 maps a to a. Similarly, γ^3 maps b to b and c to c.

Think of each application of γ as moving forward one position in the circular list defined by γ. By applying γ three times, we make a complete circuit, ending where we started. Thus, γ^3 maps each element to itself (making it the identity permutation). If we apply γ fewer than three times, we do not make a complete circuit, in which case each element is not mapped to itself. Thus, 3 is the smallest positive exponent of γ that yields the identity permutation, making 3 the order of γ.

Because γ^3 is the identity element, the composition of two or more γ^3 permutations is also the identity permutation. That is, the identity permutation equals γ^3, which equals $\gamma^3 \circ \gamma^3$, which equals $\gamma^3 \circ \gamma^3 \circ \gamma^3$, and so on.

What is the order of an n-cycle, where $n > 0$?

526

To make a complete circuit around the n-element circular list defined by an n-cycle, we have to apply the cycle n times. Thus, the order of an n-cycle is n.

Let's simplify the permutation $[(1\ 2) \circ (3\ 4)]^2 = (1\ 2) \circ (3\ 4) \circ (1\ 2) \circ (3\ 4)$. In this composition, each pair of cycles is either disjoint or equal. In either case, we can commute them without affecting the permutation defined. Commute the cycles so that the $(1\ 2)$ cycles are on the left and the $(3\ 4)$ cycles are on the right.

527

$(1\ 2) \circ (1\ 2) \circ (3\ 4) \circ (3\ 4)$

The order of any 2-cycle is 2. Thus, the preceding composition simplifies to _____.

528

$[(1\ 2) \circ (1\ 2)] \circ [(3\ 4) \circ (3\ 4)] = (1) \circ (1) = (1)$, the identity permutation.

Simplify $[(1\ 2\ 3) \circ (4\ 5\ 6)]^3$.

529

It is equal to (1), the identity permutation.

Now consider the permutation $\beta = (1\ 2) \circ (3\ 4\ 5)$. What is its order? If we square β, we get $\beta^2 = (1\ 2) \circ (3\ 4\ 5) \circ (1\ 2) \circ (3\ 4\ 5)$. Because $(1\ 2)$ and $(3\ 4\ 5)$ are disjoint, we can commute them without changing the permutation defined. We get $\beta^2 = (1\ 2) \circ (1\ 2) \circ (3\ 4\ 5) \circ (3\ 4\ 5) = (1\ 2)^2 \circ (3\ 4\ 5)^2$, which equals $(3\ 4\ 5)^2$ because $(1\ 2)^2$ is the identity permutation. By squaring β, we "get rid of" $(1\ 2)$ but not $(3\ 4\ 5)$. If, however, we cube β, we get rid of $(3\ 4\ 5)$ (because its order is 3) but not $(1\ 2)$. To get rid of both (1 2) and (3 4 5), we need β^k where k is a multiple of both 2 and 3. The smallest positive integer k that is a multiple of both 2 and 3 is the order of β. We call such an integer the **least common multiple** of 2 and 3 (denoted by lcm(2, 3)). Thus, $|\beta|$ = lcm(2, 3) = 6. In general, the order of a composition of disjoint cycles is equal to the lcm of the lengths of the cycles. For example, the order of (1 2) ∘ (3 4 5) ∘ (6 7 8 9) = lcm(2, 3, 4) = 12.

What is the order of (1 2) ∘ (3 4)?

530

|(1 2) ∘ (3 4)| = lcm[|(1 2)|, |(3, 4)|] = lcm(2, 2) = 2.

What is the order of γ = (1 2) ∘ (2 3)? *Hint*: The answer is *not* 2.

531

(1 2) and (2 3) are not disjoint. Thus, the preceding analysis does not apply. Convert γ to standard cycle form.

532

γ = (1 2 3). Thus, $|\gamma|$ = 3.

Earlier in this chapter, we learned how to convert a permutation in table form to standard cycle form. Depending on the permutation, the conversion process results in either a single cycle or multiple disjoint cycles. If we convert the permutations in S_4 this way, each permutation will be in one of the following forms: (a), $(a\ b)$, $(a\ b\ c)$, $(a\ b\ c\ d)$, or $(a\ b) \circ (c\ d)$.

What, then, is the maximum order among the elements of S_4? *Hint*: What is the order of each of the preceding forms?

533

Here are the orders for each possible form:

(a)	1
$(a\ b)$:	2
$(a\ b\ c)$:	3
$(a\ b\ c\ d)$:	4
$(a\ b) \circ (c\ d)$:	2

Thus, 4 is the maximum order.

What are all the cycle forms of the permutations in S_5?

534

$(a), (a\ b), (a\ b\ c), (a\ b\ c\ d), (a\ b\ c\ d\ e), (a\ b) \circ (c\ d), (a\ b) \circ (c\ d\ e)$

What is the maximum order among the elements in S_5?

535

$|(a\ b) \circ (c\ d\ e)|$ = lcm(2, 3) = 6

Here are two permutations from S_{10}: (1 2 3 4) ∘ (5 6 7 8 9) and (1 2 3 4) ∘ (5 6 7 8 9 10). What are their orders?

536

lcm(4, 5) = 20, lcm(4, 6) = 12

Give an element with the largest order in S_{10}. What is the order of this element?

537

|(1 2) ∘ (3 4 5) ∘ (6 7 8 9 10)| = lcm(2, 3, 5) = 30

Review Questions

1. Convert to standard cycle form and to a composition of transpositions: $\begin{pmatrix} 1 & 2 & 3 & 4 & 5 \\ 3 & 1 & 4 & 5 & 2 \end{pmatrix}$.
2. Convert to standard cycle form and to a composition of transpositions: $\begin{pmatrix} 1 & 2 & 3 & 4 & 5 & 6 & 7 \\ 2 & 3 & 1 & 7 & 6 & 4 & 5 \end{pmatrix}$.
3. Convert the following permutation in S_6 to table form: $(1\ 3) \circ (2\ 4) \circ (3\ 5) \circ (1\ 6) \circ (2\ 5)$.
4. List all the elements in S_2 in cycle form and in table form.
5. Construct the Cayley table for S_2.
6. For what values of n, if any, does $S_n = A_n$?
7. What is the order of $(1\ 2) \circ (3\ 4\ 5) \circ (6\ 7\ 8)$?
8. Give an element in S_8 that has the maximum possible order.
9. Put the inverse of $(5\ 3\ 4\ 6\ 1)$ in standard cycle form.
10. Put the inverse of $(1\ 2) \circ (2\ 3\ 4)$ in standard cycle form. *Hint*: Use the socks-shoes property of groups (see frame **453**).

Answers to Review Questions

1. $(1\ 3\ 4\ 5\ 2), (1\ 2) \circ (1\ 5) \circ (1\ 4) \circ (1\ 3)$
2. $(1\ 2\ 3) \circ (4\ 7\ 5\ 6), (1\ 3) \circ (1\ 2) \circ (4\ 6) \circ (4\ 5) \circ (4\ 7)$
3. $\begin{pmatrix} 1 & 2 & 3 & 4 & 5 & 6 \\ 6 & 1 & 5 & 2 & 4 & 3 \end{pmatrix}$
4. $(1), (1\ 2); \begin{pmatrix} 1 & 2 \\ 1 & 2 \end{pmatrix}, \begin{pmatrix} 1 & 2 \\ 2 & 1 \end{pmatrix}$
5.

*	(1)	(1 2)
(1)	(1)	(1 2)
(1 2)	(1 2)	(1)

6. $n = 1$
7. $\mathrm{lcm}(2, 3, 3) = 6$
8. $(1\ 2\ 3) \circ (4\ 5\ 6\ 7\ 8)$
9. $(1\ 6\ 4\ 3\ 5)$
10. $[(1\ 2) \circ (2\ 3\ 4)]^{-1} = (2\ 3\ 4)^{-1} \circ (1\ 2)^{-1} = (2\ 4\ 3) \circ (1\ 2) = (1\ 4\ 3\ 2)$

Homework Questions

1. Convert $(1\ 2\ 3) \circ (4\ 5\ 6\ 7)$ to a composition of transpositions.
2. Show that each permutation in S_3 is even or odd. *Hint*: Consider $(1-2) \cdot (1-3) \cdot (2-3)$.
3. Convert the permutation $(1\ 2) \circ (2\ 3) \circ (3\ 4)$ in S_4 to table form.
4. What is the order of an element with the maximum order in S_{15}?
5. What is the order of an element with the maximum order in S_{42}?
6. Convert to standard cycle form: $(1\ 2\ 3\ 4\ 5)^{-3}$.
7. Convert to standard cycle form: $[(1\ 2\ 3) \circ (4\ 5\ 6) \circ (7\ 8\ 9)]^{-2}$.
8. Give a rule to determine the evenness or oddness of an n-cycle.
9. Show that S_n is not commutative for all $n > 2$.
10. Let $\alpha = (1\ 2\ 4) \circ (3\ 5)$ and $\beta = (6\ 9\ 8) \circ (5\ 7)$. Determine a permutation γ such that $\gamma \alpha \gamma^{-1} = \beta$.

10 Divisibility Properties of the Integers

Division Algorithm

The integers are not closed under division. For example, 7 divided by 2 is 3.5, which is not an integer. However, we can define division on the integers in such a way that the result is two integers: a ***quotient*** and a ***remainder***. You probably learned this type of division in grade school.

Let's first consider the case in which the ***dividend*** (i.e., the number being divided) is nonnegative and the divisor is positive. For example, consider the division of the dividend 7 by the divisor 3:

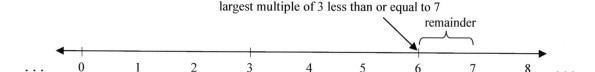

The quotient is 2, and the remainder is 1.

Let's represent this division graphically. Starting from 0 on the number line, we mark off the successive multiples of the divisor 3. We stop just before we go above 7:

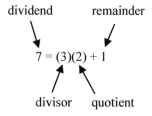

The number 6 is the last multiple of 3 we mark off because the next multiple of 3 is 9, which is greater than 7. The number of multiples we mark off in this process is the quotient. Thus, for this example, the quotient is 2. The quotient is the maximum whole number of times the divisor can "go into" the dividend. There can be only one maximum. Thus, the quotient is unique.

The number corresponding to the distance on the number line from the last multiple (6 in this example) and the dividend (7 in this example) is the remainder. That is, remainder = dividend − (last multiple). Solving for the dividend, we get dividend = (last multiple) + remainder. For this example, we have:

$$7 = (3)(2) + 1$$

with dividend = 7, divisor = 3, quotient = 2, remainder = 1.

When marking off multiplies of the divisor, we stop before we go beyond the dividend. Thus, the last multiple falls either on the dividend or to its left. But this implies that the remainder (which equals the dividend minus the last multiple) is greater than or equal to 0. Moreover, the remainder is always less than the divisor. This is true because when we compute the quotient, we take successive multiplies until doing so would take us beyond the dividend. Thus, the last multiple has to differ from the dividend by less than the divisor (otherwise, we could take at least one more multiple without going beyond the dividend). Because the remainder is uniquely determined by the quotient for a given dividend and divisor, the uniqueness of the quotient implies the uniqueness of the remainder.

Divide 15 by 4. What are the quotient and the remainder?

539

The largest multiple of 4 that is less than or equal to 15 is (4)(3). In other words, at most three 4's fit into 15. Thus, 3 is the quotient. To reach 15 from (4)(3) = 12, we have to add 3. Thus, the remainder is 3.

Divide 14 by 7. What are the quotient and remainder?

540

We can go up to the second multiple of 7 without going beyond 14. Thus, 2 is the quotient. This multiple brings us all the way to 14. Thus, the remainder is 0.

When dividing a negative integer by a positive divisor, we take negative multiples of the divisor until we either land on the dividend or go just beyond it so we end up to its left. For example, when we divide -7 by 4, we take the following multiples of 4: (4)(-1) and (4)(-2). We stop at (4)(-2) because it is the first multiple of 4 that is on or to the left of -7, the dividend:

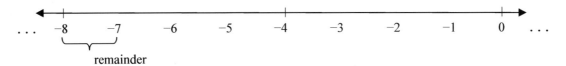

The remainder, then, is 1, the number corresponding to the distance on the number line from the last multiple (-8) to the dividend (-7). In this process, we end up on the dividend or to its left. Thus, the remainder is always greater than or equal to 0. The remainder is also less than the divisor (if it were not, then that would mean we should have taken fewer multiples). The quotient is -2 because we need no more than two multiples of 4 in the negative direction to land on or to the left of the dividend, -7. For this example, we have that

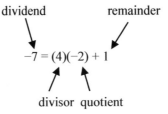

In both cases—dividend negative or dividend nonnegative—the quotient and remainder are unique, and $0 \leq$ remainder $<$ divisor.

What are the quotient and remainder when -5 is divided by 2?

541

$-5 = (2)(-3) + 1$. The quotient is -3; the remainder is 1.

What are the quotient and remainder when -1 is divided by 2?

542

$-1 = (2)(-1) + 1$. The quotient is -1; the remainder is 1.

What are the quotient and remainder when 1 is divided by 2?

543

$1 = (2)(0) + 1$. The quotient is 0; the remainder is 1.

The following theorem summarizes our observations on division. We call this theorem the ***division algorithm***:

For any integers $d > 0$ and a, there exist integers q and r such that $a = dq + r$ with $0 \leq r < d$. Moreover, q and r are unique.

We call a, d, q, and r the **dividend,** the **divisor,** the **quotient,** and the **remainder,** respectively. We will use $a \bmod d$ to represent the remainder that results when a is divided by d. For example, the remainder when 5 divided by 2 is 1. Accordingly, we write $5 \bmod 2 = 1$. We call "mod" the **remainder operator**.

What is $10 \bmod 5$?

544

0

What is $-10 \bmod 3$?

545

$-10 = (3)(-4) + 2$. Thus, the remainder is 2.

An *algorithm* is a step-by-step procedure that ultimately halts and provides an answer to some problem. Is the division algorithm in frame **543** an algorithm?

546

No. The name "division algorithm" is a misnomer. The division algorithm stated in frame **543** is not a procedure for solving a problem. Thus, it is not an algorithm. However, there is an algorithm for dividing integers—it is the long division procedure that you learned in grade school. We used it in frame **538** to divide 7 by 3.

Some Divisibility Proofs

547

Let's prove the following theorems on the divisibility of integers (recall that "|" is the "divides" operator: $a \mid b$ if and only if $a \neq 0$ and there is an integer k such that $b = ak$, or more simply, a is a factor of b):

D1: $a \mid b$ and $b \mid c \Rightarrow a \mid c$
D2: $1 \mid a$ for all $a \in \mathbb{Z}$
D3: $a \mid 0$ for all $a \in \mathbb{Z} - \{0\}$
D4: $d \mid a$ and $d \mid b \Rightarrow d \mid (ax + by)$ for all $x, y \in \mathbb{Z}$
D5: $a \mid b$ and $b \neq 0 \Rightarrow |a| \leq |b|$
D6: $a \mid b$ and $b > 0 \Rightarrow a \leq b$

Let's prove D1. What do we know, given that $a \mid b$ and $b \mid c$?

548

$b = ja$ and $c = kb$ for some $j, k \in \mathbb{Z}$.

Using these equalities, form an equality that connects a and c.

549

Substituting ja for b in $c = kb$, we get that $c = kja \Rightarrow a \mid c$. ∎

Prove D2: $1 \mid a$ for all $a \in \mathbb{Z}$.

550

$a = 1 \cdot a \Rightarrow 1 \mid a$. ∎

Prove D3: $a \mid 0$ for all $a \neq 0$.

551

Chapter 10: Divisibility Properties of the Integers

$a \neq 0$ and $0 = 0a \Rightarrow a \mid 0$. ∎

Prove D4: If $d \mid a$ and $d \mid b$, then $d \mid (ax + by)$ for all $x, y \in \mathbb{Z}$.

552

$d \mid a \Rightarrow a = kd$ for some $k \in \mathbb{Z}$, and $d \mid b \Rightarrow b = jd$ for some $j \in \mathbb{Z}$. Thus, $ax + by = kdx + jdy$. Factoring out d, we get $ax + by = d(kx + jy) \Rightarrow d \mid (ax + by)$. ∎

If $a, b \in \mathbb{Z}$ and $a \mid b$, can we conclude that $a < b$?

553

No. Counterexample: if $a = b = 5$, then $a \mid b$, but a is not less than b.

If $a \mid b$, can we conclude that $a \leq b$?

554

No. Counterexample: if $a = 5$ and $b = -10$, then $a \mid b$, but a is not less than or equal to b.

Suppose $a \mid b$ and $b \neq 0$. What can we say about the absolute values of a and b (denoted by $|a|$ and $|b|$, respectively)? *Hint*: a goes into b at least once because $b \neq 0$.

555

$|a| \leq |b|$

This assertion is the theorem D5 in frame **547**: $a \mid b$ and $b \neq 0 \Rightarrow |a| \leq |b|$. You probably feel it is obvious. However, in mathematics, what seems obvious sometimes is wrong. So to be sure we are correct, let's prove D5. The proof is important to study because it illustrates how to prove an inequality. We start our proof with

$$a \mid b \Rightarrow b = ka \text{ for some } k \in \mathbb{Z} \Rightarrow |b| = |ka| = |k| \cdot |a|.$$

If we replace $|k|$ in $|k| \cdot |a|$ with something less than or equal to $|k|$, then the result has to be less than or equal to $|k| \cdot |a|$. $k \neq 0 \Rightarrow |k| \geq 1$. Thus, if we replace k with 1 in $|k| \cdot |a|$, the result is less than or equal to $|k| \cdot |a|$. That is, $|b| = |k| \cdot |a| \geq |1| \cdot |a| = |a|$. The preceding sequence starts with $|b|$, ends with $|a|$, and has $\geq$ in between. Thus, $|b| \geq |a|$, or, equivalently, $|a| \leq |b|$. ∎

$a \mid b$ and $b \neq 0$ together imply $|a| \leq |b|$. The condition $a \mid b$ *by itself* implies $a \neq 0$ (because a divisor by definition is nonzero), but it does *not* imply $|a| \leq |b|$. To see this, consider $a = 5$ and $b = 0$. Then $a \mid b$, but $|a| \nleq |b|$. To prove D5, we have to use the condition $b \neq 0$ as well as $a \mid b$. Thus, if the preceding proof is correct, it *must* somewhere use the condition $b \neq 0$.

Does the preceding proof use the condition $b \neq 0$? *Hint*: On what basis does it assume $k \neq 0$?

556

When it asserts that $k \neq 0$, it is using the condition $b \neq 0$, which implies $k \neq 0$ (because $b = ka$ and $a \neq 0$). If b were 0, then k would also be 0. The writer of the proof is assuming that the justification for the assertion $k \neq 0$ is obvious and therefore omits it. To most readers, it is probably not that obvious. So the proof would be better if the assertion

$$k \neq 0 \Rightarrow |k| \geq 1$$

were replaced by

$$a \neq 0, b \neq 0, \text{ and } b = ka \Rightarrow k \neq 0 \Rightarrow |k| \geq 1.$$

If both a and b are positive, and $a \mid b$, what can you conclude about a and b?

557

By the preceding theorem, $|a| \leq |b|$. If a and b are both positive, then $a = |a|$ and $b = |b|$. Thus, $a \leq b$.

Prove D6: If $a \mid b$ and $b > 0$, then $a \leq b$. *Hint*: Consider two cases: $a < 0$ and $a > 0$.

558

If a is negative, then $a \leq b$, because every negative number is less than every positive number. If a is positive, then by the preceding frame, $a \leq b$.

Greatest Common Divisor

559

If $d \mid m$ and $d \mid n$, we say that d is a *common divisor* of m and n.

Does every pair of integers have at least one positive common divisor?

560

Yes. 1 divides any pair of integers.

If $m = n = 0$, then any nonzero integer is a common divisor of m and n. For example, 1 is a common divisor of 0 and 0. So are 2, 3, 4, and so on. Thus, there is no greatest common divisor of m and n for $m = n = 0$. However, if either m or n is nonzero, then a greatest common divisor can be no larger than the larger of $|m|$ and $|n|$ (by D5 in frame **547**). Thus, if at least one of m and n is nonzero, the greatest common divisor of m and n, denoted by $\gcd(m, n)$, exists.

List all the positive common divisors of 10 and 18. Also determine $\gcd(10, 18)$.

561

1 and 2, $\gcd(10, 18) = 2$

What is $\gcd(0, 20)$?

562

20

What is $\gcd(10, 20)$?

563

10

What is $\gcd(4, 9)$?

564

1

You probably have found nothing unexpected in the divisibility properties we have discussed so far. This is not surprising—you have been using these properties ever since you learned long division in grade school. However, we are now going to discuss a property that is quite unexpected—a property of the greatest common divisor.

What is the smallest positive integer of the form $10x + 18y$, where $x, y \in \mathbb{Z}$? What is $\gcd(10, 18)$?

565

2 (when $x = 2$ and $y = -1$), $\gcd(10, 18) = 2$

Let's try some more examples. What is the smallest positive integer of the form $10x + 15y$? What is $\gcd(10, 15)$?

566

5 (when $x = -1$ and $y = 1$), $\gcd(10, 15) = 5$

What is the smallest positive integer of the form $9x + 15y$? What is $\gcd(9, 15)$?

567

3 (when $x = 2$ and $y = -1$), $\gcd(9, 15) = 3$

In all three cases, the smallest positive integer value of $ax + by$ is equal to $\gcd(a, b)$. This behavior is true for any integers a and b unless $a = b = 0$, in which case the $\gcd(a, b)$ is not defined. Let's formally state this result:

Let a and b be integers with not both equal to zero. Then $\gcd(a, b)$ is the least element in the set of all positive integers of the form $ax + by$, where $x, y \in \mathbb{Z}$.

Here is an outline of the proof of this important theorem (you prove this theorem in homework question 9):

1. For integers a and b with not both equal to zero, let $M = \{ax + by : ax + by > 0 \text{ and } x, y \in \mathbb{Z}\}$.

2. Show that M is nonempty. Then, by the well-ordering principle, M has a least element denoted by g; $g = ai + bj$ for some $i, j \in \mathbb{Z}$.

3. Using the division algorithm, divide a by g. Show that the remainder is 0. *Hint*: A contradiction results if the remainder is not 0. Similarly, divide b by g, and show that the remainder is 0. Thus, g is a common divisor of a and b.

4. Show that if d is a common divisor of a and b, then $d \mid g$. Then, by D6 in frame **547**, $d \le g$. Thus, among all the common divisors of a and b, g is the greatest. That is, $g = \gcd(a, b)$.

Recall that an algorithm is a step-by-step procedure that solves a problem. The **Euclidean algorithm** is an algorithm for computing $\gcd(a, b)$ for $b > 0$ by means of repeated divisions. Let's look an example. To compute $\gcd(164, 72)$, we perform a sequence of divisions, starting with the division of 164 by 72, until a division produces a 0 remainder:

row	dividend		divisor		quotient		remainder
1	164	=	72	·	2	+	20
2	72	=	20	·	3	+	12
3	20	=	12	·	1	+	8
4	12	=	8	·	1	+	4
5	8	=	4	·	2	+	0

algorithm halts when it produces a 0 remainder

$\gcd(164, 72)$ is the divisor used in the last division

In the first division, we divide 164 by 72, giving a quotient of 2 and a remainder of 20 (row 1 in the table above). In the second row, we divide 72 (the divisor in the first row) by 20 (the remainder in the first row). In each row after the first, the dividend and the divisor is the divisor and remainder, respectively, in the *preceding* row, as indicated by the arrows in the table above. The divisor used in the last division (i.e., the division that produces a 0 remainder) is the desired gcd. Thus, $\gcd(164, 72) = 4$.

To confirm that 4 is a common divisor of 164 and 72, observe that $4 \mid 8$ (because the remainder in row 5 is 0). Moreover, $4 \mid 4$. Then by D4 in frame **547**, 4 divides the right side of the equation $12 = 8 \cdot 1 + 4$ in row 4. Thus, 4 divides 12, the left side of this equation. We now have that $4 \mid 12$ and $4 \mid 8$. Then by D4, 4 divides the right side of the equation $20 = 12 \cdot 1 + 8$ in row 3. Thus, 4 divides 20, the left side of this equation. At this point, this process has shown that 4 divides 8, 12, and 20—the dividends and divisors in the last three rows. If we continue this process back to the first row, it shows that 4 divides every dividend and divisor in the table. Thus, 4 is a common divisor of the dividend and divisor in the first row (164 and 72).

To show that 4 is the *greatest* common divisor of 164 and 72, we have to show that 4 is greater than or equal to any common divisor of 164 and 72. To do this, we use D4 repeatedly, working from the first row to the last row. Let d be a common divisor of 164 and 72. Suppose we rewrite the equation in row 1 as $164 - 72 \cdot 2 = 20$. Then by D4, d divides the left side of this equation. Thus, d also divides 20, the right side of this equation. We now have that $d \mid 72$ (because d is a common divisor of 164 and 72) and $d \mid 20$. We can rewrite the equation in row 2 as $72 - 20 \cdot 3 = 12$. Applying D4 to this equation, we get that d divides the left side of this equation. Thus, d also divides 12, the right side of this equation. At this point, this process has shown that d divides 72, 20, and 12—the divisors in the first three rows. If we continue this process to the last row, it shows that d divides 4, the divisor in the last row. Then by D6, $d \le 4$. Thus, 4 is the greatest common divisor.

From our table, we can now determine an x and y such that $4 = 164x + 72y$ by starting with the equation for 4 from row 4, and substituting successively for the remainders 8, 12, and 20 using the equations in rows 3 to 1. We get

```
4
= 12 − 8                                    this is the row 4 equation
= 12 − (20 − 12) = 12 · 2 − 20              substituting for 8 using the row 3 equation (8 = 20 − 12)
= (72 − 20 · 3) · 2 − 20 = 72 · 2 − 20 · 7  substituting for 12 using the row 2 equation (12 = 72 − 20 · 3)
= 72 · 2 − (164 − 72 · 2) · 7 = 164(−7) + 72(16).   substituting for 20 using the row 1 equation (20 = 164 − 72 · 2)
```

Prime Numbers

568

A *prime number* p is an integer greater than 1 whose only positive divisors are 1 and itself. For example, the only positive divisors of 3 are 1 and 3. Thus, 3 is a prime number. The first five prime numbers are 2, 3, 5, 7, and 11. The integer 1 is *not* a prime number (by definition, a prime number is greater than 1). A *composite number* is an integer greater than 1 that is not a prime number. That is, it is a positive integer that has a positive divisor in addition to 1 and itself. An integer n is composite if and only if there are integers j and k such that $n = j \cdot k$ and $1 < j, k < n$.

Is 15 prime or composite? Justify your answer.

569

It is composite, because $15 = 3 \cdot 5$ and $1 < 3, 5 < 15$.

Is 52 prime or composite?

570

It is composite, because $52 = 4 \cdot 13$ and $1 < 4, 13 < 52$. We can also establish that 52 is composite by observing that $52 = 2 \cdot 26$ and $1 < 2, 26 < 52$.

Is 5 a prime?

571

Yes. The only positive divisors of 5 are 1 and 5.

List the first 10 prime numbers.

572

2, 3, 5, 7, 11, 13, 17, 19, 23, 29

If an integer greater than 1 is not prime, we can factor it until all its factors are prime (we will prove this shortly). Thus, every integer greater than 1 is either a prime or a product of primes. For example, $100 = 2 \cdot 50 = 2 \cdot 2 \cdot 25 = 2 \cdot 2 \cdot 5 \cdot 5$.

    A *square root of n,* where n is a nonnegative real number, is a real number that, when squared, yields n. For example, 9 has two square roots, 3 and −3, because the square of 3 and the square of −3 both equal 9. Every positive real number has two square roots. For positive reals, its *principal square root* is its positive square root. For zero, its principal square root is its one and only square root, namely, 0. We use the radical symbol $\sqrt{\ }$ to denote only the principal square root. For example, $\sqrt{9}$ equals 3, not −3. To represent the negative square root of 9, we write $-\sqrt{9}$.

    Let's determine some of the values of p and q such that $pq = 36$. If p is 6, then q is also 6, because 6 is a square root of the 36. Now suppose we increase p to 9. Clearly we must decrease q (to 4) to maintain pq equal to 36. If, on the other hand, we decrease p, we must then increase q.

Key idea: If n is a positive integer that has a positive divisor greater than $\sqrt{n}$, then n also has a positive divisor less than $\sqrt{n}$.

To determine if a positive integer n is prime, we have to check if it has any positive divisors other than 1 and n. If it does, then it is composite. From the preceding key idea, we know that n has a positive divisor less than $\sqrt{n}$ if n has a positive divisor greater than $\sqrt{n}$. Thus, we need to check for potential divisors only in the range of 2 to $\sqrt{n}$. If there are no such divisors, then the only positive divisors of n are 1 and n, in which case n is prime. In fact, we need to check *only the primes* from 2 to $\sqrt{n}$. If a composite is a divisor, then so are its prime factors. Thus, by checking only the primes from 2 to $\sqrt{n}$, we are guaranteed to find a divisor of n if any exist. For example, $\sqrt{31}$ is between 5 and 6. Thus, to determine if 31 is prime, we need to check only 2, 3, and 5. If none are divisors, then 31 is prime.

Is 31 prime?

573

The primes 2, 3, and 5 all do not divide 31; thus, 31 is prime.

Determine if 101 is prime.

574

The primes less than or equal to $\sqrt{101}$ are 2, 3, 5, and 7. None divides 101. Thus, 101 is prime.

Determine if 45 prime.

575

The primes less than or equal to $\sqrt{45}$ are 2, 3, and 5; 45 is divisible by 3, so 45 is composite.

Euclid's Lemma

576

A *lemma* is a "little" theorem used in a proof of a main theorem. A *corollary* is a theorem that follows from the main theorem. In mathematical texts, we often see several lemmas leading to a main theorem, followed by several corollaries.

Euclid's lemma is a theorem that is used in the proofs of many theorems. For this reason, it is called a lemma rather than a theorem. Here is *Euclid's lemma*:

>Let p, a, and b be integers with p prime. If $p \mid ab$, then $p \mid a$ or $p \mid b$.

Let's prove it. The consequent of this implication is a disjunction. So we will use the technique for proving a disjunction described in frame **170**: We assume $p \mid ab$ and $p \nmid a$, and show that $p \mid b$.

Suppose $p \nmid a$. Because p is prime, its only positive divisors are 1 and p. Thus, $\gcd(a, p)$ can equal only 1 or p. But $\gcd(a, p)$ cannot be p because $p \nmid a$. Thus, $\gcd(a, p) = 1$. From frame **567**, we know that there exist integers x and y such that $ax + py = 1$. Multiplying both sides of this equation by b, we get $b = abx + pby$.

Does p divide ab? Does p divide pb?

577

Yes (we are given that $p \mid ab$), yes (p is a factor of pb).

How do we know that $p \mid (abx + pby)$?

578

By D4 in frame **547**, $p \mid (abx + pby)$. b is equal to $abx + pby$. Thus, $p \mid b$. ∎

Euclid's lemma easily generalizes to the following theorem (you prove this theorem in homework question 10):

>Let $p, a_1, a_2, \ldots, a_n$ be integers with p prime. If $p \mid a_1 a_2 \ldots a_n$, then $p \mid a_i$ for some i, $1 \leq i \leq n$.

Is it true that $6 \mid (8 \cdot 15)$?

579

Yes. $8 \cdot 15 = 120 = 6 \cdot 20$. Thus, $6 \mid (8 \cdot 15)$.

Because $6 \mid (8 \cdot 15)$, can we conclude by Euclid's lemma that $6 \mid 8$ or $6 \mid 15$?

580

No. Euclid's lemma does not apply here because 6 is not a prime.

Fundamental Theorem of Arithmetic

581

Here are all the factorizations of 12 into primes: $12 = 2 \cdot 2 \cdot 3 = 2 \cdot 3 \cdot 2 = 3 \cdot 2 \cdot 2$. All the factorizations of 12 into primes are the same, apart from the order of the factors. They all have two 2's and one 3 and no other prime factors. Is this uniqueness property of the factorization into primes true for every integer greater than 1? Let's try another example. Factor 72 into primes.

582

$72 = 2 \cdot 2 \cdot 2 \cdot 3 \cdot 3$

All the factorizations of 72 into primes have three 2's and two 3's and no other prime factors.

That the factorizations of 12 and 72 into primes are unique, apart from the order of their factors is, of course, not a proof that unique factorization into primes holds for every integer greater than 1. However, unique factorization into primes does indeed hold for every integer greater than 1. This is such an important theorem that it is called the ***fundamental theorem of arithmetic*** (abbreviated FTOA).

The proof of the FTOA consists of two parts:

1. Existence: We prove that every integer greater than 1 can be factored into primes.

2. Uniqueness: We prove that the factorization of any integer greater than 1 into primes is unique, apart from the order of the prime factors.

The existence proof of the FTOA illustrates a type of proof by induction that differs from the type we discussed in chapter 4. Specifically, the inductive hypothesis differs in the two versions. In the version in chapter 4, the inductive hypothesis is S_n. In the version we use here, the inductive hypothesis is S_n *and* all the statements preceding S_n (i.e., $S_{n-1}, S_{n-2}, \ldots$) down to the basis case.

To prove the existence part of the FTOA, we want to show that the S_n is true for all integers $n \geq 2$, where S_n is the statement "n can be factored into primes." We will prove this using our new type of proof by induction. Specifically, we will show that S_2 is true (the basis), and if S_i is true for $2 \leq i \leq n$ (the inductive hypothesis), then S_{n+1} is true.

Basis step:

2 is a prime $\Rightarrow$ 2 itself is its prime factorization $\Rightarrow S_2$ is true.

Induction step: Assume $S_2, S_3, \ldots, S_n$ are true.

Case 1: $n + 1$ is a prime $\Rightarrow n + 1$ itself is its prime factorization $\Rightarrow S_{n+1}$ is true for this case.
Case 2: $n + 1$ is composite $\Rightarrow n + 1 = pq$, where $1 < p, q, < n + 1$. By our inductive hypothesis, S_p and S_q are true $\Rightarrow p$ and q can be factored into primes $\Rightarrow pq = n + 1$ can be factored into primes (the prime factors of $n + 1$ are all the primes in p together with all the primes in q) $\Rightarrow S_{n+1}$ is true for this case. ∎

Why did we have to use a new type of proof by induction?

583

We need S_p and S_q to show S_{n+1}. We cannot do this with only S_n. In the chapter 4 version of proof by induction, we assume only S_n for the inductive hypothesis.

We now do the uniqueness part of the proof of the FTOA. We use a proof by contradiction in conjunction with the well-ordering principle.

Let S be the set of integers greater than 1 that has more than one factorizations into primes. Suppose S is nonempty. Then, by the well-ordering principle, S has a least element n. Thus, n has at least two distinct factorizations into primes, which we represent with $n = p_1 p_2 \ldots p_r$ and $n = q_1 q_2 \ldots q_s$, where $r, s \geq 1$, and the p's and q's are primes.

Can r and s both equal 1?

584

No. Otherwise, each factorization would consist of only one prime. These two primes would then have to be equal, in which case the two factorizations would not be distinct.

$p_1 \mid p_1p_2...p_r \Rightarrow p_1 \mid n \Rightarrow p_1 \mid q_1q_2...q_s$. Then by the generalized form of Euclid's lemma (frame **578**), $p_1 \mid q_i$ for some i.

From $p_1 \mid q_i$, what can we conclude about p_1 and q_i?

585

Because q_i is a prime, its only positive divisors are 1 and q_i; p_1 is a prime, so it is not equal to 1. Thus, $p_1 = q_i$.

Let's now cancel this common prime factor in the two factorizations from frame **583**:

$$n = p_1p_2...p_r \text{ and } n = q_1q_2...q_s, \text{ where } r, s \geq 1.$$

If $r = 1$ and $s > 1$, we get $1 = q_1q_2...q_{i-1}q_{i+1}...q_s$. If $r > 1$ and $s = 1$, we get $p_2p_3...p_r = 1$. If $r > 1$ and $s > 1$, we get $p_2p_3...p_r = q_1q_2...q_{i-1}q_{i+1}...q_s$. In all three cases, we have a contradiction. In the first two cases, we have 1 equal to a prime or a product of primes, which is impossible. In the third case, we get two distinct prime factorizations for an integer greater than 1 that is less than n. But n is the least such integer. We resolve these contradictions by concluding that our initial assumption—that S is nonempty—is incorrect. Recall that S is the set of integers greater than 1 that has more than one factorization into primes. We have shown that S is empty. Thus, the factorization of any integer greater than 1 into primes is unique. ∎

We will frequently use the FTOA in the following way: suppose a prime factor appears k times on one side of an equality. Then, by the FTOA, it must appear k times on the other side of the equality. For example, if $2 \cdot 2 \cdot 3 = ab$, then there must be two 2 factors and one 3 factor in ab.

Irrationality of $\sqrt{2}$

586

If the numerator and the denominator of a fraction have a prime factor in common, we can get an equal fraction by dividing the numerator and the denominator by that prime factor. We can continue this process until we get a fraction equal in value to the original fraction whose numerator and denominator have no prime factors in common.

Consider the fraction $\frac{6}{14}$. Do 6 and 14 have any prime factors in common?

587

Yes. 2 is a factor of both 6 and 14. By dividing the top and bottom of $\frac{6}{14}$ by the common prime factor 2, we get the fraction $\frac{3}{7}$, which is equal to $\frac{6}{14}$.

Eliminate the prime factors in common to the numerator and denominator of $\frac{36}{20}$.

588

The prime factor 2 appears twice in both 36 and 20. Thus, we reduce the factor by dividing through by 2^2 to get $\frac{9}{5}$.

We now prove that $\sqrt{2}$ is irrational. That is, we prove that it cannot be expressed as the ratio of two integers. We will use the FTOA and Euclid's lemma.

Assume that $\sqrt{2}$ can be expressed as the ratio of two integers. If the two integers have any common prime factors, we can divide them out, as discussed earlier. Thus, without loss of generality, we can assume that the numerator and denominator of our fraction representing $\sqrt{2}$ have no prime factors in common. That is, $\sqrt{2} = \frac{i}{j}$ where i and j are integers that have no prime factors in common.

Square both sides of the preceding equality and multiply through by j^2.

589

$i^2 = 2j^2$

The prime 2 is a factor of the right side, so by the FTOA it must also be a factor of the left side. Thus,

$2 \mid i^2$. Because $2 \mid i^2$, by Euclid's lemma, $2 \mid i$. We can also conclude that $2 \mid i$ by observing that the square of any odd number is odd. Thus, if i^2 is even, i must also be even, and, therefore, divisible by 2.

How many 2 factors are on the left side of $i^2 = 2j^2$?

590

At least two. Because $2 \mid i$, there is a 2 factor in i. Thus, there are at least two 2 factors in i^2. Because $i^2 = 2j^2$, there also are at least two 2 factors in $2j^2$.

Can we then conclude that there is a 2 factor in j^2 on the right side of $i^2 = 2j^2$?

591

Yes. One 2 factor is the initial 2 on the right side. By FTOA, there must be another 2 factor in j^2. That is, $2 \mid j^2$.

Using Euclid's lemma, what can we conclude about j?

592

$2 \mid j$

But now we have a contradiction: we have shown that $2 \mid i$ and $2 \mid j$. Thus, 2 is a common prime factor of i and j. But i and j have no prime factors in common. We resolve this contradiction by concluding that our initial assumption—that $\sqrt{2}$ can be expressed as the ratio of integers—is incorrect. ∎

Factorization of Squares into Primes

593

Whenever we show the prime factorization of an integer, we will show it in the following standard form: we will list the primes in ascending order, with each prime listed only once. We specify the multiplicity of each prime (i.e., the number of times that prime appears in the factorization) with an exponent on that prime (we omit the exponent if the multiplicity of the prime is 1). For example, we show the prime factorization of $360 = 2 \cdot 2 \cdot 2 \cdot 3 \cdot 3 \cdot 5$ as $2^3 3^2 5$.

Show the prime factorization of 40 in standard form.

594

$2^3 5$

The integers 4, 16, 64, 100, and 10,000 are all squares (i.e., they equal the square of some integer). Give their prime factorizations in standard form.

595

$4 = 2^2$, $16 = 2^4$, $64 = 2^6$, $100 = 2^2 5^2$, $10,000 = 2^4 5^4$

How would you characterize the exponents—2, 4, 6—in these factorizations?

596

They are all even numbers.

Why are the exponents in the standard form factorization of a square into primes always even?

597

Suppose n is a square. That is, $n = k^2$ for some integer k. Each prime factor in k appears twice in n because the factor k appears twice in n. Thus, the multiplicity of each prime in n must be an even number.

Trailing Zeros in *n*!

598

Characterize the number of 2 factors in 10! = 10 · 9 · ⋯ · 2 · 1. Are there none, one, or many?

599

Many. All the even numbers from 2 to 10 have 2 factors: 2 has one 2 factor; 4 has two; 6 has one; 8 has three; 10 has one. There is a total of eight 2 factors.

How many 5 factors are in 10!?

600

Two: 10 has one 5 factor; 5 itself is one 5 factor.

How many 10 factors are there in 10! Think carefully about this question. The correct answer is not 1.

601

The prime factorization of 10! has only two 5 factors but plenty of 2 factors. Thus, it has exactly two occurrences of 5 · 2. That is, it has exactly two 10 factors.

In every factorial, every 5 factor can be paired with a 2 factor (because there are plenty of 2 factors). Thus, the number of 5 factors in a factorial is equal to the number of _____ factors.

602

10

To multiply a decimal integer by 10, simply add trailing zeros. For example, 10 · 23 is 23 with a trailing zero: 230. Each 10 factor in an integer corresponds to a trailing zero in its decimal representation. If an integer has n 10 factors, it representation has n trailing zeros. Conversely, each trailing zero in the decimal representation of an integer corresponds to a 10 factor in the number. If a number has n trailing zeros, it has n 10 factors.

How many trailing zeros are in the decimal representation of an integer that has exactly three 10 factors?

603

three

The decimal representation of 10! ends with how many zeros? *Hint*: How many 5 factors are in 10!?

604

Two: 10! has two 5 factors plus plenty of 2 factors. Thus, it has exactly two 10 factors.

Check this answer by computing the value of 10!

605

3,628,800

The decimal representation of 20! ends with how many zeros?

606

Four. There is a 5 factor in 5, 10, 15, and 20. All the other factors in 20! (2, 3, 4, 6, …, 19) do not have any 5 factors.

The decimal representation of 25! ends with how many zeros? Be careful!

607

Six: 25 has two 5 factors. Thus, 25! has six 5 factors: one in 5, 10 15, and 20, and two in 25.

Relatively Prime Numbers

608

Two integers a and b are **relatively prime** if and only if $\gcd(a, b) = 1$. Equivalently, a and b are relatively prime if and only if a and b have no prime factors in common (see homework question 6).

Are 4 and 9 relatively prime?

609

Yes. They have no prime factors in common.

Suppose $a, b, c \in \mathbb{Z}$ and $a \mid bc$. Can we conclude that $a \mid b$ or $a \mid c$?

610

No. A counterexample is $4 \mid (2 \cdot 2)$, but $4 \nmid 2$. However, if $a \mid bc$ and $\gcd(a, b) = 1$, then $a \mid c$. We call this theorem the **relatively prime divisor theorem.**

Try proving this theorem. *Hint*: If a and b are relatively prime, then by frame **567**, there are integers x and y such that $ax + by = 1$.

611

$ax + by = 1$. Multiplying both sides of this equation by c, we get $cax + bcy = c$. The integer a clearly divides ca. Moreover, we are given that $a \mid bc$. Thus, by D4 in frame **547**, a divides $cax + bcy = c$. ∎

Let's apply the relatively prime divisor theorem. Because $6 \mid (25 \cdot 12)$, can we conclude that $6 \mid 12$?

612

Yes, because $\gcd(6, 25) = 1$. We can also come to this conclusion using the FTOA: Because $\gcd(6, 25) = 1$, 6 and 25 have no prime factors in common. Because $6 \mid (25 \cdot 12)$, all the prime factors in 6 are in $25 \cdot 12$. Because none of the prime factors in 6 are in 25, they all must be in 12. Thus, $6 \mid 12$.

Some review:

- Euclid's lemma: If $p \mid ab$ and p is prime, then $p \mid a$ or $p \mid b$.
- Relatively prime divisor theorem: If $d \mid ab$ and d and a are relatively prime, then $d \mid b$.

U(n)

613

$U(n)$ is defined as the set of positive integers less than n that are relatively prime to n. Equivalently, $U(n)$ is the set of positive integers less than n that have no prime factors in common with n. For example, $U(6) = \{1, 5\}$. $U(6)$ does not include 2, 3, or 4 because these integers are not relatively prime to 6.

List the elements in $U(10)$. *Hint*: 10 has a 2 factor and a 5 factor. Thus, eliminate any element that is a multiple of 2 or a multiple of 5.

614

$U(10) = \{1, 3, 7, 9\}$

List the elements in $U(30)$.

615

$U(30) = \{1, 7, 11, 13, 17, 19, 23, 29\}$

List the elements in $U(7)$.

616

$U(7) = \{1, 2, 3, 4, 5, 6\}$

Specify the elements in $U(p)$ where p is a prime.

617

$U(p) = \{1, 2, ..., p-1\}$

Least Common Multiple

618

In everyday English, we use the word *multiple* to mean "more than one." For example, "I have multiple injuries from playing football" means "I have more than one injury." However, in mathematics, *multiple* has a different meaning: a **multiple** of a nonzero number n is any number in the set $\{kn : k \in \mathbb{Z}\}$. For example, if n is 5, then 10, 15, 20, and so on, are multiples of 5, but so are 5, 0, -5, -10, and so on.

The **least common multiple** of **m** and **n** (denoted by lcm(m, n)) is the smallest *positive* integer that is a multiple of both m and n. For example, 6 is a multiple of 2 and 3. Moreover, there is no smaller positive multiple of 2 and 3. Thus, lcm(2, 3) = 6.

What is lcm (6, 9)?

619

18.

What is lcm(5, -7)?

620

35.

Is $|mn|$ a positive common multiple of m and n?

621

Yes, except when $mn = 0$. Zero has no positive multiples. Thus, if $m = 0$ or $n = 0$, then m and n have no common positive multiples. For this reason, lcm(m, n) is undefined if $m = 0$ or $n = 0$, or, equivalently, if $mn = 0$. If, on the other hand, $mn \neq 0$, then $|mn|$ is a positive common multiple of m and n. Then by the well-ordering principle, there must be a least common multiple of m and n.

What is the lcm of 0 and -7?

622

undefined

What is the lcm of 5 and 10?

623

10

What is the lcm of 1 and -1?

624

1

Determining gcd(*m*, *n*) and lcm(*m*, *n*)

625

Can a positive divisor of $2^2 3$ have more than two 2 factors? *Hint*: Consider the FTOA.

626

No. If d is a positive divisor of $2^2 3$, then $2^2 3 = dk$ for some $k \in \mathbb{Z}$. Then, by the FTOA, the total number of 2 factors in d and k is two. d can have two, one, or zero 2 factors, in which case k would have zero, one, or two 2 factors, respectively. Similarly, d can have one or zero 3 factors, in which case k would have zero or one 3 factor, respectively.

What are the possible values of i and j if $d = 2^i 3^j$ is a positive divisor of $2^2 3$?

627

The exponent i can equal 0, 1, or 2 (corresponding to the three possible multiplicities of 2 in d), and j can equal 0 or 1 (corresponding to the two possible multiplicities of 3 in d). Thus, in every positive divisor of $2^2 3$, the multiplicity of 2 is either 0, 1, or 2; the multiplicity of 3 is either 0 or 1. We have three choices for the prime 2 and two choices for the prime 3. Thus, we can construct $3 \cdot 2 = 6$ positive divisors:

Multiplicity of 2	Multiplicity of 3	Divisor
0	0	$2^0 3^0 = 1$
1	0	$2^1 3^0 = 2$
2	0	$2^2 3^0 = 4$
0	1	$2^0 3^1 = 3$
1	1	$2^1 3^1 = 6$
2	1	$2^2 3^1 = 12$

What is the greatest common divisor of $m = 2^3 3^2 5^1$ and $n = 2^1 3^3 5^2 7^1$? *Hint*: By the FTOA, the multiplicity of 2 in a divisor of m is at most 3; in a divisor of n, it is at most 1. Thus, in a *common* divisor of m and n, the multiplicity of 2 is at most 1 (the *smaller* of the two multiplicities). Because gcd(m, n) is the greatest common divisor of m and n, in gcd(m, n), the multiplicity of 2 is 1, its maximum value for a common divisor of m and n. .

628

$2^1 3^2 5^1 7^0 = 90$

The integer m does not have a 7 factor, but we can think of m as equal to $2^3 3^2 5^1 7^0$. Because $7^0 = 1$, including 7^0 in the factorization of m does not affect the value represented. Between $m = 2^3 3^2 5^1 7^0$ and $n = 2^1 3^3 5^2 7^1$, 0 is the smaller multiplicity of the 7 factor. Thus, the multiplicity of 7 in gcd(m, n) is 0.

By the FTOA, any positive multiple of $m = 2^3 3^2 5^1$ must have at least three 2 factors, because m has three 2 factors. Similarly, a positive multiple of $n = 2^1 3^3 5^2 7^1$ must have at least one 2 factor. Thus, lcm(m, n) must have exactly three 2 factors—the *larger* multiplicity of 2 in m and n. We can similarly determine the multiplicity of the 3, 5, and 7 factors in lcm(m, n).

Determine the lcm(m, n) where $m = 2^3 3^2 5^1$ and $n = 2^1 3^3 5^2 7^1$. *Hint*: view m as $2^3 3^2 5^1 7^0$. Thus, the larger multiplicity of 7 in m and n is 1.

629

$2^3 3^3 5^2 7^1 = 37{,}800$

Key idea: The multiplicity of a prime in gcd(m, n) is the smaller multiplicity of that prime in m and n. The multiplicity of a prime in lcm(m, n) is the larger multiplicity of that prime in m and n.

What are the gcd and lcm of 50 and 75?

630

$50 = 2^1 3^0 5^2$, $75 = 2^0 3^1 5^2$. Thus, gcd(50, 75) $= 2^0 3^0 5^2 = 25$, lcm(50, 75) $= 2^1 3^1 5^2 = 150$.

What are the gcd and lcm of -30 and 180?

631

$-30 = -2^1 3^1 5^1$, $180 = 2^2 3^2 5^1$. Thus, gcd(30, 180) = $2^1 3^1 5^1$ = 30, lcm(30, 180) = $2^2 3^2 5^1$ = 180.

In the remaining frames in this section, assume that m and n are integers and $mn \neq 0$. That is, both m and n are nonzero integers (otherwise lcm(m, n) is undefined).

What can you say about the multiplicity of some prime p in mn?

632

The multiplicity of p is the sum of its multiplicities in m and n.

What can you say about the multiplicity of some prime p in the product gcd(m, n) · lcm(m, n)?

633

In the product of gcd(m, n) and lcm(m, n), the multiplicity of p is the sum of its multiplicity in the gcd(m, n) (which is the smaller of its multiplicities in m and n) and its multiplicity in the lcm(m, n) (which is the larger of its multiplicities in m and n).

What can you conclude about the multiplicity of a prime p in mn and the multiplicity of p in gcd(m, n) · lcm(m, n)?

634

They are the same ⇒ $|mn|$ = gcd(m, n) · lcm(m, n).

Compute $|-12 \cdot 10|$, gcd(-12, 10), lcm(-12, 10), and gcd(-12, 10) · lcm(-12, 10).

635

$|-12 \cdot 10|$ = 120, gcd(-12, 10) = 2, lcm(-12, 10) = 60, gcd(-12, 10) · lcm(-12, 10) = 2 · 60 = 120. As we expect from our result in the previous frame, $|-12 \cdot 10|$ = gcd(-12, 10) · lcm(-12, 10).

Suppose gcd(m, n) = 6 and lcm(m, n) = 36. What does $|mn|$ equal?

636

6 · 36 = 216

If mn = 375 and gcd(m, n) = 5, what is lcm(m, n)?

637

mn/gcd(m, n) = 375/5 = 75

$|mn|$ = lcm(m, n) if and only if gcd(m, n) = _____. *Hint*: The answer is obvious from the equation in frame **634**.

638

1

Key idea: For $mn \neq 0$, $|mn|$ = gcd(m, n) · lcm(m, n).
Key idea: For $mn \neq 0$, $|mn|$ = lcm(m, n) if and only if gcd(m, n) = 1.

Finding Multipliers

639

Consider the integers 66 and 420. What is the smallest positive multiplier of 66 that yields a multiple of 420? *Hint*: Factor 66 and 420 into primes.

640

$66 = 2 \cdot 3 \cdot 11$ and $420 = 2 \cdot 2 \cdot 3 \cdot 5 \cdot 7$. Any positive multiple of 420 has to have all the prime factors of 420; 66 has two of the prime factors of 420: 2 and 3. Thus, to get the smallest multiplier that works, we have to multiply 66 only by those prime factors in 420 that are not in 66: the second 2 factor, 5, and 7. Thus, the smallest multiplier of 66 that yields a multiple of 420 is $2 \cdot 5 \cdot 7 = 70$. Let's confirm that $70 \cdot 66$ is a multiple of 420: $70 \cdot 66 = 4620 = 11 \cdot 420$.

What does gcd(66, 420) equal?

641

66 and 420 have prime factors 2 and 3 in common. Thus, $\gcd(66, 420) = 2 \cdot 3 = 6$.

If we divide 420 by gcd(66, 420), we divide out all the factors in 420 that are also in 66. Thus, 420/gcd(66, 420) is precisely the multiplier that we need in the preceding frame. It is the smallest multiplier of 66 that yields a multiple of 420.

In general, if k and n are positive integers, $n/\gcd(k, n)$ is the smallest positive multiplier of k that yields a multiple of n. We call this result the ***smallest positive multiplier theorem.***

If k and n are relatively prime, what is $n/\gcd(k, n)$?

642

If k and n are relatively prime, then $\gcd(k, n) = 1$. Thus, $n/\gcd(k, n) = n/1 = n$.

What is the smallest positive multiplier of 42 that yields a multiple of 72?

643

$72/\gcd(42,72) = 72/6 = 12$

What is the smallest positive multiplier of 4 that yields a multiple of 81?

644

81 because 4 and 81 are relatively prime. That is, $\gcd(4, 81) = 1$. Thus, the smallest multiplier is $81/\gcd(81, 4) = 81/1 = 81$.

Review Questions

1. What are the quotient and remainder when 32 is divided by 5?
2. What are the quotient and remainder when -32 is divided by 5?
3. Restate the division algorithm so that it applies to both positive and negative divisors.
4. What are the quotient and remainder when -32 is divided by -5? See review question 3.
5. Using the Euclidean algorithm, determine gcd(184, 80), and compute x and y such that $184x + 80y = \gcd(184, 80)$.
6. List the elements in $U(12)$.
7. Show that the square of any integer n is equal to $3k$ or $3k + 1$ for some $k \in \mathbb{Z}$. *Hint*: By the division algorithm, $n = 3q$, $n = 3q + 1$, or $n = 3q + 2$ for some integer q.
8. What is the smallest positive multiplier of 12 that is a multiple of 42?
9. Give an informal argument that $x \mid z$ if $x \mid (y + z)$ and $x \mid y$.
10. 50! expressed as a decimal number ends in how many zeros?

Answers to the Review Questions

1. 6, 2
2. -7, 3
3. For any integers $d \neq 0$ and a, there exist integers q and r such that $a = dq + r$ with $0 \leq r < |d|$. Moreover, q and r are unique.
4. 7, 3

5.

row	dividend		divisor		quotient		remainder
1	184	=	80	·	2	+	24
2	80	=	24	·	3	+	8
3	24	=	8	·	3	+	0

$$\gcd(184, 80) = 8 = 80 - 24 \cdot 3 = 80 - (184 - 80 \cdot 2) \cdot 3$$
$$= 184(-3) + 80(7)$$

6. $U(12) = \{1, 5, 7, 11\}$
7. *Case 1*: $(3q)^2 = 3(3q^2)$. *Case 2*: $(3q+1)^2 = 3(3q^2 + 2q) + 1$. *Case 3*: $(3q+2)^2 = 3(3q^2 + 4q + 1) + 1$.
8. $42/\gcd(42, 12) = 42/6 = 7$
9. Adding a multiple of x to another multiple of x gives a multiple of x. But adding a non-multiple of x to a multiple of x gives a non-multiple of x. Accordingly, because y is a multiple of x (i.e., $x \mid y$), for $y + z$ to be a multiple of x (i.e., $x \mid (y+z)$, z must be a multiple of x (i.e., $x \mid z$).
10. 50! has a total of twelve 5 factors. Thus, it has a total of twelve 10 factors and ends with twelve zeros.

Homework Questions

1. What are the remainder and the quotient when -50 is divided by 11?
2. Determine the prime factorization of 12345.
3. List the elements in $U(210)$.
4. Prove that the sum of any three consecutive integers is divisible by 3.
5. Prove that $\gcd(n, n+1) = 1$ for any $n \in \mathbb{Z}$. *Hint*: Use the following in a proof by contradiction: If $d > 1$, then $d \nmid 1$.
6. Prove: a and b have a prime factor in common $\Leftrightarrow \gcd(a, b) \neq 1$. Or equivalently (by frame **168**), prove: a and b have no prime factors in common $\Leftrightarrow \gcd(a, b) = 1$.
7. In this chapter, we presented and informally justified the division algorithm. Here is an outline of a rigorous proof of the division algorithm: Let $S = \{a - di : a - di \geq 0 \text{ and } i \in \mathbb{Z}\}$. Show that S is nonempty. Then S has a least element. Denote this least element with r. Thus, $r = a - dq$ for some $q \in \mathbb{Z}$. Show that $0 \leq r$. Show that $r < d$. Show that r and q are unique. Write out the proof in detail.
8. Specify the Euclidean algorithm in the form of a sequence of primitive steps. *Hint*: Your algorithm should include a *loop* (a sequence of steps that is performed repeatedly).
9. Complete the proof outlined in frame **567**.
10. Prove by induction the generalization of Euclid's theorem in frame **578**.

11 Equivalence Relations

Review of Relations

645

A relation on a set A is any subset of $A \times A$. For example, suppose $A = \{1, 2, 3\}$. Then $A \times A = \{(1, 1), (1, 2), (1, 3), (2, 1), (2, 2), (2, 3), (3,1), (3, 2), (3, 3)\}$. Any subset of this set is a relation on A. For example, $R = \{(1, 1), (1, 2), (2, 3)\}$ is a relation on A. To indicate that $(2, 3)$ is in R, we can write any of the following:

- $(2, 3) \in R$
- $2 \, R \, 3$
- 2 is R-related to 3
- 2 is related to 3 (if context indicates that the relation is R)

How many relations are there on $A = \{1, 2, 3\}$? *Hint*: See frame **45**.

646

The number of relations is equal to the number of subsets of $A \times A$. $A \times A$ has nine ordered pairs. Thus, there are $2^9 = 512$ subsets.

If a relation has a standard name or symbol, we generally use that name or symbol to indicate that an element is related to another. For example, the relation consisting of all ordered pairs of integers in which the first component of each pair is less than or equal to the second component is the "less than or equal" relation, whose symbol is $\leq$. Thus, to indicate that $(1, 5)$ is in this relation, we generally write $1 \leq 5$.

A convenient way to represent a relation is with an arrow diagram. For example, the arrow diagram corresponding to $R = \{(1, 1), (1, 2), (2, 3)\}$ is

Each arrow corresponds to an ordered pair in R. A **self-loop** is an arrow whose head points to its tail. For example, the arrow from element 1 back to element 1 in the preceding arrow diagram is a self-loop.

Special Types of Relations

647

A relation R on a set A is **reflexive** if and only if every element in A is related to itself, that is, $a \, R \, a$ for all $a \in A$, or, equivalently, $(a, a) \in R$ for all $a \in A$. For example, the relation $\leq$ on the integers is a reflexive relation because $a \leq a$ for all $a \in \mathbb{Z}$.

It is easy to determine from an arrow diagram if a relation is reflexive. Simply check if each element has a self-loop. For example, here is an arrow diagram for a reflexive relation:

Is $A \times A$ a reflexive relation on $A = \{1, 2, 3\}$?

648

Yes. Because $A \times A$ is a subset of itself, it is a relation on A. $A \times A$ contains all the ordered pairs that can be constructed from the elements of A. Thus, it contains (a, a) for all $a \in A$.

Add the minimum number of arrows needed to make the following arrow diagram correspond to a reflexive relation:

649

We call the relation that results when the minimum number of arrows is added to the arrow diagram of the given relation to make it correspond to a reflexive relation the ***reflexive closure*** of the given relation.

A relation R on a set A is ***symmetric*** if and only if, for all $a, b \in A$, $a\,R\,b \Rightarrow b\,R\,a$. That is, for every ordered pair (a, b) in R, there is an ordered pair (b, a) in R.

In an arrow diagram for a symmetric relation, wherever there is an arrow connecting two distinct elements, there is an arrow in the opposite direction connecting the same two elements. Thus, if there is a direct connection between two elements, the connection is a "two-way street."

Is the relation represented by the following arrow diagram symmetric?

650

Yes. The single self-loop on 3 represents not only the arrow from the first component of (3, 3) to the second component but also the arrow from the second component to the first component (because both components are 3). Thus, the single self-loop on 3 is consistent with a symmetric relation.

Is $\emptyset$ (i.e., the empty set) a symmetric relation on $A = \{1, 2, 3\}$?

651

Yes. $\emptyset$ is a subset of $A \times A$, and therefore, it is a relation on A. $\emptyset$ does not have any ordered pairs. Thus, it is vacuously true that for each ordered pair (a, b), there is the ordered paired (b, a).

Key idea: A universally quanfied statement of the form $(\forall x \in A)(p(x) \Rightarrow q(x))$ is ***vacuously true*** if and only if for all $x \in A$, $p(x)$ is false, in which case the implication $p(x) \Rightarrow q(x)$ is true for all x.

Is "$\leq$" a symmetric relation?

652

No. We need just one counterexample to show it is not symmetric. For example, $2 \leq 3$ but $3 \nleq 2$.

Add the minimum number of arrows needed to make the following arrow diagram correspond to a symmetric relation:

653

We call the relation that results when the minimum number of arrows is added to the arrow diagram of the given relation to make it correspond to a symmetric relation the **symmetric closure** of the given relation.

A relation R on a set A is **transitive** if and only if, for all $a, b, c \in A$, $a\,R\,b$ and $b\,R\,c \Rightarrow a\,R\,c$. In an arrow diagram for a transitive relation, if an arrow goes from a to b, and an arrow goes from b to c, then an arrow also goes directly from a to c. For example, in the following arrow diagram, there is an arrow from a to b and from b to c:

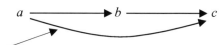

Present if relation is transitive

If the relation is transitive, there must be an arrow from a to c.

Add the minimum number of arrows needed to make the following arrow diagram correspond to a transitive relation:

654

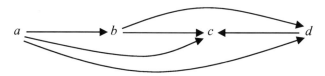

We call the relation that results when the minimum number of arrows is added to the arrow diagram of the given relation to make it correspond to a transitive relation the **transitive closure** of the given relation.

Suppose there is a path of n arrows from element p to element r all going in the p-to-r direction in the arrow diagram of some transitive relation, where $n \geq 1$. Is there necessarily an arrow going directly from p to r?

655

Yes

Let's prove this by induction. Let S_n be the following statement: if there is a path of n arrows from some element p to some element r all going in the p-to-r direction in the arrow diagram for a transitive relation, then there is an arrow going directly from p to r. Is S_1 true?

656

Yes. If there is one arrow from p to r, then obviously this arrow goes directly from p to r. This is the basis step of our induction proof. Now for the induction step. We assume S_n is true and then show that S_{n+1} is true.

Write out the statement S_{n+1}.

657

If there is a path of $n + 1$ arrows from some element p to some element r all going in the p-to-r direction in the arrow diagram of a transitive relation, then there is an arrow going directly from p to r.

Suppose there is a path of $n + 1$ arrows from p to r. This sequence of arrows determines a sequence of elements—the sequence of elements visited by following the arrows. Let q be the element in this sequence just before the final r:

$$p \to \cdots \to q \to r$$

How many arrows in the given sequence are there from p to q?

658

n, because there are $n + 1$ arrows from p to r.

Is there an arrow from p directly to q?

659

Yes

What is the justification for this assertion?

660

Our inductive hypothesis: S_n is true.

What is the justification for now concluding that there is an arrow from p directly to r?

661

The relation is transitive. Because an arrow goes directly from p to q, and an arrow goes directly from q to r, by transitivity, an arrow must go directly from p to r. ∎

An *equivalence relation* is a relation that is reflexive, symmetric, and transitive.

Add as few arrows as possible to the following arrow diagram so that it corresponds to an equivalence relation:

662

Under an equivalence relation, the elements of a set "clump" into disjoint subsets. These disjoint subsets are called *equivalence classes.* For example, in the equivalence relation represented in the preceding diagram, we have two equivalence classes: {2} and {1, 3}. In each equivalence class, every element is related to every element in that equivalence class. For example, within the equivalence class {1, 3}, 1 is related to 1 and 3, and 3 is related to 1 and 3. Is this interconnectedness within each equivalence class true for every equivalence relation? Let's answer this question.

Suppose in an arrow diagram for an equivalence relation, p and q are connected by some sequence of arrows, *not* necessarily all going in the p-to-q direction or all going in the q-to-p direction. For example, suppose p and q are connected as follows:

If this is a part of an arrow diagram of an equivalence relation and, therefore, a symmetric relation, then each arrow must be paired with an arrow going in the opposite direction. Thus, there must be a sequence of arrows—all pointing in the p-to-q direction—from p to q. Similarly, there must be a sequence of arrows—all pointing in the q-to-p direction—from q to p. Then, by frame **655**, there must be an arrow directly from p to q and from q to p. Thus, if there is *any* connection between p and q, then there must be *a direct connection from p to q and from q to p.* Thus, the elements clump into disjoint subsets. Moreover, because an equivalence relation is reflexive, every element must have an arrow pointing to itself. Thus, within each clump, every element is related to itself as well as to every other element in that clump.

We denote the equivalence class to which p belongs with $[p]$. That is, $[p]$ is the set of elements that are related to p. If $p \, R \, q$ for some equivalence relation R, what can you say about $[p]$ and $[q]$?

663

If p and q are related, then they are in the same "clump." That is, $[p] = [q]$.

Is the converse also true? That is, does $[p] = [q]$ imply $p\,R\,q$?

664

Yes. If $[p] = [q]$, then p and q are in the same "clump," and are, therefore, related. Thus, if R is an equivalence relation, whenever we write $p\,R\,q$, we can equivalently write $[p] = [q]$, and vice versa.

If p is not R-related to q, what is the strongest statement we can say about $[p]$ and $[q]$? *Hint*: $[p] \neq [q]$ is not the strongest statement.

665

$[p] \cap [q] = \emptyset$

If p and q are not related, then they must be in different "clumps" (i.e., equivalence classes), in which case, $[p]$ and $[q]$ are disjoint.

Partitions

666

In mathematics, we frequently encounter partitions. A partition is like an pie that we cut into pieces. The pieces of the pie have the following properties:

i. Each piece consists of some of the pie.

ii. Each piece is completely distinct from the other pieces of pie. In other words, no two pieces share some part of the pie.

iii. The set of all the pieces makes up the given pie.

We call the set consisting of all the pieces of the cut-up pie a ***partition*** of the pie.

As we can partition a pie, so can we partition a set. For example, consider the set of $A = \{1, 2, 3\}$. Suppose we cut up A as follows:

We get two "pieces": the disjoint subsets $\{2\}$ and $\{1, 3\}$. The set of these two subsets—that is, $\{\{2\}, \{1, 3\}\}$—has properties i, ii, and iii above: Each subset is nonempty (property i). Each pair of subsets is disjoint (property ii). The union of all the subsets is the set A (property iii). Thus, $\{\{2\}, \{1, 3\}\}$ is one of the several partitions of A.

Key idea: To partition a set, we break it up into disjoint subsets such that each subset is nonempty and the union of all the subsets is the given set.

Here are sets that are *not* partitions of A:

- $\{\{\ \}, \{2\}, \{1, 3\}\}$ does not have property i
- $\{\{1, 2\}, \{2, 3\}\}$ does not have property ii
- $\{\{1\}, \{2\}\}$ does not have property iii

We call the pieces of a partition ***blocks***. For example, the partition $\{\{2\}, \{1, 3\}\}$ of A has two blocks: $\{2\}$ and $\{1, 3\}$.

$A = \{1, 2, 3\}$ can be partitioned in five different ways. List these five partitions.

667

$\{\{1\}, \{2\}, \{3\}\}, \{\{1\}, \{2, 3\}\}, \{\{1, 3\}, \{2\}\}, \{\{1, 2\}, \{3\}\}, \{\{1, 2, 3\}\}$

Refer back to the arrow diagram at the beginning of frame **662**. The elements of A form two distinct clumps of elements under the equivalence relation. Is this set of clumps a partition of A?

668

Yes. Specifically, it is the partition $\{\{1, 3\}, \{2\}\}$.

Here is the arrow diagram of another equivalence relation on A:

What is the partition associated with this equivalence relation?

669

$\{\{1\}, \{2\}, \{3\}\}$

Because of the "clumping" effect of an equivalence relation, associated with every equivalence relation on a nonempty set is a partition of that set. The "clumps" of the equivalence relation are the blocks of the partition.

Conversely, for any partition, "in the same block" is an equivalence relation. For any partition, each element obviously is in the same block as itself. Thus, "in the same block" is reflexive. If an element a is in the same block as b, then b is in the same block as a. Thus, "in the same block" is symmetric. If a is in the same block as b, and b is in the same block as c, then a is in the same block as c. Thus, "in the same block" is transitive.

An equivalence relation on a set and the corresponding partition represent two ways of viewing the same mathematical structure. When we view this structure as an equivalence relation, we call the "clumps" equivalence classes. When we view this structure as a partition, we call the "clumps" blocks.

In mathematics, we often investigate properties of blocks of elements rather than individual elements. When we do, we are working with a partition. For example, the real numbers can be partitioned into the rationals and the irrationals. We can then investigate the rationals and the irrationals separately.

We frequently encounter equivalence relations and partitions in everyday life as well as in mathematics. For example, we often view people and objects in terms of categories to which they belong. Each category is a block in some partition. For example, on a college campus, we have four blocks of people: students, professors, administrators, and service personnel. The set of these blocks is a partition. If, however, there is any overlap between these categories, then we would not have a partition (for example, if a professor is also an administrator).

Quotient Structures

670

In everyday life, when dealing with categories of people, we sometimes carry over an activity in which individuals engage to the block of individuals to which they belong. Two individuals can fight. Similarly, two blocks of people can fight. For example, we can say the Revolutionary War was a fight between the Americans and the British. In mathematics, we can similarly extend an operation on individual elements of a set to the blocks in the partition of that set. For example, let's partition the integers into two blocks: the even integers and the odd integers. Let's also extend the operation of addition so that it applies to the blocks of our partition. An even integer plus an odd integer always yields an odd integer. Thus, it makes sense that we make the sum of the even block and the odd block equal to the odd block.

Here is the Cayley table that represents the addition operation on the even and odd blocks of the integers:

+	even	odd
even	?	odd
odd	?	?

Fill in the missing entries.

671

+	even	odd
even	even	odd
odd	odd	even

The even and odd blocks along with our new addition operation is a new algebraic structure obtained from a more complicated structure (the integers under addition). Our new structure has only two elements, the even block and the odd block, whereas the structure from which it is derived, the integers under addition, is an infinite set. Our new structure is, in a sense, a highly simplified form of the structure from which it is derived.

Whenever we obtain a new structure by partitioning a set into blocks and defining an operation on these blocks, we call the new structure a *quotient structure*. The term *quotient* conveys the idea that the elements of our structure are obtained by "dividing" a set into blocks. The term *structure* conveys the idea that there is some kind of relation on its elements, typically in the form of one or more operations defined on the blocks of the quotient structure. The even and odd blocks under addition are an example of a quotient structure. The even and odd blocks are the elements of this quotient structure; the addition operation gives it its structure.

What is a convenient way to represent the structure of the even and odd blocks under addition?

Their structure is nicely captured by a Cayley table.

When we continue our study of groups in the chapters to come, we will encounter some remarkable quotient structures.

Review Questions

1. Draw the arrow diagram corresponding to the relation $\{(1, 1), (1, 2), (2, 1)\}$ on the set $\{1, 2, 3\}$.
2. Add the minimum number of arrows to your answer for review question 1 so that the represented relation is reflexive.
3. Add the minimum number of arrows to your answer for review question 2 so that the represented relation is symmetric.
4. Add the minimum number of arrows to your answer for review question 3 so that the represented relation is transitive. What is the partition corresponding to this equivalence relation?
5. Let R be a relation on the integers defined by $a \mathrel{R} b$ if and only if $a - b$ is even. Is R reflexive? Is R symmetric? Is R transitive?
6. Let R be a relation on the set of people defined by $a \mathrel{R} b$ if and only if a and b are born within one day of each other. Is R reflexive? Is R symmetric? Is R transitive?
7. Let R be a relation on the integers defined by $a \mathrel{R} b$ if and only if $ab \geq 0$. Is R reflexive? Is R symmetric? Is R transitive?
8. Let R be the relation on the integers defined by $a \mathrel{R} b$ if and only if a and b have the same remainder when divided by 3. Describe the equivalence classes for this relation.
9. A relation R is *antisymmetric* if $a \mathrel{R} b$ and $b \mathrel{R} a$ imply $a = b$. Thus, in an arrow diagram, no two distinct elements point to each other. Is the relation $\leq$ on $\mathbb{Z}$ antisymmetric? Is the relation $\subseteq$ on the subsets of a set A antisymmetric? Is the relation in review question 1 antisymmetric? Is the relation "divides" antisymmetric on $\mathbb{Z}^+$? Is "divides" antisymmetric on $\mathbb{Z}$? Is the relation $<$ antisymmetric on $\mathbb{Z}$?
10. List all the partitions of $\{1, 2\}$.

Answers to the Review Questions

1.

2.

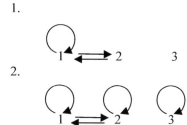

3. It is already symmetric.

4. It is already transitive. The corresponding partition is {{1, 2}, {3}}.
5. Yes, for all three questions.
6. Reflexive, symmetric, but not transitive (consider a born on Monday, b on Tuesday, and c on Wednesday)
7. Reflexive, symmetric, not transitive (consider $a = 1, b = 0, c = -1$)
8. [0] = set of integers that are a multiple of 3, [1] = set of integers that are 1 more than a multiple of 3, [2] = set of integers that are 2 more than a multiple of 3
9. $\leq$ and $\subseteq$ are antisymmetric. The relation in review question 1 is not antisymmetric (1 and 2 point to each other). "Divides" is antisymmetric on $\mathbb{Z}^+$, but not on $\mathbb{Z}$ (counterexample: $-2 \mid 2$ and $2 \mid -2$ but $-2 \neq 2$). $<$ is antisymmetric (it is vacuously true because there are no integers a and b such that $a < b$ and $b < a$).
10. {{1}, {2}} and {{1, 2}}.

Homework Questions

1. What is the equivalence relation (give its set of ordered pairs) that corresponds to the following partition: {{1, 2}, {3}, {4, 5, 6}}?
2. What is the partition that corresponds to the following equivalence relation on $A = \{1, 2, 3\}$: $\{(1, 1), (2, 2), (3, 3)\}$?
3. What is the partition that corresponds to the following equivalence relation on $A = \{1, 2\}$: $\{(1, 1), (2, 2), (1, 2), (2, 1)\}$?
4. If a set has n elements, how many partitions does that set have?
5. Let A be the set of points on a plane with the standard x and y axes. Let R be the equivalence relation defined by $a \, R \, b$ if and only if a and b are the same distance from the origin. Describe the equivalence classes of R.
6. Let A be the set of points on a plane with the standard x and y axes. Let R be the relation defined by $a \, R \, b$ if and only if a and b are *not* the same distance from the origin. Is R an equivalence relation?
7. Let R be a relation on $\mathbb{Z}^+ \times \mathbb{Z}^+$ defined by $(a, b) \, R \, (c, d)$ if and only if $ad = bc$. Show that R is an equivalence relation.
8. Is the power set of a set A a partition of A?
9. Let R be a relation on $\mathbb{R}$ defined by $a \, R \, b$ if and only if $a - b \in \mathbb{Z}$. Show that R is an equivalence relation. Describe its equivalence classes.
10. Define an equivalence relation on $\mathbb{R} \times \mathbb{R}$ whose equivalence classes correspond to the horizontal lines on a plane with the standard x and y axes.

12 Congruence

Basics

673

Congruence modulo n is a relation on the integers. We say ***a is congruent to b modulo n*** if and only if $n \mid (a - b)$, or, equivalently, $a - b$ is some multiple of n. That is, $a - b = kn$ for some $k \in \mathbb{Z}$. For example, 8 is congruent to 2 modulo 3 because $8 - 2$ is a multiple of the 3.

In a congruence modulo n, n is called the ***modulus*** of the congruence. The modulus can be any positive integer. We indicate that a is congruent to b modulo n by writing $a \equiv b \pmod{n}$. For example, to indicate that 8 is congruent to 2 modulo 3, we write $8 \equiv 2 \pmod{3}$.

Is $10 \equiv 4 \pmod{2}$?

674

Yes, because $10 - 4 = 6$ is a multiple of 2

Is $4 \equiv 10 \pmod{2}$?

675

Yes, because $4 - 10 = -6$ is a multiple of 2.

Is $10 \equiv 4 \pmod{3}$?

676

Yes, because $10 - 4 = 6$ is a multiple of 3.

Is $10 \equiv 4 \pmod{5}$?

677

No, because $10 - 4 = 6$ is not a multiple of 5.

Is $-12 \equiv 3 \pmod{5}$?

678

Yes, because $-12 - 3 = -15$ is a multiple of 5.

In mathematics, "mod" has two distinct uses:

1. To indicate the modulus in a congruence relation. For example, $18 \equiv 3 \pmod{5}$.

2. To indicate the remainder in a division operation. For example, 5 mod 3 = 2 (i.e., 5 divided by 2 yields a remainder of 2). When "mod" is used in this way, we call it the ***remainder operator***.

We use "mod" in both ways in the next section. Be careful to interpret each use of "mod" correctly. It is easy to distinguish the two: the first use of "mod" above always has a left parenthesis immediately preceding it; the second use never has a left parenthesis immediately preceding it.

Thinking about Congruence Modulo *n*

679

Let's investigate congruence modulo *n* using the approach a mathematician typically uses to investigate a new area of math. First, we try to figure out how "things work" using a variety of informal techniques such as guessing, intuition, and testing specific cases to see if a pattern emerges. Next, we formulate some conjectures based on our informal analysis. Finally, we prove or disprove our conjectures.

Consider the following number line on which multiples of *n* have been marked off. Suppose $(a \bmod n) = (b \bmod n) = r$. That is, *a* and *b* produce the same remainder, denoted by *r*, when divided by *n*. By the division algorithm, $a = nj + r$ and $b = nk + r$ for some $j, k \in \mathbb{Z}$:

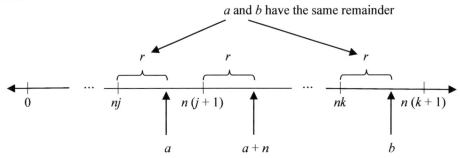

Suppose we start from *a* and move right *n* units. Because $a = nj + r$, moving *n* units to the right brings us to $(nj + r) + n = n(j + 1) + r$. Thus, we end up _____ units to the right of $n(j + 1)$.

680

r

Each time we move *n* units to the right, we end up *r* units to the right of the next multiple of *n*; *b* is *r* units to the right of some multiple of *n* (because we are given that *a* and *b* have the same remainder). Thus, we eventually end up on *b*. That means *a* and *b* differ by a multiple of *n* or, equivalently, $a \equiv b \pmod{n}$. From this informal analysis, it appears that equal remainders imply congruence. We can state this assertion symbolically:

$$(a \bmod n = b \bmod n) \Rightarrow a \equiv b \pmod{n}.$$

Conversely, suppose $a \equiv b \pmod{n}$ (i.e., *a* and *b* differ by some multiple of *n*). Let *r* be the remainder when *a* is divided by *n*. As we observed earlier, each time we move *n* units to the right starting from *a*, we end up *r* units to the right of the next multiple of *n*. We eventually land on *b* (because *a* and *b* differ by a multiple of *n*). Thus, *b* must be *r* units to the right of a multiple of *n*. But that means *b* has the remainder *r*—the same remainder as *a*.

Based on the preceding analysis, it appears that *a* and *b* have the same remainders when divided by *n* if and only if *a* and *b* are congruent modulo *n*. Stated symbolically, we get

$$(a \bmod n = b \bmod n) \text{ if and only if } a \equiv b \pmod{n}.$$

However, at this point, the preceding statement is still a conjecture we have yet to prove. Give a reason why the preceding analysis does not constitute a proof.

681

In the diagram in the preceding frames, *a* and *b* are not arbitrary integers: *a* is positive and $a < b$. Thus, our conclusion does not necessarily apply to all integers *a* and *b*. However, our analysis does provide us with some insight into congruence modulo *n* that a rigorous proof might not. Let's now prove our conjecture.

Suppose $a \bmod n = b \bmod n$. We want to show that $a \equiv b \pmod{n}$. Using the division algorithm, divide both *a* and *b* by *n*. Denote the common remainder by *r*.

682

$a = nq_a + r$, $b = nq_b + r$

Solve for *r* in both equations.

683

$r = a - nq_a$, $r = b - nq_b$, from which we get $a - nq_a = b - nq_b$.

Finish the proof. *Hint*: Solve for $a - b$.

684

Solving for $a - b$, we get $a - b = nq_a - nq_b = n(q_a - q_b) \Rightarrow n \mid (a - b) \Rightarrow a \equiv b \pmod{n}$.

Let's now prove the converse—namely, if $a \equiv b \pmod{n}$, then $a \bmod n = b \bmod n$. Using the division algorithm, divide a by n. Let q_a and r_a be the quotient and remainder, respectively. Similarly, divide b by n. Let q_b and r_b be the quotient and remainder, respectively.

685

$a = nq_a + r_a$, $b = nq_b + r_b$

Solve for $r_a - r_b$ and factor out n (we want to show that $r_a = r_b$ or, equivalently, that $r_a - r_b = 0$).

686

$r_a - r_b = (a - nq_a) - (b - nq_b) = (a - b) + nq_b - nq_a = (a - b) + n(q_b - q_a)$

We are given that $a \equiv b \pmod{n} \Rightarrow$ the difference between a and b is some multiple of n $\Rightarrow a - b = jn$ for some $j \in \mathbb{Z}$. Substituting jn for $(a - b)$, we get $r_a - r_b = (a - b) + n(q_b - q_a) = jn + n(q_b - q_a) = n(j + q_b - q_a)$. Thus, $r_a - r_b$ is a multiple of n.

What are the largest and smallest values possible for r_a and r_b? *Hint*: Consider the division algorithm.

687

By the division algorithm, $0 \leq r_a < n$, $0 \leq r_b < n$

What is the largest value possible for $r_a - r_b$?

688

$n - 1$ (when $r_a = n - 1$ and $r_b = 0$)

What is the smallest value possible for $r_a - r_b$?

689

$-(n - 1)$ (when $r_a = 0$ and $r_b = n - 1$)

What is the range of values possible for $r_a - r_b$?

690

$-(n - 1) \leq r_a - r_b \leq n - 1$

From frame **686**, we know that $r_a - r_b$ is a multiple of n. There is only one multiple of n in the range of values possible for $r_a - r_b$. What is it?

691

0

Thus, $r_a - r_b = 0 \Rightarrow r_a = r_b \Rightarrow a \bmod n = b \bmod n$. ∎

Key idea: a and b produce the same remainder when divided by n if and only if $a \equiv b \pmod{n}$.

Let's illustrate this key idea with an example: 10 and 16 have the same remainder when divided by 3. Thus, $10 \equiv 16 \pmod{3}$. Conversely, because $10 \equiv 16 \pmod{3}$, 10 and 16 have the same remainder when divided by 3.

136   Chapter 12: Congruence

Congruence Classes

692

Show that congruence modulo n is a reflexive relation. That is, show that $x \equiv x \pmod{n}$ for all $x \in \mathbb{Z}$. *Hint*: By D3 in frame **547**, $n \mid 0$.

693

For all $n \in \mathbb{Z}^+$, $n \mid 0 \Rightarrow n \mid (x - x) \Rightarrow x \equiv x \pmod{n}$. ∎

Show that congruence modulo n is a symmetric relation.

694

$x \equiv y \pmod{n} \Rightarrow x - y = kn$ for some $k \in \mathbb{Z} \Rightarrow -(x - y) = -(kn) \Rightarrow y - x = (-k)n \Rightarrow n \mid (y - x) \Rightarrow y \equiv x \pmod{n}$. ∎

Show that congruence modulo n is a transitive relation.

695

$x \equiv y \pmod{n}$ and $y \equiv z \pmod{n} \Rightarrow x - y = jn$, and $y - z = kn$ for some $j, k \in \mathbb{Z}$. Adding the corresponding sides of the two preceding equations, we get $x - z = jn + kn = (j + k)n \Rightarrow x \equiv z \pmod{n}$. ∎

Congruence modulo n is reflexive, symmetric, and transitive. Thus, it is an _____ relation.

696

equivalence

Congruence modulo n is an equivalence relation. We generally refer to its equivalence classes with the more specific name **congruence classes**. Recall from frame **662** that we denote the equivalence class to which p belongs with $[p]$. That is, $[p]$ is the set of elements that are related to p under the equivalence relation. For example, $[0]$, the congruence class (i.e., equivalence class) under congruence modulo 3, contains all the integers congruent to 0 modulo 3. Thus, $[0] = \{\ldots, -3, 0, 3, 6, \ldots\}$.

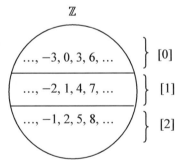

Modulo 3 congruence classes

Within each congruence class, each integer is congruent to every integer in that class. Thus, by frame **691**, *every integer within a congruence class yields the same remainder when divided by the modulus*. For example, every integer in the modulo 3 congruence class $[0] = \{\ldots, -3, 0, 3, 6, \ldots\}$ yields 0 on division by 3; every integer in $[1] = \{\ldots, -2, 1, 4, 7, \ldots\}$ yields 1 on division by 3; every integer in $[2] = \{\ldots, -1, 2, 5, 8, \ldots\}$ yields 2 on division by 3.

If $a \equiv b \pmod{n}$, then a and b are in the same "clump" induced by the modulo n equivalence relation. That is, $[a] = [b]$. Conversely, if $[a] = [b]$ (i.e., a and b are in the same "clump"), then a and b must be related under the modulo n equivalence relation. That is, $a \equiv b \pmod{n}$.

Key idea: $a \equiv b \pmod{n}$ if and only if $[a] = [b]$.

What integers are in the modulo 5 congruence class $[1]$?

697

The set of integers congruent to 1 modulo 5 = the set of integers that differ from 1 by some multiple of 5 = $\{5k + 1 : k \in \mathbb{Z}\}$ = $\{\ldots, -4, 1, 6, 11, \ldots\}$.

Using the division algorithm, represent the division of a by n.

698

$a = nq + r$, where $0 \leq r < n$

The integers a and r differ by qn, which is a multiple of n. Thus, $a \equiv r \pmod{n}$. For example, 10 divided by 4 produces the remainder 2. Thus, $10 \equiv 2 \pmod{4}$.

Key idea: Every integer is congruent to its remainder. More precisely, for all integers a and positive n, a divided by n yields a remainder that is congruent to a modulo n.

What is the set of remainders possible in a division by n?

699

$\{0, 1, ..., n-1\}$

By the preceding frame, every integer is congruent modulo n to its remainder on division by n. The set of possible remainders on division by n is $\{0, 1, ..., n-1\}$. Thus, every integer is congruent to some element in $\{0, 1, ..., n-1\}$. This, in turn, implies that every integer must be in one of the modulo n congruence classes $[0], [1], ..., [n-1]$.

Is it possible that some of the congruence classes among $[0], [1], ..., [n-1]$ might be the same? That is, might $[i] = [j]$ for some $i, j \in \{0, 1, ..., n-1\}$, where $i \neq j$? If this were the case, then i would be congruent to j. But two distinct remainders cannot be congruent. To see this, note that the absolute value of the difference between two distinct remainders is at least 1 (because they are distinct) and at most $n-1$ (when one is equal to $n-1$ and the other is equal to 0). Thus, they cannot differ by a multiple of n, and, therefore, cannot be congruent. We conclude that $[i]$ and $[j]$ are distinct congruence classes if $i \neq j$.

From the preceding analysis, we can conclude that every integer is in one and _____ one of the modulo n congruence classes $[0], [1], ..., [n-1]$.

700

only

Key idea: Under congruence modulo n, every integer is in exactly one of the congruence classes in $\{[0], [1], ..., [n-1]\}$.

We typically use $[0], [1], ..., [n-1]$ for the modulo n congruence classes. Zero through $n-1$ are the smallest nonnegative representatives of the n congruence classes, so generally they are the most convenient representatives to use. However, we could certainly use other representatives. For example, instead of the representatives 0, 1, and 2 for the modulo 3 congruence classes, we could use -1, 0, and 1. That is, we could denote the three congruence classes with $[-1], [0]$, and $[1]$. The congruence classes $[-1]$ and $[2]$ are equal because $-1 \equiv 2 \pmod{3}$. Thus, $\{[-1], [0], [1]\}$ is the same set of congruence classes as $\{[0], [1], [2]\}$.

List all the congruence classes under congruence modulo 7.

701

$[0], [1], [2], [3], [4], [5]$, and $[6]$

Replacement Rules

702

Suppose $b \equiv c \pmod{n}$. Then b and c are separated on the number line by some multiple of n units. If we add a to b, we move a units from b on the number line in either the positive or negative direction, depending on the sign of a. If we also add a to c, we move a units from c in the same direction. Thus, the difference between $a+b$ and $a+c$ is the same as the difference between b and c. The difference between b and c is a multiple of n (because we are given that $b \equiv c \pmod{n}$). Thus, the difference between $a+b$ and $a+c$ is also a multiple of n. That is, $a+b \equiv a+c \pmod{n}$. State this result as an implication.

703

$b \equiv c \pmod{n} \Rightarrow a+b \equiv a+c \pmod{n}$ or, equivalently, $b \equiv c \pmod{n} \Rightarrow [a+b] = [a+c]$.

We have a similar rule for multiplication:

$b \equiv c \pmod{n} \Rightarrow ab \equiv ac \pmod{n}$ or, equivalently, $b \equiv c \pmod{n} \Rightarrow [ab] = [ac]$.

Prove this multiplication rule.

704

$b \equiv c \pmod{n}$	
$\Rightarrow b - c = kn$ for some $k \in \mathbb{Z}$	definition of congruence modulo n
$\Rightarrow a(b - c) = a(kn)$	multiply left and right sides of equation by a
$\Rightarrow ab - ac = (ak)n$	distributive and associative laws
$\Rightarrow ab \equiv ac \pmod{n}$	definition of congruence modulo n ∎

The implications in the preceding frame can be viewed as replacement rules—the first one for addition, the second for multiplication. The addition rule tells us that if $b \equiv c \pmod{n}$, we can replace b with c in $[a + b]$ *without affecting the congruence class specified.* That is, if $b \equiv c \pmod{n}$, then $[a + b] = [a + c]$. For example, in the modulo 5 congruence class $[2 + 4]$, we can replace 4 with 9 because $4 \equiv 9 \pmod{5}$. The modification does not affect the congruence class specified. That is, $[2 + 4] = [2 + 9]$. The multiplication rule allows a similar replacement. For example, we can replace 4 with 9 in $[2 \cdot 4]$ without affecting the congruence class specified. That is, $[2 \cdot 4] = [2 \cdot 9]$.

By frame **698**, every integer is congruent modulo n to its remainder on division by n. Thus, the preceding replacement rules *allow us to replace an integer with its remainder*. For example, in the modulo 5 congruence class $[101 + 2]$, we can replace 101 with 1 (the remainder when 101 is divided by 5) without affect the congruence class specified. Thus, $[101 + 2] = [1 + 2]$.

Cancellation Laws

705

If $a + c \equiv b + c \pmod{n}$, we can cancel the c's. That is, the following cancellation law for addition holds:

If $a + c \equiv b + c \pmod{n}$, then $a \equiv b \pmod{n}$.

Show that this law holds using the same type of reasoning we used in frame **702**. *Hint*: What is the movement on the number line corresponding to subtracting c from $a + c$ and from $b + c$.

706

To go from $a + c$ to a on the number line, we move c units on the number line. Similarly, to go from $b + c$ to b, we move c units on the number line in the same direction. Thus, the distance between $a + c$ and $b + c$ is the same as the distance between b and c ⇒ if the distance between $a + c$ and $b + c$ is a multiple of n so is the distance between a and b ⇒ the cancellation in the preceding frame.

Although we have a cancellation law for addition, the corresponding cancellation for multiplication does *not* hold. That is, $ac \equiv bc \pmod{n}$ does *not* imply $a \equiv b \pmod{n}$.

What is wrong with the following "proof" of the cancellation law for multiplication?

1. $ac \equiv bc \pmod{n}$
2. $\Rightarrow ac - bc = kn$ for some $k \in \mathbb{Z}$ definition of congruence
3. $\Rightarrow (a - b)c = kn$ factoring c on left side
4. $\Rightarrow n \mid (a - b)c$ definition of "divides"
5. $\Rightarrow n \mid (a - b)$ Euclid's lemma
6. $\Rightarrow a \equiv b \pmod{n}$ definition of congruence

---

707

Statement 4 does not imply statement 5. If n is prime, and $n \mid (a - b)c$, then by Euclid's lemma, $n \mid (a - b)$ or $n \mid c$. Thus, n does not necessarily divide $a - b$. Moreover, n here is not necessarily prime, so we cannot even apply Euclid's lemma. However, if c and n are relatively prime, then by the relatively prime divisor theorem in frame **610**, statement 4 would imply statement 5. Thus, the general cancellation law for multiplication does not hold, but the following restricted law does:

If $ac \equiv bc \pmod{n}$ and c and n are relatively prime, then $a \equiv b \pmod{n}$.

For example, $2 \cdot 3 \equiv (22 \cdot 3) \pmod{10}$; 3 and 10 are relatively prime. Thus, we can cancel the 3's to get $2 \equiv 22 \pmod{10}$.

$12 \cdot 3 \equiv (8 \cdot 3) \pmod 6$. Can we cancel the 3's?

708

No; 3 and 6 are not relatively prime.

The general cancellation law does not hold for multiplication. This means that there is at least one set of values of a, b, c, and n for which we cannot cancel the c's in the equation $ac \equiv bc \pmod{n}$. It does *not* mean that for *every* set of values of a, b, c, and n, we cannot cancel the c. If c and n are relatively prime, we can *always* cancel. If c and n are not relatively prime, we may or may not be able to cancel, depending on the values of a, b, c, and n.

$2 \cdot 3 \equiv (8 \cdot 3) \pmod 6$. Can we cancel the 3's?

709

In this case we can, even though 3 and 6 are not relatively prime.

Quotient Structure under Congruence Modulo *n*

710

Congruence modulo *n* breaks up the integers into *n* congruence classes. We can define addition on these congruence classes by specifying a procedure for determining the sum of any two congruence classes. Here is the procedure:

> To add two congruence classes, take an arbitrary element from each of the two classes. Suppose the two elements are a and b. Compute $a + b$. Determine the congruence class to which $a + b$ belongs. This congruence class is the result of the addition operation.

More simply stated, $[a] + [b] = [a + b]$. This procedure is quite simple. For example, consider the congruence classes under congruence modulo 3: [0], [1], and [2]. To determine $[1] + [2]$, take an arbitrary element from [1] (let's use 1) and an arbitrary element from [2] (let's use 2). $1 + 2 = 3$. Thus, $[1] + [2] = [1 + 2] = [3]$. $[3] = [0]$. Thus, $[1] + [2] = [0]$. Alternatively, we can use the representatives 7 and 14 as the representatives of [1] and [2], respectively. We get $[1] + [2] = [7 + 14] = [21]$. $[21] = [0]$. Thus, again $[1] + [2] = [0]$.

Suppose $[a] = [c]$ and $[b] = [d]$. Then a is in the same congruence class as c. Similarly, b is in the same congruence class as d. The result of the addition of two congruence classes should not depend on the representatives we choose. Thus, $[a] + [b] = [a + b]$ should equal $[c] + [d] = [c + d]$. By our replacement rule for addition in frame **703**, we can replace a with c and b with d in $a + b$ without affecting the congruence class of the result. Thus, $[a + b] = [c + d]$. Using a and b as representatives in congruence class addition gives the same result as using c and d. If this were not the case, then our addition procedure for congruence classes defined earlier would not define a binary operation. But our addition procedure does, indeed, yield a unique result—the result does not depend on the representatives we choose. Accordingly, we say that the addition of congruence classes is ***well defined***.

Multiplication of the modulo *n* congruence classes is defined as follows: $[a] \cdot [b] = [ab]$. For example, under congruence modulo 3, $[2] \cdot [2] = [2 \cdot 2] = [4]$, and $[4] = [1]$. Thus, $[2] \cdot [2] = [1]$. Like addition, the multiplication of congruence classes is well defined. That is, the result of multiplication does not depend on the representatives we choose (this follows from the two applications of the replacement rule for multiplication in frame **703**).

Construct the Cayley table for the modulo 3 congruence classes under addition and multiplication. Represent the congruence classes with [0], [1], and [2].

711

·	[0]	[1]	[2]
[0]	[0]	[0]	[0]
[1]	[0]	[1]	[2]
[2]	[0]	[2]	[1]

+	[0]	[1]	[2]
[0]	[0]	[1]	[2]
[1]	[1]	[2]	[0]
[2]	[2]	[0]	[1]

Are the modulo 3 congruence classes under multiplication a group?

712

140   Chapter 12: Congruence

No. In a Cayley table for a group, each row is a permutation of its column labels, and each column is a permutation of its row labels. The Cayley table for multiplication in the preceding frame does not have this property.

Quotient Group of Congruence Classes

713

What are the four requirements for a set and a binary operation to be a group?

714

Closure, associativity, identity, and inverses

Let's show that the set of modulo n congruence classes forms a group under addition. First, show that the modulo n congruence classes are closed under addition.

715

Let $[a]$ and $[b]$ be two modulo n congruence classes. Then, by definition, $[a] + [b] = [a + b]$. $[a + b]$ is a modulo n congruence class. Thus, the set of congruence classes is closed under addition.

Show that addition on modulo n congruence classes is associative. *Hint*: Create a sequence of equalities that starts with $([a] + [b]) + [c]$ and ends with $[a] + ([b] + [c])$.

716

$([a] + [b]) + [c]$	
$= [a + b] + [c]$	definition of congruence class addition
$= [(a + b) + c]$	definition of congruence class addition
$= [a + (b + c)]$	associativity of addition on $\mathbb{Z}$
$= [a] + [b + c]$	definition of congruence class addition
$= [a] + ([b] + [c])$	definition of congruence class addition

What is the identity element in the set of modulo n congruence classes under addition?

717

$[0] + [a] = [0 + a] = [a]$. $[a] + [0] = [a + 0] = [a]$. Thus, $[0]$ is the identity element.

Show that every congruence class has an inverse under addition.

718

$[a] + [-a] = [a + (-a)] = [0]$. $[-a] + [a] = [-a + a] = [0]$. Thus, $[-a]$ is the inverse of $[a]$.

The set of modulo n congruence classes under addition has the closure, associativity, identity, and inverses properties. It is a group. ∎

What is the inverse of the modulo 100 congruence class $[70]$ under addition?

719

$[30]$ because $[70] + [30] = [100] = [0]$. The inverse is also $[-70]$ because $[70] + [-70] = [0]$.

What is the inverse of the modulo n congruence class $[x]$ under addition?

720

$[n - x]$ and $[-x]$.

Because the set of modulo n congruence classes under addition have the properties of closure, associativity, identity, and inverses, it is a group under addition. Thus, this set necessarily has all the properties a group has. For example, the left and right cancellation laws hold for any group. Thus, they also hold for modulo n congruence classes under addition. For example, consider the following equation of modulo 7 congruence classes: [3] + [6] = [3] + [13]. We can cancel [3] on both sides to get [6] = [13].

Does the following equation have a unique solution, where [5] and [3] are modulo 7 congruence classes: [5] + x = [3]?

721

Yes. In a group, equations of this form have a unique solution (see frame **438**).

Determine the modulo 7 congruence class equal to x in the preceding frame.

722

Using the switching-sides rule (see frame **440**), we get $x = -[5] + [3] = [-5] + [3] = [-5 + 3] = [-2]$.

Show that [−2] is a solution by substituting [−2] for x in the original equation.

723

We get [5] + [−2] = [5 − 2] = [3]. Thus, the modulo 7 congruence class [−2] is a solution.

The modulo 7 congruence class [5] is also a solution: [5] + [5] = [10] = [3]. But according to group theory, the equation should have a unique solution. What is wrong here?

724

The equation does have only one solution: [−2] and [5] are the same congruence class under congruence modulo 7.

The set of modulo n congruence classes under addition is a ***quotient structure.*** Its elements are obtained from $\mathbb{Z}$ by dividing $\mathbb{Z}$ into congruence classes (hence use of the term *quotient*). It has structure imposed on it by the addition operation (hence use of the term *structure*). This quotient structure is special because it is also a group. Accordingly, we call it a ***quotient group*** rather than the more general "quotient structure."

We observed in frame **712** that the set of modulo n congruence classes under multiplication is not a group. Which of the four group properties—closure, associativity, identity, inverses—does it have? Which does it not have?

725

It has closure (follows from the definition of multiplication). It has associativity (follows from the associativity of multiplication on $\mathbb{Z}$). [1] is the identity ([1] · [a] = [a] · [1] = [a]). But it does not have the inverses property.

Show that [0] has no inverse.

726

[0] · [x] = [$0x$] = [0] for all x. Thus, there is no congruence class [x] such that [0] · [x] = [1].

Our argument that [0] has no inverse has one flaw: we are assuming something about our set of congruence classes that is not always true. What is this faulty assumption?

727

We are assuming that there is a congruence class [1] distinct from [0]. But under congruence modulo 1, there is only one congruence class.

Is $a \equiv b \pmod{1}$ for all $a, b \in \mathbb{Z}$?

728

Yes, because $1 \mid (a - b)$ for all $a, b \in \mathbb{Z}$. Thus, under congruence modulo 1, every element in $\mathbb{Z}$ is congruent to every element in $\mathbb{Z}$.

Describe the congruence classes under congruence modulo 1.

729

The integers in $\mathbb{Z}$ form a single "clump." $\mathbb{Z}$ itself is the only congruence class.

Under congruence modulo 1, does [0] = ℤ?

730

Yes. [0] is the set of all the integers that differ from 0 by some multiple of 1. This set is ℤ.

Under congruence modulo 1, does [1] = ℤ?

731

Yes. [1] is the set of all the integers that differ from 1 by some multiple of 1, which is the set ℤ.

Under congruence modulo 1, does $[k]$ = ℤ for every $k \in$ ℤ?

732

Yes. $[k]$ contains all the integers that differ from k by some multiple of 1, which is the set ℤ. To represent the single congruence class under congruence modulo 1, let's use [0].

Construct the Cayley table for the set of modulo 1 congruence classes under multiplication. Use [0] to represent its single element.

733

·	[0]
[0]	[0]

This is the Cayley table for the cyclic group of order 1.

Is the following statement true: for all positive n, the set of modulo n congruence classes under multiplication is not a group?

734

No. It is not true for n = 1.

Qualify the statement in the preceding frame so that it is true.

735

For all $n > 1$, the set of modulo n congruence classes under multiplication is not a group.

Modular Arithmetic

736

Recall that x mod n is the remainder produced when x is divided by n. Using this notation, we define addition modulo n (denoted by $+_n$) as follows:

$$a +_n b = (a + b) \bmod n, \text{ where } n \in \mathbb{Z}^+.$$

That is, $a +_n b$ is the remainder produced when $a + b$ is divided by n. For example, $2 +_3 5 = (2 + 5)$ mod $3 = 7$ mod $3 = 1$.

In modular arithmetic, we use the regular addition and multiplication operations on the integers. However, at any step in which a value is less than zero or greater than or equal to the modulus, we replace that value with its remainder. For example, we get the value of $2 +_5 4 +_5 9$ by computing $[(2 + 4) \bmod 5 + (9 \bmod 5)] \bmod 5$. We add 2 and 4 to get 6; 6 is greater than or equal to the modulus 5, so we replace 6 with its remainder, 1; 9 is also greater than or equal to the modulus, so we replace 9 with its remainder, 4. We then add 1 and 4 to get 5. Finally, we replace 5 by its remainder to get 0.

When we perform modular arithmetic, we replace values with their remainders as we perform the computation. Alternatively, we can take the remainder only once, at the very end of the computation. Either way, we get the same result. For example, we can get the value of $2 +_5 4 +_5 9$ by computing $(2 + 4 + 9)$ mod 5. We add 2, 4, and 9 to get 15 and then take the remainder produced when 15 is divided by 5. We get 0.

In modular arithmetic, replacing a value with its remainder does not affect the congruence class of the result (we know this from frame **703**). Thus, the two approaches—taking remainders during the computation or only once at the end—produce final answers that are congruent. Moreover, because the final answer with either approach is a remainder, and distinct remainders cannot be congruent, the final answers with the two approaches must be equal.

What is the value of $2 +_7 14 +_7 10$?

737

The integers 14 and 10 are greater than or equal to the modulus 7. So we replace them with their remainders 0 and 3, respectively. We then add 2, 0, and 3 to get 5; 5 is less than the modulus 7, so 5 is our answer. Alternatively, we can compute $(2 + 14 + 10) \bmod 7 = 26 \bmod 7 = 5$.

What is the value of $10 +_7 10$?

738

$10 +_7 10 = 3 +_7 3 = 6$

Show that addition modulo n is associative.

739

$(a +_n b) +_n c = ((a + b) + c) \bmod n = (a + (b + c)) \bmod n = a +_n (b +_n c)$

Multiplication modulo n (denoted by $\cdot_n$) is defined as follows:

$$a \cdot_n b = (ab) \bmod n.$$

That is, $a \cdot_n b$ is the remainder produced when ab is divided by n. For example, $2 \cdot_3 5 = (2 \cdot 5) \bmod 3 = 10 \bmod 3 = 1$. As with modular addition, we can compute the value of a modular multiplication by replacing values with their remainders either during the computation or at the end. Compute the value of $100 \cdot_3 5$ by first replacing 100 by its remainder on division by 3.

740

$100 \cdot_3 5 = 1 \cdot_3 5 = 5 \bmod 3 = 2$

Show that multiplication modulo n is associative.

741

$(a \cdot_n b) \cdot_n c = ((ab)c) \bmod n = (a(bc)) \bmod n = a \cdot_n (b \cdot_n c)$

When performing computations in modular arithmetic, it is sometimes advantageous to determine remainders at every step rather than at the end. For example, which approach is easier for $8 \cdot_7 8 \cdot_7 8 \cdot_7 8 \cdot_7 8 \cdot_7 8 \cdot_7 8 \cdot_7 8 \cdot_7 8 \cdot_7 8$?

742

Computing 8^{10} and then dividing by 7 requires more work than replacing 8 with 1 (the remainder when 8 is divided by 7) and then computing 1^{10}.

Show that $a \cdot_n b = ab + kn$ for some $k \in \mathbb{Z}$. *Hint*: Use frame **698**.

743

$a \cdot_n b$
$= ab \bmod n$ definition of multiplication modulo n
$\equiv ab \pmod{n}$ frame **698**

Because $a \cdot_n b$ is congruent modulo n to ab, $a \cdot_n b$ and ab differ by some multiple of n. That is, $a \cdot_n b = ab + kn$ for some $k \in \mathbb{Z}$.

Group of Integers Modulo *n*

744

The set of possible remainders when an integer is divided by the positive integer n is $\{0, 1, ..., n-1\}$. We denote this set with $\mathbb{Z}_n$. For any modulus n, $\mathbb{Z}_n$ under the operation $+_n$ is a group. We call it the ***group of integers modulo n.***

For example, $\mathbb{Z}_3 = \{0, 1, 2\}$ under $+_3$ is a group. Construct its Cayley table.

745

$+_3$	0	1	2
0	0	1	2
1	1	2	0
2	2	0	1

Prove that $\langle \mathbb{Z}_n, +_n \rangle$ is a group for all $n \geq 1$.

746

Closure: By definition, $+_n$ always produces a result that is one of the remainders on division by n. $\mathbb{Z}_n$ is the set of all these remainders. Thus, $a +_n b \in \mathbb{Z}_n$ for $a, b \in \mathbb{Z}_n$.
Associativity: $(a +_n b) +_n c = ((a + b) + c) \bmod n = (a + (b + c)) \bmod n = a +_n (b +_n c)$.
Identity: $0 +_n a = a +_n 0 = a$ for all $a \in \mathbb{Z}_n$; 0 is the identity.
Inverses: Let $a \in \mathbb{Z}_n$. If $a = 0$, then a is its own inverse. If $a > 0$, then $0 < n - a < n$, which implies $n - a \in \mathbb{Z}_n$. We then have that $a +_n (n - a) = (n + a - a) \bmod n = n \bmod n = 0$. Similarly, $(n - a) +_n a = 0$. Thus, for $a > 0$, the inverse of a is $n - a$. ∎

In $\langle \mathbb{Z}_{10}, +_{10} \rangle$, what is the inverse of 3.

747

$10 - 3 = 7$.

Isomorphic Groups

748

Compare the Cayley table for $\mathbb{Z}_4$ under modular addition with the Cayley table for the modulo 4 congruence classes under addition as defined in frame **710**:

$\mathbb{Z}_4$

$+_4$	0	1	2	3
0	0	1	2	3
1	1	2	3	0
2	2	3	0	1
3	3	0	1	2

Modulo 4 congruence classes

+	[0]	[1]	[2]	[3]
[0]	[0]	[1]	[2]	[3]
[1]	[1]	[2]	[3]	[0]
[1]	[2]	[3]	[0]	[1]
[1]	[3]	[0]	[1]	[2]

The two tables are essentially the same. Thus, these two groups are essentially the same. We need to use the word "essentially" here because the two groups are not identical. The two sets are different: $\{0, 1, 2, 3\}$ is not the same set as $\{[0], [1], [2], [3]\}$. Each element in the former is an integer; each element in the latter is an infinite set of integers.

However, the two groups have the same size and structure as defined by their Cayley tables. Suppose we replace the elements of the Cayley table for $\mathbb{Z}_4$ according to the following bijection: $0 \to [0], 1 \to [1], 2 \to [2], 3 \to [3]$. That is, we replace 0, 1, 2, and 3 with [0], [1], [2], and [3], respectively. We then get the Cayley table for the modulo 4 congruence classes. If we do the reverse replacement—replace the elements in the Cayley table for the modulo 4 congruence classes according to the inverse of the given bijection—we get the Cayley table for $\mathbb{Z}_4$. Thus, the two tables differ only in the names of the elements.

We say that a group G_1 is ***isomorphic*** to a group G_2 (denoted by $G_1 \cong G_2$) if the Cayley table for G_1 can be converted to the Cayley table for G_2 by renaming the elements of G_1 according to some bijection from G_1 to G_2. A bijection that converts the Cayley table for G_1 to the Cayley table for G_2 is called an ***isomorphism*** from G_1 to G_2.

If φ is an isomorphism from G_1 to G_2, then φ^{-1} is an isomorphism from G_2 to G_1. In other words, isomorphism is a symmetric relation on the set of all groups. Moreover, isomorphism is also a reflexive and transitive relation. Thus, it is an equivalence relation. Like all equivalence relations, it corresponds to a partition. In each block of this partition, all the groups in that block are isomorphic to each other.

The bijection given earlier converts the Cayley table for $\mathbb{Z}_4$ to the Cayley table for the modulo 4 congruence classes. Thus, the corresponding groups are isomorphic. The general case also holds. That is, $\mathbb{Z}_n$ is isomorphic to the group of modulo n congruence classes.

The groups corresponding to the two preceding Cayley tables are not only isomorphic; they have Cayley tables that are constructed using *exactly* the same computation. To compute $a +_n b$, we add a and b and take the remainder. Similarly, to compute $[a] + [b]$, we add a and b to get the congruence class $[a + b]$. But we then represent this congruence class in the Cayley table using the remainder of $a + b$ as its representative. Thus, in both cases, we add a and b and take the remainder. The remainder represents the result.

Because $\mathbb{Z}_n$ and the modulo n congruence classes parallel each other so closely, it is not surprising that mathematicians often do not distinguish between the two groups and denote both with $\mathbb{Z}_n$. However, in this book, we use $\mathbb{Z}_n$ strictly to denote the integers modulo n. That is, $\mathbb{Z}_n = \{0, 1, \ldots, n-1\}$.

Now let's turn our attention to a group that, at first, appears quite different from $<\mathbb{Z}_4, +_4>$. Recall from high school algebra that i represents the imaginary number $\sqrt{-1}$. Thus, $i \cdot i = \sqrt{-1}\sqrt{-1} = -1$. The set $U_4 = \{1, i, -1, -i\}$ under multiplication is a group. The fourth power of any element in U_4 equals 1:

$$1 \cdot 1 \cdot 1 \cdot 1 = 1,$$
$$i \cdot i \cdot i \cdot i = (i \cdot i) \cdot (i \cdot i) = (-1) \cdot (-1) = 1,$$
$$(-1) \cdot (-1) \cdot (-1) \cdot (-1) = [(-1) \cdot (-1)] \cdot [(-1) \cdot (-1)] = 1 \cdot 1 = 1,$$
$$(-i) \cdot (-i) \cdot (-i) \cdot (-i) = [(-i) \cdot (-i)] \cdot [(-i) \cdot (-i)] = (i \cdot i) \cdot (i \cdot i) = (-1) \cdot (-1) = 1.$$

Thus, each element is a fourth root of 1. For this reason, we call U_4 the ***fourth roots of unity***. Do not confuse this set with $U(4)$, the set of positive integers less than 4 that are relatively prime to 4 (see frame **613**).

Construct the Cayley table for U_4 under multiplication. Also construct the Cayley table for $\mathbb{Z}_4$ under addition. Use row and column labels listed in 1, i, -1, and $-i$ order.

749

$\cdot$	1	i	-1	$-i$
1	1	i	-1	$-i$
i	i	-1	$-i$	1
-1	-1	$-i$	1	i
$-i$	$-i$	1	i	-1

U_4

$+_4$	0	1	2	3
0	0	1	2	3
1	1	2	3	0
2	2	3	0	1
3	3	0	1	2

$\mathbb{Z}_4$

The two tables have the same size, but do they also have the same structure? Are these groups isomorphic? Rename the elements in the Cayley table for U_4 according to the following bijection: $1 \to 0$, $i \to 1$, $-1 \to 2$, $-i \to 3$. Do we get the Cayley table for $\mathbb{Z}_4$?

750

Yes. Thus, these groups are isomorphic.

Are U_4, $\mathbb{Z}_4$, and the modulo 4 classes all isomorphic to each other?

751

Yes

Review Questions

1. Using set-builder notation, specify the elements in the modulo 20 congruence class [7].
2. Solve for x in this modulo 7 equation: $[3] + x = [-1]$.
3. Solve for x in this modulo 5 equation: $[3] + x = [3] + [4]$.
4. Determine the elements of $\mathbb{Z}_4$ that have a multiplicative inverse.
5. What is the additive inverse of [5] in the group of the modulo 8 congruence classes?
6. Solve for x in $3x \equiv 33 \pmod{14}$. *Hint*: Factor 33 and then use the left cancellation law.
7. Solve for x in $3x \equiv 33 \pmod{6}$. *Hint*: There are three solutions.

8. If $a \equiv c$ (mod n) and $b \equiv d$ (mod n), are the following statements true: $a + b \equiv (c + d)$ (mod n) and $ab \equiv (cd)$ (mod n).
9. Determine 6^2 mod 7. Determine 6^{1000} mod 7.
10. Show that if $n \mid m$ and $a \equiv b$ (mod m), then $a \equiv b$ (mod n).

Answers to Review Questions

1. $\{7 + 20k : k \in \mathbb{Z}\}$
2. Using the switching-sides rule for groups (see frame **440**), we get $x = -[3] + [-1] = [-3] + [-1] = [-4] = [3]$.
3. Apply the left cancellation law for groups to get $x = [4]$.
4. $1 \cdot_4 1 = 1, 3 \cdot_4 3 = 1$.
5. $[5] + [3] = [8] = [0]$. Thus, $[3]$ is the additive +inverse of $[5]$.
6. Write 33 as $3 \cdot 11$ and then cancel the 3's to get $x \equiv 11$ (mod 14). We can cancel the 3's because 3 and 14 are relatively prime.
7. 1, 3, and 5.
8. Both are true. Use the replacement rules from frame **703**: replace a with c and b with d.
9. Because -1 and 6 have the same remainder when divided by 7, we can replace 6 with -1 without affecting the result. Thus, 6^2 mod $7 = (-1)^2$ mod $7 = 1$ mod $7 = 1$. 6^{1000} mod $7 = (-1)^{1000}$ mod $7 = 1$ mod $7 = 1$.
10. $n \mid m \Rightarrow m = nk$ for $k \in \mathbb{Z}$. $a \equiv b$ (mod m) $\Rightarrow a = b + mj$ for $j \in \mathbb{Z}$. Substituting, we get $a = b + mj = b + n(kj) \Rightarrow a \equiv b$ (mod n).

Homework Questions

1. Determine the elements of $\mathbb{Z}_5$ that have a multiplicative inverse.
2. Show: $x \equiv y$ (mod n) $\Rightarrow \gcd(x, n) = \gcd(y, n)$. *Observation we can make based on this problem*: Suppose the gcd of n and some element in a modulo n congruence class is equal to k. Then the gcd of n and *any* element in that congruence class is also equal to k. That is, all the elements within any one modulo n congruence class have the same gcd (as well as the same remainder on division by n).
3. Show that $10^n \equiv 1$ (mod 9) for all $n \in \mathbb{Z}^+$. *Hint*: 10^n consists of a 1 followed by n zeros.
4. Find $a, b \in \mathbb{Z}$ and $n \in \mathbb{Z}^+$ such that $a^2 \equiv b^2$ (mod n) and $a \not\equiv b$ (mod n).
5. Is $\mathbb{Z}_3$ under multiplication a group? Justify your answer.
6. Is $\mathbb{Z}_3 - \{0\}$ under multiplication a group? Justify your answer.
7. Is $\mathbb{Z}_4 - \{0\}$ under multiplication a group? Justify your answer.
8. Does $j \equiv k$ (mod n) imply $a^j \equiv a^k$ (mod n) for all $a \in \mathbb{Z}$? Justify your answer.
9. Does $a \equiv b$ (mod n) imply $a^i \equiv b^i$ (mod n) for all $a, b \in \mathbb{Z}$, $i \in \mathbb{Z}^+$? Justify your answer.
10. An isomorphism between two groups is not necessarily unique. Show that there are two isomorphisms between the integers modulo 3 and the modulo 3 congruence classes.

13 Symmetries of a Regular Polygon

Symmetries of an Equilateral Triangle

In chapter 8, we investigated an unusual group—a group whose elements are permutations on the set consisting of the integers from 1 to n. In this chapter, we investigate another unusual group. For our investigation, you will need a triangle with which to experiment. On a piece of paper, draw a picture of an equilateral triangle (i.e., a triangle whose sides are of equal length). Label the vertices in counterclockwise order, starting with 1 on the upper vertex, as shown below. Write a T (for "top") in the middle of the triangle:

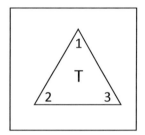

Next, cut out the triangle. Flip the triangle, keeping vertex 1 the upper vertex, as shown below. On the flip side, write a B (for "bottom"), and label each vertex of the triangle in clockwise order, as illustrated. Each vertex then will be labeled with the same number on each side. For example, the vertex labeled with "2" on the T side should also be labeled with "2" on the B side.

We now consider motions of the triangle that position the triangle so that its edges are aligned with the edges of the cutout area. In such an orientation, the triangle can be placed back into the cutout area. We call these motions **symmetries.** For example, if we rotate the triangle 120° clockwise, the triangle in its new position can be placed back into the cutout area. Thus, this motion is a symmetry. In this section, we restrict ourselves to symmetries that *start and end with the T side up*.

Remove the triangle from its original orientation in the cutout area, rotate it 120° clockwise, and place it back into the cutout area:

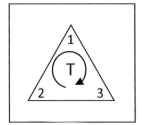 *r* motion ⟹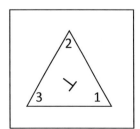

148   Chapter 13: Symmetries of a Regular Polygon

The change in the orientation of the T shows the effect of this motion. Note that in the "after" picture above, we show the vertex numbers in the upright position (to make them easier to read). Of course, if the vertex numbers are actually written on the triangle, then they, like the T, would change orientation when the triangle is rotated. We can summarize the effect of this motion with the following two-row, three-column table: $\begin{pmatrix} 1 & 2 & 3 \\ 2 & 3 & 1 \end{pmatrix}$. The first column indicates that the vertex 1 is "replaced by" the vertex 2. That is, the position occupied by vertex 1 before the rotation is occupied by vertex 2 after the rotation. The second and third columns provide the corresponding information for vertices 2 and 3, respectively.

Using a table, represent a clockwise rotation of 480°. *Hint*: 480 = 360 + 120.

753

$\begin{pmatrix} 1 & 2 & 3 \\ 2 & 3 & 1 \end{pmatrix}$ This table defines the *r* motion.

A 480° rotation is one full circle plus 120°. Thus, its net effect is the same as a 120° rotation. We are interested in only the net effect of motions. Thus, we view the set of all motions that have the same net effect as a single motion. We call any motion whose net effect is described by the preceding table an ***r* motion** (*r* is for "rotation").

What is the table representation of a 240° clockwise rotation?

754

$\begin{pmatrix} 1 & 2 & 3 \\ 3 & 1 & 2 \end{pmatrix}$ This table defines the r^2 motion.

A net 240° clockwise rotation has the same net effect as two *r* motions. Accordingly, we call a 240° clockwise rotation an r^2 motion.

What is the table representation of a 0° clockwise rotation?

755

$\begin{pmatrix} 1 & 2 & 3 \\ 1 & 2 & 3 \end{pmatrix}$ This table defines the *e* motion.

We call this motion an *e* motion. Now consider the set of motions $\{e, r, r^2\}$ under the operation "followed by." For example, *r* followed by *r* is the motion consisting of two *r* motions, one after the other. To denote this sequence of motions, we will write *rr* (we use juxtaposition to represent the "followed by" operation). *rr* has the same net effect as the r^2 motion (because two 120° clockwise rotations have the same effect as one clockwise 240° rotation). Thus, $rr = r^2$.

rrr = _____.

756

e. Three clockwise 120° rotations bring the triangle back to its original position. Thus, its net effect is the same as an *e* motion.

When any pair of elements in $\{e, r, r^2\}$ is operated on by the "followed by" operation, the motion has the same net effect as one of the elements in $\{e, r, r^2\}$. For example, rr^2 (an *r* motion followed by an r^2 motion) has the same net effect as an *e* motion. Thus, we write $rr^2 = e$. r^2 followed by r^2 has the same effect as one *r* motion. Thus, $r^2r^2 = r$.

We can summarize the "followed by" operation on $\{e, r, r^2\}$ with a Cayley table. Complete the following Cayley table:

	e	r	r^2
e		r	
r		r^2	e
r^2			r

757

	e	r	r^2
e	e	r	r^2
r	r	r^2	e
r^2	r^2	e	r

$C_3 = \{e, r, r^2\}$

Because all the entries in the table are e, r, or r^2, we can conclude that the set $\{e, r, r^2\}$ is closed under "followed by."

Is there an identity element?

758

e is the identity.

e is any motion that has the same net effect as no motion. Thus, $ex = xe$ for all $x \in \{e, r, r^2\}$. Now you know why we give the no-motion the name e: e is the name we usually give to the identity element, and the e motion is the identity element in the set of symmetries of our triangle.

Does each element have an inverse?

759

Yes. Because $ee = e$, and $rr^2 = r^2r = e$, the inverse of e is e, and r and r^2 are inverses of each other.

Suppose $a, b, c \in \{e, r, r^2\}$. In what order are the motions a, b, and c performed corresponding to $(ab)c$? Corresponding to $a(bc)$?

760

In both cases, the order is a then b then c. Thus, $(ab)c = a(bc)$. That is, "followed by" is an associative operation on $\{e, r, r^2\}$.

The set $\{e, r, r^2\}$ under the "followed by" operation has the closure, associativity, identity, and inverses properties. It is a group! This group is the set of symmetries of an equilateral triangle that *start and end with* T *on top*.

A ***regular polygon*** is a polygon whose sides are equal and whose interior angles are equal. An equilateral triangle is a three-sided regular polygon. Thus, we can equivalently say that our group is the set of symmetries of a three-sided regular polygon that starts and ends with T on top. Let's denote this group with C_3. The "3" in C_3 is for the number of elements in the group. We will explain the "C" in C_3 shortly.

What motion is the inverse of an r motion?

761

$rr^2 = e$. Thus, $r^{-1} = r^2 = 240°$ *clockwise* rotation. r^2 has the same net effect as one 120° *counterclockwise* rotation. Thus, we can think of r^{-1} as either a 240° clockwise rotation or as a 120° counterclockwise rotation.

r^{-1} is one 120° counterclockwise rotation. Thus, r^{-2}, which by definition is equal to $(r^{-1})^2$, is two 120° counterclockwise rotations, which has the same net effect as a single r motion. That is, $r^{-2} = r$. r^{-3} has the net effect of three 120° counterclockwise rotations, which is equivalent to no motion. That is, $r^{-3} = e$. Each power of r with a negative exponent is a motion whose net effect is the same as one of the motions in $\{e, r, r^2\}$.

Consider the sequence consisting of the powers of r:

$$\ldots, r^{-3} = e, r^{-2} = r, r^{-1} = r^2, r^0 = e, r^1 = r, r^2, r^3 = e, r^4 = r, r^5 = r^2, r^6 = e, \ldots$$

The powers of r—the rotation symmetries—cycle through the three elements of the group. Accordingly, we call this group the ***cyclic rotation group of order 3***, and denote it with C_3. "C" in C_3 stands for "cyclic"; "3" stands for the number of elements in the group. Because the powers of r cycle through the elements of C_3, we say that C_3 is ***generated by r***.

Is C_3 generated by e?

762

No. All the powers of e equal e.

Show the cycle associated with the powers of r^2. Is C_3 generated by r^2?

763

$\ldots, (r^2)^{-3} = e, (r^2)^{-2} = r^2, (r^2)^{-1} = r, (r^2)^0 = e, (r^2)^1 = r^2, (r^2)^2 = r^4 = r, (r^2)^3 = r^6 = e, \ldots$. The powers of r^2 cycle through all the elements of C_3. Thus, C_3 is generated by r^2, as well as by r, but not by e.

How many symmetries does a square have in which the T side starts and ends on top?

764

Four. The symmetries are e, r, r^2, and r^3, where r is a 90° clockwise rotation.

This set of rotation symmetries of a square under the "followed by" operation is a group. Like C_3, it has the cyclic property (it is the cyclic rotation group of order 4). We call this group C_4. Construct the Cayley table for C_4.

765

	e	r	r^2	r^3
e	e	r	r^2	r^3
r	r	r^2	r^3	e
r^2	r^2	r^3	e	r
r^3	r^3	e	r	r^2

$C_4 = \{e, r, r^2, r^3\}$

Four successive r motions (90° clockwise rotations) of a square bring it back to its original position. Thus, $r^4 = e$. The powers of r that follow r^4 then cycle through r, r^2, r^3, and back to e again. This table, like the tables for all finite cyclic groups appropriately organized, has an obvious pattern: each row after the first row is the preceding row circularly left shifted. For example, if we take the first row (e, r, r^2, r^3) and shift it left with its leftmost element (e) wrapping around to the right side, we get the second row (r, r^2, r^3, e).

What is the inverse of r in C_4?

766

$rr^3 = r^3r = r^4 = e$. Thus, $r^{-1} = r^3$.

The Cayley table for C_5, the symmetries of a regular five-sided polygon (i.e., a regular pentagon) that start and end with T on top, has a similar pattern. Construct it.

767

	e	r	r^2	r^3	r^4
e	e	r	r^2	r^3	r^4
r	r	r^2	r^3	r^4	e
r^2	r^2	r^3	r^4	e	r
r^3	r^3	r^4	e	r	r^2
r^4	r^4	e	r	r^2	r^3

$C_5 = \{e, r, r^2, r^3, r^4\}$

How many symmetries does an n-sided regular polygon have in which the T side starts and ends on top?

768

n. The symmetries are e, r, r^2, ..., r^{n-1}, where r is a $360°/n$ clockwise rotation. It is also a group. Specifically, it is the cyclic rotation group of order n. We call this group C_n.

How many symmetries does a line segment have?

769

Two: no motion and r, where r is a 180° rotation.

$$1 \longrightarrow 2 \quad \overset{r}{\Longrightarrow} \quad 2 \longrightarrow 1$$

Construct its Cayley table. *Hint*: Two 180° rotations have the same net effect as no motion.

770

	e	r
e	e	r
r	r	e

$C_2 = \{e, r\}$

This Cayley table corresponds to a group of order 2. We call this group C_2.

How many symmetries does an infinitesimal point have?

771

One. A point has no dimensions, so rotating it has no effect. Do not confuse an infinitesimal point with a circle or a sphere. A circle and a sphere both have an infinite number of symmetries.

Construct the Cayley table corresponding to a point.

772

	e
e	e

$C_1 = \{e\}$ This Cayley table corresponds to a group of order 1. We call this group C_1.

Let's summarize: for each $n > 0$, C_n is a group of symmetries. For $n > 2$, C_n is the group of symmetries of an n-sided regular polygon that start and end with the T side on top. C_2 is the group of symmetries of a line segment. C_1 is the group of symmetries of an infinitesimal point. All these groups are cyclic. For $n > 1$, C_n is generated by r where r is a $360°/n$ clockwise rotation. C_1 is generated by e.

Groups of Orders 3, 2, and 1

773

Compare the following Cayley tables:

	e	r	r^2
e	e	r	r^2
r	r	r^2	e
r^2	r^2	e	r

$\langle C_3,$ followed by$\rangle$

$+_3$	0	1	2
0	0	1	2
1	1	2	0
2	2	0	1

$\langle \mathbb{Z}_3, +_3 \rangle$

The two tables are essentially the same; e, r, and r^2 in C_3 correspond to 0, 1, 2 in $\mathbb{Z}_3$, respectively. Thus, $C_3 \cong \mathbb{Z}_3$. That is, C_3 is _____ to $\mathbb{Z}_3$.

774

isomorphic

Because $C_3 \cong \mathbb{Z}_3$ and C_3 is cyclic, $\mathbb{Z}_3$ must also be cyclic. The groups of order 3 that we have seen so far are isomorphic to each other. Is it possible that all groups of order 3 are isomorphic to each other? To answer this question, let's see how many Cayley tables with a distinct structure we can construct for a group that has order 3 and elements e, x, and y in which the row and column labels are listed in e-x-y order. If only one table is possible, then all groups of order 3 must have essentially the same Cayley table. If that is the case, then there is only one group of order 3 in the isomorphic sense.

Fill in the first row and the first column of the following Cayley table for a group that has order 3. Assume e is the identity element.

	e	x	y
e			
x			
y			

775

	e	x	y
e	e	x	y
x	x		
y	y		

← Determine this element next (circled cell at row x, column y)

Because e is the identity, row e has to be identical to the column labels, and column e has to be identical to the row labels.

Recall that each row in a Cayley table for a group has to be a permutation of the column labels; each column has to be a permutation of the row labels. Moreover, the positions of the identity are symmetrical with respect to the main diagonal. Using these constraints, determine the unspecified entries in the preceding Cayley table. *Hint*: Determine the entry in the circled position next.

776

	e	x	y
e	e	x	y
x	x	y	e
y	y	e	x

The position marked with a circle is in a row that contains x and is in a column that contains y. Thus, this position cannot be occupied by either x or y. The only possible alternative is e. The remaining entries can be determined in a similar fashion. This is the only Cayley table consistent with group properties for the set $\{e, x, y\}$ in which the rows and columns are in e-x-y order and e is the identity. Thus, every group of order 3 must have a Cayley table that is essentially the same as this table. What can we conclude?

777

All groups of order 3 are isomorphic to each other.

Because all groups of order 3 are isomorphic to each other, we say that **up to isomorphism,** there is only one group of order 3. The phrase "up to isomorphism" indicates that in our count, we do not treat groups isomorphic to each other as separate groups. Moreover, because C_3, one of the groups of order 3 is cyclic, all groups of order 3 must be cyclic.

Construct the Cayley tables for the group $\{e\}$ and the group $\{e, x\}$ as you did for the group $\{e, x, y\}$ in the preceding frame.

778

These are the only possible Cayley tables for groups of orders 1 and 2:

	e
e	e

	e	x
e	e	x
x	x	e

The first table is the only Cayley table consistent with group properties for the set $\{e\}$. Thus, up to isomorphism, this is the only group of order 1. Because C_1, one of the groups of order 1, is cyclic, all groups of order 1 must be cyclic. Similarly, the second table is the only table consistent with group properties for the set $\{e, x\}$ in which the rows and columns are in e-x order and e is the identity. Thus, up to isomorphism, this is the only group of order 2. Because C_2, one of the groups of order 2 is cyclic, all groups of order 2 must be cyclic.

Show that the two groups defined by the preceding tables are cyclic.

779

$\ldots, e^{-2} = e, e^{-1} = e, e^0 = e, e^1 = e, e^2 = e, \ldots$
$\ldots, x^{-2} = e, x^{-1} = x, x^0 = e, x^1 = x, x^2 = e, \ldots$

The following table summarizes what we have learned about groups of orders 1, 2, and 3:

Order	Groups
1	$C_1 \cong \mathbb{Z}_1 \cong S_1 \cong A_1 \cong A_2$
2	$C_2 \cong \mathbb{Z}_2 \cong S_2$
3	$C_3 \cong \mathbb{Z}_3 \cong A_3$

All groups of order 3 have essentially the same Cayley table. Thus, they are all isomorphic to each other. Similarly, all groups of order 2 are isomorphic to each other, and all groups of order 1 are isomorphic to each other. Thus, up to isomorphism, there is only one group of order 1, one group of order 2, and one group of order 3. All these groups are cyclic.

Of order 1, we have C_1, $\mathbb{Z}_1$, S_1 (the symmetric group of degree 1), and A_1 and A_2 (the alternating groups of degree 1 and 2). Of order 2, we have C_2, $\mathbb{Z}_2$, and S_2. Of order 3, we have C_3, $\mathbb{Z}_3$, and A_3. There is no symmetric group of order 3 (the order of S_3 is $3! = 6$).

Generating Sets

780

Let X be a nonempty subset of a group G. Let $<X>$ be the set consisting of

- all the elements in X, plus
- all the inverses of the elements in X, plus
- all the products that can be constructed using the elements of X and their inverses, with repeats allowed.

By the closure and inverses properties of G, all the elements in $<X>$ are in G. If $<X>$ equals G, we say that X **generates** G or, equivalently, that X is a **generating set** for G. For example, in C_3, $\{r\}$ is a generating set. That is, if $X = \{r\}$, then $<X> = C_3$. But $\{e\}$ is not a generating set. If $X = \{e\}$, then $<X> = \{e\} \neq C_3$.

If elements are added to a set that is already a generating set for a group, the new set will also be a generating set. For example, in C_3, $\{r\}$ is a generating set, and so are $\{e, r\}$, $\{r, r^2\}$, and C_3 itself. Thus, a generating set for a group is not unique. We will generally use a generating set for a group that has the minimum number of elements. For example, for C_3, we will use $\{r\}$ or $\{r^2\}$ rather than $\{e, r\}$, $\{e, r^2\}$, $\{r, r^2\}$, or C_3.

Constructing the Graph of a Group

781

It is often said that a picture is worth a thousand words. So is a graph of a group. A **graph** of a group is a picture that represents both its elements and its structure. A graph of a group has a node for each element in the group. Connecting the nodes are labeled arrows.

Suppose a and c are elements in a group. In the corresponding graph, an arrow from a to c labeled with b indicates that $a * b = c$, where $*$ is the group operation:

$$a \xrightarrow{b} c \qquad \text{means } a * b = c$$

We name arrows according to the label on them. For example, we call the preceding arrow a **b arrow** because it is labeled with b.

The graph of a group consists of a node for every element in the group. Every node should have an outgoing arrow for every element in a generating set for the group. For example, if the generating set is $\{r, f\}$, then the corresponding graph should have at every node an outgoing r arrow and an outgoing f arrow. To avoid unnecessary complexity in a graph, we generally construct a graph based on a minimal generating set. For example, $\{r, r^2\}$ is a generating set for C_3, but so is $\{r\}$. If the graph is based on $\{r, r^2\}$, the graph will have two outgoing arrows at every node. But if it is based on $\{r\}$, it will have only one outgoing arrow at every node.

Let's look at the graph of C_3 based on the generating set $\{r\}$. Each node has an outgoing r arrow:

Because $er = r$ in C_3, its graph has an r arrow from the e node to the r node. Because $rr = r^2$, the graph has an r arrow from the r node to the r^2 node. Because $r^2 r = e$, the graph has an r arrow from the r^2 node to the e node.

When we draw a graph, it is convenient to omit the labels on the arrows but provide the same information via a key. For example, the graph of C_3 without labels on the arrows is

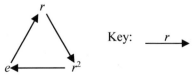

The key here indicates that every arrow is an r arrow.

154   Chapter 13: Symmetries of a Regular Polygon

Construct the graph of C_4, the symmetries of a four-sided regular polygon (i.e., a square) that start and end with T on top.

782

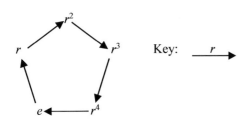

Key: $\underline{\quad r \quad}\rightarrow$

Construct the graph of C_5.

783

[Pentagon graph with vertices e, r, r^2, r^3, r^4 connected by arrows]

Key: $\underline{\quad r \quad}\rightarrow$

What does the graph of C_n look like, where $n > 2$?

784

It is an n-sided polygon, with each arrow pointing to the tail of the next arrow, and with each arrow labeled with r.

Draw the graph corresponding to C_2. *Hint*: In C_2, $er = r$ and $rr = e$.

785

$e \rightleftarrows r$ Key: $\underline{\quad r \quad}\rightarrow$

In this graph, we have two arrows labeled with the *same* label between two nodes going in opposite directions. To simplify our graphs, we will represent such pairs of arrows with a single line without any arrowheads. With this modification, the graph of C_2 simplifies to

$$e \text{———} r \qquad \text{Key: } \underline{\quad r \quad}$$

Draw the graph corresponding to C_1.

786

$\circlearrowleft e$ Key: $\underline{\quad e \quad}$

Show enough of the graph of $\mathbb{Z}$ under addition so that it is clear what the entire graph looks like (the graph is infinite, so it is impossible to show the entire graph). *Hint*: Its generating set is $\{1\}$. Thus, each node should have an outgoing 1 arrow.

787

$\cdots \xrightarrow{1} -2 \xrightarrow{1} -1 \xrightarrow{1} 0 \xrightarrow{1} 1 \xrightarrow{1} 2 \xrightarrow{1} \cdots$

Getting the Cayley Table from the Graph

788

What does the following arrow indicate in a graph of a group?

$a * b = c$

In this equation, move b to the right side.

$a = c * b^{-1}$. Interchanging sides, we get $c * b^{-1} = a$.

Show the arrow that corresponds to $c * b^{-1} = a$. *Hint*: The arrow is from c to a.

$c \xrightarrow{b^{-1}} a$ rotated 180° is $a \xleftarrow{b^{-1}} c$

Thus, the arrow from a to c labeled with b implies an arrow from c to a labeled with the inverse of b.

Key idea: In a graph for a group, an arrow going in one direction labeled with x implies an arrow going in the opposite direction labeled with x^{-1}.

When constructing a graph of a group, we draw a single outgoing arrow at every node for each element in the generating set. If we do this, the resulting graph will necessarily have exactly one incoming arrow at every node for each element in the generating set (you prove this in homework questions 1 and 2). For example, here is the graph of C_3:

We constructed it by drawing an outgoing r arrow at every node. Note that every node also has exactly one incoming r arrow.

Suppose a group has the generating set $\{x, y\}$. The graph based on this generating set has at each node exactly one outgoing x arrow and exactly one outgoing y arrow. Then, by homework questions 1 and 2, the graph will necessarily also have at each node exactly one incoming x arrow and exactly one incoming y arrow, as illustrated by the left side of the following diagram:

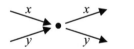   is equivalent to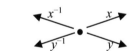

We know that an x arrow is equivalent to an x^{-1} arrow in the reverse direction, and a y arrow is equivalent to a y^{-1} arrow in the reverse direction. Thus, we can view the incoming x and y arrows as outgoing x^{-1} and y^{-1} arrows, respectively, as illustrated by the right side of the preceding diagram.

Suppose a group has a generating set $\{x, y\}$ and we want to determine the product of the elements yy and $xy^{-1}x$. Given the graph of the group, we can easily determine this product by using the ***follow-the-arrows*** technique. We start at the yy node and then follow the arrows corresponding to the components of $xy^{-1}x$. That is, we follow the outgoing x arrow, then the outgoing y^{-1} arrow, then the outgoing x arrow. The final node on which we land corresponds to the product of yy and $xy^{-1}x$.

From our preceding analysis, we know that at each node, there is exactly one outgoing arrow for x, one for x^{-1}, one for y, and one for y^{-1}. Thus, when we apply the follow-the-arrows technique, at every node, the arrow we need to follow is always available at that node, and there is only one such arrow. We never hit a "dead-end" or have a choice of arrows to follow.

The following graph is based on the generating set $\{x, y\}$. Using the follow-the-arrows technique, determine the products $(x) * (xy)$ and $(y) * (xy)$.

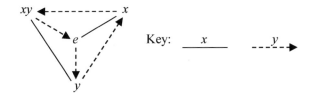

792

$(x) * (xy) = y$, $(y) * (xy) = e$

The group represented by the graph in the preceding frame is called the **dicyclic group of order 4** (denoted Dic_4). In chapter 20, we will investigate this type of group.

The follow-the-arrows technique works for any group. That is, given a graph of any group, we can use the follow-the-arrows technique to determine the product of any two elements in that group. The graph provides us with all the information that the Cayley table for the group provides. Thus, the graph of a group *fully defines the group*.

Dihedral Groups

793

We saw earlier that $C_3 = \{e, r, r^2\}$ under the operation "followed by" is a group. Each element in C_3 is a symmetry—a motion that positions the equilateral triangle so that it can be placed back into the cutout area. Are e, r, and r^2 the only symmetries of the equilateral triangle?

794

No. The triangle can be flipped so that the B side is up. With the B side up, as with the T side up, there are three positions that fit into the cutout. Thus, there are a total of six symmetries—three that end with the T side up, and three that end with the B side up.

Suppose we flip the triangle around the altitude line through vertex 1 (i.e., the line from vertex 1 perpendicular to the opposite side). We call this motion f ("f" is for "flip"). An f motion leaves vertex 1 where it is but switches vertices 2 and 3. Here are the before and after pictures for f:

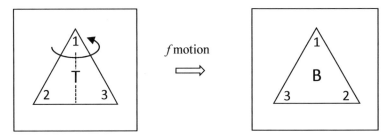

Represent this motion with a table.

795

$\begin{pmatrix} 1 & 2 & 3 \\ 1 & 3 & 2 \end{pmatrix}$ This table defines the f motion.

There are three symmetries that leave the B side up. One corresponds to f. A second corresponds to f followed by r (let's denote this symmetry with fr). The third corresponds to f followed by r followed by another r (let's denote this symmetry with fr^2). Keep in mind that we are interested in only the net effect of motions. For example, there are other motions that have the same net effect as the fr motion (e.g., try rrf with your cutout). We use fr to represent all such motions.

With your triangle cutout, try to determine the effect of an r motion when B is on top. In an r motion, 1 is replaced by 2, 2 is replaced by 3, and 3 is replaced by 1. You will find that an r motion is a 120° *clockwise* rotation when the T is on top, but it is a 120° *counterclockwise* rotation when B is on top.

Using triangle pictures, show the effect of an fr motion. Specifically, show the f motion followed by the r motion.

796

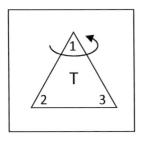 $\xrightarrow{f \text{ motion}}$ 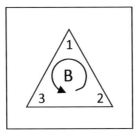 $\xrightarrow{r \text{ motion}}$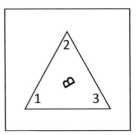

Give the table representation of the *fr* motion. *Hint*: Compare the vertices in the first triangle in the preceding diagram with the vertices in the third triangle.

797

$\begin{pmatrix} 1 & 2 & 3 \\ 2 & 1 & 3 \end{pmatrix}$ This table defines the *fr* motion.

Show the final configuration corresponding to the fr^2 motion. *Hint*: Rotate counterclockwise the final configuration for the *fr* motion in the preceding frame.

798

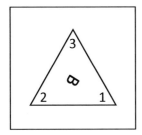

What is the table representation of the fr^2 motion?

799

$\begin{pmatrix} 1 & 2 & 3 \\ 3 & 2 & 1 \end{pmatrix}$

There are exactly six symmetries: $e, r, r^2, f, fr,$ and fr^2. Any two of these six motions performed in succession is a symmetry. Thus, any two performed in succession must have the same net effect as one of the six symmetries listed. In other words, the set of these six motions is closed under the operation "followed by."

Using triangle diagrams, show the effect of an *rf* motion, that is, a *r* motion followed by an *f* motion. Recall that in an *f* motion, we flip the triangle around the altitude line through vertex 1.

800

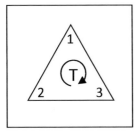 $\xrightarrow{r \text{ motion}}$ 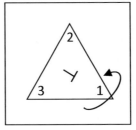 $\xrightarrow{f \text{ motion}}$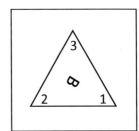

Which of our six symmetries—e, r, r^2, f, fr, fr^2—has the same net effect as *rf*?

801

fr^2

Thus, $rf = fr^2$. This equality tells us that an r before an f is equivalent to two r's after an f.

Apply this equality to r^2f, using it to move the f to the far left. Thus, the result will be an f followed by some sequence of r's. Hint: $r^2f = rrf$.

802

Replacing r^2 with the equivalent rr, we get $r^2f = r(rf)$. Next, substitute fr^2 for rf to get $r(fr^2)$. Re-parenthesize to get $(rf)(r^2)$. Substitute fr^2 for rf to get $(fr^2)(r^2)$. Re-parenthesize to get $f(r^2r^2) = fr^4$.

Three r's is the identity motion. Thus, $fr^4 = fr^3r = fer =$ _____ .

803

fr

Using $rf = fr^2$, we determined that $r^2f = fr$. Based on these two equalities, we get the following rule:

> An r moved across an f changes into an r^2. An r^2 moved across an f changes into an r.

Let's call this rule the ***flip−rotate commutation rule*** for an equilateral triangle. With this rule, along with $r^3 = f^2 = e$, we can easily determine the Cayley table for the symmetries of an equilateral triangle. We simply move all f's to the far left. The result is some sequence of f's followed by some sequence of r's. We then simplify using $r^3 = f^2 = e$. The result will be one of our six symmetries: $e, r, r^2, f, fr,$ or fr^2.

Determine $(fr^2)(fr^2)$.

804

$(fr^2)(fr^2) = (f)(r^2f)(r^2) = (f)(fr)(r^2) = f^2r^3 = ee = e$

Determine $(fr)(fr)$.

805

$(fr)(fr) = f(rf)r = f(fr^2)r = f^2r^3 = ee = e$

Let's confirm the flip-rotate commutation rule by visualizing the motions rf and fr^2. In the starting position, vertex 1 is in the 12 o'clock position. An r motion (a four-hour clockwise rotation) moves vertex 1 to the 4 o'clock position. A flip then puts the B side up but leaves vertex 1 in the 4 o'clock position. Now consider the fr^2 motion: f puts the B side up and leaves vertex 1 in the 12 o'clock position. An r^2 motion consists of two r motions. Each r motion is a four-hour *counterclockwise* rotation (recall that an r motion is clockwise when T is on top but counterclockwise when B is on top). The first r motion brings vertex 1 to the 8 o'clock position. The second r motion brings vertex 1 to the 4 o'clock position. Both rf and fr^2 place vertex 1 in the 4 o'clock position with the B side up. Thus, $rf = fr^2$.

Using the same visualization technique, determine the effect of the motions fr and r^2f.

806

fr: B side up then vertex 1 to the 8 o'clock position; r^2f: vertex 1 to the 8 o'clock position then B side up. Motions fr and r^2f both place vertex 1 in the 8 o'clock position with the B side up. Thus, $fr = r^2f$.

Here is the Cayley table for all the symmetries of an equilateral triangle:

	e	r	r^2	f	fr	fr^2
e	e	r	r^2	f	fr	fr^2
r	r	r^2	e	fr^2	f	fr
r^2	r^2	e	r	fr	fr^2	f
f	f	fr	fr^2	e	r	r^2
fr	fr	fr^2	f	r^2	e	r
fr^2	fr^2	f	fr	r	r^2	e

D_3

Is $\{e, r, r^2, f, fr, fr^2\}$ a group under the operation "followed by"? As we observed in frame **760**, "followed by" is associative. The preceding Cayley table confirms that our set of motions is closed under "followed by." The identity element is e. Notice the pattern of e's in the Cayley table: every row has an e that is either on the main diagonal or has a mirror image with respect to the main diagonal. From frame **404**, we know that this pattern of e's confirms that every element has an inverse. Our set of six symmetries under the "followed by" operation has the closure, associative, identity, and inverses properties. Thus, it is a group.

The graph of this group has six nodes—one for each element. No single element generates this group. However, r and f together generate the group. Accordingly, the graph has two types of arrows: r arrows and f arrows. Each node has two outgoing arrows: one r arrow and one f arrow. The graph looks like two triangles, one in front of the other:

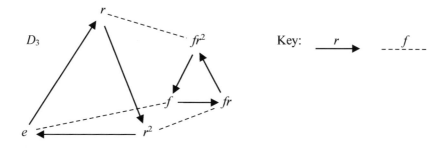

Each solid arrow is an r arrow; each dotted line represents two f arrows going in opposite directions.

Multiplying repeatedly by r, starting with e, generates the elements in the front triangle cyclically. Multiplying by f switches triangles. Then multiplying repeatedly by r generates the elements in the back triangle cyclically. In the top triangle, the r arrows are in a clockwise configuration, but in the bottom triangle, the r arrows are in a counterclockwise configuration.

In geometry, *dihedral* refers to a geometric structure that has two planes. Because the graph of our group of symmetries of an equilateral triangle has two planes—one triangle in front of another—we call this group a ***dihedral group.*** This specific dihedral group is denoted with D_3. "D" is for "dihedral," and "3" is for the three-sided regular polygon (i.e., the equilateral triangle) whose symmetries form the group.

Draw the graph of D_4: the group of all the symmetries of a square. *Hint*: It has two planes, each with four elements. It is similar to the graph of D_3. The r motion is a rotation of 90° clockwise. To simplify the drawing of the graph, show it in two dimensions rather than three. Specifically, place the back square inside the front square.

807

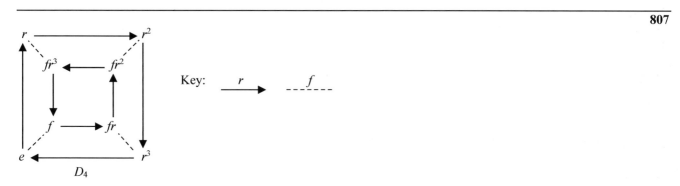

The symmetries for every regular polygon comprise a dihedral group. A regular polygon with n sides corresponds to the dihedral group D_n. A triangle corresponds to D_3, a square corresponds to D_4, a pentagon corresponds to D_5, a hexagon to D_6, a heptagon to D_7, an octagon to D_8, and so on.

What is the order of D_n?

808

$2n$. The graph of D_n contains two planes, each containing n elements.

D_n is called the **dihedral group of order 2n.** D_4 has a special name: it is called the ***octic group.*** The graphs of D_n for $n > 2$ have the same general structure: two planes, each a polygon. Successive rotations correspond to moving around one of the two planes in the graph. A flip corresponds to moving from one plane to the other.

To convert the graph of D_4 to the graph of D_3, we remove one element from each square (r^3 from the outer square and fr^3 from the inner square). The effect is to collapse the two squares into two triangles. Here is a really interesting question: what do we get if we remove the elements r^2 and fr^2 from the graph of D_3? Each triangle collapses into a _____.

809

line

One line has the elements e and r. The other line has the elements _____ and _____.

810

f, fr

The resulting graph is of the group that is called the **Klein four group,** denoted by V_4. We can also denote this group with D_2, because its graph is obtained from the graph of D_3 by removing one vertex in each triangle. However, the resulting graph does not have the dihedral property (it has two lines corresponding to r arrows, not two planes). Thus, a letter D (for "dihedral") in its designator does not reflect its graphical structure. For this reason, V_4 is the preferred designator for this group.

Draw the graph of V_4.

811

Construct the Cayley table for V_4. You should be able to determine it from its graph. For example, from the graph, it is easy to see that $ff = e$, $rr = e$, and $fr = rf$. Also construct the table for C_4, the cyclic rotation group of order 4.

812

	e	r	f	fr
e	e	r	f	fr
r	r	e	fr	f
f	f	fr	e	r
fr	fr	f	r	e

V_4

	e	r	r^2	r^3
e	e	r	r^2	r^3
r	r	r^2	r^3	e
r^2	r^2	r^3	e	r
r^3	r^3	e	r	r^2

C_4

In the Cayley table for V_4, all the e's are on the main diagonal. This means that each element is its _____ _____.

813

own inverse

Compare the Cayley tables in the preceding frame for V_4 and C_4. Are V_4 and C_4 isomorphic?

814

No.

Give an obvious indication that V_4 and C_4 are not isomorphic.

815

In V_4, each element is its own inverse, but not so in C_4. So V_4 and C_4 cannot be isomorphic.

Recall from frame **748** that groups that are isomorphic to each other are essentially the same—their Cayley tables differ only in the names of their elements. For this reason, we sometimes give every group in a collection of groups isomorphic to each other the same name. For example, we call every group isomorphic the V_4 group in frame **812** the Klein four group.

D_3 and S_3

816

Here is the table representation of each motion in D_3:

$$\overset{e}{\begin{pmatrix} 1 & 2 & 3 \\ 1 & 2 & 3 \end{pmatrix}} \quad \overset{r}{\begin{pmatrix} 1 & 2 & 3 \\ 2 & 3 & 1 \end{pmatrix}} \quad \overset{r^2}{\begin{pmatrix} 1 & 2 & 3 \\ 3 & 1 & 2 \end{pmatrix}} \quad \overset{f}{\begin{pmatrix} 1 & 2 & 3 \\ 1 & 3 & 2 \end{pmatrix}} \quad \overset{fr}{\begin{pmatrix} 1 & 2 & 3 \\ 2 & 1 & 3 \end{pmatrix}} \quad \overset{fr^2}{\begin{pmatrix} 1 & 2 & 3 \\ 3 & 2 & 1 \end{pmatrix}}$$

In frame **803**, we determined that $rf = fr^2$ by observing the effect of r followed by f on our triangle. We can also the determine the value of rf from the table representations of r and f by observing that r maps 1 to 2 and f maps 2 to 3. Thus, rf maps 1 to 3. We can similarly determine to which elements rf maps 2 and 3.

Determine rf in table form from the tables for r and f.

817

$\begin{pmatrix} 1 & 2 & 3 \\ 3 & 2 & 1 \end{pmatrix}$ This is the table representation of rf. But it is also the table representation of fr^2. Thus, $rf = fr^2$.

The table representations of the r and f in D_3 are permutations on $\{1, 2, 3\}$. We obtained the table representation for rf by composing the r permutation and the f permutation in r-then-f order. The "followed by" operation, when the elements of D_3 are in table form, is simply a function composition. For example, r followed by f corresponds to the composition $f \circ r$ (recall that in the composition, the functions are applied in right-to-left order). Moreover, D_3 in table form contains all six of the permutations on $\{1, 2, 3\}$. Thus, D_3 in table form is the group of all permutations on $\{1, 2, 3\}$ under function composition. We have seen this group in a previous chapter. It is _____.

818

S_3, the symmetric group of degree 3. Thus, $D_3 \cong S_3$ (i.e., they are isomorphic groups).

Are D_4 and S_4 isomorphic? *Hint*: What are the orders of D_4 and S_4?

819

$|D_4| = 2 \cdot 4 = 8$ and $|S_4| = 4! = 24$. Because they have different orders, they cannot be isomorphic.

Show that D_{12} and S_4 are *not* isomorphic. *Hint*: See frame **533**.

820

From frame **533**, we know that the maximum order among the elements of S_4 is 4. In D_{12}, on the order hand, r has the order 12. Thus, D_{12} cannot be isomorphic to S_4.

Give an element in D_{12} that has order 12.

821

r corresponds to a clockwise rotation of 360/12 degrees. $r^1, r^2, \ldots, r^{11}$ are all distinct elements. r^{12}, a rotation of 360°, has the same net effect as no motion. Thus, $r^{12} = e$. The order of r is 12. ,

What is the order of r in D_n?

822

n

Groups for Orders 1 to 7

823

Up to isomorphism, there is only one group with order 1, one group with order 2, and one group with order 3 (see frame **779**). There are two and only two non-isomorphic groups with order 4: V_4 and C_4 (you show this in homework questions 3 and 4).

A theorem we will prove in chapter 14 states that for every prime p, there is only one group with order p up to isomorphism, and that group is the cyclic group of order p. Thus, C_2, C_3, C_5, and C_7 are the only groups up to isomorphism of order 2, 3, 5, and 7, respectively.

In order of increasing size, the groups from order 1 to order 7 are as follows:

Order	Commutative	Noncommutative
1	$C_1 \cong \mathbb{Z}_1 \cong S_1 \cong A_1 \cong A_2$	
2	$C_2 \cong \mathbb{Z}_2 \cong S_2$	
3	$C_3 \cong \mathbb{Z}_3 \cong A_3$	
4	$C_4 \cong \mathbb{Z}_4 \cong U_4 \cong Dic_4$, V_4	
5	$C_5 \cong \mathbb{Z}_5$	
6	$C_6 \cong \mathbb{Z}_6$	$D_3 \cong S_3$
7	$C_7 \cong \mathbb{Z}_7$	

We have not yet established that C_6 and D_3 are not isomorphic. The graphs for these groups suggest that they are not isomorphic. However, structurally distinct graphs do not guarantee that the corresponding groups are not isomorphic (graphs based on different generating sets for the same group can be structurally dissimilar). Let's find a difference between C_6 and D_3 that confirms that C_6 and D_3 are not isomorphic. Examine the Cayley table for D_3 in frame **806**. Is it symmetric with respect to the main diagonal?

824

No. For example, fr is in the fourth row, second column, but fr^2 is in the second row, fourth column.

What is the significance of this lack of symmetry?

825

D_3 is not commutative.

But C_6 is commutative. Let's prove this. Let $x, y \in C_6$. Then $x = r^i$ and $y = r^j$ for some $i, j \in \{0, 1, 2, 3, 4, 5\}$, where r is a generator of C_6. Show that $xy = yx$.

826

$xy = r^i r^j = r^{i+j} = r^{j+i} = r^j r^i = yx \Rightarrow C_6$ is commutative. ∎

In fact, all cyclic groups are commutative (our argument that C_6 is cyclic generalizes to every cyclic group).

D_3 is not commutative; C_6 is commutative. What can you conclude?

827

D_3 and C_6 are not isomorphic.

Is V_4 commutative? Check its table in frame **812**.

828

Yes. Its table is symmetric with respect to its main diagonal.

What is the smallest order for which there is more than one group up to isomorphism?

829

What is the smallest group that is not commutative?

830

$D_3 \cong S_3$

For which orders less than 8 is there only one group up to isomorphism?

831

1, 2, 3, 5, 7

These orders, with the exception of 1, are all primes.

Review Questions

1. Give the table representation of the r motion in D_4.
2. Give the table representation of the f motion in D_4.
3. The f motion in D_3 is a flip around the altitude through vertex 1. Flips around the altitudes through vertices 2 and 3 are also symmetries. To which elements of D_3 do they correspond?
4. What is the flip-rotate commutation rule for D_4?
5. What is the flip-rotate commutation rule for D_n?
6. A rectangle that is not a square has how many symmetries?
7. Construct the graph of the infinite cyclic group generated by r.
8. Show that $\langle \mathbb{Z}_3, + \rangle$ is a cyclic group.
9. In D_n, an r motion corresponds to how many degrees of rotation?
10. Can a cyclic group have more than one generator? Justify your answer.

Answers to Review Questions

1. $\begin{pmatrix} 1 & 2 & 3 & 4 \\ 2 & 3 & 4 & 1 \end{pmatrix}$
2. $\begin{pmatrix} 1 & 2 & 3 & 4 \\ 1 & 4 & 3 & 2 \end{pmatrix}$
3. fr is a flip around the altitude through vertex 3; fr^2 is a flip around the altitude through vertex 2. Thus, two fr motions or two fr^2 motions in succession is the identity motion. If the table of a motion is not the identity, and its ith column has i in both positions, then that table corresponds to a flip around the altitude through the ith vertex.
4. $rf = fr^3, fr = r^3f$
5. $rf = fr^{n-1}, fr = r^{n-1}f$
6. Four: 180° clockwise rotation, flip around the vertical axis through the center, flip around the horizontal axis through the center, and the identity.
7. $\ldots \longrightarrow r^{-1} \longrightarrow e \longrightarrow r \longrightarrow r^2 \longrightarrow r^3 \longrightarrow \ldots$ Key: $\xrightarrow{r}$
8. $1, 1 + 1 = 2, 1 + 1 + 1 = 0, 1 + 1 + 1 + 1 = 1, \ldots$
9. $360°/n$
10. Yes. In C_3, both r and r^2 are generators of C_3.

Homework Questions

1. Show that in a graph of a group constructed as described in this chapter, for each element x in the generating set, each node cannot have more than one incoming x arrow.
2. Show that in a graph of a group constructed as described in this chapter, for each element x in the generating set, each node must have at least one incoming x arrow. What can you conclude based on this question and homework question 1?
3. The position marked by "?" in the following table for a group whose elements are $e, a, b,$ and c can be which elements of the group?

	e	a	b	c
e	e	a	b	c
a	a		?	
b	b			
c	c			

4. Suppose the element in the position marked by "?" in homework question 3 is *e*. Determine the rest of the elements in the table. What group corresponds to the completed table? Suppose the element in the position marked by "?" in homework question 3 is *c*. Determine the rest of the elements in the table. What group corresponds to the completed table?
5. How many symmetries does the letter Z have? The letter T? The character *? An equilateral triangle with clockwise arrow sides:

6. Give a subset of S_4 that is isomorphic to V_4. *Hint*: The square of every non-identity element in V_4 is equal to the identity. The product of any two distinct non-identity elements is equal to the remaining non-identity element. The elements of S_4 whose square is the identity have the form $(a\ b)$ or $(a\ b) \circ (c\ d)$.
7. A symmetry for a three-dimensional object is a motion that positions the object so that it can occupy precisely the space occupied by the object in its original position. How many symmetries does a regular tetrahedron have? A regular tetrahedron is a four-sided pyramid, with each side an equilateral triangle.
8. How many distinct Cayley tables are possible for a set that has three elements, *x*, *y*, and *z*, in which the rows and columns are arranged in *x-y-z* order? How many of these correspond to groups?
9. Construct the graph of the infinite dihedral group.
10. In the graph of D_3, reverse the arrows of the bottom triangle and rename its elements accordingly. Construct the corresponding Cayley table. Does the Cayley table correspond to the cyclic group of order 6?

14 Subgroups

Basics

832

A subset H of a group G is a **subgroup** of G if and only if H itself is a group under the group operation of G restricted to H. For example, $\mathbb{Z}$ under addition is a group. $2\mathbb{Z} = \{2k : k \in \mathbb{Z}\}$, the set of even integers, is a subset of $\mathbb{Z}$ that is also a group under addition. Thus, $2\mathbb{Z}$ is a subgroup of $\mathbb{Z}$.

Show that $<2\mathbb{Z}, +>$ is a group and, therefore, a subgroup of $<\mathbb{Z}, +>$.

833

Closure: The sum of any two even integers is even. So $2\mathbb{Z}$ is closed under addition
Associativity: Because addition on $\mathbb{Z}$ is associative, addition on any subset of $\mathbb{Z}$ (such as $2\mathbb{Z}$) is also associative. We say that $<2\mathbb{Z}, +>$ **inherits** associativity from $<\mathbb{Z}, +>$.
Identity: The identity element 0 is even. So $2\mathbb{Z}$ contains the identity element.
Inverses: The inverse of an even integer (its negation) is also even. Thus, every even integer has an inverse in $2\mathbb{Z}$. ∎

The definition of a subgroup does not require that a subgroup of G be a proper subset of G. Thus, every group is a subgroup of itself.

If $<G, *>$ is a group with identity e, is $<\{e\}, *>$ a subgroup of $<G, *>$?

834

$\{e\}$ is a subset of G. It is also a group under the group operation of G. Thus, it is a subgroup of G.

Because $<\{e\}, *>$ has very little structure, it is called the **trivial subgroup** of G. Any subgroup of G that is a proper subset of G is called a **proper subgroup** of G. The proper subgroups of G include all the subgroups of G, except G itself.

Is $<\{1, -1\}, \cdot>$ a group, where "$\cdot$" denotes integer multiplication?

835

Yes. $(1 \cdot 1) = 1$, $(1 \cdot -1) = (-1 \cdot 1) = -1$, and $(-1 \cdot -1) = 1 \Rightarrow \{1, -1\}$ is closed under multiplication. Multiplication on the integers is associative. The element 1 is the multiplicative identity. $(1 \cdot 1) = 1, (-1 \cdot -1) = 1 \Rightarrow$ each element is its own inverse. Thus, $<\{1, -1\}, \cdot>$ is a group (it is the cyclic group of order 2 generated by -1).

Is $<\{1, -1\}, \cdot>$ a subgroup of $<\mathbb{Z}, +>$?

836

No. Although $<\{1, -1\}, \cdot>$ is a group and $\{1, -1\}$ is a subset of $\mathbb{Z}$, its operation is not the same as the operation in $<\mathbb{Z}, +>$.

Let H be a subgroup of G. Is it possible that the identity element for H is different from the identity element for G? Recall from frame **454** that in a set S with a binary operation $*$, an idempotent is an element a such that $a * a = a$. In a group, an element is an idempotent if and only if it is the identity element.

Let e_G and e_H be the identities in the group G and its subgroup H, respectively; $e_H = e_H * e_H$ because e_H is the _____ in _____.

837

identity, H

e_H is an idempotent in H. Because e_H is an element in G as well as H, e_H is also an idempotent in G. Thus, e_H must be the identity element in G. That is, $e_H = e_G$. We saw an example of this in $\langle \mathbb{Z}, +\rangle$ and $\langle 2\mathbb{Z}, +\rangle$. Zero is the identity for both.

Suppose H is a subgroup of G with identity e, and $h \in H$. Let h_H^{-1} be the inverse of h in H, so $hh_H^{-1} = e$. Let h_G^{-1} be the inverse of h in G, so $hh_G^{-1} = e$. Then $hh_H^{-1} = e = hh_G^{-1}$. What can we conclude about h_H^{-1} and h_G^{-1}?

838

$h_H^{-1} = h_G^{-1}$ by the left cancellation law.

We can also show that $h_H^{-1} = h_G^{-1}$ with the following sequence of equalities:

$$h_H^{-1} = h_H^{-1} * e = h_H^{-1} * (h * h_G^{-1}) = \cdots = h_G^{-1}$$

Give the entire sequence along with a justification for each step.

839

h_H^{-1}	
$= h_H^{-1} * e$	e is the identity
$= h_H^{-1} * (h * h_G^{-1})$	h_G^{-1} is the inverse of h
$= (h_H^{-1} * h) * h_G^{-1}$	$*$ is associative
$= e * h_G^{-1}$	h_H^{-1} is the inverse of h
$= h_G^{-1}$	e is the identity ∎

Let's summarize: the identity in a subgroup H of a group G is the same element as the identity in G. The inverse of an element h in a subgroup H of a group G is the same element as the inverse of h in G.

What are all the subgroups of $C_4 = \{e, r, r^2, r^3\}$? *Hint*: C_4 has three subgroups.

840

$\{e\}$, $\{e, r^2\}$, and C_4. The only nontrivial proper subgroup is $\{e, r^2\}$.

Show that $\{r, r^2\}$ is not a subgroup of C_4.

841

$\{r, r^2\}$ does not contain the identity element for G.

Show that $\{e, r\}$ is not a subgroup of C_4.

842

$rr = r^2$, but r^2 is not in $\{e, r\}$. This set is not closed under the group operation.

What are the subgroups of $C_3 = \{e, r, r^2\}$?

843

C_3 and $\{e\}$. Every other subset of C_3 is not closed under the group operation. C_3 does not have any nontrivial proper subgroups.

Determine all six subgroups of $D_3 = \{e, r, r^2, f, fr, fr^2\}$. *Hint*: $ff = e$, $frfr = e$, $fr^2fr^2 = e$.

844

D_3, $\{e\}$, $\{e, r, r^2\}$, $\{e, f\}$, $\{e, fr\}$, $\{e, fr^2\}$

The orders of these subgroups are 6, 1, 3, 2, 2, and 2, respectively. The order of each subgroup of D_3 divides the order of D_3. Similarly, in C_3, as well as in C_4, the order of each subgroup divides the order of the group. Perhaps for every group, the order of each subgroup divides the order of the group. As we investigate groups further, we should keep this conjecture in mind. We may, at some point, be able to prove or disprove it. What we are doing here is typical of the process that leads to mathematical discovery: we observe a pattern in a few examples. We formulate a conjecture based on that pattern. We then study the mathematical structure with the aim of either proving or disproving the conjecture.

Suppose it is true that for any group, the order of its subgroups divides the order of the group. What can we conclude about a group G whose order is p, where p is a prime?

845

The only positive divisors of p are 1 and p. Thus, all the subgroups of G have orders that are either 1 or p. The only subgroup of order 1 is the trivial subgroup. The only subgroup with the same order as G is G itself.

CI Subgroup Test

846

Suppose H is a subset of a group G. Because the operation on G is associative, it is also associative on any subset of G. In particular, it is associative on H. We say that H *inherits* associativity from G. Thus, to prove that a subset H of a group G is a subgroup, we do *not* have to show that the group operation is associative on H.

Suppose H is a nonempty subset of a group G that has the closure and inverses properties. Show that the identity element must also be in H. *Hint*: Use an element and its inverse to show that the identity is in H.

847

H is nonempty. Thus, there exists at least one element $h \in H$. We are given that H has the inverses property. Thus, there is an element $h^{-1} \in H$ that is the inverse of h. Because H has the closure property, $hh^{-1} = e$ is also in H. ∎

We see that closure and inverses imply that H has the identity property. Associativity is inherited from G. Thus, H is a subgroup of G. These observations give us the **CI subgroup test** ("CI" stands for "closure" and "inverses"):

> If H is a nonempty subset of a group G, and H has the closure and inverses properties, then H is a subgroup of G.

Let's show that $\langle 2\mathbb{Z}, +\rangle$, the set of even integers, is a subgroup of $\langle \mathbb{Z}, +\rangle$ using the CI subgroup test. In frame **833**, we showed that $\langle 2\mathbb{Z}, +\rangle$ is a group and, therefore, a subgroup of $\langle \mathbb{Z}, +\rangle$. Recall that we did this by showing that $\langle 2\mathbb{Z}, +\rangle$ has the closure, associative, identity, and inverses property. However, because $2\mathbb{Z}$ is a nonempty subset of $\mathbb{Z}$, and the two sets have the same operation, by the CI subgroup test, we need to show only closure and inverses.

In frame **833**, we did not do a very good job justifying closure and inverses. Our reasoning is circular. We stated that the "sum of any two even integers is even. So $2\mathbb{Z}$ is closed under addition." But this is essentially saying that the even numbers are closed under addition because they are closed under addition. Our justification of the inverses property is similarly flawed.

Prove that $\langle 2\mathbb{Z}, +\rangle$ is a subgroup of $\langle \mathbb{Z}, +\rangle$. This time, do it in a logically sound way. In particular, do not use circular reasoning. Use the CI subgroup test. *Hint*: Let $x = 2i$ and $y = 2j$. Then show that $x + y$ is even and that x has an inverse.

848

Closure: Let $x, y \in 2\mathbb{Z} \Rightarrow x = 2i$ and $y = 2j$ for some $i, j \in \mathbb{Z} \Rightarrow x + y = 2i + 2j = 2(i+j) \Rightarrow x + y \in 2\mathbb{Z}$.
Inverses: Let $x \in 2\mathbb{Z} \Rightarrow x = 2i$ for some $i \in \mathbb{Z} \Rightarrow -x = 2(-i) \Rightarrow -x \in 2\mathbb{Z}$. Thus, for all x in $2\mathbb{Z}$, $-x$ is also in $2\mathbb{Z}$. ∎

In the preceding proof, if we replace 2 with n, where n is a nonnegative integer, is the proof still valid?

849

Yes. By substituting n for 2, we get the proof for the following statement: $\langle n\mathbb{Z}, +\rangle$ is a subgroup of $\langle \mathbb{Z}, +\rangle$ for every nonnegative integer n.

Using the CI subgroup test, show that if H and K are subgroups of G, then $H \cap K$ is also a subgroup of G. *Hint*: Use the closure property of H and K to show that $H \cap K$ has the closure property. Similarly, use the inverses property of H and K to show that $H \cap K$ has the inverses property. Use juxtaposition to represent the group operation.

850

Closure: $x, y \in H \cap K \Rightarrow x, y \in H$ and $x, y \in K \Rightarrow xy \in H$ and $xy \in K$ (because H and K are closed) $\Rightarrow xy \in H \cap K$.
Inverses: $x \in H \cap K \Rightarrow x \in H$ and $x \in K \Rightarrow x^{-1} \in H$ and $x^{-1} \in K$ (because H, K have the inverses property) $\Rightarrow x^{-1} \in H \cap K$. ∎

Subgroup Generated by a Subset *X*

851

Let's review and expand on the concept of a generating set from chapter 13. Let X be a nonempty subset of a group $<G, *>$. Let $<X>$ be the subset of G that contains

- all the elements in X, plus
- all the inverses of the elements in X, plus
- all the products that can be constructed using the elements of X and their inverses, with repeats allowed.

For example, suppose $X = \{a, b\}$, where a and b are elements of a group $<G, *>$. Then the elements of $<X>$ are a, b, a^{-1}, and b^{-1}, plus all the products possible consisting of any number of occurrences (including zero) of the elements a, b, a^{-1}, and b^{-1}, in any order. For example, some of the elements in $<X>$ are a, b, a^{-1}, b^{-1}, $b*b$, and $a*b*a*b^{-1}$. Clearly, if $x, y \in <X>$, then xy is also a product of the elements a, b, a^{-1}, and b^{-1} because its components x and y consist of the elements a, b, a^{-1}, and b^{-1}. Thus, $xy \in <X>$. $<X>$ is _____ under the group operation $*$.

852

closed

Show that $<X>$ has the inverses property (we can then conclude by the CI subgroup test that $<X>$ is a subgroup of G). *Hint*: Use the socks-shoes property (frame **453**).

853

By the socks-shoes property, the inverse of a product is the product of the inverses in reverse order. For example, the inverse of $a^{-1}abb^{-1} = (b^{-1})^{-1}b^{-1}a^{-1}(a^{-1})^{-1} = bb^{-1}a^{-1}a$. As this example illustrates, the inverses of the products in $<X>$ are also products constructed from the elements of X and their inverses. Thus, they are also in $<X>$. So $<X>$ has the inverses property. Because $<X>$ is a nonempty subset of G, and it has the closure and inverses properties, then by the CI subgroup test, it is a subgroup of G.

The elements that we add to X to get $<X>$ are precisely the elements necessary to give the resulting set the closure and inverses properties. Because any subgroup has the closure and inverses properties, any subgroup that contains all the elements in X must also contain at least all the elements in $<X>$. Thus, $<X>$ is a subset of *any* subgroup that contains all the elements in X. In this sense, $<X>$ is the smallest subgroup that contains X.

We call $<X>$ the **subgroup of G generated by X.** $<X>$ might be a proper subgroup of the group G, or it might be G itself. For example, in D_3, if $X = \{r\}$, then $<X> = \{e, r, r^2\}$, which is a proper subgroup of D_3. If, on the other hand, $X = \{r, f\}$, then $<X> = D_3$. Suppose X is a subset of a group G. If $<X> = G$, we say that G is **generated by** X or, equivalently, that X is a **generating set** for G. For example, $\{r, f\}$ is a generating set for D_3 because $<\{r, f\}> = D_3$. The generating set for a group in general is not unique. For example, $\{r, f\}$, $\{r^2, f\}$, $\{r, r^2, f\}$, and D_3 are all generating sets for D_3.

Suppose X contains only one element, say, g. That is, $X = \{g\}$. Then $<X>$ by definition contains g, g^{-1}, and all possible products consisting of any number of occurrences of g and g^{-1}. Every element in $<X>$ is equal to some power of g. For example, the product $gg^{-1}gg^{-1}$ is the element g^0. Moreover, every power of g is in $<X>$. Thus, $<X>$ has a simple characterization: it is the subgroup of G consisting of all the powers of g. We call this subgroup the **subgroup generated by g.** We typically denote this subgroup with $<g>$.

Key idea: $<g> = \{g^k : k \in \mathbb{Z}\}$ is the subgroup generated by g.

We define the **order of g,** denoted by $|g|$, as the order of the subgroup generated by g. That is, $|g| = |<g>|$.

Suppose g is an element in the group G. Is $<g>$ the smallest subgroup of G that contains g? That is, is $<g>$ a subset of every subgroup of G that contains g?

854

Yes. In the preceding frame, we observed that $<X>$ is a subset of every subgroup of G that contains all the elements in X.

CF Subgroup Test

855

If we are given that a nonempty subset of a group is closed under the group operation, we cannot conclude that the subset is a subgroup. In other words, closure alone of a subset does not guarantee that subset is a group.

Give an example of a subset of a group that is closed under the group's operation but is not a subgroup. *Hint*: Consider subsets of infinite order.

856

$\mathbb{Z}^+$ is a subset of $\mathbb{Z}$ that is closed under addition (the sum of any two positive integers is a positive integer). But $<\mathbb{Z}^+, +>$ is not a subgroup of $<\mathbb{Z}, +>$. Why not?

857

It does not have the identity and inverses properties.

In the preceding frame, if you tried to find an example of a finite subset of a group that is closed under the group's operation but is not a subgroup, you were doomed to fail. All such subsets are subgroups. Closure *alone* guarantees that a nonempty *finite* subset of a group is a subgroup. This property of finite subsets of groups gives us the following subgroup test, which we call the **CF subgroup test** ("CF" stands for "closure and finite"):

If H is a finite, nonempty subset of a group G, and H is closed under the group's operation, then H is a subgroup of G.

Let's show that the CF subgroup test is valid using the CI subgroup test. Let H be a finite nonempty subset of a group G that is closed under the operation of G. By the CI subgroup test, we need to show only that H has the inverses property (we are given that H has the closure property, so we do not have to show that). H is nonempty. Thus, there is an element $h \in H$. Consider the successive powers of h: $h, h^2, h^3, \ldots, h^j, \ldots, h^k, \ldots$. Because H is closed, every element in the sequence is in H. H is finite. Thus, some elements must repeat in this sequence; otherwise, H would be infinite. Suppose h^k is a repetition of h^j. That is, $h^j = h^k$ with $j < k \Rightarrow h^j = h^j h^n$, where $n = k - j \geq 1$. Then by the left cancellation law h^n must be e, the identity element. If $n = 1$, then h is the identity element and, therefore, is its own inverse. If $n > 1$, then $e = h^n = h^{n-1}h = hh^{n-1}$. In this case, the product of h^{n-1} and h in either order is the identity. Thus, h^{n-1} is the inverse of h. ∎

Let's find the inverse of r in the subset $H = \{e, r, r^2\}$ of D_3 using the technique that we used in the preceding proof. Consider the successive powers of r:

$$r, r^2, r^3 = e, r^4 = rr^3 = re = r.$$

Because H is finite, repeats must occur; r^4 is a repeat of r. We then have $r = r^4 \Rightarrow r = rr^3 \Rightarrow e = r^3 \Rightarrow rr^2 = r^2 r = e \Rightarrow r^{-1} = r^2$.

Show that $\{e, f\}$ is a subgroup of D_3 using the CF subgroup test.

858

$ee = e, ef = f, fe = f, ff = e$. Thus, $\{e, f\}$ is closed. It is also finite. Then, by the CF subgroup test, it is a subgroup of D_3.

Using the CF subgroup test, we can methodically determine all the subgroups of a finite group. Let's do this for $C_3 = \{e, r, r^2\}$. Every subgroup must contain e. So we start with all the subsets of C_3 that contain e. To each of the subsets of C_3 that contains e, add only those elements necessary to impart closure to the resulting set. By the CF subgroup set, the resulting sets are the subgroups of C_3. Moreover, every subgroup must be one of these resulting sets. For example, to impart closure to $\{e, r\}$, we have to add r^2 to get $\{e, r, r^2\}$, which, by the CF subgroup test, is one of the subgroups of C_3. If we start with a subset that is already closed, we do not add any elements. For example, to $\{e\}$, we do not add any elements, because $\{e\}$ is already closed and, therefore, a subgroup of C_3.

What are the missing entries in the following table for C_3:

Subsets containing e	Smallest superset with closure property
$\{e\}$	$\{e\}$
$\{e, r\}$	$\{e, r, r^2\}$
$\{e, r^2\}$	?
$\{e, r, r^2\}$	?

859

$\{e, r, r^2\}$ in both positions. Thus, $\{e\}$ and $\{e, r, r^2\}$ are the only subgroups of C_3.

One-Step Subgroup Test

860

The CI subgroup test requires two steps: show closure and show inverses. The CF subgroup test requires only one step: show closure. But it applies only to finite subsets. The one-step subgroup test, on the other hand, requires only one step and applies to infinite as well as finite subsets. Here is the *one-step subgroup test*:

If H is a nonempty subset of a group G, and $ab^{-1} \in H$ for all $a, b \in H$, then H is a subgroup of G.

Let's confirm the validity of this test by showing that a subset of a group that satisfies it has the closure, associative, identity, and inverses property. Suppose H is a nonempty subset of a group G, and $ab^{-1} \in H$ for all $a, b \in H$. We are given that H is nonempty. Thus, there must be at least one element $h \in H$. We are given that $ab^{-1} \in H$ for all $a, b \in H$. What does this condition imply if both a and b are h?

861

$hh^{-1} = e \in H$. Thus, H has the identity property.

What does the condition $ab^{-1} \in H$ for all $a, b \in H$ imply when $a = e$?

862

$eb^{-1} = b^{-1} \in H$ for all $b \in H$

Thus, H has the inverses property.

Let $x, y \in H$. Because H has the inverses property, $y^{-1} \in H$. Now replace a and b with x and y^{-1}, respectively, in $ab^{-1} \in H$. What do we get?

863

$x(y^{-1})^{-1} = xy \in H$. Thus, H has the closure property.

Given only that $ab^{-1} \in H$ for all $a, b \in H$, we have shown that H has the closure, identity, and inverses properties. H inherits associativity from G. Thus, H is a subgroup of G. ∎

$\mathbb{R}$ (the reals) and $\mathbb{Q}$ (the rationals) under multiplication are not groups (0 has no multiplicative inverse).

Is $\mathbb{R} - \{0\}$ a group under multiplication?

864

Yes. It has the closure, associative, identity, and inverses properties.

Using the one-step test, show that $\mathbb{Q} - \{0\}$ is a subgroup of $\mathbb{R} - \{0\}$ under multiplication. *Hint*: the inverse of $\frac{r}{s}$ is $\frac{s}{r}$.

865

Let $a, b \in \mathbb{Q} - \{0\}$. Then $a = \frac{p}{q}$ and $b = \frac{r}{s}$ for some $p, q, r, s \in \mathbb{Z} - \{0\}$, and $b^{-1} = \frac{s}{r}$. Thus, $ab^{-1} = (\frac{p}{q})(\frac{s}{r}) = \frac{ps}{qr}$, and ps and qr are nonzero because p, q, r, and s are nonzero. Thus, $ab^{-1} = \frac{ps}{qr} \in \mathbb{Q} - \{0\}$. ∎

We stated the one-step subgroup test using multiplicative notation. State it using additive notation.

866

If H is a nonempty subset of a group G, and $a + (-b) \in H$ for all $a, b \in H$, then H is a subgroup of G.

Use this version of the one-step subgroup test to show that $\langle n\mathbb{Z}, +\rangle$ is a subgroup of $\langle \mathbb{Z}, +\rangle$ for all $n \geq 0$.

867

Let $a, b \in n\mathbb{Z} \Rightarrow a = ni$ and $b = nj$ for some $i, j \in \mathbb{Z} \Rightarrow a + (-b) = ni + (-nj) = n(i - j) \in n\mathbb{Z}$. ∎

For all $n \geq 0$, $n\mathbb{Z}$ is a subgroup of $\mathbb{Z}$. But is the converse true? That is, is every subgroup of $\mathbb{Z}$ equal to $n\mathbb{Z}$ for some $n \geq 0$? After pondering this question for a while, you will probably guess that the converse is also true. But how do we prove it? When we learn more about groups, we will be able to answer this question.

Here, however, is a question you should be able to answer now. What are the finite subgroups of $\langle \mathbb{Z}, +\rangle$?

868

Suppose a subgroup of $\mathbb{Z}$ has a nonzero element k. Then $k, k + k, k + k + k, \ldots$ are all in the subgroup because a subgroup has the closure property. Thus, the subgroup is infinite. The only finite subgroup of $\langle \mathbb{Z}, +\rangle$ is $\langle \{0\}, +\rangle$.

Review Questions

1. List all the subgroups of $\langle \mathbb{Z}_6, +_6\rangle$.
2. List all the subgroups of $\langle \mathbb{Z}_7, +_7\rangle$.
3. List all the subgroups of V_4 (see frame **446**).
4. Which single elements in $\langle \mathbb{Z}_6, +_6\rangle$ generate $\langle \mathbb{Z}_6, +_6\rangle$?
5. Which single elements in $\langle \mathbb{Z}_7, +_7\rangle$ generate $\langle \mathbb{Z}_7, +_7\rangle$?
6. Suppose G is a group. Let $Z(G) = \{g \in G: gx = xg \text{ for all } x \in G\}$. Show that $Z(G)$ is a subgroup of G. $Z(G)$ is called the ***center of the group G.***
7. Give an example of a group G with subgroups H and K such that $H \cup K$ is not a subgroup.
8. If $\langle K, *\rangle$ is a subgroup of $\langle H, *\rangle$, and $\langle H, *\rangle$ is a subgroup of $\langle G, *\rangle$, is $\langle K, *\rangle$ a subgroup of $\langle G, *\rangle$? Justify your answer.
9. Let g be an element of a group. Show that g and g^{-1} have the same order.
10. Why does the CI subgroup test require the subset to be nonempty?

Answers to Review Questions

1. $\{0\}, \{0, 3\}, \{0, 2, 4\}, \mathbb{Z}_6$
2. $\{0\}, \mathbb{Z}_7$
3. $\{e\}, \{e, a\}, \{e, b\}, \{e, c\}, V_4$
4. 1 and 5 (the elements relatively prime to 6)
5. 1 through 6 (the elements relatively prime to 7).
6. *Closure*: Let $a, b \in Z(G)$. Then, for all $x \in G$, $abx = axb$ (because $b \in Z(G)$) $= xab$ (because $a \in Z(G)$). Thus, $ab \in Z(G)$. *Inverses*: Suppose $a \in Z(G)$. Then for all $x \in G$, $ax = xa$. Mutiplying on the left by a^{-1} and on the right by a^{-1}, we get $a^{-1}axa^{-1} = a^{-1}xaa^{-1}$ which simplifies to $xa^{-1} = a^{-1}x \Rightarrow a^{-1} \in Z(G)$. Then by the CI subgroup test, $Z(G)$ is a subgroup. ∎
7. In D_3, $\{e, f\}$ and $\{e, r, r^2\}$.
8. Yes. Because K is a subgroup of H, K must have the closure and inverses property. Because $K \subseteq H \subseteq G$, $K \subseteq G$. K is a subgroup, so it must be nonempty. Thus, by the CI subgroup test, K is a subgroup of G.

9. To get $\langle g \rangle$ from $\{g\}$, to $\{g\}$ we add g^{-1} and all possible products constructed from g and g^{-1}. To get $\langle g^{-1} \rangle$ from $\{g^{-1}\}$, to $\{g^{-1}\}$ we add g and all possible products constructed from g and g^{-1}. Thus, $\langle g \rangle$ and $\langle g^{-1} \rangle$ have the same set of elements.
10. An empty set has the closure and inverses property. Thus, the CI subgroup test would show the empty set is a subgroup (which it is not). By definition, a group (which includes any subgroup) is a nonempty set.

Homework Questions

1. Suppose $a, b \in G$, where G is a group. Does $a \neq b$ imply $\langle a \rangle \neq \langle b \rangle$? Justify your answer.
2. Suppose that G is a group and that $a, b \in G$. If $b \in \langle a \rangle$, does $\langle a \rangle = \langle b \rangle$? Justify your answer.
3. What are the elements in $Z(D_3)$? See review question 6.
4. How many subgroups does $\langle \mathbb{Z}_{36}, + \rangle$ have? List them.
5. Suppose $\langle g \rangle$ has 100 elements. How many elements does $\langle g^{25} \rangle$ have?
6. Suppose H is the subset of a commutative group G, and H contains all the elements in G of order 1 or 2. Show that H is a subgroup of G.
7. Suppose G is a group with subgroups H and K. Is $HK = \{hk : h \in H \text{ and } k \in K\}$ a subgroup of G? Justify your answer.
8. Suppose H is a subgroup of a group G, and $g \in G$. Show that gHg^{-1} is a subgroup of G; gHg^{-1} is called the ***conjugate subgroup of H by g.***
9. How do we know that 31 and 69 have the same order in $\langle \mathbb{Z}_{100}, + \rangle$?
10. Let g be an element of a group G. Then the set of all elements in G that commute with g (denoted by $C(g)$) is called the ***centralizer of g.*** Show that $C(g)$ is a subgroup of G.

15 Isomorphic Groups and Cayley's Theorem

How to Show Two Groups are Isomorphic

869

A *group G_1 is isomorphic to a group G_2* (denoted by $G_1 \cong G_2$) if the Cayley table for G_1 is "essentially the same" as the Cayley table for G_2. That is, if the Cayley tables for G_1 and G_2 differ, they differ only in the names of the elements. Thus, if G_1 is isomorphic to G_2, we should be able to transform the G_1 table into the G_2 table simply by renaming each element in the G_1 table with its corresponding element in G_2, possibly after reordering the rows and columns of G_1. For example, suppose the Cayley tables for G_1 and G_2 are as follows:

G_1

*	a	e	b
a	b	a	e
e	a	e	b
b	e	b	a

G_2

$\otimes$	e	x	y
e	e	x	y
x	x	y	e
y	y	e	x

A renaming scheme should be a one-to-one pairing of elements in G_1 with the elements in G_2, with no element left unpaired. That is, it should be a bijection from G_1 to G_2. Let's rename elements in G_1 according to following bijection:

$$e \rightarrow e$$
$$a \rightarrow x$$
$$b \rightarrow y$$

If this bijection or some other bijection successfully transforms the G_1 table into the G_2 table, then G_1 is isomorphic to G_2.

The row and column order in the G_2 table is *e-x-y*. Under our renaming scheme, this order corresponds to *e-a-b* in the G_1 table. But the G_1 table is in *a-e-b* order. Thus, we need to switch the *a* and *e* rows and the *a* and *e* columns in the G_1 table so that they are in *e-a-b* order. Do this.

870

G_1 table in *e-a-b* order:

*	e	a	b
e	e	a	b
a	a	b	e
b	b	e	a

Switching rows and columns does not change the operation defined by the table. For example, the original G_1 table specifies that $ab = e$. The new table specifies the same. If we now rename the G_1 table using our renaming scheme, we get the G_2 table. Thus, G_1 is isomorphic to G_2.

If a bijection converts the Cayley table for G_1 into the Cayley table for G_2, the inverse of that bijection is also a bijection. Moreover, it converts the Cayley table for G_2 into the Cayley table for G_1. In other words, if G_1 is isomorphic to G_2, then G_2 is isomorphic to G_1. So, if G_1 is isomorphic to G_2, we can simply say that G_1 and G_2 are isomorphic, indicating that each group is isomorphic to the other.

Demonstrating that the Cayley table for one group can be transformed into the Cayley table for another group can be a lengthy and tedious process. We need a simple test that determines if a given bijection performs the required transformation.

We say the function φ from the group $\langle G_1, *\rangle$ to the group $\langle G_2, \otimes\rangle$ is **operation preserving** if it has the following property:

$$\varphi(a * b) = \varphi(a) \otimes \varphi(b) \text{ for all } a, b \in G_1.$$

In the preceding equality, a and b are elements in G_1; thus, we write their product in the preceding equality with an intervening $*$ (the operation symbol for G_1). Similarly, $\varphi(a)$ and $\varphi(b)$ are elements in G_2; thus, we write their product with an intervening $\otimes$ (the operation symbol for G_2).

We will see that if φ is a bijection from G_1 to G_2, it successfully transforms the G_1 table into the G_2 table if and only if φ is operation preserving. Thus, to show that G_1 and G_2 are isomorphic, we simply show that an operation-preserving bijection φ from G_1 to G_2 exists. We call such a function an ***isomorphism from G_1 to G_2***.

Suppose φ is a bijection from G_1 to G_2. Let's use φ to rename the elements in the G_1 table. Assume that we have ordered the rows and columns in the G_1 table so that they are in the same order as the corresponding rows and columns in the G_2 table.

Let's take an arbitrary position in the G_1 table—say, row a/column b—and the corresponding position in the G_2 table, as illustrated in the following diagram:

The arrows indicate the mapping of the φ bijection. In the intersection of row a/column b of the G_1 table, we have the element equal to $a * b$. In the corresponding position in the G_2 table, we have the element $\varphi(a) \otimes \varphi(b)$. Thus, if φ maps $a * b$ to $\varphi(a) \otimes \varphi(b)$, it transforms the element in row a/column b of the G_1 table to the element in the corresponding position in the G_2 table.

What must be true of φ for φ to map $a * b$ to $\varphi(a) \otimes \varphi(b)$?

871

$\varphi(a * b) = \varphi(a) \otimes \varphi(b)$

Thus, if $\varphi(a * b) = \varphi(a) \otimes \varphi(b)$ for *all* $a, b \in G_1$, then it maps each element in the G_1 table to the element in the corresponding position in the G_2 table. In other words, if φ is operation preserving, it transforms the G_1 table into the G_2 table. Thus, G_1 and G_2 are isomorphic. Conversely, if φ transforms the G_1 table into the G_2 table, then φ maps $a * b$ to $\varphi(a) \otimes \varphi(b)$ for all $a, b \in G_1$. That is, $\varphi(a * b) = \varphi(a) \otimes \varphi(b)$ for all $a, b \in G_1$. In other words, if φ transforms the G_1 table into the G_2 table, then φ is operation preserving.

Key idea: To show that the group $\langle G_1, *\rangle$ is isomorphic to a group $\langle G_2, \otimes\rangle$, show that there exists a bijection φ from G_1 to G_2 such that $\varphi(a * b) = \varphi(a) \otimes \varphi(b)$ for all $a, b \in G_1$.

The operation-preserving property is often written without specifying the operation symbols:

$$\varphi(ab) = \varphi(a)\varphi(b) \text{ for all } a, b \in G_1$$

Be sure to understand that in ab on the left, the operation is G_1's operation, but in $\varphi(a)\varphi(b)$ on the right, the operation is G_2's operation. If the groups have well-established symbols for their operations, we generally write the operation-preserving property using those symbols. For example, suppose the operation in both groups is addition, represented by the symbol "+". How is the operation-preserving property written?

872

$\varphi(a + b) = \varphi(a) + \varphi(b)$ for all $a, b \in G_1$

Let's use this condition to show that $\mathbb{Z}$ under addition is isomorphic to $2\mathbb{Z}$, the set of even integers, under addition. Let φ be a function from $\mathbb{Z}$ to $2\mathbb{Z}$ defined by $\varphi(x) = 2x$ for all $x \in \mathbb{Z}$. Show that φ is an onto function.

873

Let $j \in 2\mathbb{Z} \Rightarrow j = 2k$ for some $k \in \mathbb{Z} \Rightarrow \varphi(k) = 2k = j \Rightarrow \varphi$ is onto.

Show that φ is a one-to-one function.

874

$\varphi(a) = \varphi(b) \Rightarrow 2a = 2b \Rightarrow a = b \Rightarrow \varphi$ is one-to-one.

Because φ is onto and one-to-one, it is a bijection.

Finally, show that φ is operation preserving. That is, show that $\varphi(a + b) = \varphi(a) + \varphi(b)$ for all $a, b \in \mathbb{Z}$.

875

$\varphi(a + b) = 2(a + b) = 2a + 2b = \varphi(a) + \varphi(b)$

The function φ is an operation-preserving bijection from $\mathbb{Z}$ to $2\mathbb{Z}$. Thus, the two groups are isomorphic.

Cayley's Theorem

876

Cayley's theorem states that every group is isomorphic to a group of permutations. This theorem is significant because it tells us that we are not limiting our study of groups if we study only groups of permutations.

For any group G, it is easy to determine a group of permutations (i.e., bijections) on G that is isomorphic to G. From frame **450**, we know that each row of the Cayley table for a group G is a permutation of the column labels in the Cayley table. The set of these permutations form a group under composition that is isomorphic to to G. For example, consider the following Cayley table for a group G:

	e	a	b
e	e	a	b
a	a	b	e
b	b	e	a

The permutation in table form corresponding to the a row is

$$\begin{pmatrix} e & a & b \\ a & b & e \end{pmatrix} \begin{matrix} \leftarrow \text{column labels from the Cayley table for } G \\ \leftarrow \text{row } a \text{ of the Cayley table for } G \end{matrix}$$

We can also represent this permutation as a function of x. This function is the permutation corresponding to row a of the Cayley table for G. So let's call it p_a ("p" for permuation, and "a" for row a). Row a consists of the elements of the group that equal a times the column labels. Thus, $p_a(x) = ax$ for $x \in \{e, a, b\}$.

Show that $p_a(x)$ is a permutation.

877

onto: Let $g \in G$. Then $a^{-1}g \in G$, and $p_a(a^{-1}g) = aa^{-1}g = g$.
one-to-one: Let $p_a(x) = p_a(y) \Rightarrow ax = ay \Rightarrow x = y$.

What is $p_e(x)$ in table form and as a function of x?

878

$\begin{pmatrix} e & a & b \\ e & a & b \end{pmatrix}$, $p_e(x) = ex$

What is $p_b(x)$ in table form and as a function of x?

879

$\begin{pmatrix} e & a & b \\ b & e & a \end{pmatrix}$, $p_b(x) = bx$

Let's determine the permutation equal to $p_a \circ p_b$. Do this using $p_a = \begin{pmatrix} e & a & b \\ a & b & e \end{pmatrix}$ and $p_b = \begin{pmatrix} e & a & b \\ b & e & a \end{pmatrix}$. Hint: p_b maps e to b; p_a then maps b to e. Thus, the composition maps e to e.

880

$\begin{pmatrix} e & a & b \\ e & a & b \end{pmatrix}$

This is the table for p_e. Thus, $p_a \circ p_b = p_e$. We can also determine $p_a \circ p_b$ using the definitions of $p_a(x)$, $p_b(x)$, $p_a \circ p_b$, and the Cayley table in frame **876**: $(p_a \circ p_b)(x) = p_a(p_b(x)) = p_a(bx)) = abx$.

Determine ab from the Cayley table in frame **876**.

881

$ab = e$

Thus, $(p_a \circ p_b)(x) = abx = ex$. Which permutation corresponds to ex?

882

$p_e(x)$

Thus, $p_a \circ p_b = p_e$. For this example, we have that $ab = e$ and $p_a \circ p_b = p_e$. Notice anything interesting?

883

Here p_a and p_b under composition are behaving like their corresponding elements, a and b, in G. This behavior suggests that G under its operation is isomorphic to P under composition. Let's prove this. First show that the set of permutations in P under composition is a group.

884

Closure: $p_g \circ p_h(x) = p_g(hx) = ghx = p_{gh}(x)$. Because $g, h \in G$, and G has the closure property, $gh \in G \Rightarrow p_{gh}(x) \in P$. Thus, $p_g \circ p_h \in P$.
Associativity: Function composition is associative (see frame **462**).
Identity: $p_g \circ p_e = p_{ge} = p_g$, $p_e \circ p_g = p_{eg} = p_g \Rightarrow p_e$ is the identity.
Inverses: Let h be the inverse of g. Then $p_g \circ p_h = p_{gh} = p_e$, and $p_h \circ p_g = p_{hg} = p_e \Rightarrow p_h$ is the inverse of p_g.

Because P has the closure, associativity, identity, and inverses properties, it is a group. It is called the **left regular representation** of G. Finally, we have to show that there is an operation-preserving bijection from G to P. Let φ be a function from G to P defined by $\varphi(g) = p_g$. Show that φ is a bijection from G to P.

885

Onto: Let $p_g \in P$. Then $\varphi(g) = p_g$.
One-to-one: Let $\varphi(a) = \varphi(b) \Rightarrow p_a(x) = p_b(x) \Rightarrow ax = bx \Rightarrow a = b$.

Show that φ is operation preserving.

886

$\varphi(ab) = p_{ab} = p_a \circ p_b = \varphi(a) \circ \varphi(b)$.

φ is an operation-preserving bijection from G to P. Thus, G is isomorphic to P. ∎

Review Questions

1. What are the elements in table form in the group of permutations isomorphic to $<\mathbb{Z}_2, +_2>$?
2. What are the elements in the group of permutations isomorphic to the Klein four group V_4 (see frame **446**)?
3. Find two isomorphisms from $<\mathbb{Z}, +>$ to $<2\mathbb{Z}, +>$.
4. Define an isomorphism from $<\mathbb{Z}_2, +>$ to a subgroup of $<\mathbb{Z}_4, +>$.

5. Suppose that G_1 and G_2 are groups and that e is the identity in G_2. Show that the function $\varphi: G_1 \to G_2$ from the group G_1 to the group G_2 defined by $\varphi(g) = e$ for all $g \in G_1$ has the operation-preserving property.
6. Find an isomorphism from $\mathbb{Z}_2$ under addition to $\mathbb{Z}_3 - \{0\}$ under multiplication.
7. Show that the identity function $e(g) = g$ on a group is an isomorphism.
8. Show that the set of powers of 10 is a group under multiplication that is isomorphic to $\mathbb{Z}$ under addition.
9. Show by "brute force" (i.e., test every possible case) that $\varphi: \mathbb{Z}_3 \to \mathbb{Z}_3$ defined by $\begin{pmatrix} 0 & 1 & 2 \\ 0 & 2 & 1 \end{pmatrix}$ is operation preserving.
10. Show that an isomorphism from group G_1 to group G_2 maps the identity element in G_1 to the identity element in G_2.

Answers to the Review Questions

1. $\begin{pmatrix} 0 & 1 \\ 0 & 1 \end{pmatrix}, \begin{pmatrix} 0 & 1 \\ 1 & 0 \end{pmatrix}$
2. $\begin{pmatrix} e & a & b & c \\ e & a & b & c \end{pmatrix}, \begin{pmatrix} e & a & b & c \\ a & e & c & b \end{pmatrix}, \begin{pmatrix} e & a & b & c \\ b & c & e & a \end{pmatrix}, \begin{pmatrix} e & a & b & c \\ c & b & a & e \end{pmatrix}$
3. $\varphi_1(x) = 2x$, $\varphi_2(x) = -2x$.
4. $0 \to 0, 1 \to 2$. This is an isomorphism from $\mathbb{Z}_2$ to $\{0, 2\}$, where $\{0, 2\}$ is a subgroup of $\mathbb{Z}_4$.
5. $\varphi(g_1 g_2) = e = ee = \varphi(g_1) \varphi(g_2)$
6. $0 \to 1, 1 \to 2$
7. $e(g)$ is a bijection, and $e(g_1 g_2) = g_1 g_2 = e(g_1) e(g_2)$.
8. Define φ as $\varphi(10^n) = n$ for all $n \in \mathbb{Z} \Rightarrow \varphi$ is onto. Let $x = 10^n$ and $y = 10^m$. Then $\varphi(x) = \varphi(y) \Rightarrow \varphi(10^n) = \varphi(10^m) \Rightarrow n = m \Rightarrow x = y \Rightarrow \varphi$ is one-to-one; $\varphi(10^n 10^m) = \varphi(10^{n+m}) = n + m = \varphi(10^n) + \varphi(10^m) \Rightarrow \varphi$ is operation preserving.
9. $\varphi(0 +_3 0) = \varphi(0) = 0 = 0 +_3 0 = \varphi(0) +_3 \varphi(0)$
   $\varphi(0 +_3 1) = \varphi(1) = 2 = 0 +_3 2 = \varphi(0) +_3 \varphi(1)$
   $\varphi(0 +_3 2) = \varphi(2) = 1 = 0 +_3 1 = \varphi(0) +_3 \varphi(2)$
   $\varphi(1 +_3 0) = \varphi(1) = 2 = 2 +_3 0 = \varphi(1) +_3 \varphi(0)$
   $\varphi(1 +_3 1) = \varphi(2) = 1 = 2 +_3 2 = \varphi(1) +_3 \varphi(1)$
   $\varphi(1 +_3 2) = \varphi(0) = 0 = 2 +_3 1 = \varphi(1) +_3 \varphi(2)$
   $\varphi(2 +_3 0) = \varphi(2) = 1 = 1 +_3 0 = \varphi(2) +_3 \varphi(0)$
   $\varphi(2 +_3 1) = \varphi(0) = 0 = 1 +_3 2 = \varphi(2) +_3 \varphi(1)$
   $\varphi(2 +_3 2) = \varphi(1) = 2 = 1 +_3 1 = \varphi(2) +_3 \varphi(2)$
10. Let e be the identity in G_1; $\varphi(e) \varphi(e) = \varphi(ee) = \varphi(e) \Rightarrow \varphi(e)$ is idempotent in $G_2 \Rightarrow \varphi(e)$ is the identity in G_2.

Homework Questions

1. Find two isomorphisms from $<\mathbb{Z}_4, +_4>$ to $<U_4, \cdot>$. See frame **749**.
2. An isomorphism on a group G (i.e., from G to G) is called an **automorphism**. Show that the set of all automorphisms of a group is a group under composition.
3. Construct the group of permutations isomorphic to D_3.
4. Show that $\varphi: \mathbb{R}^+ \to \mathbb{R}$ defined by $\varphi(x) = \log x$ for all $x \in \mathbb{R}^+$ is an isomorphism.
5. Show that the composition of an isomorphism from G_1 to G_2 and an isomorphism from G_2 to G_3 is an isomorphism from G_1 to G_3.
6. Show that an isomorphism φ from G_1 to G_2 maps g^{-1} to $\varphi(g)^{-1}$ for all $g \in G_1$.
7. Determine if $<2\mathbb{Z}, +>$ is isomorphic to $<3\mathbb{Z}, +>$.
8. Are $\mathbb{Z}$ and $\mathbb{Q}$ under addition isomorphic? Justify your answer.
9. Show that $\mathbb{R}$ under addition is isomorphic to $\mathbb{R}$ under the operation $\oplus$, where $a \oplus b = a + b + 1$ for all $a, b \in \mathbb{R}$.
10. Show that the function φ on the group G defined by $\varphi(a) = a^{-1}$ for all $a \in G$ is an isomorphism from G to G if G is commutative.

16 Cyclic Groups

Definition of a Cyclic Group

887

We have already seen cyclic groups in our study of the symmetries of regular polygons. Informally, a cyclic group is a group that can be generated by a single element. Here is a more precise definition: A group G is **cyclic** if and only if it has an element g such that $G = \{g^k : k \in \mathbb{Z}\}$. Recall from frame **853** that in a group with an element g, the set $\{g^k : k \in \mathbb{Z}\}$ is denoted with $<g>$. Restating the definition above using this notation, we get that a group G is cyclic if and only if it has an element g such that $G = <g>$. If $G = <g>$, we say that G **is generated by** g.

Characterizing Cyclic Groups

888

Let's derive the basic properties of a cyclic group starting from its definition. Let G be a cyclic group generated by g with the identity e. If G is finite, then in the sequence of powers of g with increasing nonnegative exponents, an element must repeat (otherwise G would be infinite). That is, in the sequence

$$g^0 = e, g^1, g^2, \ldots, g^n, \ldots \quad \text{first repeat}$$

there must eventually be a power of g equal to a power of g earlier in the sequence. Suppose g^n is the first power in the sequence equal to a preceding power. All the elements preceding g^n must be distinct (otherwise, g^n would not be the first element to repeat in the sequence).

Suppose the first repeat is g^4. Is it possible that g^4 is a repeat of g? That is, is it possible that $g^4 = g$? Let's suppose $g^4 = g$. Then factoring out g on the left side and multiplying by e on the right side, we get $gg^3 = ge$. What does this equation imply about g^3? *Hint*: Use the left cancellation law.

889

By the left cancellation law, $g^3 = e$.

But this is impossible. Why?

890

Because then g^3, not g^4, would be the first repeat (of $g^0 = e$).

It is also not possible that $g^4 = g^2$. Show this. *Hint*: Factor out g^2 on the left side and multiply the right side by e. Then use the left cancellation law to cancel g^2.

891

We get $g^2g^2 = g^2e$. Canceling g^2, we get $g^2 = e$, in which case g^2 is the first repeat. But g^4 is the first repeat.

Show that it is not possible that $g^4 = g^3$.

892

Factoring g^3 out of the left side and multiplying the right side by e, we get $g^3g = g^3e$. Canceling g^3, we get $g = e$, in which case g is the first repeat. But g^4 is the first repeat.

The element g^4 cannot be a repeat of any of the powers of g with exponents 1 through 3. Thus, the first repeat, g^4, must be a repeat of _____.

893

$g^0 = e$

Because $g^4 = e$, the entire sequence of elements from g^0 to g^3 repeats starting with g^4:

$$\ldots, g^4 = e, g^5 = g^4g = eg = g, g^6 = g^4g^2 = eg^2 = g^2, g^7 = g^4g^3 = eg^3 = g^3, \ldots$$

g^8 is $g^4g^4 = ee = e$. The identity e reappears with g^8 and with each subsequent power of g whose exponent is a multiple of 4. From each e up to the next e are the elements e, g, g^2, g^3. In other words, the sequence

$$g^0 = e, g^1, g^2, g^3, g^4, g^5, g^6, g^7, g^8, g^9, \ldots$$

is equal to the sequence

$$g^0 = e, g^1, g^2, g^3, e, g^1, g^2, g^3, e, g^1, \ldots$$

Thus, every power of g with a nonnegative exponent is equal to one of the elements in $\{e, g, g^2, g^3\}$.

Justify the following statement: the set $\{e, g, g^2, g^3\}$ is closed under the group's operation.

894

The product of any two elements in $\{e, g, g^2, g^3\}$ equals a power of g with a nonnegative exponent, which, by the preceding frame, is equal to an element in $\{e, g, g^2, g^3\}$.

Are the powers of g with negative exponents also in $\{e, g, g^2, g^3\}$? Before we answer this question, let's determine the inverse of g. We previously concluded that if g^4 is the first repeat, then $e = g^4 \Rightarrow g^3g = e$ and $gg^3 = e$. Thus, $g^{-1} =$ _____.

895

g^3

Consider g^{-p} where $p > 0$. By definition, $g^{-p} = (g^{-1})^p$. Substituting g^3 for g^{-1}, we get $(g^3)^p = g^{3p}$. Because $p > 0$, g^{3p} is a power of g with a positive exponent. By frame **893**, all powers of g with nonnegative exponents are in $\{e, g, g^2, g^3\}$. Thus, $g^{-p} \in \{e, g, g^2, g^3\}$ for all $p > 0$.

A simpler way to show that all the powers of g—the powers with nonnegative exponents as well as the powers with negative exponents—are in $\{e, g, g^2, g^3\}$ is to use the division algorithm. Consider g^k, an arbitrary power of g (k can be any integer). Divide k, by 4. We get $k = 4q + r$. Substituting for k in g^k, we get $g^{4q+r} = g^{4q}g^r = (g^4)^q g^r = e^q g^r = eg^r = g^r$ where, by the division algorithm, $0 \leq r < 4$. Thus, g^k is in $\{e, g, g^2, g^3\}$ for any integer k.

Let's summarize what we have discovered so far. If g is a generator of a finite cyclic group, and g^4 is the first repeat among the successive powers of g with positive exponents, then $g^4 = e$, and all the elements in $\{e, g, g^2, g^3\}$ are distinct. Moreover, all the powers of g are in $\{e, g, g^2, g^3\}$. Thus, $\langle g \rangle$, the cyclic group generated by g, consists of precisely the four elements in $\{e, g, g^2, g^3\}$.

Suppose that in a cyclic group generated by g, the first power with a positive exponent to repeat is g^{10}. What does g^{10} equal, how many elements are in the group, what are they, what is the inverse of g^3, and what is the inverse of g^i where $0 \leq i < 10$? *Hint*: Use the same sort of reasoning we used for the case when g^4 is the first repeat.

896

$g^{10} = e$, 10 elements, $\langle g \rangle = \{e, g, g^2, \ldots, g^9\}$, the inverse of g^3 is g^7 because $g^3g^7 = g^{10} = e$, the inverse of g^i is g^{10-i} because $g^i g^{10-i} = g^{10} = e$.

Suppose G is a cyclic group generated by g and that the first power with a positive exponent to repeat is g^n for some $n > 0$. What does g^n equal, how many elements are in the group, what are they, what is the inverse of g^3, what is the inverse of g^i where $0 \leq i < n$, and what does g^n equal?

897

$g^n = e$, n elements, $\langle g \rangle = \{e, g, g^2, ..., g^{n-1}\}$, the inverse of g^3 is g^{n-3}, the inverse of g^i is g^{n-i}, $g^n = e$

Key idea: If G is a finite cyclic group generated by g, and the first power with a positive exponent to repeat is g^n (or, equivalently, n is the smallest positive exponent such that $g^n = e$), then the order of G is n, and its elements are $e, g, g^2, ..., g^{n-1}$.

In the cyclic group of order 5 generated by g, what is the inverse of g^3?

898

$g^{-3}g^3 = g^0 = e$, and $g^2g^3 = g^5 = e$. Both g^{-3} and g^2 are the inverse of g^5. Inverses in a group are unique. Thus, $g^{-3} = g^2$.

In an infinite cyclic group generated by g, what is the inverse of g^3?

899

g^{-3}. Unlike the cyclic group of order 5, there is no power of g with a nonnegative exponent equal to g^{-3}.

Show that all cyclic groups are commutative. *Hint*: Let $x = g^j$ and $y = g^k$, where g is the generator of the group. Then show $xy = yx$.

900

Suppose G is a cyclic group generated by g. Let $x, y \in G \Rightarrow x = g^j$ and $y = g^k$ for some $j, k \in \mathbb{Z} \Rightarrow xy = g^jg^k = g^{j+k} = g^{k+j} = g^kg^j = yx$. ∎

Our definition of a cyclic group in frame **887** uses multiplicative notation. Write the equivalent definition using additive notation.

901

A group G is **cyclic** if and only if it has an element g such that $G = \{kg : k \in \mathbb{Z}\}$. In other words, a group G is cyclic if and only if it has an element g such that G is the set of all multiples of g.

If we are working with a group that uses additive notation, we, of course, should use the preceding definition.

Is $\langle \mathbb{Z}, + \rangle$ a cyclic group?

902

Yes. It is generated by 1. The set of all multiples of 1 is equal to $\mathbb{Z}$.

What other element can generate $\mathbb{Z}$?

903

-1

Is $\langle \mathbb{Z}_4, +_4 \rangle$ cyclic?

904

Yes. $\langle \mathbb{Z}_4, +_4 \rangle$ is generated by 1.

Is $\langle \mathbb{Z}_n, +_n \rangle$ cyclic?

905

Yes. It is generated by 1.

Cyclic Subgroups of a Group

906

Suppose G is a group (not necessarily cyclic), and $g \in G$. Show that $<g> = \{g^k : k \in \mathbb{Z}\}$ is a subgroup of G. Use the CI subgroup test.

907

Closure: The product of any two powers of g is a power of g.
Inverses: The inverse of g^i is g^{-i}, which is a power of g.

The set $<g>$ is a subgroup of G. Is $<g>$ a cyclic subgroup of G?

908

Yes, because $<g>$ is the set of all the powers of g. It is the cyclic group generated by g.

Determine all the cyclic subgroups of $C_3 = \{e, r, r^2\}$. That is, determine $<e>$, $<r>$, and $<r^2>$.

909

$<e> = \{e\}$, $<r> = C_3$, $<r^2> = C_3$

Determine all the cyclic subgroups of D_3. That is, determine $<e>$, $<r>$, $<r^2>$, $<f>$, $<fr>$, and $<fr^2>$. Hint: $r^3 = f^2 = (fr)^2 = (fr^2)^2 = e$.

910

$<e> = \{e\}$, $<r> = \{e, r, r^2\}$, $<r^2> = \{e, r, r^2\}$, $<f> = \{e, f\}$, $<fr> = \{e, fr\}$, $<fr^2> = \{e, fr^2\}$

These are all the cyclic subgroups of D_3. None, however, is D_3. Thus, D_3 itself is not a cyclic group.

Does every group have at least one cyclic subgroup? Are there groups that have both cyclic and noncyclic subgroups?

911

The trivial group is a cyclic subgroup of every group. So every group has at least one cyclic subgroup. In any noncyclic group G, G is a noncyclic subgroup of itself, and the trivial subgroup is a cyclic subgroup. So every noncyclic group has both cyclic and noncyclic subgroups.

An interesting question to ponder: Every subgroup of the cyclic group C_3 is cyclic. Is this true for every cyclic group? That is, are the subgroups of a cyclic group always cyclic? We will answer this question shortly.

Isomorphism of Cyclic Groups of the Same Order

912

Let $<G, *>$ and $<H, \otimes>$ be cyclic groups of order n generated by g and h, respectively. Let's show that G and H are isomorphic.

List all the elements of G. Below that, list all the elements of H.

913

$g^0, g^1, ..., g^{n-1}$
$h^0, h^1, ..., h^{n-1}$

There is an obvious bijection φ from G to H. Define this bijection.

914

The function $\varphi(g^i) = h^i$ for $0 \leq i < n$.

182   Chapter 16: Cyclic Groups

Show that for all $g^j, g^k \in G$, $\varphi(g^j * g^k) = \varphi(g^j) \otimes \varphi(g^k)$.

915

$\varphi(g^j * g^k) = \varphi(g^{j+k}) = h^{j+k} = h^j \otimes h^k = \varphi(g^j) \otimes \varphi(g^k)$

The function φ is an operation-preserving bijection from G to H. Thus, G and H are isomorphic. ∎

Because all the cyclic groups of order n are isomorphic, in the "up to isomorphism" sense, there is only one cyclic group of order n. To represent the cyclic group of order n, we will generally use $C_n = (e, r, r^2, ..., r^{n-1})$. Recall that for $n > 2$, C_n is the group of symmetries of an n-sided regular polygon that start and end with the T side up (see frame **768**). C_2 is the group of symmetries of a line segment. C_1 is the group of symmetries of an infinitesimal point. $\langle \mathbb{Z}_n, +_n \rangle$ is also commonly used to represent the cyclic group of order n. However, we will use C_n instead.

Let G and H be infinite cyclic groups. Are they necessarily isomorphic?

916

Yes. The proof for the finite case works for the infinite case. Simply extend φ so that it maps g^i to h^i for all $i \in \mathbb{Z}$.

Order of r^k in C_n

917

We often speak of the ***order of an element g*** (denoted by $|g|$) in a group. One definition of $|g|$ is that it is the smallest positive n such that $g^n = e$, if such an n exists. If no such n exists, then $|g|$ is infinite. We can also define $|g|$ simply as the order of the subgroup $\langle g \rangle$. As we learned in frame **897**, $|\langle g \rangle|$ is the smallest positive n such that $g^n = e$, if such an n exists. Otherwise, $|\langle g \rangle|$ is infinite. Thus, the two definitions of $|g|$ are equivalent.

Let's investigate the orders of the elements of the cyclic group C_n. Because cyclic groups of the same order are isomorphic (see frame **912**), the results we get for C_n apply to any cyclic group of order n. Before reading further, you may want to review the section "Finding Multipliers" starting at frame **639**.

In $C_3 = \{e, r, r^2\}$, what is the order of r^2, and what subgroup does it generate?

918

$(r^2)^0 = e$, $(r^2)^1 = r^2$, $(r^2)^2 = r$, $(r^2)^3 = e$. The first power of r^2 that equals e is $(r^2)^3$. Thus, $|r^2| = 3$.
$\langle r^2 \rangle = \{e, r, r^2\} = C_3$

What is the order of r^2 in C_6? Does it generate C_6?

919

$|r^2| = 3$, r^2 generates $\{e, r^2, r^4\} \neq C_6$

Determine the order and the subgroup generated by each element in C_9.

920

$|e| = 1$, $\langle e \rangle = \{e\}$
$|r| = 9$, $\langle r \rangle = C_9$
$|r^2| = 9$, $\langle r^2 \rangle = C_9$
$|r^3| = 3$, $\langle r^3 \rangle = \{e, r^3, r^6\}$
$|r^4| = 9$, $\langle r^4 \rangle = C_9$
$|r^5| = 9$, $\langle r^5 \rangle = C_9$
$|r^6| = 3$, $\langle r^6 \rangle = \{e, r^3, r^6\}$
$|r^7| = 9$, $\langle r^7 \rangle = C_9$
$|r^8| = 9$, $\langle r^8 \rangle = C_9$

Based on these results, propose a simple test for determining which elements of C_9 generate C_9. *Hint*: What do the exponents 1, 2, 4, 5, 7, and 8 have in common with respect to the order of C_9? Each power of r with one of these exponents generates C_9.

921

They are all relatively prime to 9.

Let's investigate the order of r^4 in C_9. Consider the sequence of powers of r^4:

$$(r^4)^0 = e, (r^4)^1 = r^4, (r^4)^2 = r^8, \ldots$$

We get distinct elements until e reappears. The elements in this sequence are $(r^4)^i = r^{4i}$ for $i = 0, 1, 2 \ldots$. Because $r^9 = e$, e will appear whenever the exponent of r^{4i} is some multiple of 9.

What is the smallest positive value of i such that $4i$ is a multiple of 9?

922

The integers 4 and 9 have no common factors. To make $4i$ a multiple of 9, i itself must be a multiple of 9. The smallest positive value of i such that $4i$ is a multiple of 9 is 9. Thus, our sequence does not repeat until $i = 9$:

$$(r^4)^0, (r^4)^1, (r^4)^2, \ldots, (r^4)^8, (r^4)^9 = e, \ldots$$

Thus, $|r^4| = 9$. Let's apply the same analysis to the element r^6. We want to find the smallest positive i such that $(r^6)^i = r^{6i} = e$. We get e whenever the exponent is a multiple of 9. Thus, $|r^6|$ is the smallest positive value of i such that $6i$ is a multiple of 9. The integers 6 and 9 are not relatively prime; 6 has 3 as a factor, which is one of the two 3 factors of 9. Thus, to get a multiple of 9 from $6i$, i has to provide the other 3 factor. The smallest positive value of i that provides a factor of 3 is 3 itself. Thus, $|r^6| = 3$.

Let's apply the same analysis to the general case. Let's determine the order of r^k in C_n, where $k \neq 0$ (if $k = 0$, then $|r^k| = |e| = 1$). The order of r^k is the smallest positive i such that ki is a multiple of n. Thus, ki must have all the prime factors of n. Every prime factor in n that is not in k has to be provided by i. What then must i be to make ki a multiple of n? *Hint*: gcd(n, k) contains all the prime factors that n and k have in common.

923

Because gcd(n, k) has all the prime factors common to n and k, n/gcd(n, k) has all the prime factors of n *not* provided by k. Thus, $i = n$/gcd(n, k) is the smallest positive value of i that makes ki a multiple of n.

Let's summarize: to make $(r^k)^i$ equal to e in C_n, ki has to be a multiple of n. Thus, for $k \neq 0$, i has to provide whatever factors in n that are not in k; k provides the factors in n given by gcd(n, k). Thus, i has to provide the factors in n/gcd(n, k); i can be any positive multiple of n/gcd(n, k). But we want the smallest positive value of i that provides the required factors. Thus, we want $i = n$/gcd(n, k). From this analysis, we get the following rules:

1. $|r^k| = n/\gcd(k, n)$ in C_n.
2. If $k \mid n$, then $|r^k| = n/k$.
3. If k and n are relatively prime, then $|r^k| = n$.

The second rule is a special case of the first rule: if $k \mid n$, then gcd(n, k) = k, making n/gcd(n, k) = n/k. The third rule is also a special case of the first rule. If k and n are relatively prime, then gcd(n, k) = 1, making n/gcd(n, k) = n.

What are the orders of each element in the cyclic group C_{12} generated by r?

924

$|e| = 1, |r| = 12, |r^2| = 6, |r^3| = 4, |r^4| = 3, |r^5| = 12, |r^6| = 2, |r^7| = 12, |r^8| = 3, |r^9| = 4, |r^{10}| = 6, |r^{11}| = 12$

What is the order of each element in C_{101} not equal to e? *Hint*: 101 is a prime number.

925

$101/\gcd(101, k) = 101/1 = 101$

What is the order of r^{20} in C_{100}?

184   Chapter 16: Cyclic Groups

926

$100/\gcd(100, 20) = 100/20 = 5$

What is the order of r^{16} in C_{81}?

927

$81/\gcd(81, 16) = 81/1 = 81$

What is the order of r^4 in C_6?

928

$6/\gcd(6, 4) = 6/2 = 3$

What is the order of 4 in $<\mathbb{Z}_6, +_6>$? *Hint*: This question is essentially the same as the question in the preceding frame.

929

r is the generator for C_6; 1 is the generator for $\mathbb{Z}_6$. Thus, r^4 in C_6 (using multiplicative notation here) corresponds to $4 \cdot 1 = 4$ in $\mathbb{Z}_6$ (using additive notation here). $|4|$ is the smallest positive integer n such that $4n$ is a multiple of 6. Thus, $|4| = 3$.

What is the order of 4 in $<\mathbb{Z}_{15}, +_{15}>$?

930

$15/\gcd(4, 15) = 15/1 = 15$

Suppose $\gcd(k, n) = \gcd(j, n)$. What can we say about the orders of r^k and r^j in C_n?

931

$|r^k| = n/\gcd(k, n) = n/\gcd(j, n) = |r^j|$

Subgroups of Cyclic Groups

932

In this section we show that cyclic groups have a very interesting property: every subgroup of a cyclic group is cyclic.

List all the subgroups of C_6.

933

$\{e\}, \{e, r^3\}, \{e, r^2, r^4\}, C_6$

Are all these subgroups cyclic?

934

Yes. They are generated by e, r^3, r^2, and r, respectively.

List all the subgroups of the cyclic group C_{12}.

935

$\{e\}, \{e, r^6\}, \{e, r^4, r^8\}, \{e, r^3, r^6, r^9\}, \{e, r^2, r^4, r^6, r^8, r^{10}\}, C_{12}$

Are they all cyclic?

936

Yes. They are generated by e, r^6, r^4, r^3, r^2, and r, respectively.

Moreover, each subgroup, except for $\{e\}$, is generated by the power of r it contains with the smallest positive exponent. For example, $\{e, r^4, r^8\}$ is generated by r^4, the power in that subgroup with the smallest positive exponent. Let's see if these properties are hold for all subgroups of cyclic groups.

Let H be a subgroup of a cyclic group $G = <g>$. Suppose g^4 is the power of g in H with the smallest positive exponent. Thus, H does not contain g, g^2, or g^3. By the closure property of groups, H has to contain all the positive powers of g^4 (i.e., $(g^4)^i = g^{4i} \in H$ for all $i > 0$). By the inverses property, H also has to contain the inverses of the positive powers of g^4. Finally, H has to contain $(g^4)^0 = e$, the identity. In short, H has to contain all the powers of g^4:

$$\ldots, (g^4)^{-2} = g^{-8}, (g^4)^{-1} = g^{-4}, (g^4)^0 = e, (g^4)^1 = g^4, (g^4)^2 = g^8, \ldots$$

Is it possible for H to *also* contain an element that is not a power of g^4? That is, can H contain a power of g whose exponent is not a multiple of 4? For example, can it contain g^{10}, which is between g^8 and g^{12}? Suppose H contains g^{10}. Then $g^{10}g^{-8} = g^2$ would necessarily be in H. Why?

937

g^{-8} is in H. Thus, if g^{10} is also in H, then by the closure property, $g^{10}g^{-8} = g^2$ is also in H.

But g^2 cannot be in the subgroup. Why not?

938

Because g^4 is the power of g in the subgroup with the smallest positive exponent. Thus, H contains all and *only* the powers of $g^4 \Rightarrow H$ is a cyclic subgroup of G generated by g^4.

We can easily extend the argument that we used to show that g^4 generates H (by using g^k rather in place of g^4) to prove the following theorem:

Suppose $H \neq \{e\}$ is a subgroup of a cyclic group $G = <g>$. If g^k is the power of g in H with the smallest positive exponent, then H is a cyclic subgroup of G generated by g^k.

What if $H = \{e\}$? Then $H = <e>$. That is, H is a cyclic subgroup of G generated by e.

The foregoing analysis gives us the following theorem:

If H is a subgroup of the cyclic group, then H itself is cyclic.

Determine the subgroup generated by each element of C_8.

939

$<e> = \{e\}$, $<r> = C_8$, $<r^2> = \{e, r^2, r^4, r^6\}$, $<r^3> = C_8$, $<r^4> = \{e, r^4\}$, $<r^5> = C_8$, $<r^6> = \{e, r^6, r^4, r^2\}$, $<r^7> = C_8$

Answer the following question based on the subgroups of C_8 listed above: if a subgroup H of C_8 is generated by r^k, is r^k the power of r in H with the smallest positive exponent?

940

Not necessarily. Counterexample: $<r^6>$, the subgroup generated by r^6, contains r^2.

This counterexample does *not* contradict our theorem in frame **938**. The subgroup $\{e, r^6, r^4, r^2\}$ is generated by the power of r with the smallest positive exponent, namely, r^2. It is also generated by r^6. The theorem does not imply that r^2 is the *only* power of r that generates this subgroup.

Our theorem in frame **938** tells us that every subgroup of $<\mathbb{Z}, +>$ is cyclic because $<\mathbb{Z}, +>$ is a cyclic group. Thus, if H is a subgroup of $<\mathbb{Z}, +>$, it is generated by some $n \in \mathbb{Z}$. That is, H is the set of all the multiples of n (recall that in additive notation, we use multiples, not powers, to define a cyclic group). We denote the set of all multiples of n with $n\mathbb{Z}$. In other words, every subgroup of $<\mathbb{Z}, +>$ is equal to $n\mathbb{Z}$ for some integer $n \geq 0$. In frame **848**, we learned that for $n \geq 0$, $<n\mathbb{Z}, +>$ is a subgroup of $<\mathbb{Z}, +>$. We now know that these are the only subgroups of $<\mathbb{Z}, +>$.

186   Chapter 16: Cyclic Groups

Order of a Subgroup of a Cyclic Group

941

Does the order of every subgroup of C_8 divide the order of C_8 (see frame **939**)?

942

Yes

In this section, we show that out observation in the preceding frame is true in general. That is, if G is a finite cyclic group with subgroup H, then the order of H divides the order of G. We also show that if k divides the order of a cyclic group, then there exists a subgroup of order k. For example, consider the cyclic group C_6 generated by r. Because 2 divides $|C_6|$, there must be a subgroup of C_6 of order 2 (it is $\{e, r^3\}$).

If H is the subgroup $\{e\}$ of a finite group G, does the order of H divide the order of G?

943

Yes. If $H = \{e\}$, then $|H| = 1$. By D2 in frame **547**, $|H|$ divides $|G|$.

If $H \neq \{e\}$, what do we know about H?

944

By the theorem in frame **938**, $H = <g^k>$, where g^k is the power of g in H with the smallest positive exponent.

What is the order of $<g^k>$?

945

We know from frame **923** that $|H| = |<g^k>| = |G|/\gcd(|G|, k)$.

Solve for $|G|$ in terms of $|H|$ and $\gcd(|G|, k)$.

946

$|G| = |H| \cdot \gcd(|G|, k) \Rightarrow |H|$ divides $|G|$. ∎

Suppose G is a cyclic group and $|G| = n$. We have just shown that if H is a subgroup of G and $|H| = m$, then m divides n. Let's now show the converse: if $m > 0$ and m divides n, then there is a subgroup H such that $|H| = m$.

What are positive divisors of 30?

947

1, 2, 3, 5, 6, 10, 15, and 30

For each positive divisor of 30, is there a subgroup in C_{30} with that order?

948

Yes. The subgroups are $\{e\}$, $<r^{15}>$, $<r^{10}>$, $<r^6>$, $<r^5>$, $<r^3>$, $<r^2>$, C_{30}, respectively.

3 is one of the divisors of 30. To determine the generator of the subgroup of order 3, we simply divide 30 by 3 to get 10. 10 is the exponent of the power of r that generates the subgroup of order 3. That is, $|<r^{10}>| = 3$. What are the elements in $<r^{10}>$?

949

$\{e, r^{10}, r^{20}\}$

This procedure makes sense: $3 \cdot 10 = 30$. Thus, it takes three 10's to get to 30. Correspondingly, it takes three successive powers of r^{10} ($[r^{10}]^1$, $[r^{10}]^2$, $[r^{10}]^3$) to get a power of r with the exponent of 30 (which, therefore, is equal to e). Thus, $|<r^{10}>| = 3$.

What is the generator of the subgroup of order 15 in C_{30}?

950

r^2

What is the generator of the subgroup of order 5 in C_{30}?

951

r^6

In each of the preceding cases, we get the exponent of the generator of the subgroup by dividing the desired order into n. Suppose m divides n. What is the generator for the subgroup of order m in C_n?

952

$r^{n/m}$

The following theorem summarizes the preceding analysis:

Suppose G is a cyclic group of order n and $m > 0$. Then there is a subgroup H of G of order m if and only if $m \mid n$.

Order of a Subgroup of a Finite Group

953

What are the subgroups of D_3?

954

$\{e\}, \{e, r, r^2\}, \{e, f\}$

Does the order of each subgroup of D_3 divide $|D_3|$?

955

Yes

This example suggests the following conjecture: for every finite group G (not just the finite cyclic groups), if H is a subgroup of G, then $|H|$ divides $|G|$. We will investigate this conjecture in the next chapter.

Review Questions

1. Find all the subgroups of C_{16}.
2. Find all the subgroups of C_{17}.
3. What is the order of r^4 in C_{16}?
4. What is the order of r^6 in C_{16}?
5. What is the order of 6 in $\mathbb{Z}_{16}$? *Hint*: See review question 4.
6. Find all the generators of $\mathbb{Z}_{20}$.
7. List all the elements in the subgroup $<r^6>$ of C_{15}.
8. Under what circumstances is $|g| = |g^2|$?
9. List all the elements in the subgroup $<6>$ of $\mathbb{Z}_{10}$.
10. List all the elements in the subgroup $<\alpha>$ of S_6, where $\alpha = \begin{pmatrix} 1 & 2 & 3 & 4 & 5 & 6 \\ 2 & 1 & 6 & 4 & 3 & 5 \end{pmatrix}$.

Answers to the Review Questions

1. $C_{16}, \{e, r^2, r^4, r^6, r^8, r^{10}, r^{12}, r^{14}\}, \{e, r^4, r^8, r^{12}\}, \{e, r^8\}, \{e\}$
2. Because 17 is prime, only $\{e\}$ and C_{17}

3. $16/\gcd(4, 16) = 16/4 = 4$
4. $16/\gcd(6, 16) = 8$
5. Same as for review question 4
6. All the elements that are relatively prime to 20: 1, 3, 7, 9, 11, 13, 17, 19
7. $r^6, r^{12}, r^3, r^9, r^{15} = e$
8. $|g^2| = |g|/\gcd(|g|, 2)$. If $|g|$ is odd, then $\gcd(|g|, 2) = 1$, so $|g^2| = |g|$. But if $|g|$ is even, then $|g^2| = |g|/2$. $|g^2|$ also equals $|g|$ if $|g|$ is infinite.
9. $\{6, 2, 8, 4, 0\}$
10. $\alpha = (1\ 2) \circ (3\ 6\ 5)$, $\alpha^2 = (3\ 5\ 6)$, $\alpha^3 = (1\ 2)$, $\alpha^4 = (3\ 6\ 5)$, $\alpha^5 = (1\ 2) \circ (3\ 5\ 6)$, $\alpha^6 = (1)$

Homework Questions

1. List all the subgroups of C_{120}.
2. What is the order of r^{18} in C_{120}?
3. What is the order of 6 in $\mathbb{Z}_{46}$?
4. List all the elements in $<\beta>$ where $\beta = (1\ 2) \circ (3\ 5\ 7) \circ (4\ 6)$.
5. List all the cyclic subgroups of D_4.
6. Are all the subgroups of D_3 cyclic? Justify your answer.
7. Suppose every proper subgroup of a group G is cyclic. Does it follow that G is cyclic? Justify your answer.
8. Is the intersection of two cyclic subgroups of the same group a cyclic subgroup? Justify your answer.
9. Is the union of two cyclic subgroups of the same group a cyclic subgroup? Justify your answer.
10. In a group G with $x, y \in G$, show that x and yxy^{-1} have the same order.

17 Coset Decomposition and Lagrange's Theorem

Properties of Cosets

956

Suppose G is a group with subgroup $H = \{h_1, h_2, h_3, \ldots\}$ and $g \in G$. Then gH, the ***left coset of H containing g***, is the set of elements obtained by multiplying g by each element of H, with g as the left operand of each product. That is, $gH = \{gh_1, gh_2, gh_3, \ldots\} = \{gh : h \in H\}$. Hg, the ***right coset of H containing g***, is the set of elements obtained by multiplying each element of H by g, with g as the right operand of each product. That is, $Hg = \{h_1g, h_2g, h_3g, \ldots\} = \{hg : h \in H\}$. The element g is called a ***representative*** of the cosets gH and Hg. The ***order of a coset*** H (denoted by $|H|$), like the order of a group, is the number of elements it contains.

Determine all the left and right cosets of $H = \{e, f\}$ in $D_3 = \{e, r, r^2, f, fr, fr^2\}$. Use only e, r, r^2, f, fr, or fr^2 to represent the elements of each coset. For example, $fr^2H = \{fr^2e, fr^2f\}$. By the flip-rotate commutation rule (see frame **803**), $fr^2f = ffr$, which equals r because $ff = e$. Thus, $fr^2H = \{fr^2e, fr^2f\} = \{fr^2, r\}$.

957

	Left Cosets				Right Cosets	
eH	=	$\{ee, ef\}$	=	$\{e, f\}$	He =	$\{e, f\}$
rH	=	$\{re, rf\}$	=	$\{r, fr^2\}$	Hr =	$\{r, fr\}$
r^2H	=	$\{r^2e, r^2f\}$	=	$\{r^2, fr\}$	Hr^2 =	$\{r^2, fr^2\}$
fH	=	$\{fe, ff\}$	=	$\{f, e\}$	Hf =	$\{f, e\}$
frH	=	$\{fre, frf\}$	=	$\{fr, r^2\}$	Hfr =	$\{fr, r\}$
fr^2H	=	$\{fr^2e, fr^2f\}$	=	$\{fr^2, r\}$	Hfr^2 =	$\{fr^2, r^2\}$

These cosets exhibit some interesting characteristics. Of course, what is true about these cosets may not be true about cosets in general. Let's first see what characteristics these cosets have, along with some other examples. We can then make some conjectures about cosets in general based on these examples. Finally, we can try to prove our conjectures.

Does $(fr)H = f(rH)$, where $f(rH)$ is defined as the set $\{fx : x \in rH\}$? *Hint*: Use $f^2 = e$ and the flip-rotate commutation rule.

958

Yes. $(fr)H = \{fre, frf\} = \{fr, ffr^2\} = \{fr, r^2\}$; $f(rH) = f\{r, rf\} = \{fr, frf\} = \{fr, ffr^2\} = \{fr, r^2\}$.

Does every coset contain its representative? For example, f is a representative of fH. Does fH contain f?

959

Yes. $e \in H \Rightarrow fe \in fH \Rightarrow f \in fH$.

How does the order of H compare with the order of each coset? For example, the order of H is 2. Are the orders of rH, r^2H, fH, frH, and fr^2H also 2?

960

Every coset of H has the same order as H.

r is in the coset $\{fr^2, r\}$. What does rH equal?

961

$rH = \{r, fr^2\}$. In this example, the element r in the coset $\{r, fr^2\}$ can represent that coset. That is, $rH = \{r, fr^2\}$.

Can each coset be represented by *any* element in that coset? For example, can we represent the left coset $\{r, fr^2\}$ with either of its elements, r or fr^2? That is, are rH and fr^2H both equal to $\{r, fr^2\}$?

962

Yes

Characterize the left cosets of H. Are they all equal? Are they all disjoint?

963

The left cosets are either equal or disjoint. Similarly, the right cosets are either equal or disjoint.

How do the left cosets compare with their corresponding right cosets? For example, how do rH and Hr compare?

964

Some but not all of the left cosets equal their corresponding right cosets. For example, $fH = Hf$ but $rH \neq Hr$.

Does the set of left cosets partition D_3?

965

Some of the six left cosets are duplicates. Thus, the set of the six left cosets is not a partition. However, the set of *distinct* left cosets is a partition of D_3: $\{\{e, f\}, \{r, fr^2\}, \{r^2, fr\}\}$. Similarly, the set of distinct right cosets partition D_3.

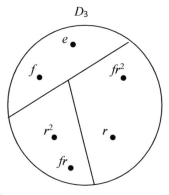 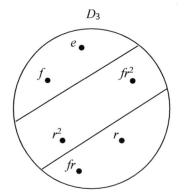

Left cosets of $\{e, f\}$ are the pieces of the D_3 pie.

Right cosets of $\{e, f\}$ cut up the D_3 pie in a different way.

We call the partition of D_3 into left cosets or into right cosets the **coset decomposition** of D_3. Of course, based on this one example, we cannot conclude that for every group G with subgroup H, the cosets of H partition G. However, for now, let's assume it is true. If it is true, and G is finite, there is a simple relationship between $|G|$, $|H|$, and the number of distinct left cosets. What is this relationship? *Hint*: The size of every coset is $|H|$.

966

$|G|$ = [number of distinct left cosets] · $|H|$, which implies that $|H|$ divides $|G|$ (i.e., the order of the subgroup H divides the order of G).

From frame **946**, we know the order of every subgroup of a finite cyclic group divides the order of the group. We see in the preceding example that the order of the subgroup H divides the order of the noncyclic group D_3. Perhaps what is true for the finite cyclic groups (i.e., the order of every subgroup of a finite group G divides the order of the group G) is true for all finite groups. Let's investigate this conjecture further with more examples.

$K = \{e, r, r^2\}$ is another subgroup of D_3. Determine the left and right cosets of K in D_3.

Left Cosets	Right Cosets
$eK = rK = r^2K = \{e, r, r^2\}$	$Ke = Kr = Kr^2 = \{e, r, r^2\}$
$fK = frK = fr^2K = \{f, fr, fr^2\}$	$Kf = Kfr = Kfr^2 = \{f, fr, fr^2\}$

Again, we see that both the distinct left cosets and the distinct right cosets partition D_3, and they all have the same order as the subgroup K. Thus, $|K|$ divides $|D_3|$. Moreover, each coset can be represented by any element in that coset. For example, the left coset $\{e, r, r^2\}$ can be represented by e, r, or r^2. That is, eK, rK, and r^2K all equal $\{e, r, r^2\}$. Unlike the preceding example (the cosets of $\{e, f\}$), in this example, every left coset equals its corresponding right coset (i.e., $xK = Kx$ for all $x \in D_3$).

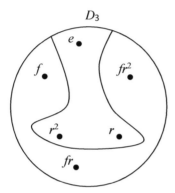 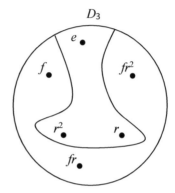

Left cosets of $\{e, r, r^2\}$ are the pieces of the D_3 pie.

Right cosets of $\{e, r, r^2\}$ cut up the D_3 pie in the *same* way.

Let's do one more example, this time with the cyclic group $C_6 = \{e, r, r^2, ..., r^5\}$. Let $J = \{e, r^3\}$. J is a subgroup of C_6. Determine the left and right cosets of J in C_6.

Left Cosets	Right Cosets
$eJ = r^3J = \{e, r^3\}$	$Je = Jr^3 = \{e, r^3\}$
$rJ = r^4J = \{r, r^4\}$	$Jr = Jr^4 = \{r, r^4\}$
$r^2J = r^5J = \{r^2, r^5\}$	$Jr^2 = Jr^5 = \{r^2, r^5\}$

In this example, the left cosets and the right cosets partition C_6, each coset can be represented by any element in that coset, and every left coset equals its corresponding right coset.

Let's formulate some conjectures about cosets based on the preceding examples. We designate each of these conjectures with the letter K followed by a sequence number. We use K because it sounds like the letter C in "coset." The letter C would be a better choice, but we are already using C in the designator for the cyclic group C_n.

Let G be a group with subgroup H.

K1: $(ab)H = a(bH)$ for all $a, b \in G$, where $a(bH) = \{ax : x \in bH\}$.

K2: $g \in gH$ for all $g \in G$ (i.e., every left coset of H contains its representative).

K3: $|gH| = |H|$ for all $g \in G$ (i.e., every left coset of H has the same order as H).

K4: *Coset absorption rule*: $hH = H$ if and only if $h \in H$ (i.e., a left coset of H equals H if and only if its representative is in H). If $hH = H$, we say that H **absorbs** h.

K5: $aH = bH$ if and only if $a \in bH$ (i.e., each left coset can be represented by any and only those elements in that coset).

K6: For all $a, b \in G$, if $aH \neq bH$, then $aH \cap bH = \emptyset$ (i.e., every pair of left cosets of H is either identical or disjoint).

K7: The set of distinct left cosets of H partition G.

K8: *Lagrange's theorem*: If G is finite, then $|G|$ = [number of distinct left cosets] $\cdot |H|$ (i.e., if G is finite, the order of a subgroup H divides the order of G).

All eight of our conjectures are true. Try proving them on your own before reading further. You will likely have trouble doing at least a few of these proofs. If so, after studying the proofs that follow, attempt all eight again without referring to the provided proofs. Repeat this process until you can easily do all of them on your own. By doing this, you will not only sharpen your proof-writing skills but also prepare yourself for the chapters that follow, in which we make frequent use of these theorems.

Prove K1: $(ab)H = a(bH)$ for all $a, b \in G$. *Hint*: The group operation on G is associative.

969

$x \in (ab)H$
$\Leftrightarrow x = (ab)h$ for some $h \in H$ definition of a coset
$\Leftrightarrow x = a(bh)$ for some $h \in H$ G's operation is associative
$\Leftrightarrow x \in a(bH)$.   definition of $a(bH)$ ∎

By K1, it makes no difference if we interpret abH as $(ab)H$ or as $a(bH)$. Either way, we get the same coset. Thus, we can write abH without the parentheses without any danger of introducing ambiguity.

Prove K2: $g \in gH$ for all $g \in G$.

970

$e \in H \Rightarrow ge \in gH \Rightarrow g \in gH$. ∎

Prove K3: $|gH| = |H|$ for all $g \in G$. *Hint*: Show that $\varphi(h) = gh$ is a bijection from H to gH.

971

Onto: Let $x \in gH \Rightarrow x = gh$ for some $h \in H \Rightarrow \varphi(h) = gh = x$.
One-to-one: Suppose $\varphi(h_1) = \varphi(h_2) \Rightarrow gh_1 = gh_2 \Rightarrow h_1 = h_2$ (by the left cancellation law for groups). ∎

Prove the "only if" part of K4: $hH = H \Rightarrow h \in H$. *Hint*: Use K2.

972

By K2, $h \in hH$. We are given that $hH = H$. Thus, $h \in H$. ∎

Now prove the "if" part of K4: $hH = H \Leftarrow h \in H$. *Hint*: The elements in hH are the elements in row h of the Cayley table for H. What do we know about each row of a Cayley table for a group (see frame **450**)?

973

hH contains all the elements in row h of the Cayley table for H. By frame **450**, every row of a Cayley table for a group contains all and only the elements of the group. Thus, $hH = H$. ∎

Here is a counterexample for K4: any group G is a subgroup of itself. Then, according to K4, $gG = G$ for all $g \in G$. Let $g = 2$ and $G = \mathbb{Z}$. We get $2\mathbb{Z} = \mathbb{Z}$. But $2\mathbb{Z} \neq \mathbb{Z}$.

Where is the error? Is it in K4 or in the counterexample?

974

The counterexample is wrong: gH in K4 is the set of elements obtained by applying the *group operation* to all the pairs of elements in which the left element is g and the right is from H. Addition is the group operation for $\mathbb{Z}$ (under multiplication, $\mathbb{Z}$

is not a group); $2\mathbb{Z}$ is the set of elements obtained by multiplying 2 by each element of $\mathbb{Z}$. Thus, K4 does not apply to $2\mathbb{Z}$. The confusion here arises because K4 is stated in multiplicative notation, but the operation for the group $\mathbb{Z}$ is addition.

State K4 in additive notation.

975

$h + H = H$ if and only if $h \in H$, which implies that $2 + \mathbb{Z} = \mathbb{Z}$. Does $2 + \mathbb{Z} = \mathbb{Z}$?

976

Yes

Prove the "only if" part of K5: $aH = bH \Rightarrow a \in bH$.

977

By K2, $a \in aH$. We are given that $aH = bH$. Thus, $a \in bH$. ∎

Prove the "if" part of K5: $aH = bH \Leftarrow a \in bH$.

978

$a \in bH \Rightarrow a = bh$ for some $h \in H \Rightarrow aH = (bh)H = b(hH) = bH$. The last equality is by K4 (the coset absorption rule). The one before that is by K1. ∎

Prove K6: for all $a, b \in G$, if $aH \neq bH$, then $aH \cap bH = \emptyset$. *Hint*: Assume aH and bH are not disjoint. Then show that $aH = bH$. Thus, aH and bH are either disjoint or equal.

979

Suppose aH and bH are not disjoint. Then there must be at least one x such that $x \in aH$ and $x \in bH \Rightarrow x = ah_1$ and $x = bh_2$ for some $h_1, h_2 \in H \Rightarrow ah_1 = bh_2$. Now solve for a.

980

$a = bh_2h_1^{-1}$. Let $h_3 = h_2h_1^{-1}$. Then $a = bh_3$. By the closure property of H, $h_3 \in H$.

To complete the proof, show that $aH = bH$.

981

$aH = (bh_3)H = b(h_3H) = bH$. The last equality is by K4 (the coset absorption rule). The one before that is by K1. ∎

K7 (the set of distinct left cosets of H partition G) follows immediately from K6, which establishes that distinct left cosets are disjoint, and K2, which establishes that each coset is nonempty and their union is equal to G.

K8 (if G is finite, $|G|$ = [number of distinct left cosets] · $|H|$) follows immediately from two of the preceding conjectures. Which two?

982

K3 (which establishes that the order of every coset is $|H|$) and K7 (which establishes that the distinct left cosets of H partition G) imply that $|G|$ = [number of distinct left cosets] · $|H|$, which in turn implies that $|H|$ divides $|G|$. ∎

Now that we have proofs for all our conjectures, we can refer to them as theorems rather than conjectures. K8 is a particularly important theorem. It is called ***Lagrange's theorem*** in honor of Joseph-Louis Lagrange, a pioneer in group theory.
  Because of the similarity between left cosets and right cosets, for every theorem on left cosets, there is a corresponding theorem on right cosets. For brevity's sake, we will usually state and prove only the left coset versions.

Show that if the group G has a prime order, then it has no subgroups other than G itself and $\{e\}$. *Hint*: Apply Lagrange's theorem.

983

Suppose G is a group and $|G|$ is a prime number. If H is a subgroup of G, then by Lagrange's theorem, $|H|$ divides $|G|$. The only positive integers that divide $|G|$ are 1 and $|G|$. Thus, $|H| = 1$ (in which case H is $\{e\}$) or $|H| = |G|$ (in which case H is G). ∎

By frame **853**, if g is an element of a group G, then $<g>$ is the subgroup of G generated by g. If $|G|$ is a prime number, what can we say about $<g>$?

984

By the preceding frame, $<g>$ has to be either $\{e\}$ or G. If $g = e$, then $<g> = \{e\}$. If $g \neq e$, then $<g> \neq \{e\}$, in which case, $<g>$ has to be G. But that means G is necessarily cyclic generated by g, where g is any non-identity element. These observations give us the following theorem:

K9: Every group of prime order is cyclic.

Moreover, by frame **912**, all cyclic groups of the same order are isomorphic. Thus, up to isomorphism, if p is prime, the only group of order p is the cyclic group of order p. For example, up to isomorphism, the only group of order 7 is C_7.

We defined cosets using multiplicative notation. In additive notation, a left coset of H containing g is $g + H = \{g + h : h \in H\}$. Similarly, a right coset of H containing g is $H + g = \{h + g : h \in H\}$. For example, $2\mathbb{Z}$ (the even integers) is a subgroup of $<\mathbb{Z}, +>$. One of its cosets of $2\mathbb{Z}$ is $1 + 2\mathbb{Z}$.

Describe the elements in the coset $1 + 2\mathbb{Z}$.

985

This is the set obtained by adding 1 to each element of $2\mathbb{Z}$, the set of even integers. The resulting set is the set of all odd integers. By K5, any element in the coset $1 + 2\mathbb{Z}$—for example, -3, -1, and 3—can represent that coset. Thus, $-3 + 2\mathbb{Z} = -1 + 2\mathbb{Z} = 1 + 2\mathbb{Z} = 3 + 2\mathbb{Z}$.

$2\mathbb{Z}$ has two distinct cosets in $<\mathbb{Z}, +>$. What are they?

986

$0 + 2\mathbb{Z} = 2\mathbb{Z}$ (the even integers) and $1 + 2\mathbb{Z}$ (the odd integers)

At this point, you are probably overwhelmed by all the theorems and proofs on cosets. We will frequently use K2 (a coset contains its representative), K4 (coset absorption), and K5 (each coset can be represented by any and only those elements in that coset). So be sure to understand these theorems before reading on

Normal Subgroups

987

Let H be a subgroup of a group G. If $gH = Hg$ for all $g \in G$, then we say that H is a ***normal subgroup.*** In other words, for normal subgroups, every left coset is equal to its corresponding right coset. For example, the subgroup $\{e, r, r^2\}$ in D_3 is normal (see frame **968**). When dealing with the cosets of a normal subgroup, we do not have to distinguish the left cosets from the right cosets. For a normal subgroup, each left coset equals its corresponding right coset. Thus, we can simply speak of the cosets of the subgroup.

A normal subgroup can be defined in a variety of ways. Here are four equivalent definitions of a normal subgroup: a subgroup H of a group G is normal if and only if

1. $gH \quad = \quad Hg$ for all $g \in G$,
2. $gHg^{-1} \quad = \quad H$ for all $g \in G$,
3. $gH \quad \subseteq \quad Hg$ for all $g \in G$,
4. $gHg^{-1} \quad \subseteq \quad H$ for all $g \in G$,

We can get definition 2 from definition 1 simply by multiplying both sides of the equality in definition 1 on the right by g^{-1}:

$gH = Hg$
$\Rightarrow gHg^{-1} = Hgg^{-1}$ multiply both sides by g^{-1}
$\Rightarrow gHg^{-1} = He$ $gg^{-1} = e$
$= H$ K4 in frame **968** (the absorption rule)

We can similarly get definition 1 from definition 2 by multiplying both sides of the equality in definition 2 on the right by g. Thus, definitions 1 and 2 are equivalent.

Surely definition 1 and definition 3 *cannot* be equivalent. Definition 1 clearly implies definition 3. But how can definition 3 imply definition 1? Do we not need both $gH \subseteq Hg$ and $Hg \subseteq gH$ to conclude that $gH = Hg$? In fact, we do not. The reason this is so is that definition 3 is not a *single* subset relation—it is a subset relation for *each* $g \in G$. It tells us that each left coset is a subset of the corresponding right coset.

If you take the inverse of each element of a group, what set do you get?

988

Every element in a group has an inverse in the group. If you take the inverse of the inverse, you get the original element. Thus, if you take the inverse of every element, you get every element. In other words, the set of the inverses of the elements in a group equals the set of elements in a group. With that in mind, determine what the following assertion tells us:

$g^{-1}H \subseteq Hg^{-1}$ for all $g \in G$.

989

It is asserting something about every inverse of a group. But the set of inverses of a group is just the set of elements in the group. So it is an assertion about each element of a group. It tells us that each left coset is a subset of the corresponding right coset. Thus, it is equivalent to definition 3 in frame **987**. We are now in a position to show that definition 3 implies definition 1:

Definition 3 $\Rightarrow g^{-1}H \subseteq Hg^{-1}$ for all $g \in G$. Now multiply both sides of this subset relation by g on the left. Then multiply both sides by g on the right. We get

$gg^{-1}Hg \subseteq gHg^{-1}g$ for all $g \in G$,

which simplifies to

$Hg \subseteq gH$ for all $g \in G$,

which, together with definition 3, implies definition 1. ∎

We can get definition 4 from definition 3 by multiplying both sides of definition 3 on the right by g^{-1}, and we can get definition 3 from definition 4 by multiplying both sides of definition 4 on the right by g. Thus, definitions 3 and 4 are equivalent. Definition 3 is equivalent to definitions 1 and 2. Thus, all four definitions are equivalent.

Let's prove some theorems about normal subgroups:

K10: If a group G is commutative, then every subgroup of G is normal.

K11: For every group G, $gG = Gg = G$ for all $g \in G$ (i.e., every group is a normal subgroup of itself).

K12: If the number of distinct left cosets of a subgroup H is 2, then H is normal.

Prove K10. *Hint*: Show $gH \subseteq Hg$.

990

Suppose H is a subgroup of a commutative group G, and $g \in G$. Let $x \in gH$. Then $x = gh$ for some $h \in H$. Because G is commutative, $x = gh = hg \Rightarrow x \in Hg$. Thus, $gH \subseteq Hg$. ∎

Prove K11. *Hint*: Use the coset absorption rule.

991

$g \in G$. Thus, by the coset absorption rule, $gG = G$. By the right coset version of the coset absorption rule, $Gg = G$. Thus, $gG = Gg = G$. ∎

Prove K12. *Hint*: If $g \in H$, then by the coset absorption rule, $gH = H = Hg$. If $g \notin H$, then by K6, gH and H are the two distinct left cosets of H.

992

Case 1: $g \in H$. Then, by the left and right coset versions of the coset absorption rule, $gH = H = Hg$.
Case 2: $g \notin H$. Then, by K6, gH and H are the two distinct left cosets of H. Because there are only two distinct cosets, gH contains all the elements in G not in H. Similarly, Hg and H are the two distinct right cosets. Hg, like gH, contains all the elements in G not in H. Thus, $gH = Hg$. ∎

Is $\{e\}$ a normal subgroup of G for any group G, where e is the identity?

993

Yes. $g\{e\} = \{e\}g = \{g\}$ for all $g \in G$.

What are the left cosets of $\{e\}$ in the group G?

994

For each $g \in G$, $g\{e\} = \{g\}$ is a left coset.

Is G a normal subgroup of G?

995

Yes, by K11.

What are the left cosets of G in the group G?

996

Only one left coset: G itself.

Normal subgroups are very important for the following reason: the cosets of a normal subgroup in a group G themselves form a group. In the next chapter, we will investigate these groups whose elements are cosets. Our investigation will culminate in a remarkable theorem aptly called the ***fundamental theorem of group homomorphisms***.

Show that $2\mathbb{Z}$ is a normal subgroup of $<\mathbb{Z}, +>$.

997

By K10 (because $\mathbb{Z}$ is commutative) or K12 (because there are only two cosets of $2\mathbb{Z}$), $2\mathbb{Z}$ is a normal subgroup of $\mathbb{Z}$.

Interesting Consequence of Lagrange's Theorem

998

The following theorem follows from Lagrange's theorem:

K13: For *any* finite group G with identity e, if $g \in G$, then $g^{|G|} = e$.

Let's test K13 on C_4. According to K13, the fourth power of any element in $C_4 = \{e, r, r^2, r^3\}$ is e. Confirm this.

999

In C_4, $r^4 = e$. Thus

$(e)^4 = e$
$(r)^4 = e$
$(r^2)^4 = (r^4)^2 = e^2 = e$
$(r^3)^4 = (r^4)^3 = e^3 = e$

If $|<g>| = n$, what are the elements in $<g>$?

1000

$e, g, g^2, ..., g^{n-1}$

What does g^n equal?

1001

e. Thus, $g^{|\langle g \rangle|} = g^n = e$.

What does Lagrange's theorem tell us about the order of a finite group G and the order of its subgroup $\langle g \rangle$?

1002

$|G| = k \cdot |\langle g \rangle|$, where k is the number of distinct left cosets of $\langle g \rangle$

Here is the start and the end of the proof for K13: $g^{|G|} = g^{k|\langle g \rangle|} = \cdots = e$. Do the whole proof.

1003

$g^{|G|}$
$= g^{k|\langle g \rangle|}$ Lagrange's theorem
$= (g^{|\langle g \rangle|})^k$ law of exponents
$= e^k$ frame **1001**
$= e$ identity property ∎

This result that is interesting because it is so general. It applies to *any* element g in *any* finite group G.

Review Questions

1. List all the left cosets of $\{e, r^4\}$ in the cyclic group of order 8 generated by r.
2. List all the left cosets of D_3 in D_3.
3. List all the left cosets of $\{e\}$ in D_3.
4. Are all the subgroups of V_4 normal (see frame **446**)? Justify your answer.
5. Show that for all $a, b \in H$, $a^{-1}b \in H$ if and only if $aH = bH$.
6. List all the left cosets of $H = \{(1), (1\ 2)\}$ in S_3.
7. List all the left cosets of $H = \{(1), (1\ 2\ 3), (1\ 3\ 2)\}$ in S_3.
8. Without comparing left and right cosets of H, how can you confirm that H in review question 7 is normal?
9. Lists the left cosets of A_3 in S_3.
10. Show that $aH = bH$ does not imply $Ha = Hb$ where H is a subgroup of G and $a, b \in G$.

Answers to the Review Questions

1. $\{e, r^4\}, \{r, r^5\}, \{r^2, r^6\}, \{r^3, r^7\}$
2. Only one: D_3
3. $\{e\}, \{r\}, \{r^2\}, \{f\}, \{fr\}, \{fr^2\}$
4. By Lagrange's theorem, each subgroup H of V_4 must have order 1, 2, or 4. If $|H| = 1$, then $H = \{e\}$. Then, by frame **993**, H is a normal subgroup. If $|H| = 4$, then $H = G$. Then, by K11 in frame **989**, H is a normal subgroup. If $|H| = 2$, then H has two cosets. Then, by K12, H is a normal subgroup. Alternatively, observe that V_4 is commutative. Then by K10, every subgroup of V_4 is normal.
5. $a^{-1}b \in H \Rightarrow a^{-1}b = h$ for some $h \in H \Rightarrow b = ah \Rightarrow ahH = bH \Rightarrow aH = bH$ (by the coset absorption rule). Conversely, $b \in bH$ and $aH = bH \Rightarrow b \in aH \Rightarrow b = ah$ for some $h \in H \Rightarrow a^{-1}b = h \Rightarrow a^{-1}b \in H$.
6. $\{(1), (1\ 2)\}, \{(1\ 3), (1\ 2\ 3)\}, \{(1\ 3\ 2), (2\ 3)\}$
7. $\{(1), (1\ 2\ 3), (1\ 3\ 2)\}, \{(1\ 2), (1\ 3), (2\ 3)\}$
8. Use K12 in frame **989**.

9. A_3 (the even permutations) and $\{(1\ 2), (1\ 3), (2\ 3)\}$ (the odd permutations)
10. Counterexample: $H = \{e, f\}$ is a subgroup of D_3. $rH = fr^2H$, but $Hr \neq Hfr^2$.

Homework Questions

1. Let H be a subgroup of a group G. Show that the only left coset of H that is a subgroup of G is H itself.
2. Suppose G is a group with subgroup H. Let $[G:H]$ be the number of distinct cosets of H in G. We call $[G:H]$ the **index** of H in G. If K is a subgroup of H, and H is a subgroup of G, show that $[G:K] = [H:K] \cdot [G:H]$.
3. Characterize the left cosets of $\{2^k : k \in \mathbb{Z}\}$ in the group $\mathbb{R} - \{0\}$ under multiplication.
4. Show that in any group of order p, where p is a prime, the number of elements of order p is $p-1$.
5. Give an example of a group G such that $k > 0$ and k divides $|G|$, but there is no subgroup of G of order k.
6. How does your example in homework question 5 relate to Lagrange's theorem?
7. Do all groups have at least two normal subgroups?
8. Show that $H = \{(1), (2\ 3)\}$ is not a normal subgroup of S_3.
9. What are all the cosets of $5\mathbb{Z}$ in $\mathbb{Z}$?
10. Suppose the number of cosets of H in G is 2. Show that if $a, b \notin H$, then $ab \in H$.

18 Quotient Groups

Quasi-Commutative Law for Normal Subgroups

1004

Suppose that H is a normal subgroup of G and that $g \in G$. Let $x \in gH$. Then $x = gh$ for some $h \in H$. Because H is normal, $gH = Hg$. Thus, x is also an element of Hg, which implies that $x = h'g$ for some $h' \in H$. Both gh and $h'g$ equal x. Thus, $gh = h'g$. This equality is a commutative law of sorts. It indicates that we can commute the group element g with the subgroup element h, if the subgroup is normal. However, after the commutation, the subgroup element is not necessarily the same. We call this rule the ***quasi-commutative law for normal subgroups.*** Because the subgroup element may change on commutation, this is not a genuine commutative law—hence the qualifier "quasi" (*quasi* means "resembling").

For example, $H = \{e, r, r^2\}$ is a normal subgroup of D_3; r is a subgroup element and f is a group element. Thus, in the product rf, we can quasi-commute the r and f. The group element f commutes without modification. The subgroup element r, however, changes into the subgroup element r^2. That is, $rf = fr^2$ (see frame **803**).

Apply the quasi-commutative law to fr.

1005

The subgroup element r changes to r^2. We get $fr = r^2 f$.

Suppose that the group element from D_3 is r^2 and that the subgroup element is r (from the subgroup $\{e, r, r^2\}$). Apply the quasi-commutative law to $r^2 r$.

1006

As always with the quasi-commutative law, the group element does not change on commutation. In this case, however, the subgroup element also does not change. We get $r^2 r = rr^2 = r^3$. We will often use the quasi-commutative law in proofs dealing with normal subgroups.

We showed above that a normal subgroup H obeys the quasi-commutative law. Let's now show the converse. Suppose that H is a subgroup of G that obeys the quasi-commutative law. Prove that H is a normal subgroup of G. *Hint*: Show that for all $g \in G$, $gH \subseteq Hg$ (this is one of the tests for the normality of a subgroup—see item 3 in frame **987**).

1007

Let $x \in gH \Rightarrow x = gh$ for some $h \in H$. By the quasi-commutative law, $x = h'g$ for some $h' \in H \Rightarrow x \in Hg$. Thus, $gH \subseteq Hg$. ∎

Because a subgroup is normal if and only if it obeys the quasi-commutative law, we can define a normal subgroup as any subgroup that obeys the quasi-commutative law. So we now have five definitions of a normal subgroup: four from the preceding chapter and a new one based on the quasi-commutative law:

1. gH = Hg for all $g \in G$.
2. gHg^{-1} = H for all $g \in G$.
3. gH ⊆ Hg for all $g \in G$.
4. gHg^{-1} ⊆ H for all $g \in G$.
5. For all $g \in G$ and $h \in H$, $gh = h'g$ for some $h' \in H$ (the quasi-commutative law).

An Operation on Cosets

1008

Suppose G is a group with a subgroup H. Let's define a binary operation on the set of left cosets of H. We are using multiplicative notation, so we will refer to the operation as **coset multiplication** and to its result as a product. Let aH and bH be two cosets of the subgroup H in a group $<G, *>$. Then $(aH) \otimes (bH)$, the product of aH and bH, is the coset obtained using the following procedure:

Take a representative from each coset. The product of the two cosets is the coset that contains the product of the two representatives.

For example, a and b are in aH and bH, respectively. Thus, we can use a and b as our representatives. Then, by our definition of coset multiplication, the product of aH and bH is the coset that contains $a * b$, where $*$ is the binary operation on G. By K2 in frame **968**, the coset that contains $a * b$ is coset $(a * b)H$. Thus, the product of the cosets aH and bH is $(a * b)H$. That is,

$$aH \otimes bH = (a * b)H.$$

There are two distinct operations in our definition of multiplication on the set of left cosets: the coset binary operation $\otimes$ and the group binary operation $*$. If we rewrite this definition using juxtaposition instead of the operation symbols, we get $aHbH = (ab)H$. Be sure to keep in mind that the two juxtapositions in this form represent different operations. On the left side, the juxtaposition of aH and bH represents the coset binary operation; on the right side, the juxtaposition of a and b represents the binary operation in the group G.

Suppose we omit the parentheses in $(ab)H$ to get abH. Then abH can be interpreted as either $(ab)H$ or $a(bH)$. Is abH therefore ambiguous? In other words, does the set specified by abH depend on how we parenthesize abH?

1009

No. By K1 in frame **968**, $(ab)H = a(bH)$. Thus, abH is not ambiguous, so we can omit the parentheses.

We say that coset multiplication is **well defined** if the product of each pair of cosets does not depend on the representatives of the cosets we use to determine the product. For example, suppose we want to multiply two cosets of H, and a and x are in one of the cosets, and b and y are in the other coset. If we use a and b as the representatives, then the product of the cosets is abH. If, alternatively, we use x and y as the representatives, then the product is xyH. If coset multiplication is well defined, abH should equal xyH. If there is at least one pair of cosets whose product depends on the representatives used, then coset multiplication is *not* well defined.

Our definition of coset multiplication states that the product of the cosets aH and bH is abH. The result, abH, is a coset of H. Thus, the set of cosets of H is closed under coset multiplication. Is coset multiplication then a binary operation on the cosets of H? Not necessarily. Closure is not enough to ensure that coset multiplication is a binary operation. Coset multiplication must also be well defined to be a binary operation. It is not well defined if there is at least one pair of cosets whose product depends on the representatives used. In that case, coset multiplication would map that pair of cosets to *more than one coset*. Then coset multiplication would not be a binary operation. By definition, a binary operation always produces one and only one result.

Is our operation on left cosets well defined? It turns out that it is *if and only if H is a normal subgroup of G*. This property of normal subgroups is the reason they are such an important category of subgroups. Let's start by proving that the operation $aHbH = abH$ is well defined if H is normal. We will then prove the converse.

Let H be a normal subgroup in the group G (so the quasi-commutative law applies). Let $x \in aH$ and $y \in bH$. By K2 in frame **968**, $a \in aH$ and $b \in bH$. Thus, a and x are in the same coset aH, and b and y are in the same coset bH. We have to show that xyH (the product when we use x and y as the representatives of aH and bH, respectively) equals abH (the product when we use a and b as the representatives). Because $x \in aH$, then $x = ah_1$ for some $h_1 \in H$. Similarly, because $y \in bH$, then $y = bh_2$ for some $h_2 \in H$. Now multiply x by y, and then quasi-commute h_1 and b in the result.

1010

$xy = ah_1bh_2 = abh_3h_2$ for some $h_3 \in H$

By the closure property of H, $h_3h_2 = h_4$ for some $h_4 \in H$. Substituting, we get $xy = abh_4$. Using this value of xy, we get that $xyH = abh_4H$. What does h_4H equal?

1011

$h_4H = H$ by the coset absorption rule (see K4 in frame **968**)

Substituting H for h_4H, we get $xyH = abh_4H = abH$. The equality $xyH = abH$ shows that the result of coset multiplication does not depend on the coset representatives used. Thus, coset multiplication is well defined.

Now let's prove the converse: if the operation defined by $aHbH = abH$ is well defined, then H is a normal subgroup. To show the normality of H, we will show $gH \subseteq Hg$ for all $g \in G$. To do this, we take an arbitrary element x in gH and show that it is also in Hg.

Suppose coset multiplication is well defined. Let $x \in gH$. Then $xH = gH$ by K5 in frame **968**. Replacing gH with xH in $(gH)(g^{-1}H)$, we get $(gH)(g^{-1}H) = (xH)(g^{-1}H)$. Performing the coset multiplication of both sides, we get $gg^{-1}H$ on the left and $xg^{-1}H$ on the right. We are given that coset multiplication is well defined. Thus, $gg^{-1}H = xg^{-1}H$, which simplifies to $H = xg^{-1}H$.

From K4 in frame **968** (which states that $hH = H$ if and only if $h \in H$), what can you conclude about xg^{-1}?

1012

$xg^{-1} \in H \Rightarrow xg^{-1} = h$ for some $h \in H$

Now solve for x and use the result to show that $x \in Hg$.

1013

$x = hg \Rightarrow x \in Hg$

We have shown that $x \in Hg$ if $x \in gH$. Thus, $gH \subseteq Hg \Rightarrow H$ is normal. ∎

$H = \{e, r, r^2\}$ is a normal subgroup of D_3. The cosets of H in D_3 are $H = \{e, r, r^2\}$ and $\{f, fr, fr^2\}$. Let's multiply these two cosets. Because H is normal, the resulting coset does not depend on the representatives from the two cosets we use. Thus, if we use r and fr as representatives, or if we use r^2 and f, the resulting coset should be the same.

Confirm that $(rH)(frH)$ and $(r^2H)(fH)$ are the same coset. *Hint*: Use the flip-rotate commutation rule from frame **803**. Also use $r^3 = e$ and $f^2 = e$.

1014

$$\left.\begin{aligned}(rH)(frH) &= rfrH = fr^2rH = fr^3H = fH = \{f, fr, fr^2\} \\ (r^2H)(fH) &= r^2fH = frH = \{fr, fr^2, fr^3\} = \{fr, fr^2, f\}\end{aligned}\right\} \text{ equal}$$

In frame **1008**, we defined a binary operation $\otimes$ on the cosets of H with $(aH) \otimes (bH) = (a * b)H$. When using multiplicative notation, we typically omit the operation symbols. So in multiplicative notation, the definition of our binary operation on cosets is $(aH)(bH) = abH$, and we refer to the binary operation on cosets as "coset multiplication." When using additive notation, we use the + symbol for both the coset binary operation and the group binary operation. So in additive notation, the definition of our binary operation on cosets is

$$(a + H) + (b + H) = (a + b) + H,$$

and we refer to the binary operation on cosets as "coset addition." All three of the preceding definitions are equivalent. They differ *only in the notation used.*

Let's now switch to a group for which we use additive notation. The operation in the group $<\mathbb{Z}, +>$ is addition. Thus, we, of course, should use additive notation when working with this group. Let's denote the set of integers that are a multiple of 3 with $3\mathbb{Z}$, which is a normal subgroup of $<\mathbb{Z}, +>$. In additive notation, the cosets of $3\mathbb{Z}$ are of the form $n + 3\mathbb{Z}$, where $n \in \mathbb{Z}$. The coset $n + 3\mathbb{Z}$ is the set of elements obtained by adding n to each element of $3\mathbb{Z}$. Thus, $0 + 3\mathbb{Z} = 3\mathbb{Z}$, and $1 + 3\mathbb{Z}$ is the set obtained by adding 1 to all the integers that are a multiple of 3. The subgroup $3\mathbb{Z}$ has three cosets: $3\mathbb{Z}$, $1 + 3\mathbb{Z}$, and $2 + 3\mathbb{Z}$.

The coset $1 + 3\mathbb{Z}$ is the set of all integers congruent to 1 modulo _____.

1015

3

Chapter 18: Quotient Groups

When we write $0 + 3\mathbb{Z}$, we are using 0 as a representative of the coset. However, by theorem K5 in frame **968**, we can use any element in $0 + 3\mathbb{Z}$ as its representative. Because the coset $0 + 3\mathbb{Z}$ contains 9, we can equally well represent the coset $0 + 3\mathbb{Z}$ with $9 + 3\mathbb{Z}$.

What does $(3 + 3\mathbb{Z}) + (6 + 2\mathbb{Z})$ equal?

1016

$(3 + 6) + 3\mathbb{Z} = 9 + 3\mathbb{Z} = 0 + 3\mathbb{Z} = 3\mathbb{Z}$

The set of cosets of a normal subgroup H of a group G is a ***quotient structure.*** Its elements are obtained by "dividing" the elements of G into cosets (hence use of the term "quotient"). The binary operation on cosets that we defined in frame **1008** (coset multiplication or coset addition, depending on the notation we are using) gives the set of cosets a structure (hence use of the term "structure"). We denote this quotient structure with G/H. That is, G/H is the set of cosets of the normal subgroup H in the group G under the operation of coset multiplication (or coset addition). For example, $\mathbb{Z}/3\mathbb{Z}$ is the set of cosets of $3\mathbb{Z}$ in $\mathbb{Z}$. $\mathbb{Z}/3\mathbb{Z}$ under coset addition is a quotient structure.

Determine the Cayley table for $\mathbb{Z}/3\mathbb{Z}$ under coset addition

---

1017

+	$3\mathbb{Z}$	$1 + 3\mathbb{Z}$	$2 + 3\mathbb{Z}$
$3\mathbb{Z}$	$3\mathbb{Z}$	$1 + 3\mathbb{Z}$	$2 + 3\mathbb{Z}$
$1 + 3\mathbb{Z}$	$1 + 3\mathbb{Z}$	$2 + 3\mathbb{Z}$	$3\mathbb{Z}$
$2 + 3\mathbb{Z}$	$2 + 3\mathbb{Z}$	$3\mathbb{Z}$	$1 + 3\mathbb{Z}$

Does this Cayley table correspond to a group?

1018

Yes. $\mathbb{Z}/3\mathbb{Z}$ under coset addition is isomorphic to C_3.

What are the elements of D_3/K, where $K = \{e, r, r^2\}$? *Hint*: See frame **967**.

1019

The cosets of K, which are $eK = K = \{e, r, r^2\}$ and $fK = \{f, fr, fr^2\}$

Determine the Cayley table for D_3/K under coset multiplication.

1020

·	K	fK
K	K	fK
fK	fK	K

Does this Cayley table correspond to a group?

1021

Yes. D_3/K under coset multiplication is isomorphic to C_2.

For all groups G, G is a normal subgroup of itself. Describe the elements of G/G.

1022

There is only one coset of G, and that is G itself.

Is G/G under coset multiplication a group?

1023

Yes. G/G is isomorphic to C_1.

Based on the preceding examples, a reasonable conjecture is that G/H under coset multiplication (or coset addition) is a group if H is a normal subgroup of G. Let's prove this. First show that G/H is closed under coset multiplication.

1024

If H is a normal subgroup of G, then by frames **1009** to **1011** the product of aH and bH is well defined and equal to the coset abH. Thus, G/H is closed under coset multiplication.

Show that coset multiplication on G/H is associative. *Hint*: Write a string of equalities that starts with $(aHbH)(cH)$ and ends with $aH(bHcH)$.

1025

$(aHbH)(cH)$
$= (abH)(cH)$ definition of coset multiplication
$= [(ab)c]H$ definition of coset multiplication
$= [a(bc)]H$ multiplication on G is associative
$= (aH)(bcH)$ definition of coset multiplication
$= (aH)(bHcH)$ definition of coset multiplication

What is the identity element in G/H given that e is the identity in G?

1026

$(eH)(aH) = eaH = aH$. $(aH)(eH) = aeH = aH$. eH (which equals H) is the identity element.

What is the inverse of aH where a^{-1} is the inverse of a in G?

1027

$(a^{-1}H)(aH) = (aH)(a^{-1}H) = eH = H$. The inverse of aH is $a^{-1}H$.

G/H has the closure, associative, identity, and inverses properties. It is a group! ∎

Because G/H is a group, we generally call it a ***quotient group*** rather than the less specific "quotient structure."

Graph of a Quotient Group

1028

The graph of the quotient group G/H is a simplified version of the graph for G. To obtain the graph for G/H, simply merge all the elements in each coset into a single element and eliminate any redundant arrows. For example, the following graph is for the group for D_3, with the two triangles drawn adjacent to each other:

   Key: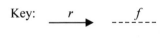

Each triangle corresponds to a coset of the normal subgroup $K = \{e, r, r^2\}$. To obtain the graph for D_3/K, we first merge the three vertices in each triangle into a single point, retaining any arrows between the two triangles. We get

We then eliminate the redundant arrows to get

which is equivalent to

This is the graph of G/K. It is also the graph of C_2, indicating that D_3/K is isomorphic to C_2.

Here is the graph of C_∞, the infinite cyclic group generated by r:

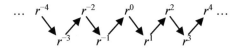

Is $<r^2>$, the subgroup generated by r^2, normal?

1029

Yes. C_∞ is cyclic $\Rightarrow C_\infty$ is commutative (by frame **900**) $\Rightarrow$ all its subgroups are normal (by K10 in frame **989**).

Construct the graph of $C_\infty/<r^2>$ from the graph C_∞. *Hint*: The two cosets of $<r^2>$ are $<r^2>$ and $r<r^2>$. The former is the set of all even powers of r (the top row in the graph of C_∞ in the preceding frame); the latter is the set of all odd powers of r (the bottom row in the graph of C_∞).

1030

Eliminating the redundant arrows, we get the graph for C_2: Thus, $C_\infty/<r^2>$ is isomorphic to C_2.

Review Questions

1. Construct the Cayley table for D_4/N, where $N = \{e, r, r^2, r^3\}$.
2. Construct the Cayley table for $<\mathbb{Z}/2\mathbb{Z}, +>$.
3. Find $|G/H|$, where $G = <\mathbb{Z}_{36}, +_{36}>$ and $H = <15>$. *Hint*: Determine $|15|$.
4. Find $|G/H|$, where $G = <\mathbb{Z}_{36}, +_{36}>$ and $H = <30>$. *Hint*: Determine $|30|$.
5. Suppose G/N is a quotient group. If G is commutative, is G/N commutative? Justify your answer.
6. Suppose H and K are two normal subgroups of G. Then $HK = \{hk : h \in H \text{ and } k \in K\}$ is a subgroup of G. Show that HK is a normal subgroup of G.
7. Suppose H is a subgroup of G. Show that gH is not a subgroup of G if $g \notin H$.
8. Determine a subgroup of S_3 that is not normal.
9. Is it true that every quotient group of a cyclic group is cyclic? Justify your answer.
10. Is it true that every quotient group of a noncyclic group is noncyclic?

Answers to the Review Questions

1.

	N	fN
N	N	fN
fN	fN	N

2.

+	$2\mathbb{Z}$	$1 + 2\mathbb{Z}$
$2\mathbb{Z}$	$2\mathbb{Z}$	$1 + 2\mathbb{Z}$
$1 + 2\mathbb{Z}$	$1 + 2\mathbb{Z}$	$2\mathbb{Z}$

3. By frame **923**, $|15| = 36/\gcd(15, 36) = 36/3 = 12 \Rightarrow |G/H| = 36/12 = 3$.
4. By frame **923**, $|30| = 36/\gcd(30, 36) = 36/6 = 6 \Rightarrow |G/H| = 36/6 = 6$.
5. Yes. $aNbN = abN = baN$ (because G is commutative) $= bNaN$.

6. Let $x \in gHK \Rightarrow x = ghk$ for some $h_1 \in H$ and $k_1 \in K$. By the quasi-commutative law, $x = gh_1k_1 = h_2gk = h_2k_2g$ for some $h_2 \in H$ and $k_2 \in K \Rightarrow x \in HKg \Rightarrow gHK \subseteq HKg \Rightarrow HK$ is normal. ∎
7. By K6 in frame **968**, gH and H are disjoint if $g \notin H$. Then because $e \in H$, $e \notin gH \Rightarrow gH$ is not a subgroup.
8. $\{e, f\}$ in D_3 is not normal. The corresponding subgroup of S_3 is $\{(1), (2\ 3)\}$.
9. Yes. Let N be a subgroup of C_n. Suppose C_n is generated by $r \Rightarrow$ every element of C_n is of the form $r^k \Rightarrow$ every coset of N is of the form $r^kN = (rN)^k \Rightarrow C_n/N = \{(rN)^k : k \in \mathbb{Z}\} \Rightarrow C_n/N$ is a cyclic group generated by rN.
10. No; $D_3/\{e, r, r^2\}$ is cyclic, but D_3 is not.

Homework Questions

1. Show that the intersection of two normal subgroups is a normal subgroup.
2. Prove or disprove: Suppose that H and K are subgroups of G. Then $HK = \{hk : h \in H$ and $k \in K\}$ is a subgroup of G if K is normal.
3. Is $g^{-1}Hg$ a normal subgroup of G if H is a normal subgroup of G? Justify your answer.
4. Show that H is a normal subgroup of G if G is finite and H is the only subgroup of order $|H|$. *Hint*: Use the result in homework question 3.
5. Show that $Z(G)$, the **center of G,** is a normal subgroup of a group G. $Z(G)$ is the set of elements in G that commute with all the elements of G.
6. Show that every element of a quotient group G/H is its own inverse if $g^2 \in H$ for all $g \in G$.
7. Is $\{(1), (1, 2)\}$ a normal subgroup of S_3?
8. Suppose that there are k cosets of the normal subgroup H in the group G. Show that $g^k \in H$ for all $g \in G$.
9. Does a quotient group G/H exist whose elements all have finite order while G has elements only of infinite order, except for the identity? Justify your answer.
10. Is the Klein four group represented as a group of permutations a normal subgroup of S_4?

19 Fundamental Theorem of Group Homomorphisms

Definition of a Group Homomorphsim

1031

Suppose φ is a function from a group $<G_1, *>$ to a group $<G_2, \otimes>$ that has the operation-preserving property

$$\varphi(a * b) = \varphi(a) \otimes \varphi(b) \text{ for } a, b \in G_1$$

We then call φ a **group homomorphism from G_1 to G_2**. A group homomorphism is simply an operation-preserving function from a group to a group. The two groups can be different groups or the same group.

If φ is operation preserving and, in addition, a bijection from G_1 to G_2, then φ is also a **group isomorphism** from G_1 to G_2, and we say that G_1 is isomorphic to G_2, or more simply, G_1 and G_2 are isomorphic (denoted by $G_1 \cong G_2$). A homomorphism φ from G_1 to G_2 requires that φ have the operation-preserving property. Isomorphism, on the other hand, requires that φ have the operation-preserving property and, in addition, requires φ to be a bijection from G_1 to G_2. Thus, every isomorphism is a homomorphism, but not every homomorphism is an isomorphism.

Homomorphism, like isomorphism, guarantees that G_2's operation parallels G_1's operation. For example, suppose G_1 and G_2 are groups with operations $*$ and $\otimes$, respectively, and φ is a homomorphism that maps a and b in G_1 to c and d, respectively, in G_2 (see the following diagram). If $a * b = x$ in G_1 and $c \otimes d = y$ in G_2, then the operation-preserving property of φ guarantees that x maps to y under φ.

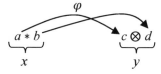

Show that the operation-preserving property of φ guarantees that x maps to y.

1032

$\varphi(x) = \varphi(a * b) = \varphi(a) \otimes \varphi(b) = c \otimes d = y$

Suppose x is the preimage of y under a homomorphism φ from a group G_1 to a group G_2. Because of the operation-preserving property of homomorphisms, the behavior of x in its group parallels the behavior of y in its group. If φ is one-to-one, then x is the only preimage of y. What if φ is not one-to-one? Then y has multiple preimages in G_1. Because of the operation-preserving property of homomorphisms, it is reasonable to conjecture that the behavior of the preimages of y *as a block* parallels the behavior of y in G_2. In this chapter, we investigate these blocks of preimages associated with a homomorphism.

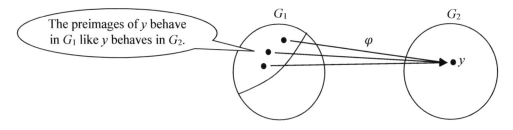

When working with groups, we often omit the symbol for the group operation in a product. Instead, we indicate a product of two elements simply by juxtaposing them. If we omit operation symbols, the operation-preserving property is written as

$$\varphi(ab) = \varphi(a)\varphi(b) \text{ for all } a, b \in G_1$$

In this form, there is potential for confusion. There are two products here, ab and $\varphi(a)\varphi(b)$. Because both products use juxtaposition, it is natural to think that the same operation is in play. However, this is not the case (unless G_1 and G_2 are the same group). In ab, a and b are elements of G_1. Thus, the operation implied is the operation in G_1. In $\varphi(a)\varphi(b)$, $\varphi(a)$ and $\varphi(b)$ are elements of G_2. Thus, the operation implied is the operation in G_2.

Properties of Group Homomorphisms

1033
Show that the function φ from $<\mathbb{Z}, +>$ to $C_1 = \{e\}$ defined by $\varphi(k) = e$ for all $k \in \mathbb{Z}$ is a homomorphism from $\mathbb{Z}$ to C_1. *Hint*: Let $i, j \in \mathbb{Z}$. Then $\varphi(i + j) = e$.

1034
Let $i, j \in \mathbb{Z}$.
$\varphi(i + j)$
$= e$ definition of φ
$= ee$ e is the identity in C_1
$\varphi(i)\varphi(j)$ definition of φ ∎

As this example illustrates, G_1 can be homomorphic to G_2 yet be quite different from G_2. $\mathbb{Z}$ is homomorphic to C_1, but the orders of $\mathbb{Z}$ and C_1 are at the opposite extremes.

Let φ be a homomorphism from G_1 to G_2. Let e_1 and e_2 be the identities in G_1 and G_2, respectively. Show that $\varphi(e_1) = e_2$ (we call this the ***identity mapping rule***). *Hint*: Use the homomorphism property to show that $\varphi(e_1)$ is idempotent, which by frame **455** implies that $\varphi(e_1)$ the identity element in G_2.

1035
$\varphi(e_1)\varphi(e_1)$
$= \varphi(e_1 e_1)$ φ is operation preserving
$= \varphi(e_1)$ e_1 is the identity in G_1
$\Rightarrow \varphi(e_1)$ is idempotent in G_2 definition of idempotent
$\Rightarrow \varphi(e_1) = e_2$ ∎ frame **455**

Show that inverses map to inverses under φ (we call this the ***inverse mapping rule***). That is, if two elements are inverses of each other in G_1, then their images in G_2 are inverses of each other. *Hint*: Show that $\varphi(a^{-1})$ is the inverse of $\varphi(a)$ in G_2.

1036
$\varphi(a)\varphi(a^{-1})$
$= \varphi(a\, a^{-1})$ φ is operation preserving.
$= \varphi(e_1)$ a^{-1} is the inverse of a
$= e_2$ identity-mapping rule

Similarly, $\varphi(a)\varphi(a^{-1}) = e_2$. Thus, $\varphi(a^{-1})$ is the inverse of $\varphi(a)$. ∎

The following diagram illustrates the identity and inverse mapping rules:

208   Chapter 19: Fundamental Theorem of Group Homomorphisms

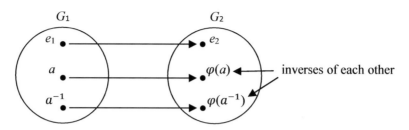

The inverse of the image of *a* is the _____ of the inverse of *a*.

1037

image

The set $\varphi(G_1)$, defined as $\{\varphi(g) : g \in G_1\}$, is called the ***image of G_1 under φ***. In words, $\varphi(G_1)$ is the set consisting of all the elements to which φ maps the elements of G_1.

Let's show that $\varphi(G_1)$ is a subgroup of G_2. We will use the CI subgroup test from frame **846**. Let's show closure:

Closure: Let $x, y \in \varphi(G_1)$. We want to show that $xy \in \varphi(G_1)$. Because $x, y \in \varphi(G_1)$, $x = \varphi(a)$, $y = \varphi(b)$ for some $a, b \in G_1$. Finish showing closure.

1038

xy
$= \varphi(a)\varphi(b)$ $x = \varphi(a), y = \varphi(b)$
$= \varphi(ab)$ φ is operation preserving
$= \varphi(c)$ for some $c \in G_1$ closure property of G_1
$\Rightarrow xy \in \varphi(G_1)$ because $xy = \varphi(c), c \in G_1$

Show that $\varphi(G_1)$ has the inverses property. *Hint*: The following diagram is essentially a pictorial proof:

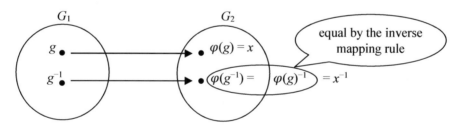

1039

Inverses: Let $x \in \varphi(G_1)$
$\Rightarrow x = \varphi(g)$ for some $g \in G_1$ definition of $\varphi(G_1)$
$\Rightarrow x^{-1} = \varphi(g)^{-1}$ inverse is unique
$= \varphi(g^{-1})$ inverse mapping rule
$\in \varphi(G_1)$ $g^{-1} \in G_1$ by inverses property of G_1

Because $\varphi(G_1)$ is closed and has the inverses property, by the CI subgroup test, $\varphi(G_1)$ is a subgroup of G_2. ∎

We have shown that if φ is a homomorphism from a group G_1 to a group G_2, then $\varphi(G_1)$ is a subgroup of G_2. Thus, whenever there is a homomorphism from G_1 to G_2, there also is a homomorphism from G_1 *onto* a group, namely, onto the subgroup $\varphi(G_1)$.

Using Logarithms to Compute a Product

1040

The following diagram shows the mapping of a homomorphism φ from a group $<G_1, *>$ to a group $<G_2, \otimes>$ on the elements a, b, and x in G_1, where $a * b = x$:

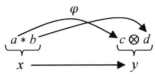

The preimage of y is x. However, y in G_2 could have multiple preimages in G_1 because φ is not necessarily one-to-one. However, if φ is an isomorphism, then φ is one-to-one. In that case, φ^{-1} maps y back to its one and only preimage, namely, x. Moreover, we can then get the answer to the computation $a * b$ (i.e., we can determine x) by mapping a and b to c and d, respectively, in G_2, performing the computation in G_2 (by computing $c \otimes d$), and then mapping the result y back to G_1. *We don't have to perform the computation in G_1*. This procedure is precisely what we do when we use logarithms to compute a product. Specifically, we map the operands of the product to a group whose operation is addition. We perform the computation (by adding) in this group. We then map the result back to the original group to get the answer to our product.

Logs are such a wonderful example of group isomorphism, let's study them in detail. We will restrict our attention to logs to the base 10. However, our results apply to logs to any base.

Let $\mathbb{R}^+$ represent the positive reals and "·" represent multiplication. Is $< \mathbb{R}^+, \cdot >$ a group?

1041

Yes. It has the closure, associative, identity (the identity is 1), and inverses properties (the inverse of a is $1/a$ for all $a \in \mathbb{R}^+$).

Is $<\mathbb{R}, +>$ a group?

1042

Yes. It has the closure, associative, identity (the identity is 0), and inverses properties (the inverse of a is $-a$ for all $a \in \mathbb{R}$).

Show that the log base 10 function from $< \mathbb{R}^+, \cdot >$ to $<\mathbb{R}, +>$ is a group homomorphism. That is, show the operation-preserving property $\log(xy) = \log(x) + \log(y)$ for all $x, y \in \mathbb{R}^+$. This equality, of course, is one of the laws of logarithms. Thus, this question is more about high school mathematics than group theory. Nevertheless, it is an important question. *Hint*: From definition of log base 10, we know that $a = 10^{\log(a)}$ for all $a \in \mathbb{R}^+$.

1043

Let $x, y \in \mathbb{R}^+$ $\mathbb{R}^+$ is the domain of log
$x = 10^{\log(x)}$, $y = 10^{\log(y)}$ definition of logs
$\Rightarrow xy = 10^{\log(x)} 10^{\log(y)}$ substitution
$= 10^{\log 10(x) + \log(y)}$ law of exponents

By definition, the log of xy is what 10 must be raised to to get xy. We have just shown that $xy = 10^{\log 10(x) + \log(y)}$. Thus, $\log(xy) = \log(x) + \log(y)$. ∎

Next, let's show that $\log(x)$ is a bijection from $< \mathbb{R}^+, \cdot >$ to $<\mathbb{R}, +>$. Start by showing that $\log(x)$ is onto $<\mathbb{R}, +>$. *Hint*: Using 10^y, show that for any $y \in \mathbb{R}$, there is an $x \in \mathbb{R}^+$ such that $\log(x) = y$.

1044

Let $y \in \mathbb{R} \Rightarrow 10^y \in \mathbb{R}^+ \Rightarrow 10^y$ is in the domain of log $\Rightarrow$ we can take the log of 10^y. We get $\log(10^y) = y$.

Show that $\log(x)$ is one-to-one.

1045

Let $\log(x) = \log(y)$.
$\Rightarrow 10^{\log(x)} = 10^{\log(y)}$ powers of 10 to the same exponent are equal
$\Rightarrow x = y$ $x = 10^{\log(x)}$, $y = 10^{\log(y)}$ ∎

By the preceding frames, log(*x*) is a bijection from < $\mathbb{R}^+$, ·> to <$\mathbb{R}$, +> that has the operation-preserving property. Thus, < $\mathbb{R}^+$, ·> is isomorphic to <$\mathbb{R}$, +>. The isomorphism between these two groups allows us to use logs to convert a multiplication problem into an addition problem. To multiply two positive reals, we map each via log to their corresponding elements in <$\mathbb{R}$, +>. We then add these elements. Because of the isomorphism, the inverse of log maps the sum back to the product in < $\mathbb{R}^+$, ·>. For example, to multiply 10 and 100, we map 10 and 100 to 1 and 2, their corresponding elements in <$\mathbb{R}$, +>, using the log function. We then add 1 and 2 to get 3. We then take the antilog of 3 (i.e., we map 3 back to <$\mathbb{R}^+$, ·> using the inverse log function) to get 10^3, the desired product:

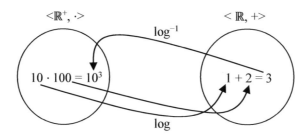

Fundamental Theorem of Group Homomorphisms

1046

Let φ be a homomorphism from a group G_1 to a group G_2, with e_2 the identity in G_2. We call the set of all the preimages of e_2 the **kernel of φ** (denoted by ker φ). That is,

$$\ker \varphi = \{g \in G_1 : \varphi(g) = e_2\}.$$

Here is our plan for this section:

1. Show that ker φ is a subgroup of G_1.
2. Show that ker φ is a normal subgroup of G_1. Then $G_1/\ker \varphi$ is a group under the coset multiplication (or coset addition if we are using additive notation).
3. Show that $G_1/\ker \varphi$ is isomorphic to $\varphi(G_1)$ (this is the fundamental theorem of group homomorphisms).

It is easy to get lost in all the mathematical symbols when studying the material in this section. The key to success is to very carefully study the diagrams that we provide—in particular, the diagrams in frames **1050** and **1055**. Let's take a quick look at these diagrams. The diagram in frame **1050** shows the mapping of a homomorphism φ from G_1 to G_2. G_1 is decomposed into the cosets of ker φ. Note that for each coset, all the elements in that coset map to a single element in G_2. Moreover, different cosets map to different elements in G_2. Thus, associated with the homomorphism φ is a *one-to-one pairing of the cosets of ker φ with elements in G_2*.

Recall that the range of a function is the set of elements to which the functions maps the elements of the domain. Thus, the range of φ in frame **1050** is the set of the three elements in G_2 to which the elements of G_1 are mapped. The homomorphism φ is not necessarily an onto function. Thus, there may be elements in G_2 in addition to the three elements in the range of φ. We denote the range of φ with $\varphi(G_1)$.

As shown in the diagram in frame **1050**, associated with the homomorphism φ is a one-to-one pairing of the cosets of ker φ with elements in G_2. This pairing is a one-to-one function from the set of cosets of ker φ onto $\varphi(G_1)$ so it is a bijection onto $\varphi(G_1)$. Let's call this bijection β. The diagram in frame **1055** shows the mapping of β. β is a bijection from the set of cosets of ker φ to the range of φ. We will show that β is also operation preserving, and, therefore, β is an isomorphism from the set of cosets to the range of φ. Thus, the cosets of ker φ under coset multiplication behave as their corresponding elements in G_2 behave under the operation of G_2. That β is an isomorphism is the fundamental theorem of group homomorphisms.

Let's start out investigation by showing that ker φ is a subgroup of G_1 using the CI subgroup test. Suppose φ is a homomorphism from G_1 to G_2 and the identities of G_1 and G_2 are e_1 and e_2, respectively. Show that ker φ is closed under the group operation of G_1.

1047

Let $x, y \in \ker \varphi$. Thus, $\varphi(x) = e_2$, $\varphi(y) = e_2$. Then $\varphi(xy) = \varphi(x)\varphi(y) = e_2 e_2 = e_2 \Rightarrow xy \in \ker \varphi$.

Show that ker φ has the inverses property. *Hint*: Let $x \in \ker \varphi$. Then show that $\varphi(x^{-1}) = e_2$, which indicates that $x^{-1} \in \ker \varphi$. Here is the start of the proof:

Let $x \in \ker \varphi$.
e_2
$= \varphi(e_1)$ identity mapping rule
$= \varphi(x^{-1}x)$ property of inverses
$= \ldots$

Finish the proof.

1048

$= \varphi(x^{-1})\varphi(x)$ φ is operation preserving
$= \varphi(x^{-1})e_2$ $x \in \ker \varphi$
$= \varphi(x^{-1})$ e_2 is the identity in G_2
$\Rightarrow x^{-1} \in \ker \varphi$ definition of ker φ ∎

Next, show that ker φ is a normal subgroup of G_1. Do this by showing that $g(\ker \varphi)g^{-1} \subseteq \ker \varphi$ for all $g \in G_1$ (this is one of the definitions of a normal subgroup given in frame **1007**).

1049

Let $z \in g(\ker \varphi)g^{-1}$. Then $z = gkg^{-1}$ for some $k \in \ker \varphi$.

$\varphi(z)$
$= \varphi(gkg^{-1})$ $z = gkg^{-1}$
$= \varphi(gk)\varphi(g^{-1})$ φ is operation preserving
$= \varphi(g)\varphi(k)\varphi(g^{-1})$ φ is operation preserving
$= \varphi(g)e_2\varphi(g^{-1})$ $k \in \ker \varphi$
$= \varphi(g)\varphi(g^{-1})$ e_2 is identity in G_2
$= \varphi(gg^{-1})$ φ is operation preserving
$= \varphi(e_1)$ g and g^{-1} are inverses
$= e_2$ identity mapping rule
$\Rightarrow z \in \ker \varphi$ definition of ker φ
$\Rightarrow g(\ker \varphi)g^{-1} \subseteq \ker \varphi$ $z \in g(\ker \varphi)g^{-1} \Rightarrow z \in \ker \varphi$
$\Rightarrow \ker \varphi$ is normal   definition of a normal subgroup ∎

What are the elements of $G_1/\ker \varphi$?

1050

the cosets of ker φ

Because ker φ is a normal subgroup of G_1, $G_1/\ker \varphi$ is a group under coset multiplication. The following diagram shows an example of how the elements in G_1 might map to G_2 under φ where e_2 is the identity in G_2:

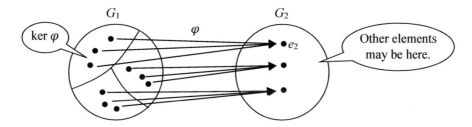

By definition, φ maps every element in ker φ to e_2 in G_2. Moreover, within each coset, φ maps each element in that coset to a single element in G_2, as illustrated in the preceding diagram. Let's prove this.

Let $x \in g\ker \varphi$. Determine $\varphi(x)$. *Hint*: $\varphi(k) = e_2$ for any $k \in \ker \varphi$.

1051

Let $x \in g\ker \varphi \Rightarrow x = gk$ for some $k \in \ker \varphi \Rightarrow$

$\varphi(x)$
$= \varphi(gk)$ substitution
$= \varphi(g)\varphi(k)$ φ is operation preserving
$= \varphi(g)e_2$ $k \in \ker \varphi$
$= \varphi(g)$ e_2 is the identity in G_2

Thus, φ maps *every* element in the coset $g\ker \varphi$ to a single element in G_2. Specifically, it maps every element to $\varphi(g)$. ∎

In the diagram in the preceding frame, φ maps the elements in different cosets to different elements in G_2. Is this always the case? Or is it possible for φ to map the elements in two or more cosets to the same element in G_2? For example, is the following structure possible:

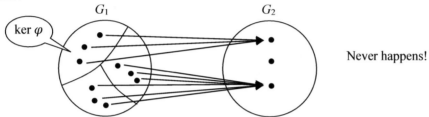

Never happens!

It is not possible; φ always maps elements from different cosets of $\ker \varphi$ to different elements in G_2. That is, if $x \in a\ker \varphi$ and $y \in b\ker \varphi$, and $a\ker \varphi \neq b\ker \varphi$, then $\varphi(x) \neq \varphi(y)$. Let's prove this statement by proving its contrapositive equivalent: if $\varphi(x) = \varphi(y)$, then x and y are in the same coset. Here is the proof:

$\varphi(x) = \varphi(y)$
$\Rightarrow \varphi(y)^{-1}\varphi(x) = e_2$ move $\varphi(y)$ to the left side
$\Rightarrow \varphi(y^{-1})\varphi(x) = e_2$ inverse mapping rule
$\Rightarrow \varphi(y^{-1}x) = e_2$ φ is operation preserving
$\Rightarrow y^{-1}x = k$ for some $k \in \ker \varphi$ definition of $\ker \varphi$
$\Rightarrow \ldots$

Finish the proof. *Hint*: Solve for x and then determine $x\ker \varphi$.

1052

$\Rightarrow x = yk$ solve for x
$\Rightarrow x\ker \varphi = yk\ker \varphi$ substitution
$\Rightarrow x\ker \varphi = y\ker \varphi$ K4 absorption rule in frame **968**

We have shown that that if x and y map to the same element in G_2, then $x\ker \varphi$ and $y\ker \varphi$ are the same coset. By K2 in frame **968**, $x \in x\ker \varphi$ and $y \in y\ker \varphi$. Thus, x and y are in the same coset. ∎

Let's summarize: φ maps every element in $g\ker \varphi$ to $\varphi(g)$. Moreover, φ maps elements in different cosets to different elements in G_2. Thus, there is a one-to-one pairing of the cosets of G_1 with the elements of some subset of G_2 (φ is not necessarily onto, so some elements of G_2 may not have a preimage among the cosets in $G_1/\ker \varphi$). We can represent this one-to-one pairing of cosets of $G_1/\ker \varphi$ and elements in G_2 with a function β, as illustrated in the following diagram:

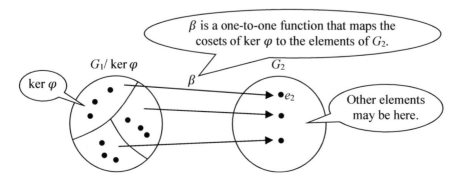

The domain of φ is G_1; φ maps the elements of G_1 to G_2. The domain of β, on the other hand, is the set of the cosets of ker φ; β maps the cosets of G_1/ker φ to G_2. The domain of β is _____.

1053

G_1/ker φ

The codomain of both φ and β is G_2. That is, both φ and β map elements in their domains to elements in G_2. Mathematicians use a shorthand notation to indicate the domain and codomain of functions. Using this notation for φ and β, we get

$$\begin{array}{lcl} \text{domains} & & \text{codomains} \\ \varphi: G_1 & \to & G_2 \\ \beta: G_1/\text{ker }\varphi & \to & G_2 \end{array}$$

In the preceding frame, we saw that φ maps every element in the coset gker φ to $\varphi(g)$. Thus, β maps the coset gker φ to $\varphi(g)$. That is, $\beta(g\text{ker }\varphi) = \varphi(g)$ for all cosets gker φ in G_1/ker φ. A concern you may have with this definition of β is that it may not be well defined. Remember that a coset can be represented by any element it contains. For example, suppose b is in $a$ker φ; then by K5 in frame **968**, $a$ker $\varphi = b$ker φ. β maps $a$ker φ to $\varphi(a)$. β maps $b$ker φ to $\varphi(b)$. If it is possible that $\varphi(a) \neq \varphi(b)$, then β is not well defined because its mapping depends on the coset representative (a or b) we use. But as we saw in frame **1051**, for each coset of ker φ, φ maps all its elements to the same element in G_2. Thus, the mapping defined by β does not depend on the representatives we use.

Is β onto G_2?

1054

Not necessarily, because φ may not be an onto function (see the diagram in frame **1052**).

Is β onto $\varphi(G_1)$?

1055

Yes. The range of β is the same as the range of φ. Because the range of φ is $\varphi(G_1)$, the range of β is also $\varphi(G_1)$.

So far, we have shown that β is a one-to-one mapping from G_1/ker φ onto $\varphi(G_1)$. Thus, β is a bijection from G_1/ker φ to $\varphi(G_1)$:

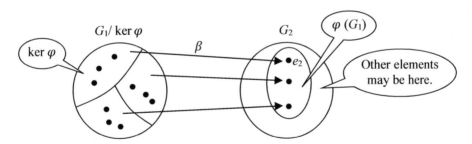

In frame **1037**, we showed that if φ is a homomorphism from G_1 to G_2, then $\varphi(G_1)$ is itself a group (it is a subgroup of G_2). Thus, β is a bijection from the group _____ onto the group _____.

1056

G_1/ker φ, $\varphi(G_1)$

Show that β is operation preserving. *Hint*: Show $\beta([a\text{ker }\varphi][b\text{ker }\varphi]) = \beta(a\text{ker }\varphi)\beta(a\text{ker }\varphi)$.

1057

$\beta([a\text{ker }\varphi][b\text{ker }\varphi])$
$= \beta(ab\text{ker }\varphi)$ definition of coset multiplication
$= \varphi(ab)$ definition of β
$= \varphi(a)\varphi(b)$ φ is operation preserving
$= \beta(a\text{ker }\varphi)\beta(a\text{ker }\varphi)$.   definition of β ∎

β is a bijection from $G_1/\ker \varphi$ onto $\varphi(G_1)$. Moreover, β is operation preserving. What can we then conclude about the groups $G_1/\ker \varphi$ and $\varphi(G_1)$?

1058

$G_1/\ker \varphi$ and $\varphi(G_1)$ are isomorphic groups.

We have just proven the following theorem:

> If φ is a homomorphism from the group G_1 to the group G_2, then the group $G_1/\ker \varphi$ is isomorphic to the group $\varphi(G_1)$. Symbolically, $G_1/\ker \varphi \cong \varphi(G_1)$.

This important theorem is called the *fundamental theorem of group homomorphisms* (FTGH). You may also find the FTGH stated this way:

> If φ is a homomorphism from the group G_1 *onto* the group G_2, then $G_1/\ker \varphi$ is isomorphic to G_2.

The two versions are equivalent. Each version implies the other.

In the diagram in frame **1055**, which groups are isomorphic?

1059

$G_1/\ker \varphi$ (i.e., the cosets of $\ker \varphi$) and $\varphi(G_1)$. By the FTGH, β is an isomorphism from $G_1/\ker \varphi$ to $\varphi(G_1)$.

Test Your Mettle

1060

We conclude this chapter with four interesting problems on the FTGH.

Problem 1: Show that if $u: G \to G_1$ and $v: G \to G_2$ are group homomorphisms onto G_1 and G_2, respectively, and $\ker u = \ker v$, then G_1 is isomorphic to G_2.

1061

$G_1 \cong G/\ker u$	FTGH
$= G/\ker v$	$\ker u = \ker v$
$\cong G_2$	FTGH ∎

Problem 2: For every homomorphism φ from the group G to a group, there is the normal subgroup $\ker \varphi$ of G that gives us the quotient group $G/\ker \varphi$. The converse is also true. That is, for every quotient group G/N, where N is some normal subgroup of G, there is a homomorphism—let's call it γ—from G to G/N.

Let N be a normal subgroup of G. Show that $\gamma: G \to G/N$ defined by $\gamma(g) = gN$ is a homomorphism from G onto G/N. γ is called the *natural map*; γ maps each element of G to the coset that contains that element. Show that γ is a homomorphism from G onto G/N.

1062

Onto: Let $gN \in G/N$. $\gamma(g) = gN \Rightarrow \varphi$ is onto.
Operation preserving:
Let $g_1, g_2 \in G$. Then
$\gamma(g_1 g_2)$
$= (g_1 g_2)N$ definition of γ
$= (g_1 N)(g_2 N)$ definition of coset multiplication
$= \gamma(g_1)\gamma(g_2)$ definition of γ ∎

If N is $\ker \varphi$, γ maps g to $g\ker \varphi$. β (as defined in frame **1052**) in turn maps $g\ker \varphi$ to $\varphi(g)$. Thus, $\beta(\gamma(g)) = \varphi(g)$. That is, $\beta \circ \gamma = \varphi$. We can represent this equality with the following diagram:

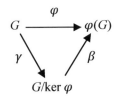

We say the diagram *commutes* because the mappings defined by the two paths from G to $\varphi(G)$—one via φ and the other via γ, then β—are the same.

Problem 3: Characterize the homomorphic image of the group G if $|G|$ is prime. *Hint*: What are the subgroups of G?

1063

By Lagrange's theorem, the order of a subgroup divides the order of the group. Thus, if $|G|$ is prime, then G itself and $\{e\}$ are the only normal subgroups of G. Then, by the FTGH, a homomorphic image of G is isomorphic to either G/G or $G/\{e\}$. G/G is isomorphic to C_1. $G/\{e\}$ is isomorphic to G. Thus, a homomorphic image of G must be isomorphic to either C_1 or G.

Problem 4: Suppose φ is a homomorphism from G_1 onto G_2. Show that $|G_2|$ divides $|G_1|$. *Hint*: Use Lagrange's theorem.

1064

The number of cosets of ker φ in G_1 is $|G_1/\text{ker }\varphi|$. $G_1/\text{ker }\varphi$ is isomorphic to G_2 by the FTGH. Thus, the number of cosets equals $|G_2|$. By Lagrange's theorem, the number of cosets of ker φ in G_1 divides $|G_1|$. Thus, $|G_2|$ divides $|G_1|$. ∎

Review Questions

1. Show that the function $\varphi(x) = x^2$ for all $x \in \mathbb{R} - \{0\}$ is a homomorphism from $<\mathbb{R} - \{0\}, \cdot>$ to $<\mathbb{R} - \{0\}, \cdot>$.
2. What is ker φ in review question 1?
3. Let $|x|$ be the absolute value of x. Show that the function $\varphi(x) = |x|$ for all $x \in \mathbb{R} - \{0\}$ is a homomorphism from $<\mathbb{R} - \{0\}, \cdot>$ to $<\mathbb{R} - \{0\}, \cdot>$.
4. What is ker φ in review question 3?
5. Let g be an element of a group G. Show that $\varphi(x)$ defined by $\varphi(x) = g^{-1}xg$ for all $x \in G$ is an isomorphism from G to G. We call an isomorphism from G to G an *automorphism* of G.
6. Show that in the quotient group G/N, it is possible to have both $aH = bH$ and $|<a>| \neq |<b>|$. *Hint*: D_3.
7. Characterize all the homomorphisms from $<\mathbb{Z}_2, +>$ to $<\mathbb{Z}_2, +>$.
8. Show that $\varphi(k) = 5k$ is a homomorphism from $<\mathbb{Z}, +>$ to $<\mathbb{Z}, +>$.
9. Show that $\varphi(k) = k + 5$ is not a homomorphism from $<\mathbb{Z}, +>$ to $<\mathbb{Z}, +>$.
10. Show that $\varphi(k) = -k$ is a homomorphism from $<\mathbb{Z}, +>$ to $<\mathbb{Z}, +>$.

Answers to the Review Questions

1. $\varphi(xy) = (xy)^2 = x^2y^2 = \varphi(x)\,\varphi(y)$
2. ker $\varphi = \{-1, 1\}$
3. $\varphi(xy) = |xy| = |x||y| = \varphi(x)\,\varphi(y)$
4. ker $\varphi = \{-1, 1\}$
5. Onto: Let $x \in G$. $\varphi(gxg^{-1}) = g^{-1}gxg^{-1}g = x$. One-to-one: Suppose $\varphi(x) = \varphi(y) \Rightarrow g^{-1}xg = g^{-1}yg \Rightarrow x = y$ (by the cancellation law for groups). Homomorphism property: $\varphi(xy) = g^{-1}xyg = g^{-1}xgg^{-1}yg = \varphi(x)\,\varphi(y)$. ∎
6. Let $H = \{e, r, r^2\}$ in D_3. $eH = rH$, but $|e| = 1$ and $|r| = 3$.
7. Corresponding to every homomorphism φ from $\mathbb{Z}_2$ to $\mathbb{Z}_2$, there is a normal subgroup ker φ in $\mathbb{Z}_2$. $\{0\}$ and $\mathbb{Z}_2$ are the only subgroups of $\mathbb{Z}_2$. The homomorphism corresponding to ker $\varphi = \{0\}$ is $\varphi(x) = x$ for all $x \in \mathbb{Z}_2$. The homomorphism corresponding to ker $\varphi = \mathbb{Z}_2$ is $\varphi(x) = 0$ for all $x \in \mathbb{Z}_2$
8. $\varphi(k + j) = 5(k + j) = 5k + 5j = \varphi(k) + \varphi(j)$
9. $\varphi(k + j) = k + j + 5 \neq k + 5 + j + 5 = \varphi(k) + \varphi(j)$. More simply, φ does not map the identity to the identity. Thus, φ cannot be a homomorphism.
10. $\varphi(k + j) = -(k + j) = (-k) + (-j) = \varphi(k) + \varphi(j)$

Homework Questions

1. Let φ be a group homomorphism from G_1 onto G_2. Show that G_1 is isomorphic to G_2 if and only if ker φ equals the trivial subgroup of G_1.
2. What are all the homomorphic images of a simple group? A ***simple group*** G is a group whose only normal subgroups are the trivial subgroup and G.
3. Show that Aut(G), the set of all automorphisms of a group G, is a group under composition (see review question 5).
4. Show that the function $\varphi(x) = x^{-1}$ for all $x \in G$ is an automorphism of a group G if and only if G is commutative (see review question 5).
5. Prove that if G_2 is the homomorphic image of G_1 and G_1 is commutative, then G_2 is commutative.
6. Let G be a group with a normal subgroup N. Show that if G is commutative, then G/N is commutative.
7. How many automorphisms does C_3 have?
8. How many automorphisms does C_∞ have? Describe them.
9. Specify three groups A, B, and C such that A is a normal subgroup of B, B is a normal subgroup of C, and A is a normal subgroup of C.
10. Show by counterexample that the following statement is false: For all groups A, B, and C, if A is a normal subgroup of B, and B is a normal subgroup of C, then A is a normal subgroup of C. *Hint*: For C, use A_4.

20 Some More Groups

$\langle \mathbb{Z}_n - \{0\}, \cdot_n \rangle$ for Prime n

1065

Recall that when performing multiplication modulo n (denoted by $\cdot_n$), we use regular multiplication and then divide the result by n and take the remainder. For example, to compute $2 \cdot_3 2$, we compute $2 \cdot 2$ to get 4 and divide by 3 to get the remainder 1. Thus, $2 \cdot_3 2 = 1$.

Construct the Cayley table for $\langle \mathbb{Z}_3, \cdot_3 \rangle$.

1066

$\cdot_3$	0	1	2
0	0	0	0
1	0	1	2
2	0	2	1

What is the identity in $\langle \mathbb{Z}_3, \cdot_3 \rangle$?

1067

1

What is the multiplicative inverse of 2 in $\langle \mathbb{Z}_3, \cdot_3 \rangle$?

1068

$2 \cdot_3 2 = 1$. Thus, 2 is its own inverse.

What is the multiplicative inverse of 0 in $\langle \mathbb{Z}_3, \cdot_3 \rangle$?

1069

$0 \cdot_3 x = 0$ for all $x \in \mathbb{Z}_3 \Rightarrow 0$ has no multiplicative inverse.

In fact, for every $n > 1$, 0 has no inverse in $\langle \mathbb{Z}_n, \cdot_n \rangle$. Thus, all of these structures are not groups. What if we eliminate the offending element 0? Do we then get a group? That is, is $\langle \mathbb{Z}_n - \{0\}, \cdot_n \rangle$ a group? Let's look at a few examples.

Construct the Cayley table for $\langle \mathbb{Z}_3 - \{0\}, \cdot_3 \rangle$. Does it represent a group?

1070

$\cdot_3$	1	2
1	1	2
2	2	1

This table represents the cyclic group of order 2.

Is $\langle \mathbb{Z}_4 - \{0\}, \cdot_4 \rangle$ a group? *Hint*: Determine the inverse of 2.

1071

$2 \cdot_4 1 = 2$, $2 \cdot_4 2 = 0$, $2 \cdot_4 3 = 2$. The element 2 has no multiplicative inverse. Moreover, because $2 \cdot_4 2 \notin \mathbb{Z}_4 - \{0\}$, $\mathbb{Z}_4 - \{0\}$ is not closed under $\cdot_4$. Thus, $\langle \mathbb{Z}_4 - \{0\}, \cdot_4 \rangle$ is not a group.

218   Chapter 20: Some More Groups

If we examine $\langle \mathbb{Z}_5 - \{0\}, \cdot_5 \rangle$, $\langle \mathbb{Z}_6 - \{0\}, \cdot_6 \rangle$, and $\langle \mathbb{Z}_9 - \{0\}, \cdot_9 \rangle$, we will find that only $\langle \mathbb{Z}_5 - \{0\}, \cdot_5 \rangle$ is a group. Let's summarize our observations: $\langle \mathbb{Z}_n - \{0\}, \cdot_n \rangle$ is a group if $n = 3$ or 5; it is not a group if $n = 4$, 6, or 9. Note that 3 and 5 are prime numbers.

Show that $\langle \mathbb{Z}_n - \{0\}, \cdot_n \rangle$ is not a group if n is composite. *Hint*: It lacks closure.

1072

n composite $\Rightarrow n = ab$ where $0 < a, b < n \Rightarrow a \cdot_n b = 0 \Rightarrow \mathbb{Z}_n - \{0\}$ is not closed under $\cdot_n$

Formulate a conjecture on the values of n for which $\langle \mathbb{Z}_n - \{0\}, \cdot_n \rangle$ is a group.

1073

n is prime $\Leftrightarrow \langle \mathbb{Z}_n - \{0\}, \cdot_n \rangle$ is a group.

This is the biconditional $p \Leftrightarrow q$, where p is "n is prime" and q is "$\langle \mathbb{Z}_n - \{0\}, \cdot_n \rangle$ is a group" Recall from frame **168** that we can prove the biconditional $p \Leftrightarrow q$ by proving $p \Rightarrow q$ and $\sim p \Rightarrow \sim q$. Let's prove $p \Rightarrow q$ first. That is, let's prove

n is prime $\Rightarrow \langle \mathbb{Z}_n - \{0\}, \cdot_n \rangle$ is a group.

Closure: Let n be prime, and $a, b \in \mathbb{Z}_n - \{0\}$. Suppose $a \cdot_n b = 0 \Rightarrow n \mid ab \Rightarrow n \mid a$ or $n \mid b$ (by Euclid's lemma), which by D6 in frame **547** implies that $n \leq a$ or $n \leq b$. But this is impossible because $a, b \in \mathbb{Z}_n - \{0\} \Rightarrow 0 < a, b < n$. Thus, $a \cdot_n b \neq 0 \Rightarrow \mathbb{Z}_n - \{0\}$ is closed under multiplication.
Associativity: $(a \cdot_n b) \cdot_n c = ((a \cdot b) \cdot c) \bmod n = (a \cdot (b \cdot c)) \bmod n = a \cdot_n (b \cdot_n c)$
Identity: $1 \in \mathbb{Z}_n - \{0\}$
Inverses: Let $a \in \mathbb{Z}_n - \{0\}$. $\gcd(a, n) = 1$ because n is prime and $0 < a < n$. Then by frame **567**, $ax + ny = 1$ for some $x, y \in \mathbb{Z} \Rightarrow ax = n(-y) + 1$. Then by the division algorithm, 1 is the remainder when ax is divided by $n \Rightarrow a \cdot_n x = 1$. If $x \in \mathbb{Z}_n - \{0\}$, then x is the inverse of a. What if $x \notin \mathbb{Z}_n - \{0\}$? Because x is congruent to its remainder on division by n, we can replace x with x mod n in $a \cdot_n x = 1$ without affecting the result of the multiplication (this is an application of the multiplication replacement rule in frame **703**). Moreover, x mod n is in $\mathbb{Z}_n - \{0\}$.

To illustrate our reasoning on the existence of inverses, let's do an example. Let's determine the multiplicative inverse of 3 in $\mathbb{Z}_7 - \{0\}$. First, we find integers x and y such that $3x + 7y = 1$. Two values for x and y that work are $x = -9$ and $y = 4$: $(3)(-9) + (7)(4) = 1$ from which we get $(3)(-9) = (7)(-4) + 1$. Then by the division algorithm, $(3)(-9)$ divided by 7 yields a remainder $1 \Rightarrow 3 \cdot_7 -9 = 1$. Because -9 is not in $\mathbb{Z}_7 - \{0\}$, the desired inverse is $(-9 \bmod 7) = 5$. Let's check if 5 is the inverse of 3. We get $5 \cdot_7 3 = 15 \bmod 7 = 1$.

Using the technique illustrated above, determine the multiplicative inverse of 4 in $\mathbb{Z}_7 - \{0\}$. Use -5 for b.

1074

$4x + 7y = 1$. Let $x = 9$ and $y = -5$. Thus, the inverse of 4 is $(9 \bmod 7) = 2$.

Let's summarize: To determine the multiplicative inverse of $a \in \mathbb{Z}_n - \{0\}$) where n is prime, we determine integers x and y such that $ax + ny = 1$. If x is in $\mathbb{Z}_n - \{0\}$, then x is the multiplicative inverse of a. If x is not in $\mathbb{Z}_n - \{0\}$, then x mod n is the inverse of a.

We have shown that for all primes n, $\langle \mathbb{Z}_n - \{0\}, \cdot_n \rangle$ has the closure, associative, identity, and inverses properties. Thus, it is a group. To complete our proof of the biconditional at the beginning of the preceding frame, we have to prove $\sim p \Rightarrow \sim q$. That is, we have to prove

n is not a prime $\Rightarrow \langle \mathbb{Z}_n - \{0\}, \cdot_n \rangle$ is not a group.

Hint: Break up proof into two cases: $n = 1$ and $n > 1$. Do the latter by showing that $\langle \mathbb{Z}_n - \{0\}, \cdot_n \rangle$ lacks the closure property if n is composite.

1075

Case 1: $n = 1$
Then $\mathbb{Z}_n - \{0\}$ is the empty set, and, therefore, cannot be a group.
Case 2: $n > 1$
$n > 1$ and n is not a prime $\Rightarrow n$ is composite $\Rightarrow n = ab$ for some $a, b \in \mathbb{Z}_n - \{0\} \Rightarrow a \cdot_n b = 0 \Rightarrow \{\mathbb{Z}_n - 0\}$ is not closed $\Rightarrow \langle \mathbb{Z}_n - \{0\}, \cdot_n \rangle$ is not a group. ■

Direct Products

1076

Let G and H be groups. Then the ***direct product of G and H***, denoted by $G \times H$, is the set $\{(g, h) : g \in G \text{ and } h \in H\}$ along with the operation defined as follows:

For all $g_1, g_2 \in G$ and $h_1, h_2 \in H$, $(g_1, h_1)(g_2, h_2) = (g_1g_2, h_1h_2)$.

g_1g_2 in the result is the product of g_1 and g_2 in G. Thus, the operation here is the operation in G. h_1h_2 is the product of h_1 and h_2 in H. Thus, the operation here is the operation in H.

$G \times H$ consists of all ordered pairs in which the first component is an element from G and the second component is an element of H. Thus, $|G \times H| = $ _____.

1077

$|G| \cdot |H|$ (i.e., the product of $|G|$ and $|H|$)

There is no requirement that the groups G and H in $G \times H$ are in some way similar or related. G and H can be *any* two groups. For example, G could be D_3, and H could be C_{100}.

What are the elements in $C_2 \times D_3$? *Hint*: Recall that C_2 (the cyclic rotation group of order 2) is $\{e, r\}$, and D_3 (the group of symmetries of an equilateral triangle) is $\{e, r, r^2, f, fr, fr^2\}$.

1078

$\{(e, e), (e, r), (e, r^2), (e, f), (e, fr), (e, fr^2), (r, e), (r, r), (r, r^2), (r, f), (r, fr), (r, fr^2)\}$

What is $(r, r)(r, r)$ in $C_2 \times D_3$? *Hint*: r^2 in C_2 is the identity, but r^2 in D_3 is not the identity.

1079

$(r^2, r^2) = (e, r^2)$

Show that the direct product of a group G and a group H is itself a group.

1080

Let $g_1, g_2, g_3 \in G$, and $h_1, h_2, h_3 \in H$. Let e_G and e_H be the identities in G and H, respectively.
Closure: $(g_1, h_1)(g_2, h_2) = (g_1g_2, h_1h_2)$; $g_1g_2 \in G$ and $h_1h_2 \in H$ because G and H are closed under their operations. Thus, $(g_1g_2, h_1h_2) \in G \times H$.
Associativity: $[(g_1, h_1)(g_2, h_2)](g_3, h_3)$
$= (g_1g_2, h_1h_2)(g_3, h_3)$ definition of multiplication
$= ([g_1g_2]g_3, [h_1h_2]h_3)$ definition of multiplication
$= (g_1[g_2g_3], h_1[h_2h_3])$ associativity property of G and H
$= (g_1, h_1)(g_2g_3, h_2h_3)$ definition of multiplication
$= (g_1, h_1)[(g_2, h_2)(g_3, h_3)]$ definition of multiplication
Identity: $(e_G, e_H)(g, h) = (g, h)(e_G, e_H) = (g, h)$. (e_G, e_H) is the identity.
Inverses: $(g, h)(g^{-1}, h^{-1}) = (gg^{-1}, hh^{-1}) = (e_G, e_H)$. Similarly, $(g^{-1}, h^{-1})(g, h) = (e_G, e_H)$. Thus, (g^{-1}, h^{-1}) is the inverse of (g, h). ∎

If we switch the order of the component groups in a direct product, we get a different but isomorphic group. That is, $G \times H$ is isomorphic to $H \times G$. Let's prove this.

Let φ be the function $\varphi: G \times H \to H \times G$ defined by $\varphi((g, h)) = (h, g)$. Show that φ is a bijection.

1081

Onto: Let $(h, g) \in H \times G \Rightarrow h \in H, g \in G \Rightarrow (g, h) \in G \times H$, and $\varphi((g, h)) = (h, g)$.
One-to-one: Let $\varphi((g_1, h_1)) = \varphi((g_2, h_2)) \Rightarrow (h_1, g_1) = (h_2, g_2) \Rightarrow h_1 = h_2$ and $g_1 = g_2 \Rightarrow (g_1, h_1) = (g_2, h_2)$.

Show that φ is operation preserving.

1082

$\varphi((g_1, h_1)(g_2, h_2)) = \varphi((g_1g_2, h_1h_2)) = (h_1h_2, g_1g_2) = (h_1, g_1)(h_2, g_2) = \varphi((g_1, h_1))\, \varphi((g_2, h_2))$. ∎

Order of an Element in a Direct Product

1083

List all the ordered pairs in the direct product $C_2 \times C_2$.

1084

$\{(e, e), (e, r), (r, e), (r, r)\}$

Every group of order 4 is isomorphic to either C_4 or V_4. $C_2 \times C_2$ is isomorphic to which one? *Hint*: Look at the squares of each element.

1085

$(e, e)^2 = (e, e)(e, e) = (e^2, e^2) = (e, e)$
$(e, r)^2 = (e, r)(e, r) = (e^2, r^2) = (e, e)$
$(r, e)^2 = (r, e)(r, e) = (r^2, e^2) = (e, e)$
$(r, r)^2 = (r, r)(r, r) = (r^2, r^2) = (e, e)$

The square of every element in $C_2 \times C_2$ is equal to the identity. This property is characteristic of V_4. Thus, $C_2 \times C_2 \cong V_4$. Because V_4 is noncyclic, so is $C_2 \times C_2$. $C_2 \times C_2$ is an example of a direct product of cyclic groups that is *not* cyclic.

$C_2 = \{e, r\}$, and $C_3 = \{e, r, r^2\}$. Note that the order of r in C_2 is 2, but the order of r in C_3 is 3. We are using the same name for two elements in different groups; r in C_2 is a 180° rotation of a line segment; r in C_3 is a 120° rotation of an equilateral triangle; (e, e) is the identity in $C_2 \times C_3$; (r, r) is an element in $C_2 \times C_3$ whose components are generators for C_2 and C_3. That is, the r in C_2 generates C_2; the r in C_3 generates C_3. Here are the first seven powers of (r, r):

$(r, r)^1 = (r, r) = (r, r)$
$(r, r)^2 = (r^2, r^2) = (e, r^2)$
$(r, r)^3 = (r^3, r^3) = (r, e)$
$(r, r)^4 = (r^4, r^4) = (e, r)$
$(r, r)^5 = (r^5, r^5) = (r, r^2)$
$(r, r)^6 = (r^6, r^6) = (e, e)$ — indicates $|(r, r)| = 6$
$(r, r)^7 = (r^7, r^7) = (r, r)$

e is the first component in every other power

The first component in the powers of (r, r) is e in every other power (because the order of r in C_2 is 2). Similarly, the second component in the powers of (r, r) is e every third power (because the order of r in C_3 is 3). Thus, to have both components of a power equal to e, the power must have an exponent that is a multiple of 2 and 3. The smallest positive common multiple of 2 and 3 is the order of (r, r).

$|(r, r)| = $ _____.

1086

lcm $(2, 3) = 6$

6 is also the order of $C_2 \times C_3$. Thus, $C_2 \times C_3$ is a _____ group generated by _____.

1087

cyclic, (r, r)

$C_2 \times C_3$ and $C_2 \times C_2$ are direct products of cyclic groups. $C_2 \times C_3$ is cyclic, but $C_2 \times C_2$ is not cyclic.

Determine the order of (r^2, r^2) in $C_3 \times C_4$ by listing successive powers of (r^2, r^2) until the identity element (e, e) appears.

1088

$(r^2, r^2)^1 = (r^2, r^2) = (r^2, r^2)$
$(r^2, r^2)^2 = (r^4, r^4) = (r, e)$
$(r^2, r^2)^3 = (r^6, r^6) = (e, r^2)$
$(r^2, r^2)^4 = (r^8, r^8) = (r^2, e)$
$(r^2, r^2)^5 = (r^{10}, r^{10}) = (r, r^2)$
$(r^2, r^2)^6 = (r^{12}, r^{12}) = (e, e)$

Because $|r^2| = 3$ in C_3, whenever the power of (r^2, r^2) has an exponent that is a multiple of 3, the first element in the pair is e. Similarly, because $|r^2| = 2$ in C_4, whenever the power of (r^2, r^2) has an exponent that is a multiple of 2, the second element in the pair is e. Thus, to get a power of (r^2, r^2) equal to (e, e), the exponent has to be a multiple of both 2 and 3. The smallest such exponent that is positive is the order of (r^2, r^2). The smallest positive exponent that is a multiple of both 2 and 3 is lcm(3, 2) = 6. In general, if $(x, y) \in G \times H$, where G and H are finite groups, then

$$|(x, y)| = \text{lcm}(|x|, |y|).$$

What is the order of (r, r) in $D_3 \times C_9$, where r is a 120° clockwise rotation in D_3 and r is a generator of C_9?

1089

In D_3: $|r| = 3$; in C_9: $|r| = 9$. Thus, $|(r, r)| = \text{lcm}(3, 9) = 9$.

The direct product of two cyclic groups is not necessarily cyclic. For example, we saw in frame **1085** that $C_2 \times C_2$ is noncyclic. Let's determine when the direct product of finite cyclic groups is cyclic.

If $C_i \times C_j$ is cyclic, then by definition there is an element (x, y) that generates $C_i \times C_j$. Thus, the order of (x, y) must equal the order of $C_i \times C_j$. What is the formula for the order of (x, y)? *Hint*: See the preceding frame. What is the order of $C_i \times C_j$?

1090

$|(x, y)| = \text{lcm}(|x|, |y|)$, $|C_i \times C_j| = ij$

Thus, if (x, y) generates $C_i \times C_j$, then $\text{lcm}(|x|, |y|) = ij$. Moreover, x and y must generate C_i and C_j, respectively. Why?

1091

If x does not generate C_i, then x in the successive powers of (x, y) would not generate all the elements in C_i, in which case, (x, y) could not generate $C_i \times C_j$. The same argument applies to y. Thus, if $C_i \times C_j$ is cyclic and generated by (x, y), then x is a generator of C_i (which means $|x| = i$) and y is a generator of C_j (which means $|y| = j$). From the preceding frame, we know that $\text{lcm}(|x|, |y|) = ij$ if $C_i \times C_j$ is cyclic. Substituting i for $|x|$ and j for $|y|$, we get

$$\text{lcm}(i, j) = ij \text{ if } C_i \times C_j \text{ is cyclic.}$$

Conversely, suppose $\text{lcm}(i, j) = ij$. Let x and y be the generators for C_i and C_j respectively. Then $|x| = i$ and $|y| = j$. From frame **1088**, $|(x, y)| = \text{lcm}(|x|, |y|)$. Substituting for $|x|$ and $|y|$, we get $|(x, y)| = \text{lcm}(i, j)$. So if $\text{lcm}(i, j) = ij$, then $|(x, y)| = ij$, which is the order of $C_i \times C_j$. Thus (x, y) generates $C_i \times C_j$, and therefore, $C_i \times C_j$ is cyclic.

In the preceding frames, we have shown that

$$C_i \times C_j \text{ is a cyclic group if and only if } \text{lcm}(i, j) = ij,$$

or, equivalently (see frame **638**),

$$C_i \times C_j \text{ is a cyclic group if and only if } \gcd(i, j) = 1,$$

or, equivalently,

$$C_i \times C_j \text{ is a cyclic group if and only if } i \text{ and } j \text{ have no common prime factor.}$$

Let's now consider direct products of more than two groups. For example, if G, H, and K are groups, then $G \times H \times K = \{(g, h, k) : g \in G, h \in H, k \in K\}$.

The properties of the direct products of two groups extend to the direct products of more than two groups. Specifically, if $G_1, G_2, \ldots, G_n$ are groups, then

- $G_1 \times G_2 \times \cdots \times G_n$ is a group under the multiplication operation defined in frame **1076** extended to n groups

- $|G_1 \times G_2 \times \cdots \times G_n| = |G_1| \cdot |G_2| \cdot \cdots \cdot |G_n|$

- $|(g_1, g_2, \ldots, g_n)| = \text{lcm}(|g_1|, |g_2|, |g_n|)$

- if $G_1, G_2, \ldots, G_n$ are cyclic groups generated by $g_1, g_2, \ldots, g_n$, respectively, then $G_1 \times G_2 \times \cdots \times G_n$ is cyclic if and only if $\text{lcm}(|g_1|, |g_2|, |g_n|) = |g_1| \cdot |g_2| \cdot \cdots \cdot |g_n|$

- if $G_1, G_2, \ldots, G_n$ are cyclic groups, then $G_1 \times G_2 \times \cdots \times G_n$ is cyclic if and only the gcd of the orders of each pair of component groups is equal to 1

- if $G_1, G_2, \ldots, G_n$ are cyclic groups, then $G_1 \times G_2 \times \cdots \times G_n$ is cyclic if and only if the orders of each pair of component groups have no common prime factor

What is the order of $C_2 \times C_3 \times C_4$?

1092

$2 \cdot 3 \cdot 4 = 24$

Is $C_2 \times C_3 \times C_4$ cyclic?

1093

No; $\text{lcm}(2, 3, 4) = 12 \neq |C_2 \times C_3 \times C_4| = 2 \cdot 3 \cdot 4 = 24$. We can also conclude this by observing that 2 and 4 have a common prime factor.

What is the order of (r, r, r) in $C_2 \times C_3 \times C_4$, where r, r, and r are the generators for $C_2, C_3,$ and C_4, respectively?

1094

$\text{lcm}(2, 3, 4) = 12$

Fundamental Theorem of Finite Abelian Groups

1095

We will now discuss but not prove (the proof is complex) a remarkable theorem on direct products. This theorem is called the ***fundamental theorem of finite Abelian groups.*** It allows us to enumerate without repeats all the Abelian (i.e., commutative) groups of any finite order. Using this theorem, let's enumerate all the commutative groups of order 72. The first step is to factor 72 into primes: $72 = 2^3 3^2$. Next, for each power of a prime in the prime factorization of 72 (i.e., for 2^3 and for 3^2), list all the distinct factorizations possible. For example, corresponding to 2^3, we list $2 \cdot 2 \cdot 2$, $2 \cdot 4$, and 8. We do not list $4 \cdot 2$ because apart from order, it is the same factorization as $2 \cdot 4$:

2^3	3^2
$2 \cdot 2 \cdot 2$	$3 \cdot 3$
$2 \cdot 4$	9
8	

Prime factorization of 72

All possible factorizations of each power

Finally, we form direct products of cyclic groups for every pair of factorizations that consists of one entry from the 2^3 column and one entry from the 3^2 column. For example, take one entry from the 2^3 column, say, $2 \cdot 4$, and one entry from the 3^2 column, say, $3 \cdot 3$. With this choice, we get four factors: 2, 4, 3, and 3. Now form the direct product of cyclic groups, one cyclic group for each of the four factors whose size is its corresponding factor:

$$\begin{array}{cc} 2 \cdot 4 & 3 \cdot 3 \\ \downarrow\downarrow & \downarrow\downarrow \\ C_2 \times C_4 \times C_3 \times C_3 \end{array}$$

This is one of the commutative groups of order 72. There are three entries in the 2^3 column and two entries in the 3^2 column. Thus, there are a total of $3 \cdot 2 = 6$ pairs, each corresponding to a distinct commutative group of order 72.

List all the commutative groups up to isomorphism of order 72.

1096

$C_2 \times C_2 \times C_2 \times C_3 \times C_3$
$C_2 \times C_4 \times C_3 \times C_3$
$C_8 \times C_3 \times C_3$
$C_2 \times C_2 \times C_2 \times C_9$
$C_2 \times C_4 \times C_9$
$C_8 \times C_9$

One of the commutative groups of order 72 is C_{72}. Because the preceding list includes all commutative groups of order 72, one must be isomorphic to C_{72}. Which one?

1097

All the direct products, except the last one, have at least one pair of component groups whose orders have a common prime factor. Thus, by frame **1091**, they are not cyclic. However, 8 and 9 have no common prime factor. Thus, $C_8 \times C_9$ is cyclic and, therefore, isomorphic to C_{72}.

Determine all commutative groups of order 8. *Hint*: Because 8 is a power of a prime, there is only one column to choose from.

1098

$C_2 \times C_2 \times C_2, C_2 \times C_4, C_8$

Determine all the commutative groups of order 4.

1099

The two factorizations of 4 are $2 \cdot 2$ and 4. The corresponding commutative groups are $C_2 \times C_2$ (which is isomorphic to V_4) and C_4.

Determine all commutative groups of order 10.

1100

$10 = 2 \cdot 5$. The 2 column has 2 as its only entry. The 5 column also 5 as its only entry. Thus, there is only one possible pair, 2 and 5, corresponding to the group $C_2 \times C_5$, which is isomorphic to C_{10}.

Is $C_2 \times C_3 \times C_5$ isomorphic to $C_6 \times C_5$? *Hint*: How do $C_2 \times C_3$ and C_6 compare?

1101

2 and 3 are relatively prime $\Rightarrow C_2 \times C_3$ is cyclic $\Rightarrow C_2 \times C_3$ and C_6 are isomorphic $\Rightarrow C_2 \times C_3 \times C_5$ isomorphic to $C_6 \times C_5$.

Subgroups of Direct Products

1102

Suppose that G_1 and G_2 are groups and that e_1 and e_2 are the identity elements in G_1 and G_2, respectively. Prove that $H = \{(g, e_2): g \in G_1\}$ is a subgroup of $G_1 \times G_2$. Use the one-step subgroup test from frame **860**.

1103

Let $(a, e_2), (b, e_2) \in H$. Then $(a, e_2)(b, e_2)^{-1} = (a, e_2)(b^{-1}, e_2) = (ab^{-1}, e_2 e_2) = (ab^{-1}, e_2) \in H$. ∎

Prove that H is a normal subgroup by showing that $(g_1, g_2)H(g_1, g_2)^{-1} \subseteq H$ (this is one of the definitions of a normal subgroup—see item 4 in frame **1007**).

1104

Let $x \in (g_1, g_2)H(g_1, g_2)^{-1}$. Then, for some $(g_3, e_2) \in H$, $x = (g_1, g_2)(g_3, e_2)(g_1, g_2)^{-1} = (g_1, g_2)(g_3, e_2)(g_1^{-1}, g_2^{-1}) = (g_1g_3g_1^{-1}, g_2e_2g_2^{-1}) = (g_1g_3g_1^{-1}, e_2) \in H$. ∎

Graphs of Direct Products of Cyclic Groups

1105

The set $X = \{(e, r), (r, e)\}$ is a generating set for $C_2 \times C_2$, where $C_2 = \{e, r\}$. Draw the graph for $C_2 \times C_2$ based on X. It should have two line types, each bidirectional: one for (e, r) and one for (r, e).

1106

```
(r, e) ——— (r, r)
  |          |
  |          |              Key:   (e, r) ———      (r, e) - - - - -
  |          |
(e, e) ——— (e, r)
```

Compare this graph with the graph in frame **811** of V_4. They are identical, except for the names of the elements. Thus, $C_2 \times C_2$ and V_4 are isomorphic.

$C_2 = \{e, r\}$ and $C_3 = \{e, r, r^2\}$. The set $Y = \{(r, e), (e, r)\}$ is a generating set for $C_2 \times C_3$. Draw the graph for $C_2 \times C_3$ based on Y.

1107

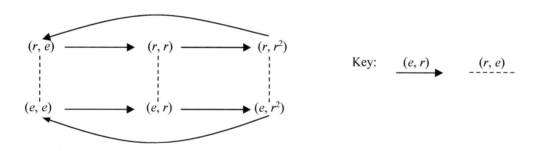

This graph can be drawn as two triangles, with one in front of the other as follows:

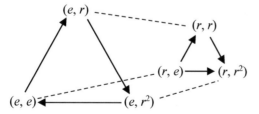

Compare this graph with the graph for D_3 (frame **806**). Are they the same apart from the names of the elements (in which case, $C_2 \times C_3$ would be isomorphic to D_3)?

1108

The graphs are different. In the preceding frame, the arrows are clockwise in both triangles. In the graph for D_3, the arrows in the top triangle are clockwise, but in the bottom triangle, the arrows are counterclockwise. That the graphs differ is to be expected: $C_2 \times C_3$ is not isomorphic to D_3. $C_2 \times C_3$ is both cyclic and commutative, but D_3 is neither.

Given that $C_2 \times C_3$ is isomorphic to C_6, why is the graph for $C_2 \times C_3$ in the preceding frame structurally different from the graph of C_6 (which has the form of a single hexagon)?

1109

They are based on different generating sets.

Symmetries of Regular Polyhedra

1110

In chapter 13, we investigated the symmetries of two-dimensional objects: the regular polygons. We can do the same with three-dimensional objects: the regular polyhedra.

There are exactly five regular polyhedra: the ***tetrahedron*** (4 sides, each an equilateral triangle), the ***cube*** (6 sides, each a square), the ***octahedron*** (8 sides, each an equilateral triangle), the ***dodecahedron*** (12 sides, each a regular pentagon), and the ***icosahedron*** (20 sides, each an equilateral triangle). The symmetries of each regular polyhedron form a group under the operation "followed by." These groups are named according their corresponding polyhedron. For example, the group of symmetries for the tetrahedron is called the ***tetrahedral group.***

Let's examine the symmetries of a regular tetrahedron. We number its vertices 1, 2, 3, and 4. Initially, vertex 1 is on top, vertex 2 is on the left, vertex 3 is in the front, and vertex 4 is on the right:

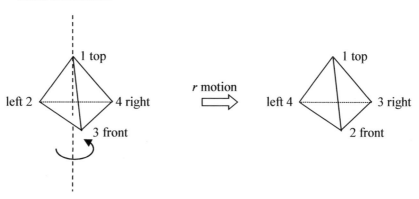

If we rotate the tetrahedron 120° counterclockwise from a position looking down on vertex 1 around an axis through vertex 1, the position vertex 2 is in originally becomes the position of vertex 4. We say vertex 2 is "replaced by" vertex 4. Vertex 4 is replaced by vertex 3, and vertex 3 is replaced by vertex 2. The tetrahedron after this motion can fit precisely in the space that it occupied when in its original orientation. Thus, this motion is a symmetry. We call this symmetry a ***rotation*** and denote it with r. We can represent this motion with the permutation $\begin{pmatrix} 1 & 2 & 3 & 4 \\ 1 & 4 & 2 & 3 \end{pmatrix}$. This permutation defines the r motion. Each column indicates what happens to the corresponding vertex. For example, in this permutation, the first column indicates that vertex 1 remains where it is. Column 2 indicates that the position of vertex 2 before the rotation is occupied by vertex 4 after the rotation. That is, vertex 2 is replaced by vertex 4.

Convert the preceding permutation to cycle form (see frame **468**).

1111

(2 4 3)

Convert this cycle to a sequence of transpositions (see frame **497**).

1112

(2 3) ∘ (2 4)

Is this an even or odd permutation?

1113

even

As with the symmetries of the regular polygons, we are interested only in the net effect of a motion. For example, if a tetrahedron is rotated 480° (one full circle plus 120°) about the axis through vertex 1, the net effect is the same as a 120° rotation about the axis through vertex 1. Thus, both motions are the same symmetry.

The motion corresponding to a 240° rotation, denoted by r^2, is also a symmetry. It is equivalent to two successive r rotations. On the first of the two rotations in an r^2 motion, the position occupied by vertex 2 becomes occupied by vertex 4. On the second rotation, this position becomes occupied by vertex 3. Thus, in the r^2 motion, vertex 2 is replaced by vertex 3:

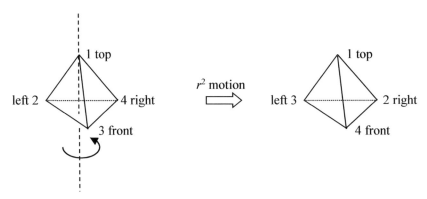

What is the permutation that represents the r^2 motion?

1114

$\begin{pmatrix} 1 & 2 & 3 & 4 \\ 1 & 3 & 4 & 2 \end{pmatrix}$. This permutation defines the r^2 motion.

Convert this permutation to cycle form and to a sequence of transpositions.

1115

$(2\ 3\ 4) = (2\ 4) \circ (2\ 3)$

This permutation, like the permutation for r, is even.

We call the symmetry whose net effect is the same as no motion the identity motion, denoted by e. This motion is the identity element in the group of symmetries. The e motion corresponds to the permutation $\begin{pmatrix} 1 & 2 & 3 & 4 \\ 1 & 2 & 3 & 4 \end{pmatrix}$, which equals $(1\ 2) \circ (1\ 2)$. It, like the permutations for r and r^2, is even.

There are three symmetries for each position vertex 1 can occupy: three when vertex 1 is in the top position, three when vertex 1 is in the left position, three when vertex 1 is in the front position, and three when vertex 1 is in the right position. Thus, the total number of symmetries is _____.

1116

12

To realize all 12 symmetries, we need another motion—one that makes it possible to move vertex 1 to the left, front, and right positions. A motion that will do this is the flip, denoted by f. An f motion is a 180° rotation around a line segment drawn from the midpoint of the line segment connecting vertices 1 and 2 to the midpoint of the line segment connecting vertices 3 and 4:

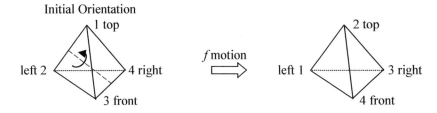

In an f motion, vertices 1 and 2 trade places, and vertices 3 and 4 trade places. Thus, its corresponding permutation is $\begin{pmatrix} 1 & 2 & 3 & 4 \\ 2 & 1 & 4 & 3 \end{pmatrix}$. This permutation defines the f motion.

Is this an even or odd permutation?

1117

In transposition form, the permutation is (1 2) ∘ (3 4). It is an even permutation.

Rather than draw a tetrahedron to show its orientation, we can simply write the numbers of its vertices in the positions they occupy. For example, we represent the orientation in which vertex 1 is on top, vertex 2 is on the left, vertex 3 is in front, and vertex 4 is on the right, with

$$\begin{array}{c} 1 \\ 2\ \ 4 \\ 3 \end{array}$$

In the following diagram, we show the orientations of the tetrahedron that result under some symmetries. The left column of orientations shows the effect of one r motion followed by a second r motion. Each orientation in the right column shows the effect of an f motion applied to its left neighbor:

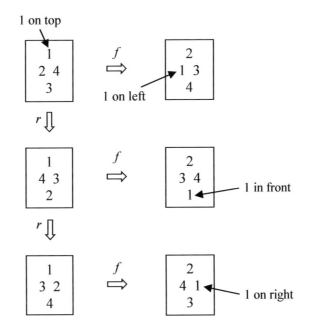

The initial orientation of the tetrahedron has vertex 1 on top. With two r motions from this starting orientation, we can realize the three symmetries with 1 on top.

An f motion on the initial orientation puts the 1 vertex in the left position (see the first row of the preceding diagram). Thus, with two r motions after the initial f motion, we can realize the three symmetries associated with vertex 1 on the left.

An rf motion puts vertex 1 in the front position. Thus, with two r motions following an rf motion, we can realize the three symmetries associated with vertex 1 in the front.

An $rrf = r^2f$ motion puts vertex 1 in the right position. Thus, with two r motions following an r^2f motion, we can realize the three symmetries associated with vertex 1 on the right.

From the preceding observations, we can conclude that every symmetry can be expressed as some product of powers of r and f. In other words, $\{r, f\}$ generates the tetrahedral group.

Express each symmetry of a tetrahedron in terms of the r and f motions.

1118

228   *Chapter 20: Some More Groups*

Vertex 1 on top: $e = r^0$ $r,\quad r^2$
Vertex 1 on left: f $fr,\quad fr^2$
Vertex 1 in front: rf $rfr,\quad rfr^2$
Vertex 1 on right: r^2f $r^2fr,\quad r^2fr^2$

$\underbrace{\hphantom{XXXXXXXXXXX}}$ $\underbrace{\hphantom{XXXXXXXXXXX}}$

        Motions in this r and then a second r
        column position   motion after positioning
        vertex 1 on top,   vertex 1 realize the other
        left, front, or right.   symmetries with vertex 1
                                           in that position.

The r and f motions are even permutations. Every symmetry of the tetrahedron can be expressed in terms of r and f motions. Thus, all _____ symmetries of a tetrahedron are _____ permutations on $\{1, 2, 3, 4\}$.

1119

12, even

Describe A_4 (see frame **518**).

1120

It is the group of the 12 even permutations on $\{1, 2, 3, 4\}$ under function composition.

Thus, the tetrahedral group in permutation form has the same set of permutations as A_4. Moreover, the "followed by" operation on the symmetries is a function composition when the symmetries are in permutation form.

What can you conclude about A_4 and the tetrahedral group?

1121

They are isomorphic.

On the basis of the information in frame **1118**, construct the graph of the tetrahedral group. *Hint*: It has four triangles with r arrows as sides. These triangles are interconnected with f lines. Each triangle corresponds to one of the four possible positions of vertex 1.

1122

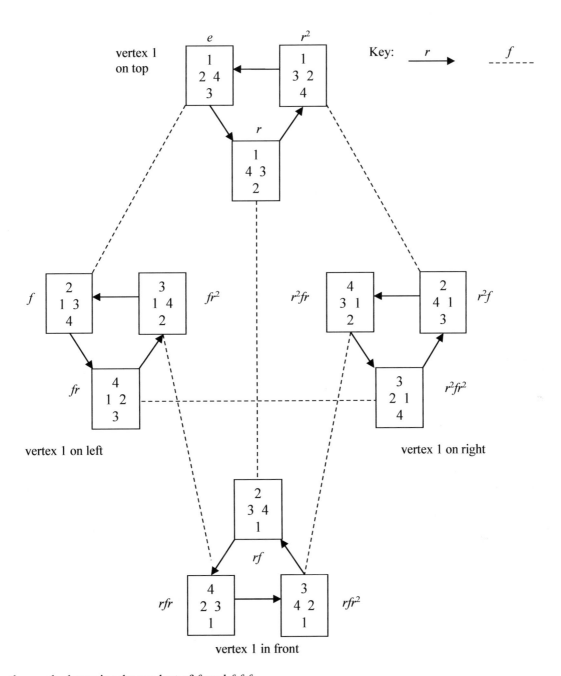

From the graph, determine the product of *fr* and *frfrf*.

1123

(*fr*)(*frfrf*) = *f*. Start at the *fr* node and follow the *f*, *r*, *f*, *r*, and *f* lines and arrows. The final node in this path is *f*.

From the graph, determine the value of *rfrfrf*.

1124

Start from the *e* node and follow the *r*, *f*, *r*, *f*, *r*, *f* arrows to *e*.

From the graph, determine if (*f*)(*r*) = (*r*)(*f*).

1125

(*f*)(*r*) ≠ (*r*)(*f*). The *f-r* and *r-f* paths from the *e* element end on different nodes.

Is the tetrahedral group commutative?

1126

No; $(f)(r) \neq (r)(f)$.

The following table summarizes the groups associated with the five regular polyhedra. The dodecahedral and the icosahedral groups are isomorphic to each other (and to A_5). This is no accident. The dodecahedron and the icosahedrons are **duals**: the centers of the 12 pentagons in a dodecahedron correspond to the 12 vertices of an icosahedron; the 20 vertices of a dodecahedron correspond to the 20 sides of an icosahedron. The cube and the octagon are also duals. Thus, their groups are isomorphic to each other (and to S_4). The tetrahedron is self-dual.

Regular Polyhedron	Type of Each Side	Number of Sides	Number of Vertices	Group of Symmetries Isomorphic to
Tetrahedron	equilateral triangle	4	4	A_4
Hexahedron (cube)	square	6	8	S_4
Octahedron	equilateral triangle	8	6	S_4
Dodecahedron	regular pentagon	12	20	A_5
Icosahedron	equilateral triangle	20	12	A_5

One of the important features of A_5 (and the icosahedral and dodecahedral groups) is that its only normal subgroups are the trivial subgroup and A_5 itself. We call a group with this property a **simple group**. A_5 is the smallest noncommutative group that is simple. In the opposite extreme is Q_8, the **quaternion group**. Q_8 is the smallest noncommutative group whose subgroups are all normal. We investigate Q_8 in the next section.

What is the smallest noncommutative group?

1127

D_3

Unlike like A_5, D_3 is not simple: $\{e, r, r_2\}$ is a normal subgroup of D_3.
A simple group G has a simple structure in that it can be decomposed into only two quotient groups: G/G and $G/\{e\}$ (because G and $\{e\}$ are its only normal subgroups).

Quaternion Group

1128

Complex numbers are of the form $a + bi$, where $a, b \in \mathbb{R}$, and $i^2 = -1$. The quaternions, denoted by $\mathbb{H}$, are an extension of the complex numbers. A quaternion is a number of the form $a + bi + cj + dk$, where

- $a, b, c, d \in \mathbb{R}$
- $i^2 = j^2 = k^2 = -1$
- $ij = k, jk = i, ki = j$
- $ji = -k, kj = -i, ik = -j$

The latter six equalities can be remembered easily with the aid of the following circle:

The product of any two adjacent elements in clockwise order is equal to the third element. For example, $ij = k$. However, the product of any two adjacent elements in counterclockwise order is equal to the negation of the third element. Thus, $ji = -k$. The subset $\{1, -1, i, -i, j, -j, k, -k\}$ of $\mathbb{H}$ forms a group under multiplication. We call this group the **quaternion group,** denoted by Q_8.

Q_8 has three subgroups of order 4: $\{1, -1, i, -i\}$, $\{1, -1, j, -j\}$, and $\{1, -1, k, -k\}$, generated by i, j, and k, respectively. The order of these subgroups is half the order of Q_8. Thus, by K12 in frame **989**, they are normal subgroups. The only subgroup of order 2 in Q_8 is $\{1, -1\}$. The coset $q\{1, -1\} = \{q, -q\} = \{1, -1\}q$ for all $q \in Q_8$. Thus, this subgroup, like the subgroups of order 4, is normal. The only remaining subgroups are the trivial subgroup and Q_8 itself, both of which are also normal. Every subgroup of Q_8 is normal.

Is Q_8 commutative?

1129

No; $ij = k \neq -k = ji$. Q_8 is the smallest noncommutative group whose subgroups are all normal.

Let's determine the order of i: $i^1 = i$, $i^2 = -1$, $i^3 = -i$, $i^4 = -i(i) = -(-1) = 1$. Thus, $|i| = 4$. List the order of each element in Q_8.

1130

1: 1, -1: 2, i: 4, j: 4, k: 4, $-i$: 4, $-j$: 4, $-k$: 4

Generators and Relations

1131

Recall from frame **780** that a subset X of a group G is a generating set for G if every element in G can be constructed from the elements of X and their inverses.

Let X be a generating set for a group G. A **word** of G is any finite sequence of elements that can be constructed from the elements in X and their inverses. For example, f, r, ff, and rfr^{-1} are words of D_3 given the generating set $\{r, f\}$.

In chapter 6, we learned that a relation is a subset of a Cartesian product. In this section, the term "relation" has a completely different meaning. Here, a **relation** for a group G is an equation whose left side is a word of G and whose right side is the identity element. For example, $rr^{-1} = e$ is a relation for $C_4 = \{e, r, r^2, r^3\}$. Because the relation $rr^{-1} = e$ is implied by the inverses property of a group, it is a relation for every group with an element r. Thus, it does not in any way define C_4.

An infinite set of relations hold for C_4. In addition to the relations that follow solely from the group properties, it has the relations $r^4 = e$, $r^8 = e$, $r^{12} = e$, $r^{16} = e$, and so on.

We say that a set of relations R **implies** a relation r if and only if r can be obtained from R using only the properties all groups share. In C_4, for example, $\{r^4 = e\}$ implies $r^8 = e$ because $r^4 = e \Rightarrow r^4 r^4 = ee \Rightarrow r^8 = e$. A set of relations R **implies** a set of relations S if and only if R implies each relation in S. For example, $\{r^4 = e\}$ implies $\{r^8 = e\}$.

A set of relations R is **stronger** than a set of relations S if and only if R implies S, but S does not imply R. For example, $\{r^4 = e\}$ implies $\{r^8 = e\}$, but $\{r^8 = e\}$ does not imply $\{r^4 = e\}$. Thus, $\{r^4 = e\}$ is stronger than $\{r^8 = e\}$.

Let R be a set of relations that hold for a group G. We say that R is a **defining set of relations** for G if every set of relations that is stronger than R does not hold for G. For example, $\{r^4 = e\}$ is a defining set of relations for C_4. The sets $\{r^2 = e\}$, $\{r^2 = e, r^3 = e\}$, and $\{r = e\}$ are stronger than $\{r^4 = e\}$. But all such stronger sets do not hold for C_4. Thus, $\{r^4 = e\}$ is a defining set of relations for C_4.

Show that a defining set of relations R for a group G implies *all* the relations that hold for G. *Hint*: Use a proof by contradiction.

1132

Assume to the contrary that there exists a relation r not implied by R that holds for G. Then $R \cup \{r\}$ is stronger than R, and it holds for G. But then R is not a defining set of relations for G. We resolve this contradiction by concluding that our initial assumption—that there exists a relation r not implied by R that holds for G—is incorrect.

What group is specified by the generating set $\{r\}$ and the defining relation $r^2 = e$?

1133

C_2

For some groups, more than one relation is required to define them. For example, D_3 is defined with the generating set $\{r, f\}$ and three relations: $r^3 = e$, $f^2 = e$, and $rfrf = e$.

Defining relations are sometimes written with something other than the identity element on the right side. For example, if we move r and f in $rfrf = e$ to the right side, we get $rf = f^{-1}r^{-1}$. From the relation $f^2 = e$, we get $f^{-1} = f$. From $r^3 = e$, we get $r^{-1} = r^2$. Thus, the relation $rf = f^{-1}r^{-1}$ can be written as $rf = fr^2$ (our flip-rotate commutation rule from frame **803**).

Let's use the defining relations of D_3 to determine its elements and Cayley table. Because $\{r, f\}$ is a generating set for D_3, each element of D_3 is either r, f, r^{-1}, f^{-1}, or some product of these elements. Given any element in D_3 represented in this way, we can replace f^{-1} with f (because $f^{-1} = f$) and r^{-1} with r^2 (because $r^{-1} = r^2$). Then using the flip-rotate commutation rule $rf = fr^2$, we can commute all the f factors to the left and all the r factors to the right. Thus, every element of D_3 can be represented as some sequence of f factors followed by some sequence of r factors. The sequence of f factors simplifies to f^i where $i = 0$ or 1 (because $f^2 = e$). The sequence of r factors simplifies to r^j, where $j = 0, 1$, or 2 (because $r^3 = e$). Thus, every element has the form $f^i r^j$ where $i = 0$ or 1, and $j = 0, 1$, or 2. When $i = j = 0$, we get the identity element e. For the other values of i and j, we get the remaining five elements in D_3: r, r^2, f, fr, and fr^2. To determine the product of any two elements, we multiply the two elements, and then convert the result to one of our six elements using the defining relations. For example, the product of f and fr is ffr. Using the relation $f^2 = e$, we get $ffr = er = r$.

Dicyclic Groups

1134

Dicyclic groups have an unusual structure. Let's define them with a generating set and a set of relations.

For $n \geq 1$, the ***dicyclic group*** of order $4n$ (denoted by Dic_{4n}) is the group generated by $\{x, y\}$, where $x^{2n} = e$, $x^n = y^2$, and $y^{-1}xy = x^{-1}$.

Although not explicitly stated in this definition, we are to assume that x and y are distinct and not equal to the identity. Thus, the generated group has at least three elements: e, x, and y.

Show that xy is not equal to x or y. *Hint*: Use a proof by contradiction.

1135

If $xy = x$, then canceling x, we get $y = e$. But y is distinct from e. Similarly, $xy \neq y$.

Show that $xy \neq e$. *Hint*: Use $y^{-1}xy = x^{-1}$.

1136

$y^{-1}xy = x^{-1} \Rightarrow xy = yx^{-1}$. If $xy = e$, then $yx^{-1} = e \Rightarrow y = x$. But $y \neq x$. Thus, dicyclic groups have at least four elements: e, x, y, and xy.

Let's investigate Dic_4, the dicyclic group Dic_{4n} for $n = 1$. According to the definition of a dicyclic group, for $n = 1$, the group is generated by x and y, where $x^2 = e$, $x = y^2$, and $y^{-1}xy = x^{-1}$.

Determine the elements generated by y.

1137

$y, y^2 = x, y^3 = y^2y = xy, y^4 = y^2y^2 = xx = x^2 = e$

The order of y is 4. The order of x is 2. What is the order of xy? *Hint*: Make substitutions using the defining relations.

1138

By the definition of a dicyclic group, $y^{-1}xy = x^{-1}$, from which we get $xy = yx^{-1}$. Using $xy = yx^{-1}$, we get

$$(xy)^1 = xy,$$
$$(xy)^2 = (xy)(xy) = (xy)(yx^{-1}) = xy^2x^{-1} = xxx^{-1} = x,$$
$$(xy)^3 = (xy)^2(xy) = x(xy) = x^2y = ey = y,$$
$$(xy)^4 = (xy)(xy)^3 = (xy)y = xy^2 = xx = e.$$

The order of xy is 4.

What is x^{-1}?

1139

$x^2 = e \Rightarrow x^{-1} = x$.

Show that $xy = yx$. *Hint*: Use the defining relation $y^{-1}xy = x^{-1}$.

1140

$y^{-1}xy = x^{-1} \Rightarrow xy = yx^{-1}$. Substituting x for x^{-1} in yx^{-1}, we get $xy = yx$. ∎

Complete the following Cayley table, where e, y, x, and xy are elements of Dic_4. *Hint*: Use $x = x^{-1}$, $xy = yx$, and the defining relations for the group. For example, to determine $y(xy)$, note that $y(xy) = y(yx) = y^2x = xx = e$.

	e	y	x	xy
e	e	y	x	xy
y	y			e
x	x			
xy	xy			

1141

	e	y	x	xy
e	e	y	x	xy
y	y	x	xy	e
x	x	xy	e	y
xy	xy	e	y	x

This table has a familiar pattern: each row is the preceding row circularly left shifted; each column is the preceding column circularly upshifted.

This is the pattern of which group?

1142

the cyclic group of order 4

What are the inverses of e, y, x, and xy?

1143

$e^{-1} = e$, $y^{-1} = xy$, $x^{-1} = x$, $(xy)^{-1} = y$

Draw the graph based on the generating set $\{x, y\}$. Use x lines and y arrows.

1144

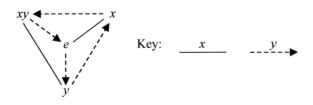

Determine from the graph if y generates $\{e, x, y, xy\}$. *Hint*: Does $yyyy$ traverse all four nodes?

1145

Applying $yyyy$ starting from the e node, we traverse the following nodes: y, $yy = x$, $yyy = xy$, $yyyy = e$. Thus, y generates $\{e, x, y, xy\}$. This confirms our observation in frame **1142** that the Cayley table in frame **1141** corresponds to the cyclic group of order 4.

The generating set for Dic_4 is $\{x, y\}$. The graph based on this generating set has nodes for e, x, y, and xy and no other nodes. What can we conclude about $\{e, x, y, xy\}$ and Dic_4?

1147

We can rule out the possibility that $\{e, x, y, xy\}$ is a proper subset of Dic_4. Thus, $Dic_4 = \{e, x, y, xy\}$.

Draw the graph for C_4.

1148

   Key:

This graph and the graph in frame **1144** are structurally dissimilar, yet they correspond to isomorphic groups (both are cyclic groups of order 4). Dissimilar graphs are not unexpected. The graph in this frame is based on a generating set with one element (r); the graph in frame **1144** is based on a generating set with two elements (x and y).

Let's now investigate Dic_8, the dicyclic group Dic_{4n} for $n = 2$. Its defining relations are $x^4 = e$, $x^2 = y^2$, and $y^{-1}xy = x^{-1}$. This group has eight elements: e, x^2, x, y, xy, x^3, y^3, and $(xy)^3$. If we match each element of Dic_8 with an element of Q_8 (the quaternion group), as follows,

Dic_8	e	x^2	x	y	xy	x^3	y^3	$(xy)^3$
Q_8	1	-1	i	j	k	$-i$	$-k$	$-j$

then the elements of Dic_8 behave exactly like their corresponding elements in Q_8. Dic_8 and Q_8 are isomorphic. For example, in the quaternion group, $jk = i$. Thus, in the dicyclic group, $y(xy)$ should equal x. Show this. *Hint*: Use the defining relations $y^2 = x^2$ and $y^{-1}xy = x^{-1}$ (or, equivalently, $xy = yx^{-1}$).

1149

$xy = yx^{-1} \Rightarrow y(xy) = y(yx^{-1}) = y^2x^{-1} = x^2x^{-1} = x$.

Groups of Orders 1 to 12

1150

Using the fundamental theorem of finite Abelian groups, determine all the commutative groups of order 12.

1151

$12 = 2^2 3$. Thus, the commutative groups of order 12 are $C_2 \times C_2 \times C_3$ and $C_4 \times C_3$.

Up to isomorphism, there are three noncommutative groups of order 12. Two are D_6 (the symmetries of a regular hexagon) and A_4 (the symmetries of a regular tetrahedron). The third noncommutative group of order 12 is Dic_{12}, the dicyclic group of order 12.

Here are all the groups of orders 1 to 12:

Order	Commutative	Noncommutative
1	$C_1 \cong \mathbb{Z}_1 \cong S_1 \cong A_1 \cong A_2$	
2	$C_2 \cong \mathbb{Z}_2 \cong S_2$	
3	$C_3 \cong \mathbb{Z}_3 \cong A_3$	
4	$C_4 \cong \mathbb{Z}_4 \cong U_4 \cong Dic_4, C_2 \times C_2 \cong V_4$	
5	$C_5 \cong \mathbb{Z}_5$	
6	$C_6 \cong \mathbb{Z}_6 \cong C_2 \times C_3$	$D_3 \cong S_3$
7	$C_7 \cong \mathbb{Z}_7$	
8	$C_8 \cong \mathbb{Z}_8, C_2 \times C_2 \times C_2, C_2 \times C_4$	D_4 (octic group), $Q_8 \cong Dic_8$
9	$C_9 \cong \mathbb{Z}_9, C_3 \times C_3$	
10	$C_{10} \cong \mathbb{Z}_{10} \cong C_2 \times C_5$	D_5
11	$C_{11} \cong \mathbb{Z}_{11}$	
12	$C_{12} \cong \mathbb{Z}_{12} \cong C_4 \times C_3, C_2 \times C_2 \times C_3$	$D_6, A_4 \cong$ tetrahedral group, Dic_{12}

Review Questions

1. Is $\langle \mathbb{Z}_5 - \{0\}, \cdot_5 \rangle$ isomorphic to C_4 or V_4?
2. Is $Q_8/\{1, -1\}$ isomorphic to C_4 or V_4?
3. In $\langle \mathbb{Z}_{30} - \{0\}, \cdot_{30} \rangle$, determine the multiplicative inverse of 7.
4. Is $\langle \mathbb{Z}_1, \cdot_1 \rangle$ a group? What is the inverse of 0?
5. List all the subgroups of $\langle \mathbb{Z}_2, +_2 \rangle \times \langle \mathbb{Z}_3, +_3 \rangle$.
6. What is the order of $\langle (a, b) \rangle$ in $C_9 \times C_{45}$, where a is a generator of C_9 and b is a generator of C_{45}?
7. Is $C_6 \times C_{15} \times C_7$ cyclic? Justify your answer.
8. List all the commutative groups of order 210.
9. What does $fr^2(r^2fr)$ equal in the tetrahedral group?
10. Show that in Dic_{4n}, $y^4 = e$.

Answers to the Review Questions

1. C_4. Only two of its elements (1 and 4) are their own inverses.
2. V_4. Each element has order 2, which is characteristic of V_4.
3. 13
4. Yes. 0 is its own inverse.
5. $\{(0, 0)\}$, $\{(0, 0), (1, 0)\}$, $\{(0, 0), (0, 1), (0, 2)\}$, $\mathbb{Z}_2 \times \mathbb{Z}_3$
6. lcm(9, 45) = 45
7. No; $|C_6|$ and $|C_{15}|$ have a common prime factor.
8. $210 = 2 \cdot 3 \cdot 5 \cdot 7$; $C_2 \times C_3 \times C_5 \times C_7 \cong C_{210}$
9. $r^2 f$
10. $y^4 = (y^2)^2 = (x^n)^2 = x^{2n} = e$

Homework Questions

1. What elements must be removed from $\mathbb{Z}_{10}$ so that the remaining elements form a group under multiplication?
2. Prove that $\langle U(n), \cdot_n \rangle$ is a group for all $n > 1$ (see frame **613**).
3. Determine all the commutative groups of order 900.
4. Does A_5 have a subgroup of order 30? Justify your answer with a simple argument.
5. Show that $1, n - 1 \in U(n)$ for all $n > 1$ (see frame **613**).
6. For $n = 2, 3, 5, 7, 11$, and 13, what groups in frame **1151** are isomorphic to $\langle \mathbb{Z}_n - \{0\} \cdot_n \rangle$?
7. Characterize the order of each element in $C_5 \times C_5 \times C_5$.
8. List all subgroups of order 4 in $C_4 \times C_4$.
9. Find an isomorphism from $\mathbb{Z}_{15}$ to $\mathbb{Z}_3 \times \mathbb{Z}_5$.
10. Using the defining relations of Dic_4, show that $y^{-1} = xy$.

Introduction to Rings

Semigroups

1152

In this chapter, we will study rings, an algebraic structure that has two binary operations. Rings are especially important because they model the algebraic structures we use every day: the integers, rationals, and reals under the operations of addition and multiplication.

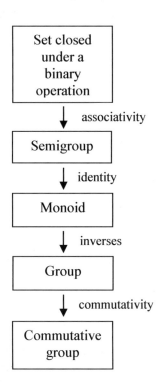

Every ring has both a semigroup and a group as component parts. For this reason, we start this chapter with a review of semigroups and groups. A *semigroup* is a set closed under an associative binary operation. For example, the set of positive integers under addition is a semigroup—it is closed under addition, and addition is associative. But the set of odd integers under addition is not a semigroup—the odd integers are not closed under addition (the sum of two odd numbers is not odd).

The defining properties of a semigroup—closure and associativity—are two of the four defining properties of a group. Thus, a semigroup in a sense is half of a group—hence the name "semigroup." Because every group has the closure and associativity properties, every group is also a semigroup. But not every semigroup is a group.

A semigroup with an identity is a *monoid*. A monoid with the inverses property (i.e., every element has an inverse) is a *group*. A group in which the operation is commutative is a *commutative* or *Abelian group*. This hierarchy of algebraic structures is illustrated by the diagram to the left.

Is $\mathbb{Z}$ under addition a group? Is it a monoid? Is it a semigroup?

1153

It is a group, a monoid, and a semigroup.

Is $\mathbb{Z}$ under multiplication a group? Is it a monoid? Is it a semigroup?

1154

It is a semigroup (it is closed under multiplication, and multiplication is associative) and a monoid (it is a semigroup with the identity 1), but it is not a group (it does not have the inverses property—for example, 2 has no multiplicative inverse in $\mathbb{Z}$).

Definition of a Ring

1155

Now that we are familiar with groups and semigroups, it is easy to define a ring: a ***ring*** is a set R with two operations. Under one operation, R is a commutative group; under the other operation, R is a semigroup. The two operations are connected by a distributive law. In particular, the semigroup operation distributes over the group operation. For example, $\mathbb{Z}$ under addition is a commutative group. Under multiplication, $\mathbb{Z}$ is a semigroup. Moreover, multiplication (the semigroup operation) distributes over addition (the group operation). That is,

$$a \cdot (b + c) = (a \cdot b) + (a \cdot c) \text{ for all } a, b, c \in \mathbb{Z}$$
$$(b + c) \cdot a = (b \cdot a) + (c \cdot a) \text{ for all } a, b, c \in \mathbb{Z}.$$

Thus, $\mathbb{Z}$ under addition and multiplication is a ring. We denote this ring with $<\mathbb{Z}, +, \cdot>$, listing the set, the group operation, and the semigroup operation, in that order.

When discussing rings in general, we will use $+$ for the group operation and 0 for the group identity. We will use $\cdot$ or juxtaposition for the semigroup operation and 1 for the semigroup identity (if it exists). We will call 0 the ***additive identity*** or simply the ***identity***, and 1 the ***multiplicative identity*** or the ***unity***. We will represent the inverse of an element a under the group operation with $-a$. Thus, $a + (-a) = (-a) + a = 0$. We will represent the inverse (if it exists) of an element a under the semigroup operation with a^{-1}. Thus, $aa^{-1} = a^{-1}a = 1$. The following table summarizes the symbols and terminology will we use for the various components of a ring:

Component	Symbol	Terminology
group operation	$+$	additive operation
group identity	0	additive identity, identity
group inverse	$-a$	additive inverse of a
semigroup operation	$\cdot$	multiplicative operation
semigroup identity	1	multiplicative identity, unity
semigroup inverse	a^{-1}	multiplicative inverse of a

Let's summarize: a ***ring*** $<R, +, \cdot>$ is an algebraic structure consisting of a set R and two operations $+$ and $\cdot$ such that

1. $<R, +>$ is a commutative group.
2. $<R, \cdot>$ is a semigroup.
3. For all $a, b, c \in R$, $a \cdot (b + c) = (a \cdot b) + (a \cdot c)$. (the ***left distributive law***)
4. For all $a, b, c \in R$, $(b + c) \cdot a = (b \cdot a) + (c \cdot a)$. (the ***right distributive law***)

A ***commutative ring*** is a ring in which the semigroup operation is commutative. By definition, the group operation in a ring is commutative. Thus, in a commutative ring, both operations—the group operation and the semigroup operation—are commutative. A ***ring with unity*** is a ring that contains a unity (i.e., a multiplicative identity). By definition, a ring contains an additive identity (because it is a group under the additive operation). Thus, in a ring with unity, there is a multiplicative identity as well as an additive identity.

Is the ring $<\mathbb{Z}, +, \cdot>$ a commutative ring? Is it a ring with unity?

1156

Because $ab = ba$ for all $a, b \in \mathbb{Z}$, and the multiplicative identity 1 is in $\mathbb{Z}$, $<\mathbb{Z}, +, \cdot>$ is a commutative ring with unity.

Let's see if $<\mathbb{Z}_6, +_6, \cdot_6>$ is a ring. Recall that $\mathbb{Z}_6 = \{0, 1, 2, 3, 4, 5\}$, $+_6$ denotes addition modulo 6, and $\cdot_6$ denotes multiplication modulo 6. $<\mathbb{Z}_6, +_6>$ is a commutative group, and $<\mathbb{Z}_6, \cdot_6>$ is a semigroup. Thus, if the distributive laws hold, $<\mathbb{Z}_6, +_6, \cdot_6>$ is a ring.

Show that $\cdot_6$ is left distributive over $+_6$. *Hint*: This follows from the distributive property of the integers under regular addition and multiplication.

1157

$a \cdot_6 (b +_6 c)$
$= a(b + c) \bmod 6$ property of modular arithmetic

= (ab + ac) mod 6   distributive property of the integers
= (a ·₆ b) +₆ (a ·₆ c)   property of modular arithmetic ∎

We can similarly show that ·₆ is right distributive over +₆. Thus, $\langle \mathbb{Z}_6, +_6, \cdot_6 \rangle$ is a ring.

Is $\langle \mathbb{Z}_6, +_6, \cdot_6 \rangle$ a commutative ring with unity?

1158

Yes. Multiplication is commutative, and the multiplicative identity 1 is in $\mathbb{Z}_6$.

Our argument that $\langle \mathbb{Z}_6, +_6, \cdot_6 \rangle$ is a commutative ring with unity easily extends to $\langle \mathbb{Z}_n, +_n, \cdot_n \rangle$ for all $n \geq 1$. Thus, for all $n \geq 1$, $\langle \mathbb{Z}_n, +_n, \cdot_n \rangle$ is a commutative ring with unity.

One and only one ring in the $\mathbb{Z}_n$ family of rings has an additive identity equal to its unity. Which ring is it?

1159

$\langle \mathbb{Z}_1, +_1, \cdot_1 \rangle$. Its only element, 0, is both the additive identity ($0 +_1 0 = 0$) and the unity ($0 \cdot_1 0 = 0$).

Basic Properties of a Ring

1160

In addition to the defining properties of a ring given in frame **1155**, rings have other properties in common. Let R be a ring with $a, b, c \in R$. Then the following are true:

- **R1:** There is one and only one additive identity in R.
- **R2:** There is at most one unity in R.
- **R3:** $a0 = 0a = 0$, where 0 is the additive identity.
- **R4:** $(-a)b = -(ab)$.
- **R5:** $(-1)a = a(-1) = -a$.
- **R6:** $a(-b) = -(ab)$.
- **R7:** $(-1)(-1) = 1$.
- **R8:** $(-a)(-b) = ab$.
- **R9:** $-(a + b) = -a - b$.
- **R10:** $-(a - b) = b - a$.
- **R11:** $a(b - c) = ab - ac$.

Most of these properties should be familiar to you either from grade school or from our study of binary operations and groups. For example, in grade school, you surely learned the property that zero times anything is zero (R3) or that the product of two negatives is a positive (R8).

In chapter 7, we learned that in a set with a binary operation and an identity, the identity is unique (see frame **377**). Thus, in a ring, the additive identity is unique, and if the multiplicative identity exists, then it too is unique. In other words, in a ring, there is one and only one additive identity (R1) and at most one multiplicative identity (R2).

Show R3: $a0 = 0a = 0$. *Hint*: Starting with $a0$, use $0 = 0 + 0$, and then the distributive law to show that $a0$ is idempotent and, therefore, the additive identity (see frame **454**). Repeat for $0a$.

1161

$a0$
$= a(0 + 0)$ 0 is the additive identity
$= a0 + a0$ left distributive law

Because $a0$ is idempotent, by frame **454**, $a0 = 0$. Similarly, $0a = (0 + 0)a = 0a + 0a$. Because $0a$ is idempotent, $0a = 0$. ∎

This theorem tells us that, in a ring,

(any element) · (additive identity) = (additive identity) · (any element) = additive identity.

It applies regardless of the symbol we use to represent the additive identity. For example, suppose i is the additive identity in some ring. What does ia equal, where a is some element in the ring? What does $i + a$ equal?

1162

$ia = i$, $i + a = a$

Show R4: $(-a)b = -(ab)$. *Hint*: If $(-a)b + ab = 0$, then $(-a)b$ is the additive inverse of ab. That is, $(-a)b = -(ab)$. Thus, to show R4, simply show that $(-a)b + ab = 0$. We do not also have to show that $ab + (-a)b = 0$ because + is commutative in a ring. *Hint*: Use the right distributive law.

1163

$(-a)b + ab$	
$= (-a + a)b$	right distributive law
$= 0b$	$-a$ is the additive inverse of a
$= 0$	R3 ∎

Another way to show this is with a string of equalities that starts with $(-a)b$ and ends with $-(ab)$:

$(-a)b$	
$= (-a)b + 0$	0 is the additive identity
$= (-a)b + ab + [-(ab)]$	$ab + [-(ab)] = 0$
$= (-a + a)b + [-(ab)]$	right distributive law
$= 0b + [-(ab)]$	$-a + a = 0$
$= 0 + [-(ab)]$	R3
$= -(ab)$	0 is the additive identity ∎

Show R5: $(-1)a = a(-1) = -a$. *Hint*: Show that $(-1)a + a = 0$. Note that $(-1)a$ and $a(-1)$ are *fundamentally different* from $-a$: $(-1)a$ and $a(-1)$ are *products* under the semigroup operation, whereas $-a$ is the additive inverse of a. Thus, R5 is an assertion that needs to be proven.

1164

$(-1)a + a$	
$= (-1)a + 1a$	1 is the multiplicative identity
$= (-1 + 1)a$	right distributive law
$= 0a$	$-1 + 1 = 0$
$= 0$	R3

Thus, $(-1)a$ is the additive inverse of a. That is, $(-1)a = -a$. We can similarly show that $a(-1) = -a$. ∎

Show R6: $a(-b) = -(ab)$. *Hint*: Use the same technique we used to prove R5.

1165

$a(-b) + ab = a(-b + b) = a0 = 0$. Thus, $a(-b)$ is the additive inverse of ab. That is, $a(-b) = -(ab)$. ∎

Show R7: $(-1)(-1) = 1$. *Hint*: The proof starts with $(-1)(-1) = (-1)(-1) + 0$. Next, substitute $-1 + 1$ for 0, use the left distributive law and R3, and ultimately derive 1.

1166

$(-1)(-1)$	
$= (-1)(-1) + 0$	0 is the additive identity
$= (-1)(-1) + (-1 + 1)$	-1 is the additive inverse of 1
$= (-1)(-1) + [(-1)(1) + 1]$	$-1 = (-1)1$
$= [(-1)(-1) + (-1)(1)] + 1$	+ is associative
$= -1(-1 + 1) + 1$	left distributive law
$= -1(0) + 1$	-1 is the inverse of 1
$= 0 + 1$	R3

Chapter 21: Introduction to Rings

= 1   0 is the additive identity ∎

Now show the general case (R8): $(-a)(-b) = ab$. *Hint*: The proof is similar the proof that $(-1)(-1) = 1$.

1167

$(-a)(-b)$	
$= (-a)(-b) + 0$	0 is the additive identity
$= (-a)(-b) + -(ab) + ab$	$-(ab)$ is the inverse of ab
$= (-a)(-b) + -a(b) + ab$	R4
$= (-a)(-b + b) + ab$	left distributive law
$= (-a)0 + ab$	$-b + b = 0$
$= 0 + ab$	R3
$= ab$	0 is the additive identity ∎

Subtraction in a ring is defined as follows: $a - b = a + (-b)$. That is, subtraction applied to a and b is defined as the addition of a and the additive inverse of b.

Using this definition, show that the minus sign distributes over addition (R9). That is, $-(a + b) = -a - b$. *Hint*: Show that $(a + b) + (-a - b) = 0$.

1168

$(a + b) + (-a - b)$	
$= (a + b) + (-a + [-b])$	definition of subtraction
$= (a + -a) + (b + [-b])$	+ is associative and commutative.
$= 0 + 0$	inverse property
$= 0$	identity property

Thus, the inverse of $a + b$ is $-a - b$. That is, $-(a + b) = -a - b$. ∎

Show R10: $-(a - b) = b - a$.

1169

$(a - b) + (b - a)$	
$= (a + [-b]) + (b + [-a])$	definition of subtraction
$= (a + [-a]) + (b + [-b])$	+ is associative and commutative.
$= 0 + 0$	inverse property
$= 0$	identity property

Thus, the inverse of $a - b$ is $b - a$. That is, $-(a - b) = b - a$. ∎

Show R11: $a(b - c) = ab - ac$. *Hint*: Use the distributive law and R6.

1170

$a(b - c)$	
$= a(b + [-c])$	definition of subtraction
$= ab + a(-c)$	left distributive law
$= ab + [-(ac)]$	R6
$= ab - ac$	definition of subtraction ∎

Generalized Distributive Laws

1171

The left distributive law of rings says that multiplication left distributes over the sum of any *pair* of elements. That is, in a ring R, $a(b + c) = ab + ac$ for all $a, b, c \in R$. Using this law, we can show that multiplication left distributes over the sum of any *triple* of elements:

$a(b + c + d)$
$= a([b + c] + d)$ + is associative
$= a(b + c) + ad$ a left distributes over the pair $[b + c]$ and d
$= (ab + ac) + ad$ a left distributes over the pair b and c
$= ab + ac + ad$ + is associative ∎

Using this result, show that multiplication left distributes over the sum of any *four* elements.

1172

$a(b + c + d + e)$
$= a([b + c + d] + e)$ + is associative
$= a(b + c + d) + ae$ a left distributes over the pair $[b + c + d]$ and e
$= (ab + ac + ad) + ae$ distributivity result from preceding frame
$= ab + ac + ad + ae$ + is associative ∎

We can continue this process indefinitely. When we establish that multiplication left distributes over the sum of any n elements, we can then use that result to show that multiplication left distributes over any $n + 1$ elements. Thus, multiplication left distributes over the sum of any number of elements. The right distributive law similarly generalizes.

The preceding analysis is essentially an informal proof by induction. Let S_n represent the statement "multiplication left distributes over the sum of any n elements." We are given that S_2 is true (S_2 is the left distributive property of a ring). By proving S_3 and then S_4, we see a pattern in these two proofs that can be repeated indefinitely. In particular, we observe that we prove S_3 using S_2, then prove S_4 using S_3, and so on. That is, we observe that for any $n \geq 2$, if S_n is true, then S_{n+1} is true. Then, by induction, S_n is true for all $n \geq 2$.

Characteristic of a Ring

1173

Recall from frame **917** that if $<G, *>$ is a group with $g \in G$, then the **order of g**, denoted by $|g|$, is the smallest positive n such that the product of n occurrences of g is the identity. If no such n exists, then $|g|$ is infinite.

In multiplicative notation, we represent the product of n occurrences of g with g^n. Thus, the order of g is the smallest positive n for which g^n equals the identity. For example, in D_3, the order of r is 3 because $r^1 = r \neq e$, $r^2 \neq e$, but $r^3 = e$ (r is a 120° clockwise rotation of an equilateral triangle). The preceding definition of the order of g uses multiplicative notation. Alternatively, we can define the order of g using additive notation. In addition notation, the order of g is defined as the smallest positive n such that the sum of n occurrences of g is the identity. In additive notation, we represent the sum of n occurrences of g with ng. Thus, in additive notation, the order of g is defined as the smallest positive n such that ng equals the identity. This is not a different definition of the order of an element—it is the same definition as the one that uses multiplicative notation, except that it is expressed using a different notation.

What is $|f|$ in D_3?

1174

$f^1 = f \neq e, f^2 = e \Rightarrow |f| = 2$ (using multiplicative notation here)

What is $|3|$ in $<\mathbb{Z}_6, +_6>$?

1175

$1(3) = 3 \neq 0, 2(3) = 3 +_6 3 = 0 \Rightarrow |3| = 2$ (using additive notation here)

When we speak of the order of an element in a ring, unless otherwise indicated, we mean the order of that element under its additive operation. For example, the order of 3 in $<\mathbb{Z}_6, +_6, \cdot_6>$ is 2.

In the standard notation for rings, we use juxtaposition in two ways:

- It is used to indicate a product of two elements. For example, we write the product of a and b as ab.
- It is used to indicate the replication of a term in a sum. For example, we write $3a$ as shorthand for $a + a + a$. Here $3a$ is not the product of 3 and a under the multiplicative operation of the ring. The 3 simply indicates that we want the sum of three a's.

Chapter 21: Introduction to Rings

Consider the ring defined by the following two Cayley tables—one for + (the addition operation), and one for ·, the multiplication operation:

+	a	b	c
a	a	b	c
b	b	c	a
c	c	a	b

·	a	b	c
a	a	a	a
b	a	b	c
c	a	c	b

The expression ab represents the product of a and b. Thus, to determine its value, we use the multiplication table. On the other hand, $2b$ represents the sum of two b's. Thus, to determine the value of $2b$, we use the addition table.

What does ab equal? What does $2b$ equal?

1176

$ab = a$, $2b = b + b = c$

Which element is the additive identity in the ring in the preceding frame? Which element is the unity?

1177

a is the additive identity, and b is the multiplicative identity.

In the ring in frame **1175**, $1b = b \neq a$ and $2b = b + b \neq a$, but $3b = b + b + b = a$. Thus, the order of b under the additive operation is 3.

What is the order of c under the additive operation?

1178

3

Why do the first row and the first column on the multiplication table in frame **1175** consist of all a's?

1179

a is the additive identity. Then, by R3 in frame **1160**, $ax = xa = a$ for all elements x in the ring.

Let R be a ring with additive identity 0. The **characteristic of R** is the smallest positive integer n such that $nx = 0$ for all $x \in R$, if such an n exists (nx here is the sum of n x's). If no such n exists, then the characteristic of R is 0.

For example, consider the ring R in frame **1175**. Its identity is a. Its characteristic is the smallest positive n such that $nx = a$ for all $x \in R$. What is its characteristic?

1180

$1a = a$, $1b = b$, $1c = c$	Only $1a$ equals the identity a.
$2a = a$, $2b = c$, $2c = b$	Only $2a$ equals the identity a.
$3a = a$, $3b = a$, $3c = a$	All equal the identity a, so 3 is the characteristic of the ring.

What is the characteristic of $\langle \mathbb{Z}, +, \cdot \rangle$? *Hint*: What is the additive order of 1? That is, what is the smallest positive n for which $n1$ (i.e., the sum of n 1's) is equal to 0?

1181

There is no such n. Because there is no positive integer n such that $n1 = 0$, there certainly is no positive integer n such that $nx = 0$ for *every* $x \in \mathbb{Z}$. Thus, the characteristic of the ring is 0.

In the ring $\langle \mathbb{Z}_6, +_6, \cdot_6 \rangle$, what is the order of 2?

1182

$2 \neq 0$ and $2 +_6 2 \neq 0$ but $2 +_6 2 +_6 2 = 0$. Thus, $|2| = 3$.

What is the order of 1?

1183

$1 \neq 0$, $1 +_6 1 \neq 0$, $1 +_6 1 +_6 1 \neq 0$, $1 +_6 1 +_6 1 +_6 1 \neq 0$, $1 +_6 1 +_6 1 +_6 1 +_6 1 \neq 0$, $1 +_6 1 +_6 1 +_6 1 +_6 1 +_6 1 = 0$. Because 6 is the smallest number of 1's that when summed equal 0, $|1| = 6$.

Could the characteristic of this ring be less than the order of 1? For example, could the characteristic be 3?

1184

No. If the characteristic were 3, then by the definition of a characteristic, the sum of three 1's would equal 0, in which case the order of 1 would be at most 3. But the order of 1 is 6. Clearly, in any ring with unity 1 and a positive characteristic, the characteristic must be greater than or equal to the order of *every* element. So it certainly must be greater than or equal to $|1|$. In other words, if the characteristic is n, then

$$n \geq |1|.$$

What does $\underbrace{1 + 1 + \cdots + 1}_{\text{the sum of } |1| \text{ 1's}}$ equal?

1185

0 (by definition, $|1|$ is the smallest positive integer k such that the sum of k 1's is 0)

Let x be an arbitrary element of a ring R with unity 1. Then

$$\underbrace{x + x + \cdots + x}_{|1| \ x\text{'s}} = \underbrace{x \cdot 1 + x \cdot 1 + \cdots + x \cdot 1}_{|1| \text{ terms}} = x \cdot \underbrace{(1 + 1 + \cdots + 1)}_{|1| \ 1\text{'s (which is equal to 0)}} = x \cdot 0 = 0$$

Thus, $|1|x$ (i.e., the sum of $|1|$ x's) $= 0$ for all $x \in R$. The characteristic of the ring is the *smallest* positive integer n such that $nx = 0$ for all $x \in R$. What can we conclude about the relative values of n and $|1|$?

1186

Both nx and $|1|x$ equal 0 for all $x \in R$. The characteristic n of R is the smallest positive integer with this property. Thus,

$$|n| \leq |1|.$$

By frame **1184**, we also know that $n \geq |1|$. Thus, $n = |1|$. That is, if the order of 1 is finite, the characteristic of a ring is equal to the order of 1.

Key idea: If the order of 1 in a ring is finite, the order of 1 is the characteristic of the ring.

If the order of 1 is infinite, then $n1 \neq 0$ for any positive n. Then, clearly, there is no positive n such that $nx = 0$ for all x in the ring. Thus, for this case, the characteristic of the ring is 0.

Let's summarize what we have learned about the characteristic of a ring with unity: To determine the characteristic of a ring with unity, we need to look at only the unity element. If its order is the integer k, then the ring has the characteristic k. If its order is infinite, then the ring has the characteristic 0. For example, we know that the characteristic of $\langle \mathbb{Z}_6, +_6, \cdot_6 \rangle$ is 6 because the additive order of 1 in this ring is 6. We know that the characteristic of $\langle \mathbb{Z}, +, \cdot \rangle$ is 0 because the additive order of 1 in $\mathbb{Z}$ is infinite.

Multiplying a Sum of 1's by a Sum of 1's

1187

Suppose 1 is the unity in a ring. What does $(1 + 1 + 1)(1 + 1)$ equal? Give your answer as a sum of 1's. *Hint*: Use the left distributive law then the right distributive law.

1188

$(1 + 1 + 1)(1 + 1) = (1 + 1 + 1)1 + (1 + 1 + 1)1 = (1 + 1 + 1) + (1 + 1 + 1)$

$(1 + 1 + 1)$ distributes over $(1 + 1)$ so, for each 1 in $(1 + 1)$, we get $(1 +1+ 1)$. There are two 1's in $(1 + 1)$. Thus, we get the sum of two occurrences of $(1 + 1 + 1)$.

Formulate a general rule to determine the result when the sum of m 1's is multiplied by the sum of n 1's.

1189

For each of the n 1's, we get m 1's. Thus, we get a sum consisting of mn 1's.

$m1$ is shorthand for the sum of m 1's. Similarly, $n1$ is shorthand for the sum of n 1's. Restate the preceding rule using $m1$ and $n1$.

1190

$(m1)(n1) = (mn)1$. That is, the sum of m 1's times the sum of n 1's is equal to the sum of mn 1's.

We use this rule in chapter 24 where we investigate the characteristic of an integral domain (a special type of ring).

Caveat: Be sure to understand the multiple use of juxtaposition in this rule. The juxtaposition of $m1$ and $n1$ on the left side represents the product of the sum of m 1's and the sum n 1's under the ring's multiplicative operation. The juxtaposition of m and n on the right side represents the product of the integers m and n under regular multiplication. Thus, the equality $(m1)(n1) = (mn)1$ should be interpreted as

$$(\text{sum of } m \text{ 1's}) \cdot (\text{sum of } n \text{ 1's}) = \text{sum of } mn \text{ 1's},$$

where $\cdot$ represents the multiplicative operation of the ring, and 1 represents the unity. Do not assume 1 here has all the properties of 1 in $<\mathbb{Z}, +, \cdot>$. For example, on the integers, $m1$ (i.e., the sum of m 1's) equals the integer m. But in a ring whose elements are not integers, the integer m is not an element, in which case $m1$ obviously cannot equal m. For example, in the ring in frame **1175**, b is the unity; $2b = b + b = c$. The element $2b$ does not equal 2.

Trivial Ring

1191

By its name, you would expect the trivial ring not to be very interesting. However, it is interesting in one respect: it is the only ring in which the additive identity is also the unity. The ***trivial ring*** is the ring that contains only one element. This element must be the additive identity because a ring is a group under its additive operation and therefore must contain an identity.

Construct the Cayley tables for the additive operation and the multiplicative operation in a trivial ring.

1192

Is it true that $0 + x = x + 0 = x$ for all elements x in a trivial ring?

1193

Yes, which means that 0 is the additive _____.

1194

identity

Is it true that $0x = x0 = x$ for all elements x in a trivial ring?

1195

Yes, which means that 0 is not only the additive identity but also the _____.

1196

unity

Caveat: In this book, we define the unity in a ring simply as the multiplicative identity. Thus, the trivial ring is a ring with unity because it has the multiplicative identity 0 (which is also the additive identity). Some textbooks, however, define the unity as a *nonzero* element that is the multiplicative identity. With this definition, the trivial ring has no unity because its multiplicative identity is 0. This variation in the definition of a unity affects the way some theorems are stated. For example, where we start a theorem with "Let R be a nontrivial ring with unity," books that use the other definition of unity start the same theorem with "Let R be a ring with unity," omitting the qualifier "nontrivial." They can omit "nontrivial" because their definition of unity requires the unity to be nonzero. Thus, any ring with unity necessarily has at least two elements, 0 (because every ring has an additive identity) and the unity, and is therefore a nontrivial ring.

Show that in a ring with unity with more than one element, 0 cannot be the unity. *Hint*: Such a ring must have an element $a \neq 0$. Consider the value of $0a$. Use a proof by contradiction.

1197

Let a be a nonzero element in the ring. Assume 0 is the unity. Then $0a = a \neq 0$. But by R3 in frame **1160**, $0a = 0$. Thus, our assumption that 0 is the unity must be incorrect. ■

Here is another interesting fact about the trivial ring. It is a commutative group under *both* its additive and multiplicative operations (under either operation, the group is the cyclic group of order 1). The trivial ring is the only ring for which this is true.

Show that for all nontrivial rings, their elements do not form a group under their multiplicative operations. *Hint*: Does 0 have a multiplicative inverse?

1198

In any ring with unity with more than one element, by the preceding frame, the additive identity 0 and the unity 1 are distinct elements. In all such rings, 0 does not have a multiplicative inverse. If, to the contrary, 0 has a multiplicative inverse x, then $0x = 1$. But $0x = 0 \neq 1$. Thus, under its multiplicative operation, a nontrivial ring cannot be a group. ■

Ring Isomorphisms

1199

Suppose G_1 and G_2 are groups. We learned in chapter 15 that G_1 is isomorphic to G_2 (denoted by $G_1 \cong G_2$) if and only if there exists a bijection φ from G_1 to G_2 that is operation preserving. φ is operation preserving if it has the following property:

$$\varphi(ab) = \varphi(a)\varphi(b), \text{ for all } a, b \in G_1.$$

Equivalently, $G_1 \cong G_2$ if and only if the Cayley table for G_1 can be converted to the Cayley table for G_2 by renaming the elements of the G_1 table according to some bijection from G_1 to G_2 (see frame **869**), in which case we say the tables are essentially the same.

We can extend the idea of isomorphism to rings. Suppose R_1 and R_2 are rings. **R_1 is isomorphic to R_2** (denoted by $R_1 \cong R_2$) if and only if there exists a bijection φ from R_1 to R_2 that preserves both the additive and multiplicative operations. That is,

$$\varphi(a + b) = \varphi(a) + \varphi(b), \text{ for all } a, b \in R_1$$

and

$$\varphi(ab) = \varphi(a)\varphi(b), \text{ for all } a, b \in R_1.$$

Equivalently, $R_1 \cong R_2$ if and only if the Cayley tables for R_1 are essentially the same as the corresponding tables for R_2. That is, the tables for R_1 can be converted to the tables for R_2 by renaming the elements according to some bijection.

Let's see if $<\mathbb{Z}, +, \cdot> \cong <2\mathbb{Z}, +, \cdot>$. $2\mathbb{Z}$ is the set of even integers. First show that $\varphi(x) = 2x$ is a bijection from $\mathbb{Z}$ to $2\mathbb{Z}$.

246 Chapter 21: Introduction to Rings

1200

Onto: Let $y \in 2\mathbb{Z} \Rightarrow y = 2x$ for some $x \in \mathbb{Z} \Rightarrow \varphi(x) = 2x = y$.
One-to-one: Let $\varphi(x) = \varphi(y) \Rightarrow 2x = 2y \Rightarrow x = y$.

Show that φ preserves addition.

1201

$\varphi(a + b) = 2(a + b) = 2a + 2b = \varphi(a) + \varphi(b)$. ∎

Show that φ preserves multiplication.

1202

$\varphi(ab) = 2(ab)$. But $\varphi(a)\varphi(b) = 2a2b = 4ab$. φ does *not* preserve multiplication. Thus, this φ is not the bijection we need to establish the isomorphism of these two rings. Perhaps these rings are *not* isomorphic.

Does $<\mathbb{Z}, +, \cdot>$ have a unity?

1203

Yes. It is 1.

Does $<2\mathbb{Z}, +, \cdot>$ have a unity?

1204

No; 1 is not in $2\mathbb{Z}$.

$<\mathbb{Z}, +, \cdot>$ and $<2\mathbb{Z}, +, \cdot>$ differ in an essential way: one has a unity, the other does not. Thus, these two rings cannot be isomorphic.

Divisors of Zero

1205

Let R be a ring with additive identity 0. A nonzero element $a \in R$ is a **divisor of zero** if and only if there is a nonzero element $b \in R$ such that $ab = 0$ or $ba = 0$.

What are the divisors of zero in $<\mathbb{Z}_6, +_6, \cdot_6>$? *Hint*: There are three.

1206

$2 \cdot_6 3 = 0$ and $3 \cdot_6 4 = 0 \Rightarrow 2, 3,$ and 4 are divisors of zero.

Show that there are no divisors of zero in $<\mathbb{Z}_p, +_p, \cdot_p>$, where p is a prime. *Hint*: Use a proof by contradiction.

1207

Assume $a \in \mathbb{Z}_p$ is a divisor of zero $\Rightarrow a$ is nonzero, and there is a nonzero $b \in \mathbb{Z}_p$ such that $a \cdot_p b = 0 \Rightarrow p \mid ab$. Thus, by Euclid's lemma, $p \mid a$ or $p \mid b$, which, by D6 in frame **547**, implies that $p \leq a$ or $p \leq b$. But this is impossible, because $a, b \in \mathbb{Z}_p \Rightarrow 0 < a, b < p$. Thus, our assumption that a is a divisor of zero is incorrect. ∎

Suppose a and b are elements in a ring $<R, +, \cdot>$ that does not have any divisors of zero. If $a \cdot b = 0$, what can you conclude about a and b?

1208

$a = 0$ or $b = 0$. Otherwise, a and b would be divisors of zero.

Do the integers have any divisors of zero. That is, are there nonzero integers a and b such that $ab = 0$?

1209

No.

We use the fact that the ring of integers has no divisors of zero when we determine the integer roots of a factorable quadratic equation. For example, to determine the roots of $x^2 + x - 2 = 0$, we first factor the equation to get

$$(x + 2)(x - 1) = 0.$$

Because there are no divisors of zero, we know that

$$x + 2 = 0 \text{ or } x - 1 = 0,$$

from which we get $x = -2$ or $x = 1$.

Suppose $<R, +, \cdot>$ is a ring in which a is not a divisor of zero and $a \neq 0$. If $ab = 0$, what can we conclude about b?

1210

$b = 0$, otherwise a would be a divisor of zero.

Units Cannot Be Divisors of Zero

1211

In a ring with unity, a ***unit*** is a nonzero element that has a multiplicative inverse. That is, if u is a unit, then there exists an element u^{-1} in the ring such that $u^{-1}u = uu^{-1} = 1$. Can a unit be a divisor of zero? Let's investigate. Let u be a unit in some ring. Assume that u is also a divisor of zero. Then there is a nonzero element v in the ring such that $uv = 0$.

Multiply on the left both sides of $uv = 0$ by u^{-1}, and simplify.

1212

$u^{-1}uv = u^{-1}0$. Simplifying, we get $v = 0$. But $v \neq 0$. Thus, our assumption that u is a divisor of zero is incorrect. A unit cannot also be a divisor of zero.

Can a divisor of zero be a unit?

1213

No. We just showed that a unit cannot be a divisor of zero. Thus, the set of units and the set of divisors of zero must be disjoint.

Cancellation Law for Multiplication in a Ring

1214

In a ring, every element has an additive inverse. If an element appears as a term in a sum on both sides of an equality, we can add its inverse to both sides to cancel that element. Thus, the cancellation law for addition holds in a ring. For example, if $a + b = c + a$, we can commute the c and a on the right side to get $a + b = a + c$ (remember that a ring under its additive operation is a commutative group). We can then add $-a$ on the left to both sides to cancel the a's. We get $b = c$.

A cancellation law also holds for multiplication in a ring. However, the cancellation law for multiplication is restricted to nonzero elements that are not divisors of zero. Here is the ***left cancellation law*** for multiplication:

Suppose $a, b, c \in <R, +, \cdot>$, where a is nonzero and not a divisor of zero. If $ab = ac$, then $b = c$.

Prove this cancellation law. *Hint*: Move ac to the left side and factor out a. Then use R6 in frame **1160** and the distributive law.

1215

248   Chapter 21: Introduction to Rings

$ab = ac$
$\Rightarrow ab + -(ac) = 0$ add inverse of ac to both sides
$\Rightarrow ab + a(-c) = 0$ R6 in frame **1160**
$\Rightarrow a(b + [-c]) = 0$ distributive law
$\Rightarrow b + (-c) = 0$ $a \neq 0$ and frame **1210**
$\Rightarrow b = c$ ∎   add c to both sides

Similarly prove the ***right cancellation law***: if $ba = ca$, then $b = c$, where a, b, and c are as in the preceding frame.

1216

$ba = ca \Rightarrow ba + [-(ca)] = 0 \Rightarrow ba + (-c)a = 0 \Rightarrow (b + [-c])a = 0 \Rightarrow b + [-c] = 0 \Rightarrow b = c$ ∎

Does $ab = ca$ imply that $b = c$ if a is nonzero and not a divisor of zero?

1217

No. Neither the left nor the right cancellation law applies. However, if the ring is commutative, we can commute ca to get $ab = ac$. We can then use the left cancellation law to get $b = c$.

In the ring $\langle \mathbb{Z}_6, +_6, \cdot_6 \rangle$, for what values of a does the left cancellation law hold in the equation $a \cdot_6 b = a \cdot_6 c$?

1218

1 and 5. These are the only nonzero elements of $\mathbb{Z}_6$ that are not divisors of zero. 2, 3, and 4 are divisors of zero.

The left cancellation does not hold for zero or for divisors of zero. Thus, it does not hold for 0, 2, 3, or 4 in $\langle \mathbb{Z}_6, +_6, \cdot_6 \rangle$. Give an example that demonstrates that the left cancellation law for multiplication does *not* hold for 2 in $\langle \mathbb{Z}_6, +_6, \cdot_6 \rangle$.

1219

$2 \cdot_6 1 = 2 \cdot_6 4$. If we cancel 2 on each side, we get $1 = 4$. But $1 \neq 4$.

Give an example that demonstrates that the left cancellation law for multiplication does *not* hold for 3 in $\langle \mathbb{Z}_6, +_6, \cdot_6 \rangle$.

1220

$3 \cdot_6 2 = 3 \cdot_6 4$. If we cancel the 3 on each side, we get $2 = 4$. But $2 \neq 4$.

Give an example that demonstrates that the right cancellation law for multiplication does *not* hold for 4 in $\langle \mathbb{Z}_6, +_6, \cdot_6 \rangle$.

1221

$1 \cdot_6 4 = 4 \cdot_6 4$. If we cancel 4 on each side, we get $1 = 4$. But $1 \neq 4$.

Give an example that demonstrates that the right cancellation law for multiplication does not hold for 0 in $\langle \mathbb{Z}_6, +_6, \cdot_6 \rangle$.

1222

$1 \cdot_6 0 = 2 \cdot_6 0$. If we cancel 0 on each side, we get $1 = 2$. But $1 \neq 2$.

Show that the cancellation laws for multiplication do not hold for *any* element a in a ring that is a divisor of zero. *Hint*: If a is a divisor of zero, then $ac = 0$ for some nonzero c. Is there an equation involving ac in which the cancellation of a shows that $c = 0$?

1223

If a is a divisor of zero, then $c \neq 0$ exists such that $ac = 0$ or $ca = 0$. If $ac = 0$, then $ac = a0$ (because $a0 = 0$ by R3 in frame **1060**). If $ca = 0$, then $ca = 0a$. In either case, if we cancel the a's, we get $c = 0$. But $c \neq 0$. ∎

In frame **434**, we observed that in the ring of real numbers, $\langle \mathbb{R}, +, \cdot \rangle$, we can cancel an element in a product as long as it is nonzero. We do not require that the element not be a divisor of zero. Why can we omit this requirement?

1224

There are no divisors of zero in $\langle \mathbb{R}, +, \cdot \rangle$. Thus, this requirement need not be stated.-

Direct Sum of Rings

1225

Suppose R_1 and R_2 are rings. Then the ***direct sum of R_1 and R_2*** (denoted by $R_1 \oplus R_2$) is the set $\{(a, b) : a \in R_1, b \in R_2\}$ together with the operations of addition and multiplication defined as follows:

Addition: $(a_1, b_1) + (a_2, b_2) = (a_1 + a_2, b_1 + b_2)$ where $+$ in the first coordinate is addition on R_1 and $+$ in the second coordinate is addition on R_2.

Multiplication: $(a_1, b_1) \cdot (a_2, b_2) = (a_1 a_2, b_1 b_2)$ where multiplication in the first coordinate is multiplication on R_1 and multiplication in the second coordinate is multiplication on R_2.

For example, the elements of $\mathbb{Z}_2 \oplus \mathbb{Z}_3$ are (0, 0), (0, 1), (0, 2), (1, 0), (1, 1), and (1, 2).

In $\mathbb{Z}_2 \oplus \mathbb{Z}_3$, what does (0, 2) + (1, 1) equal?

1226

$(0 +_2 1, 2 +_3 1) = (1, 0)$

In $\mathbb{Z}_2 \oplus \mathbb{Z}_3$, what does $(0, 1) \cdot (2, 2)$ equal?

1227

$(0 \cdot_2 2, 1 \cdot_3 2) = (0, 2)$

A direct sum of rings is itself a ring (you prove this in homework problem 5).

Construct the Cayley tables for the ring $\mathbb{Z}_2 \oplus \mathbb{Z}_2$.

1228

+	(0, 0)	(0, 1)	(1, 0)	(1, 1)
(0, 0)	(0, 0)	(0, 1)	(1, 0)	(1, 1)
(0, 1)	(0, 1)	(0, 0)	(1, 1)	(1, 0)
(1, 0)	(1, 0)	(1, 1)	(0, 0)	(0, 1)
(1, 1)	(1, 1)	(1, 0)	(0, 1)	(0, 0)

·	(0, 0)	(0, 1)	(1, 0)	(1, 1)
(0, 0)	(0, 0)	(0, 0)	(0, 0)	(0, 0)
(0, 1)	(0, 0)	(0, 1)	(0, 0)	(0, 1)
(1, 0)	(0, 0)	(0, 0)	(1, 0)	(1, 0)
(1, 1)	(0, 0)	(0, 1)	(1, 0)	(1, 1)

What is the additive identity in $\mathbb{Z}_2 \oplus \mathbb{Z}_2$?

1229

(0, 0)

The Cayley table for addition on $\mathbb{Z}_2 \oplus \mathbb{Z}_2$ corresponds to which group of order 4? C_4 or V_4?

1230

V_4 is the only group of order 4 whose diagonal elements are all the identity. The Cayley table for addition on $\mathbb{Z}_2 \oplus \mathbb{Z}_2$ has this property. Thus, it corresponds to V_4.

Direct sums of rings can be formed from two or more rings. For example, the direct sum formed from the rings $\mathbb{Z}_2$, $\mathbb{Z}_3$, and $\mathbb{Z}_4$ is $\mathbb{Z}_2 \oplus \mathbb{Z}_3 \oplus \mathbb{Z}_4$. It consists of all triples (a, b, c) where $a \in \mathbb{Z}_2$, $b \in \mathbb{Z}_3$, and $c \in \mathbb{Z}_4$.

What is the order of $\mathbb{Z}_2 \oplus \mathbb{Z}_3 \oplus \mathbb{Z}_4$?

1231

The orders of $\mathbb{Z}_2$, $\mathbb{Z}_3$, and $\mathbb{Z}_4$ are 2, 3, and 4, respectively. Thus, the order of the direct sum is $2 \cdot 3 \cdot 4 = 24$.

In $\mathbb{Z}_2 \oplus \mathbb{Z}_3 \oplus \mathbb{Z}_4$, what does $(1, 2, 3) + (1, 2, 3)$ equal?

1232

$(1 +_2 1, 2 +_3 2, 3 +_4 3) = (0, 1, 2)$

In $\mathbb{Z}_2 \oplus \mathbb{Z}_3 \oplus \mathbb{Z}_4$, what does $(1, 2, 3) \cdot (1, 2, 3)$ equal?

1233

$(1 \cdot_2 1, 2 \cdot_3 2, 3 \cdot_4 3) = (1, 1, 1)$

What are the orders of the rings $\mathbb{Z}_2$, $\mathbb{Z}_2 \oplus \mathbb{Z}_2$, $\mathbb{Z}_2 \oplus \mathbb{Z}_2 \oplus \mathbb{Z}_2$, and $\mathbb{Z}_2 \oplus \mathbb{Z}_2 \oplus \mathbb{Z}_2 \oplus \mathbb{Z}_2$?

1234

The orders are the successive powers of 2: $2^1 = 2$, $2^2 = 4$, $2^3 = 8$, and $2^4 = 16$.

Let R be the ring that is the direct sum of n occurrences of $\mathbb{Z}_2$. What is the order of R? What is the additive identity in R? What is the unity in R?

1235

$|R| = 2^n$. The identity is an n-tuple of 0's. The unity is an n-tuple of 1's.

Let R be the ring that is the direct sum of n occurrences of $\mathbb{Z}_2$, where $n \geq 2$. If $x \in R$, what does $x + x$ equal? Characterize the main diagonal of the Cayley table for the addition operation on R.

1236

If a component of x is 0, then the corresponding component in $x + x = 0 +_2 0 = 0$. If a component of x is 1, then the corresponding component in $x + x = 1 +_2 1 = 0$. Thus, $x + x$ is the identity (an n-tuple of 0's). The main diagonal of the Cayley table for addition contains exclusively the identity element (an n-tuple of 0's).

An Interesting Family of Rings

1237

Suppose A and B are sets. $A \triangle B$ is defined as $(A \cup B) - (A \cap B)$. In the following Venn diagram, indicate the region(s) that correspond to $A \triangle B$.

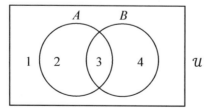

1238

Regions 2 and 4. $A \triangle B$ contains all elements that are in A or B, but not both. Think of it as the exclusive "or" of the two sets. We call $A \triangle B$ the **symmetric difference** of A and B.

Show that $\triangle$ is commutative.

1239

$\cup$ and $\cap$ are commutative. Thus, $A \triangle B = (A \cup B) - (A \cap B) = (B \cup A) - (B \cap A) = B \triangle A$.

Show the Venn diagram for $(A \triangle B) \triangle C$.

1240

$(A \triangle B) \triangle C$ corresponds to regions 2, 4, 6, and 8 in the following Venn diagram:

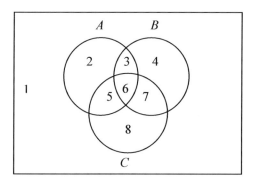

$(A \triangle B) \triangle C$ contains the elements that are in exactly one of the three sets plus the elements in all three.

Now consider the following algebraic structure: $<\mathcal{P}(S), \triangle, \cap>$, where S is a finite nonempty set, $\mathcal{P}(S)$ is the power set of S, $\triangle$ is the symmetric difference, and $\cap$ is set intersection. Let's see if this structure is a ring, where $\triangle$ is the additive operator and $\cap$ is the multiplicative operator.

The $\triangle$ operation is defined in terms of basic set operations, all of which yield subsets of S. Thus, $\mathcal{P}(S)$ is closed under $\triangle$. By the preceding frame, $\triangle$ is commutative. $\triangle$ is also associative (you prove this in homework problem 10).

Is there an identity element under $\triangle$? That is, is there an X such that $A \triangle X = X \triangle A = A$ and for all $A \in \mathcal{P}(S)$?

1241

$A \triangle \emptyset = (A \cup \emptyset) - (A \cap \emptyset) = A - \emptyset = A$ and $\emptyset \triangle A = (\emptyset \cup A) - (\emptyset \cap A) = A - \emptyset = A$. Thus, $\emptyset$ is the identity.

What is the inverse of A?

1242

$A \triangle A = (A \cup A) - (A \cap A) = A - A = \emptyset$. Every set A in $\mathcal{P}(S)$ is its own inverse.

Describe the diagonal in the Cayley table for $\mathcal{P}(S)$ under $\triangle$.

1243

Each element is its own inverse. Thus, each element on the diagonal is the identity element $\emptyset$.

$\mathcal{P}(S)$ under $\triangle$ has the closure, commutative, associative, identity, and inverses properties. Thus, it is a commutative group.

What is the multiplicative identity in $<\mathcal{P}(S), \triangle, \cap>$ (i.e., what is the identity in $\mathcal{P}(S)$ under $\cap$)? *Hint*: All the elements in $\mathcal{P}(S)$ are subsets of S.

1244

$S \cap X = X \cap S = X$ (because X is a subset of S). Thus, S is the multiplicative identity.

Is $\mathcal{P}(S)$ closed under $\cap$? Is $\cap$ associative? Is $\cap$ commutative?

1245

Because the intersection of two subsets of S is a subset of S, $\mathcal{P}(S)$ is closed under $\cap$. By frame **244**, $\cap$ is associative and commutative.

Moreover, $\cap$ distributes over $\triangle$ (you prove this in homework problem 10).

252   Chapter 21: Introduction to Rings

Let's summarize what we know: $\mathcal{P}(S)$ under Δ is a commutative group. $\mathcal{P}(S)$ under $\cap$ is closed, associative, and commutative and has the unity S. $\cap$ distributes over Δ. Thus, $<\mathcal{P}(S), \Delta, \cap>$ is a ____ ____ ____ ____.

1246

commutative ring with unity

What is the order of $<\mathcal{P}(S), \Delta, \cap>$? *Hint*: How many subsets are in $\mathcal{P}(S)$?

1247

$2^{|S|}$ (see frame **52**)

What is the order of $\mathcal{P}(S)$ when $|S| = 1$, $|S| = 2$, $|S| = 3$, and $|S| = 4$?

1248

The orders are the successive powers of 2: $2^1 = 2$, $2^2 = 4$, $2^3 = 8$, and $2^4 = 16$.

For each possible positive order of S, there is a distinct ring $<\mathcal{P}(S), \Delta, \cap>$. Each of these rings has the interesting property that all the elements on the diagonal of the Cayley table for the additive operation (i.e., Δ) are the identity element.
   We have seen two families of rings that have some striking similarities:

$$\mathbb{Z}_2, \mathbb{Z}_2 \oplus \mathbb{Z}_2, \mathbb{Z}_2 \oplus \mathbb{Z}_2 \oplus \mathbb{Z}_2, \mathbb{Z}_2 \oplus \mathbb{Z}_2 \oplus \mathbb{Z}_2 \oplus \mathbb{Z}_2, \ldots$$

$$<\mathcal{P}(S), \Delta, \cap>, |S| = 1, 2, 3, 4, \ldots$$

In both families, the order of the rings is the successive powers of 2, and all the elements on the diagonal of the Cayley table for the additive operation are the identity element. Perhaps each ring in one family is isomorphic to the ring in the other family with the same order. Let's prove this conjecture.
   Suppose S is a set with n elements. A subset of S can be represented by an *n*-tuple—a string of 1's and 0's in which each 1 indicates that the corresponding element is in the subset and each 0 indicates that the corresponding element in not in the subset. For example, suppose $S = \{a, b, c, d\}$. The subset $\{a, b, d\}$ is represented by the 4-tuple $(1, 1, 0, 1)$. The four bits in the 4-tuple in order correspond to the elements a, b, c, and d, respectively. In the subset $\{a, b, d\}$, a, b, and d are present. Thus, their corresponding bits in the 4-tuple are 1. c is not present. Thus, its corresponding bit is 0.

Represent the subset $\{b\}$ of $S = \{a, b, c, d\}$ as a 4-tuple.

1249

$(0, 1, 0, 0)$

Suppose A and B are elements in $<\mathcal{P}(S), \Delta, \cap>$ where $S = (a, b, c, d\}$. If A and B in 4-tuple form are $(0, 0, 1, 1)$ and $(0, 1, 0, 1)$, respectively, what is the 4-tuple form of $A \Delta B$? *Hint*: Recall that the Δ operation is an exclusive "or": an element is in $A \Delta B$ if and only if it is in A or B, but not both.

1250

$A = \quad (0, 0, 1, 1)$
$B = \quad (0, 1, 0, 1)$
$A \Delta B = \quad (0, 1, 1, 0)$

Now consider $\mathbb{Z}_2 \oplus \mathbb{Z}_2 \oplus \mathbb{Z}_2 \oplus \mathbb{Z}_2$. Describe what the elements in the ring $\mathbb{Z}_2 \oplus \mathbb{Z}_2 \oplus \mathbb{Z}_2 \oplus \mathbb{Z}_2$ look like.

1251

Each element has four components, each of which is an element of $\mathbb{Z}_2$. Thus, each component is either 0 or 1. For example, $(0, 0, 1, 1)$ and $(0, 1, 0, 1)$ are two of the 16 elements in $\mathbb{Z}_2 \oplus \mathbb{Z}_2 \oplus \mathbb{Z}_2 \oplus \mathbb{Z}_2$.

Suppose A and B are two elements in the ring $\mathbb{Z}_2 \oplus \mathbb{Z}_2 \oplus \mathbb{Z}_2 \oplus \mathbb{Z}_2$ with $A = (0, 0, 1, 1)$ and $B = (0, 1, 0, 1)$. What does $A + B$ equal? *Hint*: To determine each component of $A + B$, add modulo 2 the corresponding components of A and B. That is, add, divide by 2, and take the remainder.

1252

$A =$ (0, 0, 1, 1)
$B =$ (0, 1, 0, 1)
$A + B =$ (0, 1, 1, 0)

For both rings, $<\mathcal{P}(\{a, b, c, d\}, \triangle, \cap>$ and $\mathbb{Z}_2 \oplus \mathbb{Z}_2 \oplus \mathbb{Z}_2 \oplus \mathbb{Z}_2$, we get exactly the same results under their additive operations: the 4-tuple (0, 1, 1, 0). This is not unexpected because the exclusive "or" of two bits gives the same result as adding two bits modulo 2.

Let's see if the multiplicative operations in the two rings also give matching results. Again, assuming $A = (0, 0, 1, 1)$ and $B = (0, 1, 0, 1)$, what is $A \cap B$ in $<\mathcal{P}(\{a, b, c, d\}), \triangle, \cap>$ and $A \cdot B$ in $\mathbb{Z}_2 \oplus \mathbb{Z}_2 \oplus \mathbb{Z}_2 \oplus \mathbb{Z}_2$?

1253

We get matching results:

	$\mathcal{P}(S)$		$\mathbb{Z}_2 \oplus \mathbb{Z}_2 \oplus \mathbb{Z}_2 \oplus \mathbb{Z}_2$
$A =$	(0, 0, 1, 1)	$A =$	(0, 0, 1, 1)
$B =$	(0, 1, 0, 1)	$B =$	(0, 1, 0, 1)
$A \cap B =$	(0, 0, 0, 1)	$A \cdot B =$	(0, 0, 0, 1)

Again, the result is not unexpected: the "and" of two bits gives the same result as multiplying two bits modulo 2.

The exclusive "or" of two bits gives the same result as adding two bits modulo 2; the "and" of two bits gives the same result as multiplying two bits modulo 2. Thus, if we represent sets with 4-tuples, the Cayley tables for $<\mathcal{P}(\{a, b, c, d\}, \triangle, \cap>$ are identical to the corresponding tables for $\mathbb{Z}_2 \oplus \mathbb{Z}_2 \oplus \mathbb{Z}_2 \oplus \mathbb{Z}_2$. Thus, the two rings are isomorphic.

The analysis we have used to establish that $<\mathcal{P}(\{a, b, c, d\}), \triangle, \cap>$ is isomorphic to $\mathbb{Z}_2 \oplus \mathbb{Z}_2 \oplus \mathbb{Z}_2 \oplus \mathbb{Z}_2$ extends to the other rings in these two families of rings. That is, the rings $<\mathcal{P}(S), \triangle, \cap>$ for $|S| = 1, 2, 3, \ldots$ are isomorphic to the rings $\mathbb{Z}_2, \mathbb{Z}_2 \oplus \mathbb{Z}_2, \mathbb{Z}_2 \oplus \mathbb{Z}_2 \oplus \mathbb{Z}_2, \ldots$, respectively.

Because $<\mathcal{P}(\{a, b, c, d\}), \triangle, \cap>$ is isomorphic to $\mathbb{Z}_2 \oplus \mathbb{Z}_2 \oplus \mathbb{Z}_2 \oplus \mathbb{Z}_2$, there must exist an operation-preserving bijection φ between the two rings. Describe how this bijection maps the various subsets of $\{a, b, c, d\}$. For example, how would it map $\{a, b, d\}$?

1254

It maps each 4-tuple representing an element in $\mathcal{P}(\{a, b, c, d\})$ to the same 4-tuple representing an element in $\mathbb{Z}_2 \oplus \mathbb{Z}_2 \oplus \mathbb{Z}_2 \oplus \mathbb{Z}_2$.

A Caveat

1255

Because $<\mathbb{Z}, +, \cdot>$ is so often used as an example of a ring, a common mistake is to assume that a ring has all the properties of $\mathbb{Z}$. To ensure that you do not have this misconception about rings, be sure to study the list below: Each item is a property of $<\mathbb{Z}, +, \cdot>$ that a ring in general does *not* have:

- $\mathbb{Z}$ is infinite.
- $\mathbb{Z}$ has a multiplicative identity.
- Multiplication on $\mathbb{Z}$ is commutative.
- For all $a, b, c \in \mathbb{Z}$, if $ab = ac$ and $a \neq 0$, then $b = c$.
- $\mathbb{Z}$ has no divisors of zero.
- $\mathbb{Z}$ has a linear ordering of its elements: $\ldots < -2 < -1 < 0 < 1 < 2 < \ldots$

Some Specializations of a Ring

Two important specializations of a ring are integral domains and fields. We cover them in detail in chapter 24. But let's take a quick look at them here. An ***integral domain*** is a nontrivial commutative ring with unity and no divisors of zero. For example, the integers under addition and multiplication comprise an integral domain. It is nontrivial (it has nonzero elements), it has a unity (1 is the unity), it is a commutative ring (multiplication is commutative on the integers), and it has no divisors of zero (i.e., if x and y are integers and $xy = 0$, then $x = 0$ or $y = 0$). An integral domain is an abstract model of the integers. That is why it is an important algebraic structure. Another example of an integral domain is $\langle \mathbb{Z}_p, +_p, \cdot_p \rangle$ for prime p.

Why is $\langle \mathbb{Z}_4, +_4, \cdot_4 \rangle$ *not* an integral domain?

1256

It has divisors of zero. For example, $2 \cdot_4 2 = 0$.

A ***field*** is a nontrivial, commutative ring in which every nonzero element has a multiplicative inverse. For example, the real numbers under addition and multiplication is a field. It is nontrivial (the reals have nonzero elements), it is commutative (multiplication is commutative on the reals), and every nonzero element x has a multiplicative inverse (the inverse of x is $1/x$).

Are the rational numbers under addition and multiplication a field?

1257

Yes. It is nontrivial and commutative, and every nonzero element $\frac{a}{b}$ has a multiplicative inverse (the inverse of $\frac{a}{b}$ is $\frac{b}{a}$).

Fields are important because they model both the rational and the real numbers.

Review Questions

1. Prove that if a ring has a unity, the unity is unique.
2. Show that $(a + b)(c + d) = ac + bc + ad + bd$ where a, b, c, d are elements of a ring.
3. Let R be a ring. What must be true if $a^2 - b^2 = (a + b)(a - b)$ for all $a, b \in R$?
4. Find all the divisors of zero in $\mathbb{Z}_{30}$ and $\mathbb{Z}_{31}$.
5. For which values of $n > 0$ is $\langle \mathbb{Z}_n, +_n, \cdot_n \rangle$ a ring? For which values of n is $\langle \mathbb{Z}_n - \{0\}, \cdot_n \rangle$ a commutative group?
6. Suppose $a, b \in R$, and R is a nontrivial ring with unity. Show that if a and b have multiplicative inverses, then ab has a multiplicative inverse.
7. Show that if ab is a divisor of zero in a nontrivial ring, then a or b is a divisor of zero.
8. Is $\langle \mathcal{P}(\{\ \}), \Delta, \cap \rangle$ a ring? Justify your answer.
9. Determine all the divisors of zeros and units in $\mathbb{Z} \oplus \mathbb{Z}$.
10. Construct the Cayley tables for $\langle \mathcal{P}(\{a, b\}), \Delta, \cap \rangle$.

Answers to the Review Questions

1. Suppose a ring has 1 as a unity and 1' as a unity. Then $1 = 1 \cdot 1'$ (because 1' is a unity) $= 1'$ (because 1 is a unity).
2. $(a + b)(c + d) = (a + b)c + (a + b)d$ (by the left distributive law) $= ac + bc + ad + bd$ (by the right distributive law).
3. $(a + b)(a - b) = a^2 + ba + a(-b) + b(-b) = a^2 + ba - ab - b^2$ (by R6 in frame **1160**). For $a^2 + ba - ab - b^2$ to equal $a^2 - b^2$, ba must equal ab.
4. $\mathbb{Z}_{30}$: All except 1, 7, 11, 13, 17, 19, 23, 29), $\mathbb{Z}_{31}$: none.
5. $\langle \mathbb{Z}_n, +_n, \cdot_n \rangle$ a ring for all $n > 0$. $\langle \mathbb{Z}_n - \{0\}, \cdot_n \rangle$ a commutative group for n prime.
6. $(ab)(b^{-1}a^{-1}) = a(bb^{-1})a^{-1} = a1a^{-1} = aa^{-1} = 1$. Similarly, $(b^{-1}a^{-1})(ab) = 1$.
7. ab is a divisor of zero $\Rightarrow (ab)x = 0$ or $x(ab) = 0$ for some $x \neq 0$. $a \neq 0$ and $b \neq 0$ (otherwise ab would be 0 in which case ab would not be a divisor of zero). *Case 1*: $(ab)x = 0$. Then $a(bx) = 0$. If $bx \neq 0$, then a is a divisor of zero; if $bx = 0$, then

b is a divisor of zero. *Case 2*: $x(ab) = 0$. Then $(xa)b = 0$. If $xa \neq 0$, then b is a divisor of zero; if $xa = 0$, then a is a divisor of zero.

8. Yes. It has only one element. Thus, it is the trivial ring.
9. Divisors of zero: $\{(x, y) : x \text{ or } y \text{ but not both are zero}\}$. Unit: $(1, 1)$.
10.

Δ	00	01	10	11
00	00	01	10	11
01	01	00	11	10
10	10	11	00	01
11	11	10	01	00

∩	00	01	10	11
00	00	00	00	00
01	00	01	00	01
10	00	00	10	10
11	00	01	10	11

Homework Questions

1. Which of the following are rings under addition and multiplication?
   - $\{i \in \mathbb{Z} : i < 0\}$
   - $\{x \in \mathbb{R} : x \notin \mathbb{Q}\}$
   - $5\mathbb{Z}$
   - $\{x + y\sqrt{3} : x, y \in \mathbb{Z}\}$
2. Determine all the divisors of zeros and units in $\mathbb{Z}_2 \oplus \mathbb{Z}_2$.
3. Show that $\mathbb{Z}$ under the additive operation $\oplus$ defined by $a \oplus b = a + b - 1$ and the multiplicative operation $\otimes$ defined by $a \otimes b = a + b - ab$ is commutative ring with unity.
4. Show that the set of all multiplicative idempotent elements in a commutative ring is closed under the multiplicative operation.
5. Prove that the direct sum of two rings is a ring. *Hint*: Model your proof after the proof in frame **1080** that a direct product of groups is a group.
6. Suppose $<R, +>$ is a commutative group with a multiplicative operation defined by $a \cdot b = 0$ for all $a, b \in R$. Show that $<R, +, \cdot>$ is a commutative ring.
7. Show that the set of units—the elements that have a multiplicative inverse—in a ring with unity is a group under the multiplicative operation.
8. Suppose $a \in R$, and R is a nontrivial ring with unity. Show that if $a^2 = 0$, then $a + 1$ and $a - 1$ have multiplicative inverses.
9. An element a in a ring is **nilpotent** if, for some positive n, $a^n = 0$. Which elements of $\mathbb{Z}_n$ are nilpotent?
10. Prove that the operation Δ on sets defined by $A \Delta B = A \cup B - A \cap B$ is associative and that $\cap$ distributes over Δ. *Hint*: Exclusive "or" is associative.

22 Subrings and Quotient Rings

Definition of a Subring

1258

Let S be a subset of a ring R. Then S is a **subring** of R if and only if it is a ring under the operations of R, restricted to the elements of S.

Is $<2\mathbb{Z}, +, \cdot>$, the even integers, a subring of $<\mathbb{Z}, +, \cdot>$?

1259

Yes. $<2\mathbb{Z}, +, \cdot>$ has the same operations as $<\mathbb{Z}, +, \cdot>$, and has all the properties required of a ring:

- $<2\mathbb{Z}, +>$ is a commutative group.
- $<2\mathbb{Z}, \cdot>$ is a semigroup (it is closed under $\cdot$, and it inherits associativity from $<\mathbb{Z}, \cdot>$).
- $<2\mathbb{Z}, +, \cdot>$ inherits distributivity of multiplication over addition from $<\mathbb{Z}, +, \cdot>$.

Is $<\mathbb{Z}, +, \cdot>$ a subring of $<\mathbb{Q}, +, \cdot>$, where $\mathbb{Q}$ is the set of rational numbers?

1260

Yes. $<\mathbb{Z}, +>$ is a commutative group, $<\mathbb{Z}, \cdot>$ is a semigroup, and the distributive laws hold.

Is $<\mathbb{Z}^+, +, \cdot>$ a subring of $<\mathbb{Z}, +, \cdot>$, where $\mathbb{Z}^+$ is the set of positive integers.

1261

No. $<\mathbb{Z}^+, +>$ is not a group. It lacks the identity and inverses properties.

Subring Test

1262

Showing that a subset S of a ring $<R, +, \cdot>$ is a subring is a two-step procedure:

Step 1: Show that $<S, +>$ is a subgroup of $<R, +>$. To do this, we use one of the subgroup tests:

- CI subgroup test: Show that S has the closure and inverses properties under addition.
- CF subgroup test: If S is finite, show only that S has the closure property under addition.
- One-step subgroup test (stated here in additive notation): Show that $a - b \in S$ for all $a, b \in S$, where $a - b = a + (-b)$.

Step 2: Show that S is closed under multiplication.

Step 1 establishes that $<S, +>$ is a subgroup of $<R, +>$. Commutativity is inherited from R. Thus, $<S, +>$ is a commutative subgroup. Step 2, along with the associativity of multiplication (which is inherited from R), establishes that $<S, \cdot>$ is a semigroup. Distributivity on S is inherited from R. We call this two-step procedure the **SC subring test** ("SC" stands for "subgroup" and "closure" corresponding to the two steps in subring test given above).

Show that $\{0, 3\}$ is a subring of $\mathbb{Z}_6$.

1263

Step 1(using the CI subgroup test):
Closure under addition: $0 +_6 0 = 0$, $0 +_6 3 = 3 +_6 0 = 3$, $3 +_6 3 = 0$.
Additive Inverses: $0 +_6 0 = 0$, $3 +_6 3 = 0$. 0 and 3 are their own inverses.
Step 2:
Closure under multiplication: $0 \cdot_6 0 = 0$, $0 \cdot_6 3 = 3 \cdot_6 0 = 0$, $3 \cdot_6 3 = 3$. ∎

Let's show that the intersection of two subrings S_1 and S_2 of a ring $<R, +, \cdot>$ is a subring. Do step 1 using the CI subgroup test.

1264

Closure: Let $a, b \in S_1 \cap S_2 \Rightarrow a, b \in S_1$ and $a, b \in S_2 \Rightarrow a + b \in S_1$ and $a + b \in S_2$ because S_1 and S_2 are closed under $+ \Rightarrow a + b \in S_1 \cap S_2$.
Inverses: Let $a \in S_1 \cap S_2 \Rightarrow a \in S_1$ and $a \in S_2$. Because S_1 and S_2 have the inverses property, $-a \in S_1$ and $-a \in S_2 \Rightarrow -a \in S_1 \cap S_2$.

To complete the proof that $S_1 \cap S_2$ is a subring, show that $S_1 \cap S_2$ is closed under multiplication.

1265

Let $a, b \in S_1 \cap S_2 \Rightarrow a, b \in S_1$ and $a, b \in S_2 \Rightarrow ab \in S_1$ and $ab \in S_2$ because S_1 and S_2 are closed under multiplication $\Rightarrow ab \in S_1 \cap S_2$. ∎

In the example above, we used the CI subgroup test to show that $S_1 \cap S_2$ is a subgroup. Again show that $S_1 \cap S_2$ is a subgroup, but this time with the one-step subgroup test.

1266

Let $a, b \in S_1 \cap S_2$

$$\Rightarrow \begin{cases} a, b \in S_1 \Rightarrow a, -b \in S_1 \Rightarrow a + (-b) = a - b \in S_1 \\ \text{and} \\ a, b \in S_2 \Rightarrow a, -b \in S_2 \Rightarrow a + (-b) = a - b \in S_2 \end{cases} \Rightarrow a - b \in S_1 \cap S_2 \ \blacksquare$$

Properties of Subrings

1267

Suppose G is a group with subgroup H. In frame **837**, we showed that the identity for G is the same element as the identity for H. A subring is a subgroup under the additive operation of the ring. Thus, the additive identity of a ring is the additive identity of every subring of that ring. However, the unity does not have the same property. The subring of a ring with unity may not have a unity at all, or it may have a different unity. Let's look at an example of each of these cases.

$<2\mathbb{Z}, +, \cdot>$ is a subring of $<\mathbb{Z}, +, \cdot>$. What are the additive identities in $2\mathbb{Z}$ and $\mathbb{Z}$?

1268

0 is the common additive identity.

What is the unity in $2\mathbb{Z}$ and in $\mathbb{Z}$?

1269

1 is the unity in $\mathbb{Z}$. $2\mathbb{Z}$ has no unity.

Let's now look at an example in which a subring has a unity, but it is not the same element as the unity of the ring. Let R be a ring and $S = \{(r, 0) : r \in R\}$. S is a subset of the direct sum $R \oplus R$.

Show that S is a subring of $R \oplus R$ using the SC subring test.

1270

Closure under addition: Let $(r_1, 0), (r_2, 0) \in S$. Then $(r_1, 0) + (r_2, 0) = (r_1 + r_2, 0)$. $r_1 + r_2 \in R$ because R is closed under addition. Thus, $(r_1 + r_2, 0) \in S$.

Inverses under addition: Let $(r, 0) \in S$. Then $(-r, 0) \in S$, and $(r, 0) + (-r, 0) = (-r, 0) + (r, 0) = (0, 0)$. $(-r, 0)$ is the inverse of $(r, 0)$.

Closure under multiplication: Let $(r_1, 0), (r_2, 0) \in S$. Then $(r_1, 0) \cdot (r_2, 0) = (r_1 r_2, 0)$. $r_1 r_2 \in R$ because R is closed under multiplication. Thus, $(r_1 r_2, 0) \in S$. ∎

Suppose R is a ring with unity 1. What is the unity in $R \oplus R$?

1271

Let $(r_1, r_2) \in R \oplus R$. Then $(r_1, r_2) \cdot (1, 1) = (1, 1) \cdot (r_1, r_2) = (r_1, r_2)$. Thus, $(1, 1)$ is the unity in $R \oplus R$.

What is the unity in S, where $S = \{(r, 0) : r \in R\}$?

1272

$(r, 0) \cdot (1, 0) = (1, 0) \cdot (r, 0) = (r, 0)$. Thus, $(1, 0)$ is the unity in S.

$(1, 1)$ is the unity in $R \oplus R$. Why is $(1, 1)$ not the unity in S?

1273

All the elements of S have 0 as their second component. Thus, $(1, 1)$ is not an element in S. This is an unexpected result: A subring can have a unity different from the unity in the ring.

Coset Decomposition of a Ring

1274

Let start with a brief review of quotient groups (see chapter 18). Suppose $<R, +, \cdot>$ is a ring. Then $<R, +>$ (i.e., the elements in R under the additive operation of the ring) is a commutative group. Suppose N is a subgroup of $<R, +>$. Because we use additive notation for $<R, +>$, we naturally use additive notation for R/N. For example, in additive notation, we represent the coset of N containing a with $a + N$. a is called the ***representative*** of the coset $a + N$. The distinct cosets in R/N break up R into pairwise disjoint, non-empty blocks. We call this division of R into cosets the ***coset decomposition*** of R.

In chapter 18, we defined a binary operation on the cosets of a subgroup N of a group G. In *multiplicative notation*, the binary operation is defined by

$$(aN)(bN) = abN \text{ for all } a, b \in G$$

and is called ***coset multiplication***. In *additive notation*, the binary operation is defined by

$$(a + N) + (b + N) = (a + b) + N \text{ for all } a, b \in G$$

and is called ***coset addition***. Keep in mind that these are two forms of the same definition. We are simply representing the same definition using two different notations. Because we are using additive notation for the cosets of a subring, *we refer to the binary operation on these cosets as coset addition*.

We learned in Chapter 18 that if H is a normal subgroup of a group G, then the binary operation on G/H is well defined. Moreover, G/H under the binary operation on cosets forms a group. We call such a group a ***quotient group***. Because $<R, +>$ (i.e., the elements of the ring R under is additive operation) is a commutative group, every subgroup of $<R, +>$ is normal by K10 in frame **989**. Thus, if N is a subgroup of $<R, +>$, then R/N is a quotient group under coset addition. We denote this quotient group with $<R/N, +>$. Moreover, because $<R, +>$ is commutative, so is coset addition on R/N.

Show that $<R/N, +>$ is commutative.

1275

$(a + N) + (b + N)$
$= (a + b) + N$ definition of coset addition
$= (b + a) + N$ + is commutative on R.
$= (b + N) + (a + N)$ definition of coset addition ∎

What does + represent in <$R, +, \cdot$>, in <$R, +$>, and in <$R/N, +$>?

1276

In <$R, +, \cdot$> and <$R, +$>, + represents the additive operation in the ring R. In <$R/N, +$>, + represents coset addition. We are using the same symbol (+) to represent two distinct operations (addition on R and coset addition). This dual use of +, of course, can be confusing. However, the meaning of + is generally clear from the context in which it appears.

Let's examine R/N, where R is the ring <$\mathbb{Z}_6, +_6, \cdot_6$> and $N = \{0, 3\}$. N is a normal subgroup of <$\mathbb{Z}_6, +_6$>. Thus, <$\mathbb{Z}_6/N, +$> is a quotient group. Let's determine the elements of $\mathbb{Z}_6/N$. We start with N itself, which is one of the cosets. Next, we pick an element in $\mathbb{Z}_6$ that is not in N (so we do not pick 0 or 3}, and determine the coset it represents. Let's pick 1. The coset of N represented by 1 is $1 + N = 1 + \{0, 3\} = \{1, 4\}$.

Now pick an element in $\mathbb{Z}_6$ that is in neither of the two cosets we have already determined (so we do not pick 0, 3, 1, or 4). Let's pick 2. The coset of N represented by 2 is $2 + N = 2 + \{0, 3\} = \{2, 5\}$. These cosets, $\{0, 3\}$, $\{1, 4\}$, and $\{2, 5\}$, of N partition R. Thus, they are all the cosets that are in R/N.

At each step in this procedure, we pick an element of R that is not an element of any of the cosets determined up to that point. If we were to pick an element from one of the cosets already determined, we would get that coset again. For example, suppose in the last step, we picked 4, an element of the coset $\{1, 4\}$. We get the coset $\{1, 4\}$ again: $4 + \{0, 3\} = \{4 +_6 0, 4 +_6 3\} = \{4, 1\} = \{1, 4\}$. This behavior follows from K5 in frame **968**, which in additive notation states that $a + N = b + N$ if and only if $a \in b + N$ (i.e., each left coset can be represented by any and only those elements in that coset).

Let's represent the three cosets in $\mathbb{Z}_6/N$ with $0 + N$, $1 + N$, and $2 + N$. Construct the Cayley table for $\mathbb{Z}_6/N$ under coset addition. *Hint*: $0 + N$ and $3 + N$ are the same coset.

1277

+	0+N	1+N	2+N
0+N	0+N	1+N	2+N
1+N	1+N	2+N	0+N
2+N	2+N	0+N	1+N

This is the Cayley table for the cyclic group of order 3 (up to isomorphism, this is the only group of order 3).

Key idea: If N is a subring of a ring <$R, +, \cdot$>, then <$N, +$> is a normal subgroup of <$R, +$> and R/N is a commutative group under coset addition.

Absorption Rule for Cosets

1278

Suppose N is a subgroup of a ring R. By K4 in frame **968** (the coset absorption rule), if $b \in N$, then $b + N = N$. Using this property of cosets, show that $(a + b) + N = a + N$ if $b \in N$.

1279

$(a + b) + N$
$= a + (b + N)$ K1 in frame **968**
$= a + N$ K4 in frame **968**

Because $(a + b) + N = a + N$ if $b \in N$, we say that N **absorbs** b.

Simplify $a + b + c + d + N$, where $b, c, d \in N$.

1280

N absorbs b, c, and d. Thus, $a + b + c + d + N = a + N$.

Generalized Absorption Rule of Cosets: Let N be a coset of a ring R. In the coset $(a_1 + a_2 + \cdots + a_n) + N$, N absorbs every a_i that is an element of N.

Can a coset N absorb a term in its representative that is not one of its elements? For example, suppose $a \notin N$. Can $a + N = N$? It cannot by K4 in frame **968** (K4 in additive notation states that $a + N = N$ if and only if $a \in N$). Thus, a coset absorbs its own elements and *only* its own elements. However, two or more terms in a coset representative can combine into an element that is in the coset, in which case the coset can absorb all these terms, although each one individually is not absorbable. For example, consider the coset $\frac{1}{2} + \frac{3}{2} + \mathbb{Z}$ in the quotient group $\mathbb{Q}/\mathbb{Z}$. Because $\frac{1}{2}$ and $\frac{3}{2}$ are not in $\mathbb{Z}$, $\mathbb{Z}$ cannot absorb them individually. However, their sum is 2, which is in $\mathbb{Z}$. Thus, $\mathbb{Z}$ can absorb both terms: $\frac{1}{2} + \frac{3}{2} + \mathbb{Z} = 2 + \mathbb{Z} = \mathbb{Z}$.

Ideals

1281

An ideal is a special type of subring. Ideals are important because the cosets of an ideal form a ring. That is, if I is an ideal of a ring R, then R/I (the set of cosets of I in R) under the operations of coset addition and coset multiplication is itself a ring. We call R/I a *quotient ring*. An ideal plays the same role in rings as normal subgroups play in groups: The cosets of a normal subgroup form a group. Similarly, the cosets of an ideal form a ring.

An *ideal* I of a ring $<R, +, \cdot>$ is a subring of R with the following property:

$$ri \in I \text{ and } ir \in I \text{ for all } r \in R \text{ and } i \in I.$$

Because an ideal is a subring, it is closed under multiplication. That is, the product of any two elements in I produces an element in I. However, an ideal has the *superclosure property*: The product of any element in I and any element in R (in I or not) is an element in I.

Is $<2\mathbb{Z}, +, \cdot>$ an ideal of $<\mathbb{Z}, +, \cdot>$?

1282

Yes. By frame **1259**, $2\mathbb{Z}$ is a subring. Take any element of $2\mathbb{Z}$, say $2k$, and any element of $\mathbb{Z}$, say n. Their product, $2kn$, is also even, and, therefore, in $2\mathbb{Z}$. Thus, $2\mathbb{Z}$ has the superclosure property.

Is $\mathbb{Z}$ an ideal of $<\mathbb{Q}, +, \cdot>$?

1283

No. Counterexample: $\frac{1}{2} \in \mathbb{Q}$, and $5 \in \mathbb{Z}$, but $\frac{1}{2} \cdot 5 \notin \mathbb{Z}$. The subring $\mathbb{Z}$ does not have the superclosure property.

Is the subring $I = \{0, 3\}$ an ideal of $<\mathbb{Z}_6, +_6, \cdot_6>$?

1284

$0 \cdot_6 x = 0 \in I$ for all $x \in \mathbb{Z}_6$.
$1 \cdot_6 3 = 3$, $2 \cdot_6 3 = 0$, $3 \cdot_6 3 = 3$, $4 \cdot_6 3 = 0$, and $5 \cdot_6 3 = 3$.

Because all of the products above yield an element in I, I has the superclosure property. Thus, $I = \{0, 3\}$ is an ideal.

We checked above that $4 \cdot_6 3 \in I$, but we did not also check that $3 \cdot_6 4 \in I$. Is this a mistake?

1285

No. $<\mathbb{Z}_6, +_6, \cdot_6>$ is a commutative ring. Thus, $4 \cdot_6 3 = 3 \cdot_6 4$. Checking only one ordering of each product is sufficient to confirm the superclosure property in a commutative ring.

Let R be a ring with subring S. Suppose r is an element of R not in S, and s is an element of S. Where does $r + s$ reside? Is it always in S, always outside of S, or does it depend on r and s? *Hint*: Assume $r + s$ is in S, and determine what that implies.

1286

Suppose $r + s \in S \Rightarrow r + s = s'$, where $s' \in S \Rightarrow r = s' + (-s) \Rightarrow r \in S$, but $r \notin S$. Thus, $r + s$ cannot be in S. It is in R but outside of S. In other words, the *sum* of an element outside the subring and an element inside the subring is *never* inside the subring. This property of subrings (and, therefore, ideals) is called the *anti-superclosure property*.

Key idea: Under multiplication, an ideal has the superclosure property. Under addition, an ideal has the anti-superclosure property.

$2\mathbb{Z}$ is a subring and an ideal of $\mathbb{Z}$. Characterize elements in $\mathbb{Z}$ outside of $2\mathbb{Z}$. Characterize elements in $\mathbb{Z}$ inside $2\mathbb{Z}$.

1287

Outside are the odd numbers, inside are the even numbers.

Based on the anti-superclosure property of subrings, where does the sum of a number outside $2\mathbb{Z}$ and a number inside $2\mathbb{Z}$ reside? Inside or outside $2\mathbb{Z}$?

1288

Outside, which indeed is the case: the sum of an outside element (an odd number) and an inside element (an even number) is an outside element (an odd number).

To show that a subset S of a ring is an ideal, we have to show it is a subring. We do this using the SC subring test (see frame **1262**). We also have to show that S has the superclosure property. Because superclosure implies closure under multiplication, we can dispense with the test for closure under multiplication. Thus, to show that S is an ideal, we need to show only that

- S is a subgroup, and
- S has the superclosure property.

We call this two-step procedure the **SS ideal test** ("SS" stands for "subgroup" and "superclosure").

Suppose I_1 and I_2 are two ideals of a ring R. Show that $I_1 \cap I_2$ is an ideal. Use the CI subgroup test and then show superclosure.

1289

Closure under addition:
Let $a, b \in I_1 \cap I_2$
$a, b \in I_1, a, b \in I_2$ definition of intersection
$\Rightarrow a+b \in I_1, a+b \in I_1$ closure of I_1 and I_2 under $+$
$\Rightarrow a+b \in I_1 \cap I_2$ definition of intersection
Inverses under addition:
Let $a \in I_1 \cap I_2$ given
$\Rightarrow a \in I_1, a \in I_2 \Rightarrow$ definition of intersection
$\Rightarrow -a \in I_1, -a \in I_2$ inverses property of I_1 and I_2
$\Rightarrow -a \in I_1 \cap I_2$ definition of intersection
Superclosure:
Let $r \in R, a \in I_1 \cap I_2$
$\Rightarrow a \in I_1, a \in I_2$ definition of intersection
$\Rightarrow ra \in I_1$ and $ra \in I_2$ superclosure of I_1 and I_2
$\Rightarrow ra \in I_1 \cap I_2$ definition of intersection
Similarly, $ar \in I_1 \cap I_2$. ∎

Suppose I is an ideal of a ring R, and I contains 1, the unity in R. Characterize the ideal I. *Hint*: Consider the superclosure property of I.

1290

If I contains the unity 1, then by the superclosure property of I, $r1 = r \in I$ for all $r \in R$. Thus, $I = R$.

Key idea: If an ideal I of the ring R contains the unity of R, then $I = R$.

Principal Ideals

1291

Suppose a is an element of a *commutative ring R with unity*. What is the minimum number of elements we would have to add to the set $S = \{a\}$ to get an ideal? Because of the superclosure property of ideals, any ideal that contains a must also contain every element in $aR = \{ar : r \in R\}$.

Is the element a in aR, given that R is a commutative ring with unity?

1292

$a = a1 \in aR$

Let's summarize: Any ideal that contains a must contain all the elements in aR. In other words, aR is a subset of any ideal that contains a. As we will see shortly, aR itself is an ideal. Thus, aR is an ideal that contains a, and is a subset of every ideal that contains a. In this sense, aR is the smallest ideal that contains a. We call aR the ***principal ideal generated by a***, and denote it with $<a>$. Because R is commutative, $aR = Ra$. Thus, $<a>$ equals Ra as well as aR. Beware that we also use angle brackets to denote cyclic groups. Thus, $<a>$ might be the principal ideal generated by a, or the cyclic group generated by a. However, context should make the intended interpretation clear. Some books use parentheses to denote a principal ideal (for example, (a) for the principal ideal generated by a).

Show that $<a>$ is an ideal. Use the one-step subgroup test.

1293

One-step subgroup test: Let $x, y \in <a> = aR$. Then $x = ar_1$, $y = ar_2$ for some $r_1, r_2 \in R$. $x - y = ar_1 - ar_2 = a(r_1 - r_2) \in aR = <a>$.
Superclosure under multiplication: Let $x \in <a> = aR$. Then $x = ar_1$ for some $r_1 \in R$. Let $r_2 \in R$. Then $r_2 x = r_2 a r_1 = a r_2 r_1 \in aR = <a>$. Similarly, $x r_2 = a r_1 r_2 \in aR = <a>$. ∎

Let A = the intersection of all the ideals of a commutative ring with unity R that contain the element a. Show that $A \subseteq <a>$.

1294

$<a>$ is an ideal that contains a. Thus, $<a>$ is one of the ideals intersected to get A. Because the result of an intersection is a subset of each of the sets intersected, $A \subseteq <a>$.

Show that A is an ideal that contains a.

1295

From frame **1289**, we know that an intersection of two ideals is an ideal. This property of the intersection of ideals generalizes to the intersection of any number of ideals. Thus, A is an ideal. Because a is in every ideal intersected to get A, $a \in A$.

Show that $<a> \subseteq A$.

1296

From frame **1292**, we know that $<a>$ is the smallest ideal that contains a. More precisely, $<a>$ is a subset of any ideal that contains a. A is an ideal that contains a. Thus, $<a> \subseteq A$.

We have shown that $<a> \subseteq A$ and $A \subseteq <a>$. Thus, $A = <a>$.

Key ideas: If $a \in R$, where R is a commutative ring with unity, then

- $<a> = aR = \{ar : r \in R\} = Ra = \{ra : r \in R\}$. We call $<a>$ the principal ideal generated by a.
- $<a>$ is the smallest ideal that contains a. That is, $<a>$ is a subset of any ideal that contains a.
- $<a>$ is the intersection of all the ideals that contain a.

A ring in which every ideal is a principal ideal is called a ***principal ideal ring*** (abbreviated PIR).

Quotient Rings

1297

When we discussed quotient groups in chapter 18, we called the operation on cosets "coset addition" (if using additive notation) or "coset multiplication" (if using multiplicative notation). The difference between the two types of operations is strictly notational. Consider their definitions:

Coset addition: $(a + N) + (b + N) = (a + b) + N$
Coset multiplication: $aNbN = abN$

where N is a normal group. $a + b$ on the right side of the definition of coset addition is the result of the group operation applied to a and b. We get the elements of the coset $(a + b) + N$ by applying the group operation to $(a + b)$ and n for each $n \in N$. That is, $(a + b) + N = \{(a + b) + n : n \in N\}$. Similarly, ab on the right side of the definition of coset multiplication is the result of the group operation applied to a and b. We get the elements of the coset abN by applying the group operation to ab and n for each $n \in N$. That is, $abN = \{(ab)n : n \in N\}$. The two definitions *are the same* except for notation: In one we are using $+$ to represent the group operation; in the other, we are using juxtaposition to represent the group operation.

In a ring, we define coset addition and coset multiplication as follows:

Coset addition: $(a + N) + (b + N) = a + b + N$
Coset multiplication: $(a + N) \cdot (b + N) = ab + N$

A ring has two distinct operations: a group operation and a semigroup operation. In coset addition, the "$+$" in $a + b + N$ represents the *group* operation of the ring. But in coset multiplication, the juxtaposition of a and b in $ab + N$ represents the *semigroup* operation of the ring. Thus, coset multiplication and coset addition in a ring are *distinct* operations. We specify coset multiplication with either the symbol "$\cdot$" or by simply juxtaposing the two cosets in the product.

Key idea: In the context of groups, the only difference between coset addition and coset multiplication is notational (in the former, additive notation is used; in the latter multiplicative notation is used). However, in the context of rings, coset addition and coset multiplication are distinct operations.

Suppose N is a subgroup of a ring R under the addition operation of R. By definition, R is a commutative group under addition. Thus, by K10 in frame **989**, N is a normal subgroup of R. Might $<R/N, +, \cdot>$ be a ring, where $+$ is coset addition and $\cdot$ is coset multiplication as defined above? And if $<R/N, +, \cdot>$ is not a ring for every subgroup N, are there special types of subgroups N for which $<R/N, +, \cdot>$ is a ring?

According to the definition of a ring, $<R/N, +, \cdot>$ is a ring if

- $<R/N, +>$ is a commutative group.
- $<R/N, \cdot>$ is a semigroup.
- The semigroup operation (coset multiplication) distributes over the group operation (coset addition).

By frame **1275**, we know that R/N has the first property in the preceding list. Let's see if R/N has the remaining two properties. To show that $<R/N, \cdot>$ is a semigroup, we first have to show that coset multiplication on R/N is well defined. Coset multiplication on R/N, however, is *not* necessarily well defined. It depends on N. Let's determine what kind of N makes coset multiplication well defined. Coset multiplication is well defined if

$$(a + N) = (c + N) \text{ and } (b + N) = (d + N) \Rightarrow (a + N)(b + N) = (c + N)(d + N)$$

or, equivalently

$$(a + N) = (c + N) \text{ and } (b + N) = (d + N) \Rightarrow ab + N = cd + N$$

If so, we can use either a or c as the representative of the first coset, and either b or d as the representative of the second coset. The representative we use for each coset in the multiplication does not affect the result, in which case we say that coset multiplication is well defined. Let's determine what properties the subgroup N must have for the preceding implications to be true. If N has these properties, then coset multiplication is well defined. First, let's derive expressions for a and b in terms of c and d, given that $a + N = c + N$ and $b + N = d + N$:

$a \in a + N$ K2 in frame **968**

$\Rightarrow a \in c + N$ — given that $a + N = c + N$
$\Rightarrow a = c + n_1$ for some $n_1 \in N$ — definition of a coset
$b \in b + N$ — K2 in frame **968**
$\Rightarrow b \in d + N$ — given that $b + N = d + N$
$\Rightarrow b = d + n_2$ for some $n_2 \in N$ — definition of a coset

Using the equalities for a and b above, determine the product of a and b.

1298

$ab = (c + n_1)(d + n_2) = cd + n_1d + cn_2 + n_1n_2$

Thus, $ab + N = (cd + n_1d + cn_2 + n_1n_2) + N$

If N has the superclosure property, how does the preceding equality simplify?

1299

Superclosure implies that n_1d, cn_2, and n_1n_2 are all elements of N. Thus, N absorbs n_1d, cn_2 and n_1n_2 so that $ab + N = cd + N$, in which case *coset multiplication is well defined*.

Let's summarize what we have discovered. If N is a subgroup of a ring under its addition operation, and N has the superclosure property, then coset multiplication is well defined on R/N.

By definition, a subgroup N that has the superclosure property is an ideal. Thus, if N is an ideal, coset multiplication on R/N is well defined. What if N is a subgroup that is not an ideal (i.e., it is a subgroup without the superclosure property)? Might coset multiplication still be well defined? Let's investigate.

Suppose N is a subgroup that is not an ideal. Then there are elementse $r \in R$ and $n \in N$ such that $rn \notin N$. Because $rn \notin N$, N does not absorb rn. That is, $rn + N \neq N$. On the other hand, 0 (the additive identity in the subgroup N) and n are both in N. Thus, by the absorption rule, $0 + N = n + N = N$. By the definition of coset multiplication, $(r + N)(n + N) = rn + N$. Suppose we replace $n + N$ is this equality with $0 + N$. Because $n + N = 0 + N$, we should get the same result if coset multiplication is well defined. Do we? That is, does $(r + N)(n + N) = (r + N)(0 + N)$?

1300

We are given that $rn \notin N$. Then by K5 in frame **968**, $rn + N \neq N \Rightarrow (r + N)(n + N) = rn + N \neq N$. But $(r + N)(0 + N) = r0 + N = 0 + N = N$. Thus, $(r + N)(n + N) \neq (r + N)(0 + N)$ which indicates that multiplication is *not* well defined if N is not an ideal.

We have shown that

1. N an ideal $\Rightarrow$ coset multiplication is well defined
2. N not an ideal $\Rightarrow$ coset multiplication is not well defined.

Key idea: Recall from frame **168** that a proof of $p \Rightarrow q$ and $\sim p \Rightarrow \sim q$ is a proof of "p if and only if q." Implication 1 above corresponds to $p \Rightarrow q$; implication 2 above corresponds to $\sim p \Rightarrow \sim q$. Thus, the two implications above establish that coset multiplication on R/N is well defined if and only if N is an ideal.

Let's see where we are at this point in our process of determining the types of subgroups N for which $<R/N, +, \cdot>$ is a ring. We have shown that

- $<R/N, +>$ is a commutative group.
- Coset multiplication is well defined on R/N if and only if N is an ideal.

If N is an ideal, we can also show that

- $<R/N, \cdot>$ is a semigroup.
- Coset multiplication distributes over coset addition.

Thus, if N is an ideal, then $<R/N, +, \cdot>$ has all the defining properties of a ring, and, therefore, is a ring.

Let's show that $<R/N, \cdot>$ is a semigroup. Start by showing that R/N is closed under multiplication.

1301

$(a + N)(b + N) = ab + N$. $ab \in R$ because R is closed under its multiplication operation. Thus, $ab + N \in R/N$. ■

Show that $<R/N, \cdot>$ has the associative property.

1302

$([a + N][b + N])(c + N)$
$= (ab + N)(c + N)$ definition of coset multiplication
$= (ab)c + N$ definition of coset multiplication
$= a(bc) + N$ multiplication on R is associative
$= (a + N)(bc + N)$ definition of coset multiplication
$= (a + N)([b + N][c + N])$ definition of multiplication ■

Because R/N is closed under multiplication, and multiplication is associative, $<R/N, \cdot>$ is a semigroup.

Show that coset multiplication is left distributive over coset addition. That is, $(a + N)([b + N] + [c + N]) = (a + N)(b + N) + (a + N)(c + N)$

1303

$(a + N)([b + N] + [c + N])$
$= (a + N)(b + c + N)$ definition of coset addition
$= a(b + c) + N$ definition of coset multiplication
$= ab + ac + N$ left distributive law in R
$= (ab + N) + (ac + N)$ definition of coset addition
$= (a + N)(b + N) + (a + N)(c + N)$ definition of coset multiplication ■

We can similarly show that coset multiplication is right distributive over coset addition.

We have reached our principal goal for this chapter. We have shown that if N is an ideal, the coset structure $<R/N, +, \cdot>$ has all the defining properties of a ring, and therefore, is a ring. We call $<R/N, +, \cdot>$ a ***quotient ring***. Moreover, if N is not an ideal, then coset multiplication is not well defined. Thus, if N is not an ideal, then $<R/N, +, \cdot>$ is not a ring.

Key idea: $<R/N, +, \cdot>$ is a ring if and only if N is an ideal.

Ideals are to rings as normal subgroups are to groups: The cosets of ideals make quotient rings; the cosets of normal subgroups make quotient groups.
     It is interesting to compare the names "normal subgroup" and "ideal". A normal subgroup is hardly normal. To the contrary, it is quite special: It is the type of subgroup that makes quotient groups. Thus, calling it "normal" seems to undervalue its significance. An ideal, like a normal subgroup, is special. But perhaps calling it "ideal" may be a bit of an overstatement. The names "normal subgroup" and "ideal" are universally used. So we, of course, should use them, too. However, if you were to pick names for normal subgroups and ideals, what would you choose? Our choice is "quasi-commutative subgroup" (for a normal subgroup) and "superclosed subring" (for an ideal).

Review Questions

1. How do you show that a subset S of a ring R is a subring of R.
2. How do you show that a subset S of a ring R is an ideal of R.
3. Let R be a commutative ring with unity with an ideal I. If $a \in I$ and a has a multiplicative inverse, show that $I = R$.
4. Find positive integers a and b such that $<a> \cup <b>$ is not an ideal of $\mathbb{Z}$. Thus, ideals are not closed under union.
5. If $<a> = <b>$, does $a = b$? Justify your answer.
6. Let R be a ring with an ideal I. Show that R/I has a unity if I has a unity.
7. What are all the principal ideals of $\mathbb{Z}_7$?
8. What are all the principal ideals of $\mathbb{Z}_{12}$?
9. Show that $I = \{a + b : a \in A \text{ and } b \in B\}$ is an ideal of R if A and B are ideals of R.
10. Show that $I = \{0, 2, 4\}$ is an ideal of $\mathbb{Z}_6$.

Answers to the Review Questions

1. Show that it is a subgroup of R and is closed under multiplication.
2. Show that it is a subgroup of R and has the superclosure property.
3. I has the superclosure property $\Rightarrow a^{-1}a = 1 \in I \Rightarrow I = R$ by frame **1290**.
4. Let $a = 2, b = 3$. Then $a, b \in <a> \cup <b>$ but $a + b = 5 \notin <a> \cup <b>$.
5. No. Counterexample: $<1> = <-1>$ in $\mathbb{Z}$.
6. $(1 + I)(a + I) = (a + I)(1 + I) = a + I$. Thus, $1 + I$ is the unity in R/I.
7. $<x> = \mathbb{Z}_7$ for all nonzero $x \in \mathbb{Z}_7 \Rightarrow \{0\}$ and $\mathbb{Z}_7$ are the only principal ideals.
8. $\{0\}, \mathbb{Z}_{12} = <1> = <5> = <7> = <11>, <2> = <10>, <3> = <9>, <4> = <8>, <6>$
9. *Closure under addition*: Let $x, y \in I \Rightarrow x = a + b$ and $y = a' + b'$ for $a, a' \in A$ and $b, b' \in B \Rightarrow x + y = (a + b) + (a' + b') = (a + a') + (b + b')$. Because A and B are closed under addition, $a + a' \in A$ and $b + b' \in B$. Thus, $x + y = (a + a') + (b + b') \in I$.
   Inverses: Let $x \in I \Rightarrow x = a + b$ for some $a \in A$ and $b \in B \Rightarrow -a \in A, -b \in B$ (by the inverses property) $\Rightarrow -a - b \in I$ (by the closure property). $-a - b = -(a + b)$ (by R9 in frame **1160**) $= -x$. Thus, $-x \in I$.
   Superclosure: Let $r \in R. \ x \in I \Rightarrow x = a + b \Rightarrow rx = r(a + b) = ra + rb$. Because A and B have the superclosure property, $ra \in A$ and $rb \in B$. Thus, $rx \in I$. ∎
10. *Closure under addition*:
    $0 +_6 0 = 0, 0 +_6 2 = 2, 0 +_6 4 = 4$
    $2 +_6 2 = 4, 2 +_6 4 = 6,$
    $4 +_6 4 = 2.$
    Inverses: 0 is is own inverses; 2 and 4 are inverses of each other.
    Superclosure under multiplication:
    $0 \cdot_6 0 = 0, 1 \cdot_6 0 = 0, 2 \cdot_6 0 = 0, 3 \cdot_6 0 = 0, 4 \cdot_6 0 = 0, 5 \cdot_6 0 = 0$
    $0 \cdot_6 2 = 0, 1 \cdot_6 2 = 2, 2 \cdot_6 2 = 4, 3 \cdot_6 2 = 0, 4 \cdot_6 2 = 2, 5 \cdot_6 2 = 4$
    $0 \cdot_6 4 = 0, 1 \cdot_6 4 = 4, 2 \cdot_6 4 = 2, 3 \cdot_6 4 = 0, 4 \cdot_6 4 = 4, 5 \cdot_6 4 = 2$ ∎

Homework Questions

1. Determine a subring of $\mathbb{Z} \oplus \mathbb{Z}$ that is not an ideal.
2. What are the ideals of $\mathbb{Z}_{30}$?
3. Suppose a is an element in a commutative ring R that does not have a unity. What elements must be added to $\{a\}$ to get an ideal of R?
4. Show that $<r> \subseteq <s>$ if and only if $s \mid r$, where $<r>$ and $<s>$ are ideals of $\mathbb{Z}$.
5. Show that $I = \{a \in R: (\exists i \in \mathbb{Z})(a^i = 0)\}$ is an ideal of a commutative ring R.
6. Suppose b is an element of a commutative ring R. Show that $I = \{a \in R: ab = 0\}$ is an ideal of R.
7. Suppose $<R, +, \cdot>$ is a ring in which $ab = 0$ for all $a, b \in R$. Let $<H, +>$ be a subgroup of $<R, +>$. Is H always, sometimes, or never an ideal of R?
8. How many elements are in $\mathbb{Q}/H$, where H is the set of rational numbers which in reduced form have odd denominators? Is H an ideal?
9. Suppose I is an ideal of a noncommutative ring R in which $ab - ba \in I$ for all $a, b \in R$. Is multiplication commutative on I? Justify your answer.
10. Is $\mathbb{Q}$ an ideal of $\mathbb{R}$?

23 Ring Homomorphisms

Definition of Ring Homomorphisms

1304

As you read this chapter, you will likely experience *déjà vu*. Much of what we do here parallels what we did in chapter 19 on group homomorphisms. A function φ from a ring $<R_1, +, \cdot>$ to a ring $<R_2, +, \cdot>$ is a ***ring homomorphism*** if and only if for all $r, r' \in R_1$,

$$\varphi(r + r') = \varphi(r) + \varphi(r') \text{ and } \varphi(r \cdot r') = \varphi(r) \cdot \varphi(r')$$

In words, a ring homomorphism φ is a function from one ring to another that preserves both the additive and multiplicative operations. If, in addition, φ is both onto and one-to-one (i.e., a bijection from R_1 to R_2), we say that R_1 and R_1 are ***isomorphic rings***.

In the definition above of a ring homomorphism, we are using the operation symbols $+$ and $\cdot$ for both the R_1 and R_2 rings. The specific operation specified by each operation symbol depends on its operands. For example, the operation specified by the $+$ on the left side of the first equation is addition on R_1 because its operands (r and r') are in R_1. But the $+$ on the right side of the first equation is addition on R_2 because its operands ($\varphi(r)$ and $\varphi(r')$) are in R_2.

We will often use juxtaposition to represent multiplication in a ring. For example, we can write the second equation above as $\varphi(rr') = \varphi(r)\varphi(r)$.

Kernel of a Ring Homomorphism

1305

Suppose R_1 and R_2 are rings with additive identities 0_1 and 0_2, respectively. If φ is a ring homomorphism from R_1 to R_2, show that $\varphi(0_1) = 0_2$. We call this the ***identity-mapping rule***. *Hint*: Show that $\varphi(0_1)$ is idempotent and, therefore, the identity) in R_2.

1306

$\varphi(0_1) + \varphi(0_1) = \varphi(0_1 + 0_1) = \varphi(0_1) \Rightarrow \varphi(0_1)$ is idempotent $\Rightarrow \varphi(0_1)$ is the additive identity in R_2.

Let φ be a homomorphism from R_1 to R_2, with 0_2 the identity element in R_2. We call the set of all preimages in R_1 of 0_2 the ***kernel of*** φ, denoted by $\ker \varphi$. That is, $\ker \varphi = \{r \in R_1 : \varphi(r) = 0_2\}$. Let's show that $\ker \varphi$ is an ideal of R_1. To do this, we use the SS ideal test: We show that

- $\ker \varphi$ is a subgroup of R_1, and
- $\ker \varphi$ has the superclosure property.

First show that $\ker \varphi$ is a subgroup of R_1. Use the CI subgroup test.

1307

Closure: Let $x, y \in \ker \varphi \Rightarrow \varphi(x) = 0_2$ and $\varphi(y) = 0_2 \Rightarrow \varphi(x + y) = \varphi(x) + \varphi(y) = 0_2 + 0_2 = 0_2 \Rightarrow x + y \in \ker \varphi$.
Inverses: Let $x \in \ker \varphi$.

$0_2 = \varphi(0_1)$	identity mapping rule in frame 2
$= \varphi(-x + x)$	inverse property of R_1.
$= \varphi(-x) + \varphi(x)$	φ is operation preserving
$= \varphi(-x) + 0_2$	$x \in \ker \varphi$
$= \varphi(-x)$	identity property of R_2

The preceding sequence of equalities indicates that $0_2 = \varphi(-x)$. Thus, $-x \in \ker \varphi$.

Next, show that $\ker \varphi$ has the superclosure property.

1308

Let $r \in R_1$, $x \in \ker \varphi$.
$\varphi(rx)$
$= \varphi(r)\varphi(x)$ φ is operation preserving
$= \varphi(r)0_2$ because $x \in \ker \varphi$
$= 0_2$ property of rings (R3 in frame **1160**)
$\Rightarrow rx \in \ker \varphi$ definition of $\ker \varphi$

We also have to show that $xr \in \ker \varphi$ because R_1 is not necessarily commutative. Do this.

1309

$\varphi(xr)$
$= \varphi(x)\varphi(r)$ φ is operation preserving
$= 0_2\varphi(r)$ $x \in \ker \varphi$
$= 0_2$ property of rings (R3 in frame **1160**).
$\Rightarrow xr \in \ker \varphi$ definition of $\ker \varphi$ ∎

Because $\ker \varphi$ is a subgroup of R_1, and it has the superclosure property, $\ker \varphi$ is an ideal. Thus, $R_1/\ker \varphi$ is a quotient ring.

Fundamental Theorem of Ring Homomorphisms

1310

Suppose φ is a ring homomorphism from R_1 to R_2. What are the elements of $R_1/\ker \varphi$?

1311

The set of cosets of $\ker \varphi$ in R_1.

The following picture shows how the elements in R_1 map to R_2 under φ:

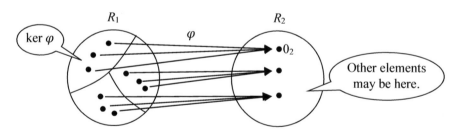

Within each coset, φ maps all the elements to the same element in R_2. Within $\ker \varphi$, φ maps all the elements to 0_2 in R_2. Moreover, elements in different cosets are mapped to different elements in R_2. Thus, there is a one-to-one correspondence between the cosets of $R_1/\ker \varphi$ and the elements in the range of φ. We proved all the preceding statements in Chapter 19, so we will not redo the proofs here.

We can represent this one-to-one correspondence with a function β. If φ maps all the elements in a coset to the element r in R_2, then β maps that coset to r in R_2 as well. For example, consider the coset $a + \ker \varphi$. φ maps all the elements of $a + \ker \varphi$ to $\varphi(a)$. Thus, β maps the coset $a + \ker \varphi$ to $\varphi(a)$. That is, $\beta(a + \ker \varphi) = \varphi(a)$:

Chapter 23: Ring Homomorphisms   269

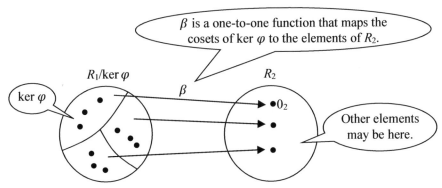

Within each coset, all the elements map to the same element in R_2. Moreover, different cosets map to different element of R_2. Thus, our β function is a one-to-one function from $R_1/\ker \varphi$ to R_2. Is β onto?

1312

Not necessarily. The range of β is the same as the range of φ. φ is not necessarily onto R_2. So neither is β (see the picture in the preceding frame).

Is β a bijection from $R_1/\ker \varphi$ onto $\varphi(R_1)$? Recall that $\varphi(R_1)$ denotes the range of φ (see frame **280**).

1313

Yes. φ and β have the same range. φ is onto $\varphi(R_1)$. Thus, β is also onto $\varphi(R_1)$. β is also one-to-one. Thus, it is a bijection from $R_1/\ker \varphi$ to $\varphi(R_1)$.

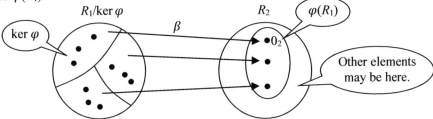

Show that β preserves the addition operation. That is, show that $\beta([a + \ker \varphi] + [b + \ker \varphi]) = \beta(a + \ker \varphi) + \beta(a + \ker \varphi)$.

1314

$\beta([a + \ker \varphi] + [b + \ker \varphi])$
$= \beta((a + b) + \ker(\varphi))$ definition of coset addition
$= \varphi(a + b)$ definition of β
$= \varphi(a) + \varphi(b)$ φ is operation preserving
$= \beta(a + \ker \varphi) + \beta(a + \ker \varphi)$ definition of β ∎

Show that β preserves the multiplication operation. That is, $\beta([a + \ker \varphi][b + \ker \varphi]) = \beta(a + \ker \varphi)\beta(a + \ker \varphi)$.

1315

$\beta([a + \ker \varphi][b + \ker \varphi])$
$= \beta(ab + \ker \varphi)$ definition of coset multiplication
$= \varphi(ab)$ definition of β
$= \varphi(a)\varphi(b)$ φ is operation preserving
$= \beta(a + \ker \varphi)\beta(a + \ker \varphi)$ definition of β ∎

β is a bijection from $R_1/\ker \varphi$ onto $\varphi(R_1)$. Moreover, β is operation preserving. What can we conclude about $R_1/\ker \varphi$ and $\varphi(R_1)$?

1316

$R_1/\ker(\varphi)$ is isomorphic to $\varphi(R_1)$

This result is called the ***fundamental theorem of ring homomorphisms***. Let's restate here:

If φ is a ring homomorphism from R_1 to R_2, then $R_1/\ker \varphi$ is isomorphic to $\varphi(R_1)$.

One of the interesting consequences of this theorem is that $\varphi(R_1)$ is a ring (because it is isomorphic to the quotient ring $R_1/\ker \varphi$), and, therefore, is a subring of R_2.

Review Questions

1. Is a ring homomorphism necessarily a group homomorphism?
2. Is $\varphi: \mathbb{Z} \to \mathbb{Z}$ defined by $\varphi(i) = 0$ for all $i \in \mathbb{Z}$ a ring homomorphism. Justify your answer.
3. Does a ring homomorphism always map the multiplicative identity, if one exists, to the multiplicative identity?
4. Is $\varphi: \mathbb{Z}_2 \to \mathbb{Z}_4$ defined by $\varphi(x) = 2x$ a ring homomorphism? Justify your answer.
5. Is $\varphi: \mathbb{Z}_2 \to \mathbb{Z}_4$ defined by $\varphi(x) = 3x$ mod 4 a ring homomorphism (recall that mod is the remainder operator)? Justify your answer.
6. Suppose φ is a ring homomorphism from R_1 to R_2. Show that $\ker \varphi = \{0\}$ if φ is one-to-one.
7. Suppose φ is a ring homomorphism from R_1 to R_2. Show that φ is one-to-one if $\ker \varphi = \{0\}$.
8. Suppose R is a commutative ring of characteristic 2. Show that $\varphi: R \to R$ defined by $\varphi(x) = x^2$ is a homomorphism.
9. Suppose R is a commutative ring. Show that the homomorphic image of R is commutative.
10. Suppose φ is a homomorphism from R_1 onto R_2. Show that if 1 is the multiplicative identity in R_1, then $\varphi(1)$ is the multiplicative identity in R_2. This is not the same question as review question 3 (φ in this question is onto).

Answers to the Review Questions

1. Yes. By definition, a ring homomorphism preserves the additive operation.
2. Yes. $\varphi(a+b) = 0 = 0 + 0 = \varphi(a) + \varphi(b)$. $\varphi(ab) = 0 = 0 \cdot 0 = \varphi(a) \cdot \varphi(b)$.
3. No. φ in review question 2 maps 1 to 0.
4. No. $\varphi(1 \cdot_2 1) = \varphi(1) = 2$. But $\varphi(1) \cdot_4 \varphi(1) = 2 \cdot_4 2 = 0$.
5. No. $\varphi(1 \cdot_2 1) = \varphi(1) = 3$. But $\varphi(1) \cdot_4 \varphi(1) = 3 \cdot_4 3 = 1$.
6. Suppose φ is one-to-one, and $x \in \ker \varphi \Rightarrow \varphi(x) = 0$. By the identity mapping rule, $\varphi(0) = 0$. Thus, $\varphi(x) = \varphi(0)$. Because φ is one-to-one, $x = 0 \Rightarrow \ker \varphi = \{0\}$. ∎
7. Suppose $\varphi(x) = \varphi(y) \Rightarrow \varphi(x) - \varphi(y) = 0 \Rightarrow 0 = \varphi(x) + \varphi(-y) = \varphi(x - y) \Rightarrow x - y \in \ker \varphi = \{0\} \Rightarrow x - y = 0 \Rightarrow x = y \Rightarrow \varphi$ is one-to-one. ∎
8. Characteristic 2 implies $1 + 1 = 0$. $\varphi(a + b) = (a + b)^2 = a^2 + ab + ab + b^2 = a^2 + b^2 + ab(1 + 1) = a^2 + b^2 + ab(0) = a^2 + b^2 = \varphi(a) + \varphi(b)$. $\varphi(ab) = (ab)^2 = a^2 b^2 = \varphi(a)\varphi(b)$.
9. $\varphi(a)\varphi(b) = \varphi(ab) = \varphi(ba) = \varphi(b)\varphi(a)$.
10. Let $y \in R_2$. Because φ is onto, there is an $x \in R_1$ such that $\varphi(x) = y \Rightarrow y\varphi(1) = \varphi(x)\varphi(1) = \varphi(x \cdot 1) = \varphi(x)$. Similarly, $\varphi(1)\varphi(x) = \varphi(1 \cdot x) = \varphi(x)$. Thus, $\varphi(1)$ is the multiplicative identity in R_2.

Homework Questions

1. Characterize all the homomorphic images of a ring R up to isomorphism if $|R|$ is prime.
2. Show that $\varphi: \mathbb{Z} \to \mathbb{Z}_n$ defined by $\varphi(i) = i$ mod n for all $i \in \mathbb{Z}$ is a ring homomorphism.
3. Show that $\varphi: \mathbb{R} \to \mathbb{R}$ defined by $\varphi(x) = |x|$ is not a ring homomorphism.
4. Show the function φ from the complex numbers to $\mathbb{R}$ defined by $\varphi(x + iy) = x$ is not a ring homomorphism.
5. Show that a ring homomorphism is one-to-one if its kernel is $\{0\}$.
6. Suppose R_1 and R_2 are rings. Show that R_1 is isomorphic to an ideal of $R_1 \oplus R_2$.
7. Suppose $<R, +, \cdot>$ is a ring with unity 1. Show that $\varphi: \mathbb{Z} \to R$ defined by $\varphi(k) = k1$ for all $k \in \mathbb{Z}$ is a ring homomorphism.
8. Suppose R is a ring with unity and a characteristic $n > 0$. Show that R contains a subring isomorphic to $\mathbb{Z}_n$.
9. Show the function φ on the complex numbers defined by $\varphi(x + yi) = x - yi$ is a ring homomorphism.
10. Let φ be a ring homomorphism from R onto S, where R and S are rings each with a unity. Show that φ maps the unity in R to the unity in S.

24 Integral Domains and Fields

Definition of an Integral Domain

1317

Recall that in a ring, an element a is a ***divisor of zero*** if and only if a is nonzero and there exists a nonzero element b in the ring such that $ab = 0$ or $ba = 0$.

Is 4 a divisor of zero in $\langle \mathbb{Z}_6, +, \cdot \rangle$?

1318

Yes, because $4 \cdot_6 3 = 0$. The elements 2 and 3 are also divisors of zero.

An integral domain is a specialization of a ring. Here is its definition: An ***integral domain*** is a nontrivial commutative ring with unity that has no divisors of zero.

Is $\langle \mathbb{Z}, +, \cdot \rangle$ an integral domain?

1319

Yes. It is a nontrivial ring, multiplication is commutative, 1 is the unity, and it has no divisors of zero.

To show that a ring R has no divisors of zero, we show that for $a, b \in R$, $ab = 0$ implies the disjunction $a = 0$ or $b = 0$.

Show that $\langle \mathbb{Z}_3, +_3, \cdot_3 \rangle$ has no divisors of zero. *Hint*: What must be true about $a, b \in \mathbb{Z}_3$ if $a \cdot_3 b = 0$.

1320

Suppose $a, b \in \mathbb{Z}_3$ and $a \cdot_3 b = 0$. Then $3 \mid ab$, and by Euclid's lemma, $3 \mid a$ or $3 \mid b$. The only element of $\mathbb{Z}_3$ divisible by 3 is 0. Thus, $a = 0$ or $b = 0 \Rightarrow \langle \mathbb{Z}_3, +_3, \cdot_3 \rangle$ has no divisors of zero. ∎

Is $\langle \mathbb{Z}_3, +_3, \cdot_3 \rangle$ an integral domain?

1321

Yes. It is a nontrivial commutative ring with unity with no divisors of zero.

Is $\langle \mathbb{Z}_4, +_4, \cdot_4 \rangle$ an integral domain?

1322

No. $2 \cdot_4 2 = 0$. The element 2 is a divisor of zero.

Is $\langle \mathbb{Z}_p, +_p, \cdot_p \rangle$ an integral domain, where p is a prime?

1323

Yes. We can use the same argument we used in frame **1320** to show that there are no divisors of zero in $\langle \mathbb{Z}_p, +_p, \cdot_p \rangle$.

Is $\langle \mathbb{Z}_n, +_n, \cdot_n \rangle$ an integral domain for n composite?

1324

No. n is composite $\Rightarrow$ there exists nonzero $r, s \in \mathbb{Z}_n$ such that $n = rs \Rightarrow r \cdot_n s = 0 \Rightarrow r$ and s are divisors of zero $\Rightarrow <\mathbb{Z}_n, +_n, \cdot_n>$ is not an integral domain.

Cancellation Law for Integral Domains

1325

Suppose in an integral domain D, $ab = ac$. Move ac in this equation to the left side.

1326

$ab + (-ac) = 0$, or equivalently by R6 in frame **1160**, $ab + a(-c) = 0$.

Now factor out a.

1327

$a(b + [-c]) = 0$.

If we are given that $a \neq 0$, what can we conclude about $b + [-c]$? *Hint*: D is an integral domain and, therefore, has no divisors of zero.

1328

Because D has no divisors of zero and $a \neq 0$, $b + (-c)$ must be equal to $0 \Rightarrow b = c$.

Thus, in an integral domain, the following ***cancellation law for multiplication*** holds: If $ab = ac$ and $a \neq 0$, then $b = c$.

Does this law apply to rings in general?

1329

No. For rings in general, the cancellation law for multiplication has two requirements on the element to be canceled (see frame **1214**): It must be nonzero and, in addition, it must not be a divisor of zero. Because an integral domain has no divisors of zero, the divisor of zero requirement need not be specified in its cancellation law.

The lack of divisors of zero in an integral domain implies the cancellation law given in the preceding frame. What about the converse? Does this cancellation law imply no divisors of zero in a nontrivial commutative ring with unity? Let's show that it does.

Suppose D is a nontrivial commutative ring with unity in which the cancellation law in the preceding frame holds. Suppose $a, b \in D$, and $ab = 0$. We want to show that D has no divisors of zero. We do this by showing that $ab = 0$ implies the disjunction $a = 0$ or $b = 0$. How do we prove the disjunction, $a = 0$ or $b = 0$? Recall from frame **170**, we do this by assuming $a \neq 0$, and then showing that $b = 0$.

What does $a0$ equal?

1330

$a0 = 0$ by R3 in frame **1160**.

We are given that $ab = 0$. Thus, $ab = a0$. If $a \neq 0$, what can we conclude from $ab = a0$?

1331

Canceling a, we get $b = 0$. Thus, D has no divisors of zero. ∎

Because the cancellation law for multiplication implies no divisors of zero, and, conversely, no divisors of zero implies the cancellation law, we can equivalently define an integral domain as follows:

An ***integral domain*** is a nontrivial commutative ring with unity in which the cancellation law in frame **1328** holds.

Another possible definition of an integral domain:

An ***integral domain*** is a nontrivial commutative ring with unity in which the set of nonzero elements is closed under multiplication.

Is this definition equivalent to the two previously given?

1332

Yes. If the nonzero elements are closed under multiplication, then the product of any two nonzero elements is nonzero. Thus, there are no divisors of zero. Conversely, suppose there are no divisors of zero. Then if $a \neq 0$ and $b \neq 0$, then $ab \neq 0$ (otherwise, a and b would be divisors of zero). Thus, the nonzero elements are closed under multiplication.

Is the ring of even integers an integral domain?

1333

No. It is nontrivial, commutative, and has no divisors of zero. But it has no unity.

Characteristic of an Integral Domain

1334

In frame **1179**, we learned that the characteristic of a ring R is the smallest positive n such that $nx = 0$ for all $x \in R$ (here nx is shorthand for the sum of n x's). If no such n exists, then the characteristic is zero. We also learned that if the ring has a unity and the characteristic is not zero, then the characteristic is the additive order of the unity.

The characteristic of an integral domain cannot be a composite. Let's prove this with a proof by contradiction. Suppose the characteristic of an integral domain is n, where n is a composite. Then there exists integers p and q such that $n = pq$, where $1 < p, q < n$. Because n is the characteristic of the integral domain, $0 = n1 = (pq)1$, which by frame **1190** equals $(p1)(q1)$ (i.e., the sum of p 1's times the sum of q 1's).

How do we know that $p1$ and $q1$ are nonzero?

1335

If $p1$ is zero, then the characteristic of the ring can be no greater than p (by frame **1186**). But we are given that the characteristic is n, which is greater than p. Thus, $p1$ must be nonzero. Similarly, $q1$ is nonzero. But it is impossible for both $p1$ and $q1$ to be nonzero. Why?

1336

Because $p1$ and $q1$ would then be divisors of zero. An integral domain has no divisors of zero. Thus, our initial assumption that the characteristic of an integral domain is a composite must be incorrect.

The characteristic of an integral domain also cannot be 1. Why? *Hint*: Use a proof by contradiction.

1337

Assume the characteristic of an integral domain D is $1 \Rightarrow 1x = x = 0$ for all $x \in D \Rightarrow D = \{0\}$. That is, D is the trivial ring. But then D cannot be an integral domain. Thus, if D is an integral domain, its characteristic cannot be 1. ∎

Key idea: Because the characteristic of an integral domain can be neither composite nor 1, it must be prime or 0.

What is the characteristic of $<\mathbb{Z}, +, \cdot>$?

1338

The additive order of 1 is infinite. Thus, the characteristic is 0.

What is the characteristic of $<\mathbb{Z}_4, +_4, \cdot_4>$?

1339

The order of 1 is 4. Thus, the characteristic is 4.

4 is not prime. Is this a contradiction of the key idea stated in frame **1337**?

1340

No because $\mathbb{Z}_4$ is not an integral domain. It has divisors of zero ($2 \cdot_4 2 = 0$).

Fields

1341

A field is specialization of an integral domain. Here is its definition:

A *field* is a nontrivial, commutative ring in which every nonzero element has a multiplicative inverse.

A field can be equivalently defined as follows (you show the equivalence of the two definitions in homework question 9):

A *field* is a commutative ring in which the set of nonzero elements form a group under multiplication.

An example of a field is $<\mathbb{R}, +, \cdot>$, the real numbers under addition and multiplication. It is a nontrivial commutative ring. Every nonzero element x in $\mathbb{R}$ has an inverse (the inverse of $x \neq 0$ is $1/x$). Another example of a field is $<\mathbb{Z}_p, +_p, \cdot_p>$ where p is a prime. Recall from frame **1073** that all the nonzero elements of $\mathbb{Z}_p$ have a multiplicative inverse if p is prime. Moreover, $<\mathbb{Z}_p, +_p, \cdot_p>$ is a nontrivial commutative ring with unity. Thus, it is a field. We call $<\mathbb{Z}_p, +_p, \cdot_p>$ for prime p the **Galois field of order p**, and denote it by $GF(p)$.

A *unit* in an element in a ring that has a multiplicative inverse. Thus, we can equivalently define a field as a nontrivial commutative ring in which every nonzero element is a unit. From frame **1211** we know that a unit cannot be a divisor of zero. Thus, a field has no divisors of zero because all its nonzero elements are units. Is a field necessarily an integral domain?

1342

Yes, because a field is a nontrivial commutative ring with unity, and it has no divisors of zero.

The categories of rings in order of increasing specialization are as follows: a nontrivial ring, a nontrivial commutative ring, a nontrivial commutative ring with unity, an integral domain, and a field.

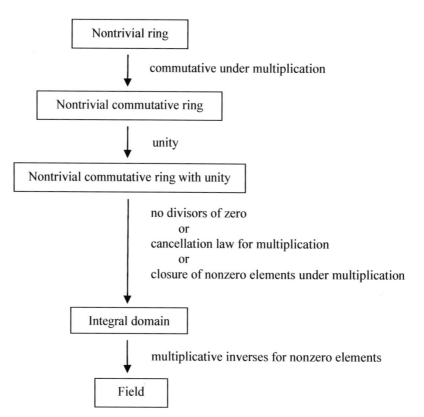

If we start with a nontrivial ring, and require commutativity under multiplication, we get a nontrivial commutative ring If we then require a unity, we get a nontrivial commutative ring with unity. If we then require any one of the following,

- no divisors of zero,
- the cancellation law for multiplication,
- closure of the nonzero elements under multiplication,

we get an integral domain. If we then require that every nonzero element has a multiplicative inverse, we get a field.

Suppose R is a nontrivial ring. What *additional* properties must R have to ensure that $R - \{0\}$ is a commutative group under the *multiplicative operation* of the ring?

1343

commutativity, unity, closure of the nonzero elements under multiplication, and each nonzero element has a multiplicative inverse

We did not forget to include associativity: By the definition of a ring, the multiplicative operation is associative in every ring.

Key idea: Each one of the four properties listed above—commutativity, unity, closure of the nonzero elements under multiplication, and each nonzero element has a multiplicative inverse—corresponds to an arrow in the preceding frame. Thus, to go from a nontrivial ring to a field, we simply add the properties, one by one, required to ensure that the nonzero elements of the ring form a commutative group under the multiplicative operation.

A field is necessarily an integral domain. Is an integral domain necessarily a field? *Hint*: Consider $\mathbb{Z}$.

1344

No. $\langle \mathbb{Z}, +, \cdot \rangle$ is an integral domain, but not all of the nonzero elements of $\mathbb{Z}$ have multiplicative inverses (only 1 and -1 have multiplicative inverses).

A Finite Integral Domain is a Field

1345

An integral domain is not necessarily a field. However, a finite integral domain is necessarily a field. Let's prove this. We need to show that each nonzero element in a finite integral domain under multiplication has an inverse (see the diagram in frame **1342**).

Suppose x is a nonzero element in a finite integral domain D. Consider the successive powers of x: $x, x^2, x^3, \ldots$. Because D is finite, there must eventually be a repeat in this sequence. That is, for some p and q with $p < q$, $x^p = x^q$. Then $x^p = x^p x^k$, where $k = q - p$. Moreover, all the powers of x are nonzero (otherwise D would have divisors of zero).

Given that $x^p = x^p x^k$, what can we conclude about x^k? *Hint*: $x^p = x^p x^k \Rightarrow x^p \cdot 1 = x^p \cdot x^k$

1346

Canceling x^p (which we can do in an integral domain), we get $x^k = 1$.

If $k = 1$, what is the multiplicative inverse of x?

1347

If $k = 1$ then $x = 1$, and x is its own inverse.

If $k > 1$, what is the multiplicative inverse of x?

1348

$x^k = 1 \Rightarrow x x^{k-1} = 1$. Thus, x^{k-1} is the inverse of x.

This analysis demonstrates that every nonzero element of in a finite integral domain D has an inverse. Thus, D is a field. ∎

Chapter 24: Integral Domains and Fields

Why does the reasoning we used above to show that a finite integral domain is a field does not work for an infinite integral domain?

1349

If the integral domain is not finite, a repeat does not necessarily occur among the powers of d.

Is $\langle \mathbb{Z}_{11}, +_{11}, \cdot_{11} \rangle$ a field?

1350

Yes. By frame **1322**, it is an integral domain. $\mathbb{Z}_{11}$ is finite. Thus, it is a field.

Units in Ideals

1351

A **unit** is an element in a ring that has a multiplicative inverse. Suppose u is a unit in a ring R with unity. Because u is a unit, there must be an element $u^{-1} \in R$ such that $u^{-1}u = 1$.

Show that if R has an ideal I that contains a unit u, then $1 \in I$.

1352

Because we are given that u is in I, by the superclosure property of I, $u^{-1}u = 1$ must also be in I.

Then by the superclosure property of I, $r1 = r \in I$ for all $r \in R$. Thus, $I = R$.

Key idea: In frame **1290**, we showed that if an ideal I of a ring R contains 1, then $I = R$. But our analysis above shows that we can make a stronger statement: If an ideal I of a ring R contains a unit (which includes 1), then $I = R$.

This result has an interesting consequence for fields. Every nonzero element in a field has a multiplicative inverse. That is, every nonzero element is a unit. Thus, if I is an ideal of a field F, and I has any nonzero elements, then $I = F$.

What is the only proper ideal that a field can have?

1353

The ideal that contains no nonzero elements. There is only one: $\{0\}$.

As we will see in chapter 26, our study of the quotient rings formed from integral domains provides some very interesting results. But we do not similarly study the quotient rings formed from fields. Why not?

1354

A field F has only two ideals—F and $\{0\}$—and, therefore, only two quotient rings: F/F, which is isomorphic to the trivial ring $\{0\}$, and $F/\{0\}$, which is isomorphic to F.

Prime Ideals

1355

In chapter 22, we showed that R/I under coset addition and multiplication is itself a ring if and only if I is an ideal of the ring R. We did this by showing that R/I has all the properties required of a ring—namely, under coset addition, R/I is a commutative group, coset multiplication is well defined, closed and associative, and the distributive laws hold.

In this section, we investigate what properties a ring R and an ideal I of R are necessary and sufficient for R/I to be an integral domain. We will see that in a commutative ring R with unity,

> R/I is an integral domain if and only if the ideal I is not equal to R, and I has the $oo = o$ property (i.e., the product of any two elements in R that are "outside" I equals some element outside I),

Let R be a ring with ideal I. Show that quotient ring R/I is a commutative ring if R is a commutative ring.

1356

$(a + I)(b + I)$
$= ab + I$ definition of coset multiplication
$= ba + I$ R is commutative
$= (b + I)(a + I)$ definition of coset multiplication ∎

Show that R/I is a ring with unity if R is a ring with unity.

1357

$(1 + I)(a + I)$
$= 1a + I$ definition of coset multiplication
$= a + I$ 1 is the unity in R

Similarly, $(a + I)(1 + I) = a + I$. Thus, $1 + I$ in the unity in R/I. ∎

What would guarantee that the quotient ring R/I is a nontrivial ring?

1358

R/I is a nontrivial ring if I is a proper ideal of R (i.e., I is proper subset of R).

Let's summarize what we know at this point: Suppose R is a commutative ring with unity, and I is a proper ideal of R. Then R/I is a nontrivial commutative ring with unity. By definition, an integral domain is a nontrivial commutative ring with unity with no divisors of zero. Thus, R/I is an integral domain if and only if it has no divisors of zero.

Let's determine what property of I guarantees that R/I has no divisors of zero. In R/I, what is the additive identity.

1359

$0 + I = I$ is the additive identity, where 0 is the additive identity in R.

It follows from the definition of a divisor of zero that if R has no divisors of zero, then whenever both a and b are nonzero, then $ab \neq 0$. Conversely, if $ab \neq 0$ whenever a and b are both nonzero, then R has no divisors of zero. Let's restate this observation more concisely:

$$ab \neq 0 \text{ whenever } a \neq 0 \text{ and } b \neq 0$$
$$\text{if and only if}$$
$$R \text{ has no divisors of zero.}$$

Suppose R is a ring with additive identity 0 and ideal I. Let's rephrase the statement above on divisors of zero so that it corresponds to the quotient ring R/I. The "zero" element in R/I (i.e., additive identity in R/I) is the coset I. The elements of R/I are the cosets of I. Thus, the cosets $a + I \neq I$ and $b + I \neq I$ are "nonzero" elements of R/I. Our rephrasing requires us to replace a and b with $a + I$ and $b + I$, respectively, and 0 with I. We get

$$(a + I)(b + I) \neq I \text{ whenever } a + I \neq I \text{ and } b + I \neq I$$
$$\text{if and only if}$$
$$R/I \text{ has no divisors of zero.}$$

By the definition of coset multiplication, we can also replace $(a + I)(b + I)$ with $ab + I$. We get

$$ab + I \neq I \text{ whenever } a + I \neq I \text{ and } b + I \neq I$$
$$\text{if and only if}$$
$$R/I \text{ has no divisors of zero.}$$

Let's simplify this statement using K4 in frame **968,** which in additive notation states that for any coset H, $h + H = H$ if and only if $h \in H$, or, equivalently, $h + H \neq H$ if and only if $h \notin H$. Thus, we can replace "$ab+I \neq I$" with "$ab \notin I$", "$a + I \neq I$" with "$a \notin I$", and "$b + I \neq I$" with "$b \notin I$". We get

$ab \notin I$ whenever $a \notin I$ and $b \notin I$
if and only if
R/I has no divisors of zero.

The top part of this statement specifies a property of I. Specifically, it specifies that whenever a and b are both "outside" I (i.e., not in I), ab is also outside I. That is, a product of an outside element and an outside element is an outside element. If we let "o" represent any outside element, we can state this property simply as ***oo = o***. Using this terminology, we have that

I has the $oo = o$ property
if and only if
R/I has no divisors of zero.

Show that the ideal $2\mathbb{Z}$ in the ring $\mathbb{Z}$ has the $oo = o$ property.

1360

$2\mathbb{Z}$ is the set of even integers. Thus, the elements outside $2\mathbb{Z}$ are the odd numbers. The product of two odd numbers is an odd number. Thus, $2\mathbb{Z}$ has the $oo = o$ property.

Does the ideal $\{0, 2\}$ in $\mathbb{Z}_4$ have the $oo = o$ property?

1361

Yes. The outside elements are 1 and 3. $1 \cdot_4 1 = 1$, $1 \cdot_4 3 = 3$, $3 \cdot_4 3 = 1$. The product of two outside elements is an outside element.

Does $\{0, 4\}$ in $\mathbb{Z}_8$ have the $oo = o$ property?

1362

No. 2 is an outside element. $2 \cdot_8 2 = 4$, an inside element.

If R is a commutative ring with unity with a proper ideal I, can we conclude that I has the $oo = o$ property if and only if R/I is an integral domain?

1363

Yes. Because I is a proper ideal of R, R/I is a nontrivial ring. By frames **1356** and **1357**, R/I is a commutative ring with unity. If I has the $oo = o$ property, then by frame **1359**, R/I has no divisors of zeros. Thus, R/I has all the properties required of an integral domain. Conversely, if R/I is an integral domain, by definition, R/I has no divisors of zero. Then by frame **1359**, I has the $oo = o$ property.

To simplify the statement of the conditions under which R/I is an integral domain, let's define a ***prime ideal*** of a ring R as a proper ideal I of R that has the $oo = o$ property. We can then say that if a ring R is a commutative ring with unity with an ideal I, then

I is prime
if and only if
R/I is an integral domain.

Is $\mathbb{Z}/3\mathbb{Z}$ an integral domain?

1364

The outside elements are integers that are not multiples of 3. That is, they do not have any 3 factors. Then by the fundamental theorem of arithmetic, the product of any two such integers cannot have a 3 factor. That means the product is an outside integer. Thus, $3\mathbb{Z}$ is a prime ideal, and $\mathbb{Z}/3\mathbb{Z}$ is an integral domain.

Is $\mathbb{Z}/4\mathbb{Z}$ an integral domain?

1365

2 is an outside integer (because it is not a multiple of 4). But $2 \cdot 2 = 4$, an inside integer. Thus, $4\mathbb{Z}$ is not a prime ideal, and $\mathbb{Z}/4\mathbb{Z}$ is not an integral domain.

As confirmation, determine (2 + 4ℤ)(2 + 4ℤ).

1366

(2 + 4ℤ)(2 + 4ℤ) = 4 + 4ℤ = 4ℤ. 4ℤ is the zero element in ℤ/4ℤ. Thus, 2 + 4ℤ is a divisor of zero, and ℤ/4ℤ is not an integral domain.

Here is an alternate but equivalent definition of a prime ideal (it is essentially the contrapositive of the definition in frame **1363**):

If *I* is a proper ideal of a ring *R*, then *I* is **prime** if and only if $ab \in I \Rightarrow a \in I$ or $b \in I$.

In other words, if a product is "inside" then at least one of its factors has to be inside. This is the definition you will most often encounter in textbooks.

Maximal Ideals

1367

Here is a natural question to ask at this point: Suppose *R* is a commutative ring with unity. What kind of ideal *I* is necessary and sufficient for *R*/*I* to be a field? We know a prime ideal *I* guarantees that *R*/*I* is an integral domain. However, a prime ideal does not guarantee that *R*/*I* is a field. The kind of ideal that guarantees *R*/*I* is a field is a maximal ideal.

A **maximal ideal** of a commutative ring *R* is an ideal not equal to *R* that is not a proper subset of any ideal of *R* except for *R*. Equivalently, an ideal *M* of *R* is a maximal ideal if and only if for all ideals *I* of *R*, if $M \subset I \subseteq R$, then $I = R$. By definition, a maximal ideal *M* of a ring *R* is not equal to *R*. Thus, *R*/*M* is not the trivial ring.

$\{0, 2, 4\}$ is an ideal of $<\mathbb{Z}_6, +_6, \cdot_6>$. Show that $\{0, 2, 4\}$ is a maximal ideal. *Hint*: By Lagrange's theorem, the order of a subgroup divides the order of the group.

1368

An ideal is a subgroup under addition. Thus, the order of an ideal of a ring divides the order of the ring. Any ideal that properly contains $\{0, 2, 4\}$ has to have order 6 (it cannot have orders 4 or 5 because these orders do not divide 6). But that means the only ideal that properly contains $\{0, 2, 4\}$ is $\mathbb{Z}_6$. Thus, $\{0, 2, 4\}$ is maximal. Then, according to the following biconditional we are about to prove, $\mathbb{Z}_6/\{0, 2, 4\}$ is a field:

Let *M* be an ideal of a nontrivial commutative ring *R* with unity. Then

R/*M* is a field ⇔ *M* is maximal.

Let start by proving the "⇐" part of the preceding biconditional:

R/*M* is a field if *M* is maximal.

Suppose *M* is maximal. We know that *R*/*M* is a nontrivial ring (by the preceding frame), and it is a commutative ring (by frame **1356**). Thus, if every nonzero element in *R*/*M* has a multiplicative inverse, then *R*/*M* is a field (recall that a field is a nontrivial commutative ring in which every nonzero element has a multiplicative inverse). Thus, to prove that *R*/*M* is a field, we have to show only that every nonzero element in *R*/*M* has a multiplicative inverse if *M* is maximal. Before we do this, let's review some basics.

What is the zero element (i.e., the additive identity) in *R*/*M*?

1369

0 + *M* = *M*

What is the multiplicative identity in *R*/*M*?

1370

1 + *M*

$0 + M = M$ is the zero element in R/M. Thus, $a + M$ is a nonzero element if and only if $a + M \neq M$. By K4 in frame 17.13, $a + M \neq M$ if and only if $a \notin M$. Thus, $a + M$ is a nonzero element in R/M if and only if $a \notin M$.

Now on to our proof. Here is our plan: Let $a + M$ be a nonzero element in R/M. Because $a + M$ is nonzero, $a \notin M$. Corresponding to a, we will define a subset J_a of R, and show that it is an ideal that properly contains M. Because M is maximal, J_a must equal R, and, therefore, have the element 1. It will then follow easily that $a + M$ has a multiplicative inverse.

Suppose $a + M$ is a nonzero element in R/M. Let $J_a = \{ar + m : r \in R$ and $m \in M\}$. Show that J_a is an ideal using the SS ideal test (see frame **1288**).

1371

Closure under addition: Let $r_1, r_2 \in R$, and $m_1, m_2 \in M$. Then $ar_1 + m_1, ar_2 + m_2 \in J_a$, and $(ar_1 + m_1) + (ar_2 + m_2) = a(r_1 + r_2) + (m_1 + m_2) \in J_a$.
Inverses under addition: Let $r \in R$, and $m \in M$. Then $ar + m \in J_a$. By the inverses property of R and M, $-r \in R$, and $-m \in M$. Thus, $a(-r) + (-m) = -ar + (-m) \in J_a$. Because $(ar + m) + [-ar + (-m)] = 0$, $-ar + (-m)$ is the additive inverse of $ar + m$.
Superclosure: Let $r_1, r_2 \in R$, and $m \in M$. Then $ar_1 + m \in J_a$. $r_2(ar_1 + m) = ar_1r_2 + r_2m$. $r_1r_2 \in R$ by the closure property of R. $r_2m \in M$ by the superclosure property of M. Thus, $ar_1r_2 + r_2m \in J_a$. ∎

Show that $M \subseteq J_a$.

1372

Let $m \in M \Rightarrow m = a0 + m \Rightarrow m \in J_a \Rightarrow M \subseteq J_a$. ∎

Show that $M \subset J_a$. *Hint*: We have just shown that $M \subseteq J_a$. Thus, to show that M is a proper subset of J_a, we need to show only that there is some element in J_a not in M. Try a.

1373

$ar + m = a$ when $r = 1$ and $m = 0 \Rightarrow a \in J_a$. $a + M$ is a nonzero element in $R/M \Rightarrow a + M \neq M$. Thus, by K4 in frame **968**, $a \notin M \Rightarrow M$ is a proper subset of J_a. ∎

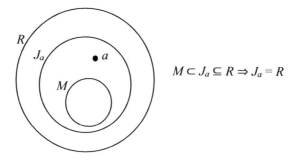

$M \subset J_a \subseteq R \Rightarrow J_a = R$

What can we conclude about J_a? *Hint*: We have shown that $M \subset J_a \subseteq R$.

1374

Because M is maximal, it is not properly contained by any ideal except for R. Thus, $R = J_a$.

Is 1 an element of J_a?

1375

Yes. $1 \in R$ and $R = J_a \Rightarrow 1 \in J_a$.

Thus, there exists elements $b \in R$ and $m \in M$ such that $1 = ab + m$. Using $1 = ab + m$, we get the following sequence of equalities:

$1 + M$
$= ab + m + M$ substitute $ab + m$ for 1
$= ab + M$ M absorbs m
$= (a + M)(b + M)$ definition of coset multiplication

Thus, $b + M$ is the multiplicative inverse of $a + M$. We have shown that R/M is a nontrivial commutative ring in which every nonzero element has an inverse. Thus, R/M is a field. ∎

Let's now prove the "⇒" part of the biconditional in frame **1368**:

If R/M is a field, then M is maximal.

To show that M is maximal, we have to show that R is the only ideal that properly contains M. Here is our plan: Show that any ideal J that properly contains M has the element 1. Then by frame **1352**, $J = R$.

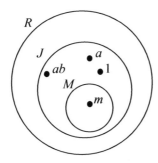

$1 = ab + m \Rightarrow 1 \in J \Rightarrow J = R \Rightarrow M$ is maximal.

Suppose J is an ideal that properly contains M. Then there must be some element a in J that is not in M. Because $a \notin M$, $a + M \neq M$. That is, $a + M$ is a nonzero element in R/M. Because R/M is a field, $a + M$ must have a multiplicative inverse. That is, there must exist a coset $b + M$ such that $(a + M)(b + M) = 1 + M$, or, equivalently $ab + M = 1 + M$. Because M contains the additive identity, $1 \in 1 + M$, and $1 + M = ab + M$. Thus, $1 \in ab + M \Rightarrow 1 = ab + m$ for some $m \in M$.

Let's examine the right side of $1 = ab + m$. Is $m \in J$?

1376

Yes, because $m \in M$ and $M \subseteq J$.

Is $ab \in J$? Hint $a \in J$.

1377

Because $a \in J$ and J has the superclosure property, $ab \in J$.

Because $m, ab \in J$, and $1 = ab + m$, what can we conclude about 1?

1378

$1 \in J$. Then by frame **1290**, $J = R$. Thus, any ideal J that properly contains M is equal to R, and, therefore, M is maximal. ∎

In this section, we have proven the following theorem:

> Let M be an ideal of a ring R where R is a commutative ring with unity. Then R/M is a field if and only if M is maximal.

Field of Quotients of an Integral Domain

1379

The rational numbers are numbers of the form $\frac{a}{b}$ or a/b where a is any integer and b is any nonzero integer. Rational numbers are really ordered pairs of integers in which the second component is nonzero. $\frac{a}{b}$ and a/b are just other ways of representing the ordered pair (a, b).

Addition and multiplication on the rationals are defined as follows:

Chapter 24: Integral Domains and Fields

$$\text{Addition} \qquad\qquad \text{Multiplication}$$

$$\frac{a}{b}+\frac{c}{d}=\frac{ad+bc}{bd} \qquad\qquad \left(\frac{a}{b}\right)\left(\frac{c}{d}\right)=\frac{ac}{bd}$$

Equality is defined as follows: $\frac{a}{b}=\frac{c}{d}$ if and only if $ad = bc$.

An important question about the rationals that you probably have never considered is this: Is addition well defined? That is, in an addition, does the sum depend on the rational numbers used to represent the values that are being added. For example, in $\frac{1}{2}+\frac{3}{5}$, if we replace $\frac{1}{2}$ with $\frac{4}{8}$, do we get the same sum? Perform both additions. Do you get the same sum?

1380

$$\frac{1}{2}+\frac{3}{5}=\frac{11}{10}, \quad \frac{4}{8}+\frac{3}{5}=\frac{20+24}{40}=\frac{44}{40}=\frac{11}{10}$$

We get the same sum. However, this one example in which the sum does not depend on the representatives used does not prove the general case. Let's prove the general case.

Suppose $\frac{a}{b}=\frac{e}{f}$ and $\frac{c}{d}=\frac{g}{h}$, or, equivalently, $af = be$ and $ch = dg$. If the two sums, $\frac{a}{b}+\frac{c}{d}=\frac{ad+bc}{bd}$ and $\frac{e}{f}+\frac{g}{h}=\frac{eh+fg}{fh}$, are equal then addition is well defined. Show that the two additions produce equal sums. *Hint*: Show that $\frac{ad+bc}{bd}=\frac{eh+fg}{fh}$, or, equivalently ($ad + bc)fh = bd(eh + fg)$, given that $\frac{a}{b}=\frac{e}{f}$ and $\frac{c}{d}=\frac{g}{h}$, or, equivalently, $af = be$ and $ch = dg$.

1381

$(ad + bc)fh$	
$= adfh + bcfh$	right distributive law
$= afdh + chbf$	multiplication is commutative.
$= bedh + dgbf$	given that $af = be$ and $ch = dg$
$= bdeh + bdfg$	multiplication is commutative.
$= bd(eh + fg)$	left distributive law ∎

In a similar fashion, show multiplication on the rationals is well defined. That is, show that $\frac{ac}{bd}=\frac{eg}{fh}$, or equivalently, $acfh = bdeg$, given that $\frac{a}{b}=\frac{e}{f}$ and $\frac{c}{d}=\frac{g}{h}$, or, equivalently, $af = be$ and $ch = dg$.

1382

$acfh$	
$= afch$	multiplication is commutative.
$= bedg$	given that $af = be$ and $ch = dg$
$= bdeg$	multiplication is commutative. ∎

Equality is an equivalence relation on the rational numbers. Each equivalence class corresponds to some value, and it contains all the rational numbers that represent that value. One equivalence class is *OneHalf*, the set of all rational numbers equal to $\frac{1}{2}$: *OneHalf* = $\{\frac{1}{2}, \frac{-1}{-2}, \frac{2}{4}, \frac{-2}{-4}, \frac{3}{6}, \frac{-3}{-6}, \ldots\}$. Recall from frame **1379** that two rational numbers, $\frac{a}{b}$ and $\frac{c}{d}$ are equal if and only if $ad = bc$. Any two elements from *OneHalf* are equal, and, therefore, in the same equivalence class. For example, $\frac{1}{2}=\frac{-2}{-4}$ because $1 \cdot -4 = 2 \cdot -2$.

Describe the equivalence classes corresponding to the equality relation on the rational numbers.

1383

Associated with each equivalence class is a value. Each equivalence class contains all the rational numbers that represent that value.

An alternate view of the set of rational numbers is to view it as the set of the equivalence classes of the equality relation given in frame **1379** rather than the set of the individual ratios of integers. To add two of these classes, simply take a representative from each class, and add them. The equivalence class of the result is then the sum of the two equivalence classes. To multiply two of these classes, simply take a representative from each class, and multiply them. The equivalence class of the result is then the product of the two equivalence classes.

Let's review what we have done. From $\mathbb{Z}$, which is a ring that is not a field, we have constructed $\mathbb{Q}$.

Is $\mathbb{Q}$ a field under addition and multiplication?

1384

Yes.

How can $\mathbb{Q}$ be a field? A field has a unique additive identity, and every nonzero element has a unique multiplicative inverse. But $\mathbb{Q}$ has an infinite number of additive identities: $\frac{0}{1}, \frac{0}{-1}, \frac{0}{2}, \frac{0}{-2}, \frac{0}{3}, \frac{0}{-3}, \ldots$, and every nonzero element has an infinite number of multiplicative inverses. For example, the inverses of $\frac{1}{2}$ are $\frac{2}{1}, \frac{-2}{-1}, \frac{4}{2}, \frac{-4}{-2}, \frac{6}{3}, \frac{-6}{-3}, \ldots$. So is $\mathbb{Q}$ a field?

1385

Yes. $\frac{0}{1}, \frac{0}{-1}, \frac{0}{2}, \frac{0}{-2}, \frac{0}{3}, \frac{0}{-3}$ are different representations of the *same* rational number (i.e., the rational number whose value is 0). Thus, $\mathbb{Q}$ has only one additive identity. Similarly, $\frac{1}{2}$ has only one multiplicative inverse (the rational number whose value is 2).

An equivalent way to answer the preceding question is to say that the elements in the field $\mathbb{Q}$ are not the individual rational numbers, but the equivalence classes under the equality relation. We can then say that the equivalence class $\{\frac{0}{1}, \frac{0}{-1}, \frac{0}{2}, \frac{0}{-2}, \frac{0}{3}, \frac{0}{-3}, \ldots\}$ is the unique additive identity of $\mathbb{Q}$, and the equivalence class $\{\frac{2}{1}, \frac{-2}{-1}, \frac{4}{2}, \frac{-4}{-2}, \frac{6}{3}, \frac{-6}{-3}, \ldots\}$ is the unique multiplicative inverse of the equivalence class $\{\frac{1}{2}, \frac{-1}{-2}, \frac{2}{4}, \frac{-2}{-4}, \frac{3}{6}, \frac{-3}{-6}, \ldots\}$.

A good question to ask at this point is this: Using the same approach that we used to construct $\mathbb{Q}$ from $\mathbb{Z}$, can we construct a field from *any* ring. Specifically, suppose R is a ring, and $F_R = \{\frac{p}{q} : p, q \in R \text{ and } q \neq 0\}$. Do the equivalence classes of F_R under the equality relation defined in frame **1379** form a field?

It is easy to show that F_R (or, more precisely the equivalence classes of F_R, but for brevity's sake, we will just say F_R) is a commutative ring with unity if and only if R is a commutative ring with unity. Thus, R has to be a commutative ring with unity for F_R to be a field. Moreover, R has to satisfy two more conditions for F_R to be a field.

If F_R is a field, why is $R \neq \{0\}$? *Hint*: Use a proof by contradiction.

1386

Suppose F_R is a field and $R = \{0\}$. The bottom number in each element of F_R has to be nonzero. Thus, if 0 is the only element in R, then F_R is empty. But F_R is nonempty because it is a field. Thus, $R \neq \{0\}$. ∎

There is one more condition that R must satisfy for F_R to be a field. Look at the denominator bd in the definitions of addition and multiplication in frame **1379**. If bd is zero, then addition and multiplication can produce elements not in F_R (all the elements of F_R have a nonzero denominator). In that case, addition and multiplication would not be closed on F_R.

What condition must R satisfy to guarantee that bd is not zero?

1387

To guarantee bd is not zero, R should have no divisors of zero.

Let's review: If the equivalence classes of $F_R = \{\frac{p}{q} : p, q \in R \text{ and } q \neq 0\}$ under the equality relation form a field, R must be a nontrivial commutative ring with unity and have no divisors of zero. In other words, R must be an integral domain. The converse is also true. That is, if R is an integral domain then F_R is a field. We call F_R a ***field of quotients***.

What are the elements of the field of quotients $F_{\mathbb{Z}_3}$ constructed from $\langle \mathbb{Z}_3, +_3, \cdot_3 \rangle$?

1388

$F_{\mathbb{Z}_3} = \{\frac{0}{1}, \frac{0}{2}, \frac{1}{1}, \frac{1}{2}, \frac{2}{1}, \frac{2}{2}\}$

There are repeats among the elements listed for $F_{\mathbb{Z}_3}$ above. For example, $\frac{0}{1} = \frac{0}{2}$. Specify $F_{\mathbb{Z}_3}$ without any repeats.

1389

$F_{\mathbb{Z}_3} = \{\frac{0}{1}, \frac{1}{1}, \frac{1}{2}\}$. Note that $\frac{1}{2} = \frac{2}{1}$ because $1 \cdot_3 1 = 2 \cdot_3 2$. Similarly, $\frac{1}{1} = \frac{2}{2}$ because $1 \cdot_3 2 = 1 \cdot_3 2$.

Construct the Cayley table for multiplication on $F_{\mathbb{Z}_3}$. From the table, confirm that each nonzero element in $F_{\mathbb{Z}_3}$ has a multiplicative inverse.

1390

·	$\frac{0}{1}$	$\frac{1}{1}$	$\frac{1}{2}$
$\frac{0}{1}$	$\frac{0}{1}$	$\frac{0}{1}$	$\frac{0}{1}$
$\frac{1}{1}$	$\frac{0}{1}$	$\frac{1}{1}$	$\frac{1}{2}$
$\frac{1}{2}$	$\frac{0}{1}$	$\frac{1}{2}$	$\frac{1}{1}$

$(\frac{1}{1})(\frac{1}{1}) = \frac{1}{1}$, $(\frac{1}{2})(\frac{1}{2}) = \frac{1}{1}$. Each nonzero element is its own inverse.

Construct the Cayley table for $F_{\mathbb{Z}_3}$ under addition.

1391

+	$\frac{0}{1}$	$\frac{1}{1}$	$\frac{1}{2}$
$\frac{0}{1}$	$\frac{0}{1}$	$\frac{1}{1}$	$\frac{1}{2}$
$\frac{1}{1}$	$\frac{1}{1}$	$\frac{1}{2}$	$\frac{0}{1}$
$\frac{1}{2}$	$\frac{1}{2}$	$\frac{0}{1}$	$\frac{1}{1}$

Ring Extensions

1392

If a ring R is isomorphic to a subring of a ring E, we say that R is **embedded** in E, and E is an **extension** of R. A quotient field F_D constructed from an integral domain D contains the subset $S = \{\frac{a}{1} : a \in D\}$. S is a subring of F_D. Moreover, D is isomorphic to S. The obvious isomorphism from D to S is $\varphi(a) = \frac{a}{1}$ for all $a \in D$. Thus, D is embedded in F_D. For example, the integral domain $\mathbb{Z}$ from which we constructed the quotient field $\mathbb{Q}$ is embedded in $\mathbb{Q}$. $\mathbb{Q}$ is an extension of $\mathbb{Z}$.

Review Questions

1. For which x and y is it true that "x is a y" where x and y can be a ring, an integral domain, a finite integral domain, or a field. For example is it true that a finite integral domain is a field?
2. "Integral" in "integral domain" indicates that an integral domain is an abstraction of the integers. However, an integral domain does not necessarily have all the properties of the integers. Give one property of the integers that is not necessarily a property of an integral domain.
3. List all the elements in each equivalence class of $F_{\mathbb{Z}_3}$ under equality, where $F_{\mathbb{Z}_3}$ is as given in frame **1388**.
4. Show why 0 cannot have a multiplicative inverse in a field.

5. Prove that a nonzero element a in an integral domain is the multiplicative identity if $aa = a$.
6. What are the elements in the field of quotients constructed from $\mathbb{Z}_2$?
7. Find all maximal ideals of $\mathbb{Z}_{17}$.
8. If a and b are divisors of zero in a commutative ring, is ab a divisor of zero?
9. Are all commutative rings with unity and with no divisors of zero integral domains? Justify your answer.
10. Suppose F is a field and $|F| = 2^n$ for some $n > 0$. Show that the characteristic of F is 2. *Hint*: Use Lagrange's theorem.

Answers to the Review Questions

1. An integral domain is a ring. A field is a ring and an integral domain. A finite integral domain is a ring, an integral domain, and a field.
2. $\mathbb{Z}$ is infinite.
3. $\{\frac{0}{1}, \frac{0}{2}\}, \{\frac{1}{1}, \frac{2}{2}\}, \{\frac{1}{2}, \frac{2}{1}\}$
4. Suppose x is the multiplicative inverse of 0. Then $0x = 1$. But $0x = 0 \neq 1$.
5. $aa = a \Rightarrow aa = a1$. Canceling a on each side, we get $a = 1$.
6. $\frac{0}{1}$ and $\frac{1}{1}$.
7. Because 17 is prime, the only ideals of $\mathbb{Z}_{17}$ are $\{0\}$ and $\mathbb{Z}_{17}$. Only $\{0\}$ is maximal.
8. Not necessarily. ab might equal 0 (because a and b are divisors of zero) in which case ab by definition is not a divisor of zero.
9. No. The trivial ring is a commutative ring with unity (which is 0), and no divisors of zero, but it is not an integral domain. By definition, an integral domain is a nontrivial ring.
10. By frame **1337**, the characteristic of a finite field is a prime number. It is also the order of the subgroup generated by 1. Thus, by Lagrange's theorem, the characteristic of a field F divides $|F|$. The only prime that divides 2^n is 2. Thus, the characteristic of F is 2.

Homework Questions

1. If S is a subring of an integral domain D, and $1 \in S$, show that S is an integral domain.
2. Show that every ideal that is a proper subset of a field is maximal.
3. Construct the quotient field from $<\mathbb{Z}_5, +, \cdot_5>$. Give its Cayley tables.
4. Let S be a subset with at least two elements of a field F. Show that S is a field if for all $a, b \in S$ with $b \neq 0$, $a - b \in S$, and $ab^{-1} \in S$.
5. Suppose u is a unit in an integral domain D. Show that for every $x \in D$, there is a $y \in D$ such that $x = uy$.
6. List the prime ideals of $\mathbb{Z}_{12}$.
7. List the maximal ideals of $\mathbb{Z}_{12}$.
8. Show that $6\mathbb{Z}$ is a maximal ideal of $<2\mathbb{Z}, +, \cdot>$.
9. Show that the two definitions of a field in frame **1341** are equivalent.
10. Is the subset $\{0, 2, 4, 6, 8, 10, 12, 14, 16\}$ of $\mathbb{Z}_{18}$ an integral domain? Justify your answer.

25 Polynomials Part 1

A Different Way of Viewing Polynomials

1393

Polynomials play an important role in ring theory, particularly field theory. Recall that every group is isomorphic to a group of permutations. There is an analogous result for finite fields and polynomials: Every finite field is isomorphic to a field of polynomials.

In this section, we introduce the concept of a polynomial in a way that likely is different from the way they were introduced to you in high school. We do not view a polynomial as an expression with various powers of x, such as $x^2 + 3x + 5$. Instead, we view a polynomial as an infinite sequence of elements from a ring.

A *polynomial over a ring R* is an infinite sequence of elements from R in which only a finite number of these elements is nonzero. An example of a polynomial over $\mathbb{Z}$ is (..., 0, 0, 0, 15, 7, 3). This polynomial consists of all zeros except for its three rightmost components.

A *constant polynomial* over the ring R is a polynomial of the form (..., 0, 0, 0, r) where $r \in R$, and 0 is the additive identity in R. The *zero polynomial* over the ring R is the polynomial whose components are all 0. A *monic polynomial* is a polynomial whose leftmost nonzero component is 1 (the multiplicative identity in R). If a ring does not have a multiplicative identity (i.e., it is not a ring with unity), then, of course, none of the polynomials over that ring can be monic.

Be sure to understand that when we are discussing polynomials in general, we use 0 for the additive identity, 1 for the multiplicative identity (if it exists), and the term "nonzero" for any element that is not equal to the additive identity. Thus, when we say that a monic polynomial is a polynomial whose leftmost nonzero component is 1, we mean a polynomial whose leftmost component not equal to the additive identity is the multiplicative identity.

We number the positions in a polynomial from right to left, starting with 0. For example, in the polynomial

$$\text{position 2} \quad \text{position 1} \quad \text{position 0}$$
$$(\ldots,\ 0,\ 0,\ 0,\ 15,\ 7,\ 3)$$

3 is in position 0, 7 is in position 1, and 15 is in position 2. To the left of 15, all the positions have 0.

A polynomial has only a finite number of nonzero elements. Thus, except for the zero polynomial, every polynomial has a leftmost nonzero component. We call this component of a polynomial the *leading term*. We can represent a polynomial simply by listing all its components from the leftmost nonzero component to rightmost component, with the understanding that all the elements to the left of the leftmost component listed are 0. For example, in place of (..., 0, 0, 0, 15, 7, 3), we simply write (15, 7, 3). To represent the zero polynomial, we write (0), or, more simply, 0. We call this type of representation of polynomials *n-tuple representation* because every polynomial over a ring R is represented with an n-tuple of elements from R. An *n-tuple* is an ordered collection of n objects. For example, a 2-tuple is an ordered pair, and a 3-tuple is an ordered triple.

The *degree* of a nonzero polynomial is the position of its leftmost, nonzero component. For example, the degree of (15, 7, 3) is 2. We denote the degree of a polynomial p with deg p.

What is the degree of the polynomial (5)?

1394

5 is in position 0 so its degree is 0.

The degree of all constant polynomials, except the zero polynomial, is 0. The zero polynomial has no leftmost nonzero component. For this reason, a degree is not defined for the zero polynomial. Two polynomials are equal if and only if their corresponding components are equal.

Does (10, 20, 30) equal (20, 30)?

1395

10 is in position 2 of the left polynomial. Because an element is not specified in position 2 in the right polynomial, we assume it is 0. The two polynomials differ in position 2. Thus, they are not equal.

Addition and Multiplication of Polynomials

1396

The addition of two polynomials is componentwise. That is, each component of the result polynomial is the sum of the corresponding components of the two given polynomials.

What is $(1, 2) + (3, 4, 5)$, where $(1, 2)$ and $(3, 4, 5)$ are polynomials over $\mathbb{Z}$?

1397

$(1, 2) + (3, 4, 5) = (0, 1, 2) + (3, 4, 5) = (0 + 3, 1 + 4, 2 + 5) = (3, 5, 7)$

Multiplication of polynomials is more complicated than addition. We call the polynomial that results from the multiplication of two polynomials the ***product polynomial***.

A convenient way to describe polynomial multiplication is with a series of ***product diagrams***—one diagram for each position of the product polynomial. To construct the product diagram for position i of p_1p_2 (the product of the polynomials p_1 and p_2), list the two polynomials horizontally, with p_2 on top of p_1 so that corresponding positions are in the same column. Then draw arrows from p_1 to p_2 such that each arrow connects positions whose position numbers sum to i. For example, draw arrows from position i of p_1 to position 0 of the p_2, from position $i-1$ of p_1 to position 1 of p_2, from position $i-2$ of p_1 to position 2 of p_2, as so on. Each arrow represents the product of the two elements at its endpoints. For example, an arrow pointing from a to b represents the product ab. The sum of all these products is the component in position i of the product polynomial.

Let's determine the product p_1p_2 where $p_1 = (4, 6, 7)$ and $p_2 = (0, 2, 3)$ are polynomials over $\mathbb{Z}$. First we list the polynomials horizontally with p_2 on top. Which one is on top makes a difference if the polynomials are over a non-commutative ring. In this example, it does not matter because $\mathbb{Z}$ is a commutative ring.

$(0, 2, 3) \quad p_2$

$(4, 6, 7) \quad p_1$

Next, draw arrows that connect positions whose position numbers sum to 0. Only one arrow is possible: from position 0 of the bottom polynomial to position 0 of the top polynomial:

$(0, 2, 3)$
↑
$(4, 6, 7)$ Product diagram for position 0

Thus, the element in position 0 in the product polynomial is $7 \cdot 3 = 21$.

Using a product diagram, determine the element in position 1 of the product of $(4, 6, 7)$ and $(0, 2, 3)$.

1398

Two arrows sum to 1: from position 0 on the bottom to position 1 on the top, and from position 1 on the bottom to position 0 on the top:

$(0, 2, 3)$

$(4, 6, 7)$ Product diagram for position 1

According to this product diagram, the element in position 1 of the product polynomial $= 7 \cdot 2 + 6 \cdot 3 = 32$.

Using a product diagram, determine the element in position 2 of the product.

1399

Product diagram for position 2

The element in position 2 of the product polynomial = $7 \cdot 0 + 6 \cdot 2 + 4 \cdot 3 = 24$.

Using a product diagram, determine the value in position 3 of the product. *Hint*: Extend the two polynomials on the left with 0's.

1400

Product diagram for position 3

The element in position 3 of the product polynomial = $7 \cdot 0 + 6 \cdot 0 + 4 \cdot 2 + 0 \cdot 3 = 8$.

Using a product diagram, determine the value in position 4 of the product.

1401

Product diagram for position 4

Each arrows connects with one 0. Thus, the element in position 4 is 0.

In the product diagrams for position 0 of the product, the element 7 in the bottom polynomial connects with 3 in the top polynomial. In the next product diagram (for position 1 of the product), 7 connects with 2. In the next diagram, 7 connects with 0. In the succession of product diagrams, 7 connects with the elements in the top polynomial in right-to-left order. We have the same right-to-left behavior with 6 in the bottom polynomial. In fact, every component, in both the top and bottom polynomials displays this right-to-left order in the succession of product diagrams. For example, 6 in the bottom polynomial connects to 3, then 2, then 0 (the components of the top polynomial in right-to-left order).

In the product diagram for position 3 in frame **1400**, there is an arrow from 4 and 2, both of which are the leftmost nonzero components of their respective polynomials. Because 4 and 2 are the leftmost nonzero components, they have 0's exclusively to their left. All the components to the right of 4 connect with components to the left of 2, all of which are 0. Similarly, all the components to the right of 2 connect with components to the left of 4, all of which are 0. Thus, only one arrow—the arrow from 4 to 2—contributes to the component in position 3 of the product polynomial. In other words, the component in position 3 of the product polynomial contains the product of the two leftmost nonzero components of the two given polynomials. Moreover, in the next product diagram, 4 and 2 connect with the components to the left of 2 and 4, respectively (because of the right-to-left behavior discussed above). But that means 2 and 4 both connect with 0's. The other nonzero components to the right connect with components even further to the left, all of which are 0. Thus, in the product polynomial, position 4 (and all higher positions) contain 0. Generalizing, we get the following rule:

Rule: Suppose a and b in positions i and j, respectively, are the leftmost nonzero components of two polynomials. Then the position $i+j$ in the product polynomial contains ab. Moreover, all positions to the left of position $i+j$ in the product polynomial contain 0. Thus, the degree of the product polynomial is at most $i+j$.

The easiest way to compute the product of two polynomials is to use essentially the same technique we use to multiply two decimal numbers: We multiply one polynomial by each component of the other polynomial, placing the partial products in separate rows, with each partial product shifted one position to the left of the previous partial product. We then add up the columns. Each column corresponds to a position in the product polynomial. Let's determine $(4, 6, 7)(2, 3)$ using this technique:

```
                2    3
         4      6    7
         ─────────────
               14   21  ⎫
         12    18       ⎬  partial products
   8     12             ⎭
   ─────────────────────
   8    24    32   21       product polynomial
```

The partial products are obtained by multiplying the polynomial (2, 3) first by 7, then by 6, and finally by 4. When we multiply 3 (in position 0) by 7 (in position 0), the product 21 goes in column 0 (i.e., the rightmost column) because the sum of the two position numbers is 0. Thus, the first partial product starts in column 0. However, when we multiply 3 (in position 0) by 6 (in position 1), the product 18 goes into column 1 because the sum of the two position numbers is 1. Thus, the second partial product starts in column 1, not column 0. The third partial product is similarly shifted because the multiplier 4 is in column 2. Each partial product starts in same column as the multiplier for that product.

Determine the polynomial equal to (4, 1)(4, 3), where (4, 1) and (4, 3) are polynomials over $\mathbb{Z}$.

1402

```
              4    3
              4    1
         ─────────────
              4    3
        16   12
        ─────────────
        16   16    3
```

Note that the leftmost nonzero components of (4, 3) and (4, 1)) are both 4, and they both are in position 1. Thus, according to our rule in frame **1401**, $4 \cdot 4 = 16$ is in position $1 + 1 = 2$ in the product polynomial. Moreover, all the components in the product polynomial to the left of position 2 are 0. The degree of the product polynomial in this example is 2. It is the sum of the degrees of the given polynomials (1 and 1).

Determining the Degree of a Product and a Sum

1403

What is deg [(4, 1)(4, 3)], where (4, 1) and (4, 3) are polynomials over $\mathbb{Z}$?

1404

The positions of the two leftmost nonzero components are 1 and 1. According to the rule in frame **1401**, position $1 + 1 = 2$ in the product polynomial contains 4, and all the components to the left are 0. Thus, the degree of the product polynomial is 2.

What is deg[(4, 1)(4, 3)], where (4, 1) and (4, 3) are polynomials over $\mathbb{Z}_8$?

1405

$(4, 1)(4, 3) = (4 \cdot_8 4, 1 \cdot_8 4 +_8 4 \cdot_8 3, 1 \cdot_8 3) = (0, 0, 3)$. The degree of the product polynomial is 0.

What is deg [(4, 1) + (4, 3)], where (4, 1) and (4, 3) are polynomials over $\mathbb{Z}$?

1406

$[(4, 1) + (4, 3) = (4 + 4, 1 + 3) = (8, 4)$. Its degree is 1.

What is deg [(4, 4) + (4, 4)], where (4, 4) is a polynomial over $\mathbb{Z}_8$?

1407

$(4, 4) + (4, 4) = (4 +_8 4, 4 +_8 4) = (0)$. It has no degree.

Given the preceding examples, you should be able to answer the following questions: Suppose p_1 and p_2 are polynomials over a ring R. Suppose a in position i is the leftmost nonzero component of p_1, and b in position j is the leftmost nonzero component of p_2. What are the degrees of p_1 and p_2?

1408

290   Chapter 25: Polynomials Part 1

deg $p_1 = i$, deg $p_2 = j$

According to the rule in frame **1401**, in the product polynomial $p_1 p_2$, position $i + j$ contains ab, and all positions to the left of position $i + j$ contain 0. Can we conclude that the degree of the product polynomial is $i + j$?

1409

No. ab could be equal to 0 if R has divisors of zero. In that case, the degree of the product polynomial is less than $i + j$ (as in frame **1405**). If, however, R has no divisors of zero, then the degree of the product polynomial is $i + j$.

Rule: If p_1, p_2, and $p_1 p_2$ are nonzero polynomials, then deg $p_1 p_2 \le$ deg p_1 + deg p_2. If R has no divisors of zero, then deg $p_1 p_2 =$ deg p_1 + deg p_2.

If p_1 and p_2 are nonzero and deg $p_1 \ne$ deg p_2, what is deg$(p_1 + p_2)$?

1410

The larger of deg p_1 and deg p_2.

If deg $p_1 =$ deg p_2, what is deg$(p_1 + p_2)$?

1411

If the leftmost components of p_1 and p_2 are additive inverses of each other, then their sum is 0. In that case, if $p_1 + p_2$ is nonzero, deg $(p_1 + p_2) <$ deg p_1. If $p_1 + p_2$ is zero, then $p_1 + p_2$ has no degree. These observations give us the following rule for nonzero polynomials p_1 and p_2:

Rule: If deg $p_1 \ne$ deg p_2, deg$(p_1 + p_2) = \max($deg p_1, deg $p_2)$. If deg $p_1 =$ deg p_2 and $p_1 + p_2$ is nonzero, deg$(p_1 + p_2) \le$ deg p_1.

Give an example of two polynomials of degree 1 whose product and sum are the zero polynomial. Each polynomial should have two nonzero components. *Hint*: Consider polynomials over $\mathbb{Z}_8$.

1412

Polynomials over $\mathbb{Z}_8$: $(4, 4)(4, 4) = (0)$, $(4, 4) + (4, 4) = (0)$

Properties of *R* than Carry over to *R*[*x*]

1413

If R is a ring, the polynomials over R under addition and multiplication as defined above form a ring. To show this, we have to show that the polynomials over R have all the properties required of a ring. The proof is long and tedious so we will omit it. Most of the steps of the proof are straightforward. For example, suppose $p = (r_n, r_{n-1}, ..., r_0)$ is a polynomial over R. Every element in R has an additive inverse in R. Thus, $q = (-r_n, -r_{n-1}, ..., -r_0)$ is also a polynomial over R. Clearly $p + q = (0)$. q is the additive inverse of p. Thus, the polynomials over R have the inverses property under addition.

Let's multiply (1 2) by (3 4), two polynomials over $\mathbb{Z}$, showing the factors in each individual multiplication:

$$
\begin{array}{rrrl}
 & 1 & 2 & \\
 & 3 & 4 & \\
\hline
 & 4 \cdot 1 & 4 \cdot 2 & \\
3 \cdot 1 & 3 \cdot 2 & & \\
\hline
3 \cdot 1 & 3 \cdot 2 + 4 \cdot 1 & 4 \cdot 2 & = \quad (3, 10, 8)
\end{array}
$$

Now let's do the same, but reverse the order of the two polynomials in the product:

$$
\begin{array}{rrrl}
 & 3 & 4 & \\
 & 1 & 2 & \\
\hline
 & 2 \cdot 3 & 2 \cdot 4 & \\
1 \cdot 3 & 1 \cdot 4 & & \\
\hline
1 \cdot 3 & 1 \cdot 4 + 2 \cdot 3 & 2 \cdot 4 & = \quad (3, 10, 8)
\end{array}
$$

If R is a commutative ring, are the polynomials over R also a commutative ring? *Hint*: Consider the preceding two multiplications.

1414

Yes. The sum of products that give the value of each component of p_1p_2 differ from the corresponding sum of products for p_2p_1 only in the order of the terms in the sums and the order of the factors in the terms. In a commutative ring, both addition and multiplication are commutative. Thus, if R is a commutative ring, the terms and factors in p_1p_2 can be commuted to give p_2p_1. Thus, the polynomials over R are commutative.

If R is a ring with unity, are the polynomials over R also a ring with unity?

1415

Yes. The polynomial (1) is the multiplicative identity in the ring of polynomials over R.

If the product of two nonzero elements a and b in a ring R is 0, then the product of the polynomials (a) and (b) over R is $(ab) = (0)$. Thus, divisors of zero in R imply divisors of zero in the polynomials over R. Conversely, if the product of two nonzero polynomials over a ring R is the zero polynomial, then the product of their two leftmost nonzero components must be 0. Thus, these two components are divisors of zero in R. In other words, R has divisors of zero if and only if the polynomials over R have divisors of zero, or equivalently (by frame **145**), R has no divisors of zero if and only if the polynomials over R have no divisors of zero.

If D is an integral domain, are the polynomials over D also an integral domain?

1416

Yes. An integral domain is a nontrivial commutative ring with unity with no divisors of zero. Thus, by the two preceding frames, the polynomials over an integral domain D is a commutative ring with unity. Moreover, the polynomials over an integral domain have no divisors of zero because an integral domain has no divisors of zero. Finally, the additive and multiplicative identities, (0) and (1), are distinct elements. Thus, the ring of polynomials over D is a nontrivial ring.

If F is a field, are the polynomials over F also a field? *Hint*: What is the multiplicative inverse of (1 0)?

1417

No. The product of (1 0) and any polynomial has 0 in position 0. Thus, the product cannot equal the unity polynomial (1).

Powers-of-*x* Representation

1418

Let's take another look at polynomials. This time we will introduce polynomials as they were introduced to you in high school, with one important exception that we will explain later.

Let R be a ring. Then a ***polynomial over R*** is an infinite sum of the form

$$\cdots + c_2x^2 + c_1x^1 + c_0x^0$$

where $c_i \in R$ for all $i \geq 0$, and only a finite number of the c_i elements are nonzero. Polynomials in this form are in ***powers-of-x representation***. We call each c_ix^i a ***term*** and c_i the ***coefficient*** of that term. There is no value associated with x. In particular, it is not an element of R. For this reason, we call x an ***indeterminate***. We generally write the c_0x^0 term simply as c_0, and the c_1x^1 term as c_1x. We also omit the coefficient in a term if the coefficient is the multiplicative identity. For example, we write the polynomial $1x^2 + 1x^1 + 2$ simply as $x^2 + x + 2$. We denote the set of all polynomials over R that use the indeterminate x with $R[x]$. The indeterminate does not have to be the symbol x. We, for example, can use t as the indeterminate, in which case we would denote the set of all polynomials over R with $R[t]$.

Because only a finite number of terms have nonzero coefficients, we can represent a polynomial with a finite sum that includes only the terms with nonzero coefficients, with the understanding that any term not included has a zero coefficient. For example, in the polynomial $7x^5 + 3x^2 + 5$ over $\mathbb{Z}$, the x, x^3, x^4, and x^i terms for $i > 5$ have zero coefficients.

We call the term in a polynomial with the largest exponent that has a nonzero coefficient the ***leading term***. For example, in $7x^5 + 3x^2 + 5$, the leading term is $7x^5$.

Convert $7x^5 + 3x^2 + 5$ to *n*-tuple representation.

1419

(7, 0, 0, 3, 0, 5)

Convert (2, 0, 3, 0) to powers-of-x representation.

1420

$2x^3 + 3x$

For polynomials in n-tuple representation, we have to include all the zero components to the right of the leftmost nonzero component. However, in the powers-of-x representation, we typically omit all terms with a zero coefficient. Thus, (2, 0, 3, 0) in powers-of-x representation is simply $2x^3 + 3x$.

In powers-of-x representation, the powers of x indicate the position of each coefficient. Specifically, the exponent in each power of x is the position of the corresponding coefficient. For example, in $7x^5 + 3x^2 + 5$, $7x^5$ indicates that 7 is in position 5 in the polynomial, $3x^2$ indicates that 3 is in position 2, and 5 without a power of x (or with x^0) indicates 5 is in position 0. Because the exponents in each term indicate the position of the coefficient in that term, we can reorder the terms without affecting the polynomial represented. For example, $7x^5 + 3x^2 + 5$ is the same polynomial as $5 + 3x^2 + 7x^5$.

When using powers-of-x representation, we typically denote a polynomial with an identifier followed by the indeterminate within parentheses. For example, we write, $p(x) = x^2 + 2x + 1$, and then use $p(x)$ to represent $x^2 + 2x + 1$.

To multiply two polynomials in powers-of-x representation, we multiply each term in one polynomial by each term in the other polynomial, and then collect and add like terms (***like terms*** are terms with the same exponent). For example, let's multiply $7x^2 + 3x + 5$ by $2x + 1$:

$$\begin{array}{r} 7x^2 + 3x + 5 \\ 2x + 1 \\ \hline 7x^2 + 3x + 5 \\ 14x^3 + 6x^2 + 10x \\ \hline 14x^3 + 13x^2 + 13x + 5 \end{array}$$

By collecting and adding like terms, we are adding exactly the same products that we add when we use product diagrams to perform the multiplication. For example, $13x$ in position 1 of the product polynomial is the sum of $3x$ and $10x$. 3 is the product of 1 and 3 (two coefficients whose positions add up to 1). Similarly, 10 is the product of 2 and 5 (the other two coefficients whose positions add up to 1).

When we multiply $7x^2$ by $2x$, we get $(7x^2)(2x) = (7 \cdot 2)x^3 = 14x^3$. To get $(7 \cdot 2)x^3$ from $(7x^2)(2x)$, we have to commute x^2 and 2. On what basis are we allowed to commute x^2 and 2? Remember that the powers of x are placeholders. Their only function is to indicate the position of their corresponding coefficients. For example, in $7x^2$, the x^2 indicates that 7 is in position 2. Similarly, in $2x$, the x (which is the abbreviated form of x^1) indicates that 2 is in position 1. From the definition of polynomial multiplication, we know that the product of a component in position 2 and a component in position 1 contributes to the component in position $2 + 1 = 3$ in the product polynomial. Thus, we should represent the product of $7x^2$ and $2x$ using x^3 as the power of x. This is the justification for equating $(7x^2)(2x)$ to $14x^3$.

In high school, you probably viewed polynomials as functions of x. For example, the polynomial x^2 is the square function. When you substitute 5 for x, you get back 25, the square of 5. However, in ring theory, we generally do not view polynomials as functions.

Consider the polynomial $x^2 + x$ over $\mathbb{Z}_2$. There are only two elements in $\mathbb{Z}_2$: 0 and 1. What is the value of $x^2 + x$ for $x = 0$ and $x = 1$?

1421

For $x = 0$, the value is $0^2 + 0 = 0$
For $x = 1$, the value is $1^2 + 1 = 1 + 1 = 0$

What is the value of the polynomial $x^4 + x^3 + x^2 + x$ over $\mathbb{Z}_2$ for $x = 0$ and $x = 1$?

1422

0 in both cases.

Thus, the polynomials $x^2 + x$ and $x^4 + x^3 + x^2 + x$ over $\mathbb{Z}_2$ represent the *same* function. However, they are *not* the same polynomials. In n-tuple representation, the former is (1, 1, 0); the latter is (1, 1, 1, 1, 0). Remember polynomials are equal if and only if each pair of corresponding coefficients is equal.

In ring theory, why do we generally not view polynomials as functions?

1423

Equal polynomial functions are not necessarily equal polynomials.

$x^4 + x$ and $x^2 + x$ as functions over $\mathbb{Z}_3$ have what values for $x = 0, 1$, and 2. Are they the same function? Represent each polynomial in n-tuple representation. Are they the same polynomial?

1424

x	$x^4 + x$	$x^2 + x$
0	0	0
1	2	2
2	0	0

same function but different polynomials

$x^4 + x = (1, 0, 0, 1, 0)$, $x^2 + x = (1, 1, 0)$.

Adjoining *x* to a Ring

1425

$\mathbb{Z}$ under addition and multiplication is a ring. Suppose we add $\sqrt{2}$ to $\mathbb{Z}$ to get the set $\mathbb{Z} \cup \{\sqrt{2}\}$. This set consists of all the integers, and, in addition, $\sqrt{2}$. Is our new set still a ring under addition and multiplication?

1426

No. It is not closed under addition. For example, 2 and $\sqrt{2}$ are in the $\mathbb{Z} \cup \{\sqrt{2}\}$, but $2 + \sqrt{2} \notin \mathbb{Z} \cup \{\sqrt{2}\}$. $\mathbb{Z} \cup \{\sqrt{2}\}$ is also not closed under multiplication. For example, $2 \cdot \sqrt{2} \notin \mathbb{Z} \cup \{\sqrt{2}\}$.

Show that $\mathbb{Z} \cup \{\sqrt{2}\}$ does not have the additive inverses property.

1427

$\sqrt{2}$ has no additive inverse ($-\sqrt{2}$ is not in $\mathbb{Z} \cup \{\sqrt{2}\}$).

Now suppose we add to $\mathbb{Z} \cup \{\sqrt{2}\}$ all and only those elements needed to get back a ring. Clearly, to maintain the closure property, we have to add $5 + \sqrt{2}$ and $7 + \sqrt{2}$ (because 5, 7 and $\sqrt{2}$ are all in $\mathbb{Z} \cup \{\sqrt{2}\}$). But then we have to add the sum and product of these two elements. What is the sum and product of these two elements?

1428

$5 + \sqrt{2} + 7 + \sqrt{2} = 12 + 2\sqrt{2}$
$(5 + \sqrt{2})(7 + \sqrt{2}) = 35 + 7\sqrt{2} + 5\sqrt{2} + \sqrt{2}\sqrt{2} = 35 + 12\sqrt{2} + 2 = 37 + 12\sqrt{2}$

In both cases, we get more elements of the form $a + b\sqrt{2}$. In fact, every element in the set after we get back a ring is of the form or can be put into the form of $a + b\sqrt{2}$, where $a, b \in \mathbb{Z}$, and all the elements of this form are in the ring. We call this process of adding an element and then restoring the ring properties by adding additional elements *adjoining an element to a ring*. In this example, we adjoin $\sqrt{2}$ to $\mathbb{Z}$. The resulting ring is denoted by $\mathbb{Z}[\sqrt{2}]$.

The set $S = \{a + b\sqrt{2} : a \in \mathbb{Z} \text{ and } b = 0\}$ under addition and multiplication is a subring of $\mathbb{Z}[\sqrt{2}]$. Moreover, $\mathbb{Z}$ is isomorphic to S. We say that $\mathbb{Z}$ is *embedded* in $\mathbb{Z}[\sqrt{2}]$, or $\mathbb{Z}[\sqrt{2}]$ is an *extension* of $\mathbb{Z}$ (see frame **1392**).

What does all this have to do with polynomials? Suppose R is a commutative ring, and x is an element not in R. Assume that x can be manipulated and operated on in the same way the elements in R can be manipulated and operated on. For example, $ab = ba$ for all $a, b \in R$ because R is a commutative ring. Thus, $ax = xa$ for all $a \in R$. What is $R[x]$, the ring that results when we when add to $R \cup \{x\}$ all and only those elements to get back a ring? In other words, what ring do we get if we adjoin x to R?

1429

We get the ring of polynomials over R. We added x, so to restore closure under multiplication, we also have to add $xx = x^2$, $xx^2 = x^3$, $xx^3 = x^4$, and so on. We also have to add all the products whose factors are x and/or elements of R. Because x commutes with every element in R, we can put each of these products in the form cx^i where $c \in R$ by commuting all the x factors to the right (replacing them with a single power of x) and all the factors that are elements of R to the left (replacing them with their product). To restore closure under addition, we have to include all the sums of terms of the form cx^i. Adding these sums yields a sum of the same form. Multiplying these sums yields products of the same form. Thus, if we adjoin x to R, then every element in $R[x]$ can be put in the form

$$c_n x^n + c_{n-1} x^{n-1} + \cdots + c_2 x^2 + c_1 x + c_0$$

where all the coefficients $c_0, c_1, \ldots, c_n$ are elements of R. This set, of course, is the set of polynomials over R.

The set of all the constant polynomials in $R[x]$ is a subring of $R[x]$. Moreover, R is isomorphic to this subring of $R[x]$. Define a function φ from R to the set of constant polynomials in $R[x]$ that is an isomorphism.

1430

$\varphi(c) = cx^0 = (c)$ for all $c \in R$.

Because R is "inside" $R[x]$ (in the isomorphism sense), we say that R is embedded in $R[x]$, and $R[x]$ is an extension of R.

We have four distinct ways of viewing polynomials over a ring R:

1. As a sequence of elements from R
2. As a sum of terms of the form $c_i x^i$, where $c_i \in R$
3. As the extension $R[x]$ obtained by adjoining x to R
4. As a function of the indeterminate

Division of Polynomials

1431

You probably learned how to divide polynomials in high school. Let's review the procedure. Let's divide $2x^3 + 3x^2 - 5x + 1$ by $x + 4$, both of which are polynomials over $\mathbb{Z}$:

$$\text{divisor} \longrightarrow x+4 \overline{)2x^3 + 3x^2 - 5x + 7} \longleftarrow \text{dividend}$$

We make the first term of the quotient $2x^2$ so that the product of it and the divisor produces $2x^3$, the first term of the dividend. Then on subtraction, the leading term of the dividend is zeroed out:

$$\begin{array}{r} 2x^2 \\ x+4 \overline{)2x^3 + 3x^2 - 5x + 7} \\ \underline{2x^3 + 8x^2 } \\ 0 -5x^2 - 5x + 7 \end{array}$$

We then repeat this step but with the result of the subtraction rather than with the original dividend:

$$\begin{array}{r} 2x^2 - 5x \\ x+4 \overline{)2x^3 + 3x^2 - 5x + 7} \\ \underline{2x^3 + 8x^2 } \\ -5x^2 - 5x + 7 \\ \underline{-5x^2 - 20x } \\ 15x + 7 \end{array}$$

We again repeat with the result of the subtraction:

$$
\begin{array}{r}
2x^2 - 5x + 15 \\
x + 4 \overline{\smash{)} 2x^3 + 3x^2 - 5x + 7 } \\
\underline{2x^3 + 8x^2 } \\
-5x^2 - 5x + 7 \\
\underline{-5x^2 - 20x } \\
15x + 7 \\
\underline{15x + 60 } \\
-53
\end{array}
$$
←— remainder polynomial

The result of the subtraction, -53, has a degree less than the degree of the divisor, $x + 4$. Multiplying $x + 4$ by any nonzero polynomial produces a polynomial of degree at least 1, the degree of $x + 4$. There is nothing we can multiply $x + 4$ by to get -53. Thus, our division process now stops, giving us the quotient $2x^2 - 5x + 15$ and the remainder -53.

Do the division above again, but with the polynomials in n-tuple representation (omit the parentheses and commas).

1432

$$
\begin{array}{r}
2 -5 15 \\
1 4 \overline{\smash{)} 2 3 -5 7} \\
\underline{2 8 } \\
-5 -5 7 \\
\underline{-5 -20 } \\
15 7 \\
\underline{15 60} \\
-53
\end{array}
$$

Divide $x + 1$ by $x^2 + 1$.

1433

If we start the quotient with any nonzero term, the product of it and the divisor will produce a term with degree 2 or greater. But then we cannot get x, the first term in the dividend, when we multiply by the divisor $x^2 + 1$. Thus, the quotient must be 0:

$$
\begin{array}{r}
0 \\
x^2 + 1 \overline{\smash{)} x + 1} \\
\underline{0 } \\
x + 1
\end{array}
$$

The subtraction of 0 from the dividend gives $x + 1$, whose degree is less than the degree of the divisor. Thus, the division process terminates. The quotient is 0; the remainder is $x + 1$.

Rule: The long division process stops when the polynomial produced by the subtraction step is either the zero polynomial or has a degree less than the degree of the divisor. If the original dividend has degree less than the degree of the divisor, the quotient is 0 and the remainder is the given dividend.

Rule: The long division process produces a remainder that is either 0 or has a degree less than the degree of the divisor.

Divide $x^2 + 3x + 5$ by $x^4 + x^2$.

1434

The degree of the divisor is greater than the degree of the dividend. Thus, the quotient is 0, and the remainder is the given dividend, $x^2 + 3x + 5$.

Divide $x^2 - 1$ by $x - 1$.

1435

```
         x + 1
       _____
x - 1 ) x²     - 1
        x² - x
        _____
             x - 1
             x - 1
             _____
                 0
```

Division by the zero polynomial is not defined. Thus, we cannot do any divisions in $R[x]$, where R is the trivial ring because its only polynomial is the zero polynomial.

In the examples of division in the preceding frames, the divisor is a monic polynomial (recall that a monic polynomial is a polynomial whose leading term has a coefficient of 1). Let's now consider the most general case:

- Our polynomials are over some ring R.
- The leading term of the divisor is ax^m.
- The leading term of the dividend is bx^n and $m \leq n$.

We need a term in the quotient which when multiplied by the divisor term ax^m produces the term bx^n.

```
                    ?  ←——————— What term do we need here?
              _____
ax^m + ... )  bx^n    + ...
              bx^n    + ...
```

1436

$a^{-1}bx^{n-m}$ because $(ax^m)(a^{-1}bx^{n-m}) = bx^n$

```
                 a^{-1}bx^{n-m}
              _____
ax^m + ... )  bx^n    + ...
              bx^n    + ...
              _____
               0      + ...
```

We use the inverse of a (the coefficient of the leading term in the divisor) in the quotient term to eliminate the a coefficient in the divisor. The b factor in the quotient term then produces the required b coefficient in the product. If $b = 1$, then the coefficient of the quotient term is a^{-1}. Thus, division on the polynomials over R *requires* all nonzero elements to be units (i.e., to have a multiplicative inverse), otherwise some divisions involving polynomials over R are not possible.

What coefficient is required here?
↓

```
               ?x^{n-m}  + ...
              _____
ax^m + ... )  ax^n    + ...
```

1437

1

Thus, the division of polynomials over a ring R requires that R be a ring with unity. Let's summarize our observations: To divide polynomials over R, R has to have the following properties:

- R is a nontrivial ring with unity.
- Every nonzero element is a unit (i.e., has a multiplicative inverse).

A ring with these properties is called a ***division ring*** or a ***skew field***. In a division ring R, every nonzero element is a unit (i.e., have a multiplicative inverse). Recall from frame **1341** that a field is a nontrivial commutative ring in which every nonzero element has a multiplicative inverse. Thus, the only difference between a division ring and a field is that a division ring *does not have to be commutative*.

Is $\mathbb{Z}$ a division ring?

1438

No. Only 1 and −1 are units.

The rationals, the reals, and the complex numbers are all commutative division rings. Thus, they are also fields.

$\mathbb{Z}$ is *not* a division ring, but we can divide polynomials over $\mathbb{Z}$ (we divided such polynomials in our high school math classes). Why then do we need polynomials over a division ring to divide?

1439

Division is restricted for polynomials over $\mathbb{Z}$. Some divisions are possible; some are not. For example, we can divide $3x$, by $x − 2$, but we cannot divide $3x$ by $2x$. That is, there is no polynomial over $\mathbb{Z}$ which when multiplied by $2x$ gives $3x$ ($\frac{3}{2}$ works, but it is not a polynomial over $\mathbb{Z}$).

Have we seen any division rings that are not fields? Recall from frame **1128** that a quaternion is a generalization of a complex number. A quaternion has the form $a + bi + cj + dk$, where $a, b, c, d \in \mathbb{R}$. The set of all quaternions under addition and multiplication is a noncommutative division ring—that is, it is a division ring that is not a field.

Divide $3x^2 + x + 1$ by $2x + 1$, where both polynomials are over $\mathbb{Q}$.

1440

$$
\begin{array}{r}
\frac{3}{2}x - \frac{1}{4} \\
2x + 1 \overline{\smash{)}\, 3x^2 + x + 1} \\
3x^2 + \frac{3}{2}x \\
\hline
-x + 1 \\
-\frac{1}{2}x - \frac{1}{4} \\
\hline
\frac{5}{4}
\end{array}
$$

Check this answer by multiplying $\frac{3}{2}x - \frac{1}{4}$ by $2x + 1$, and adding $\frac{5}{4}$. You should get the dividend, $3x^2 + x + 1$.

1441

$$
\begin{array}{r}
\frac{3}{2}x - \frac{1}{4} \\
2x + 1 \\
\hline
\frac{3}{2}x - \frac{1}{4} \\
3x^2 - \frac{1}{2}x \\
\hline
3x^2 + x - \frac{1}{4}
\end{array}
$$

Adding the remainder $\frac{5}{4}$, we get $3x^2 + x - \frac{1}{4} + \frac{5}{4} = 3x^2 + x + 1$.

Divide $3x^2 + 2x + 2$ by $2x + 1$, where both polynomials are over $\mathbb{Z}_7$. *Hint*: The first term of the divisor ($2x$) times the first term of the quotient should equal the first term of the dividend ($3x^2$). Thus, the first term of the quotient should be the multiplicative inverse of 2 (to eliminate the 2 in $2x$) times 3 (to get the 3 coefficient in $3x^2$) times x (to get the x^2 in $3x^2$). The multiplicative inverse of 2 is 4.

1442

$$5 = 4 \cdot_5 3 \qquad \text{(4 is the multiplicative inverse of 2, and 3 is the leading coefficient in the dividend)}$$

$$
\begin{array}{r}
5x + 2 \\
2x + 1 \overline{\smash{)}\, 3x^2 + 2x + 2} \\
3x^2 + 5x \longleftarrow \text{subtract } 5x, \text{ or, equivalently, add its additive inverse } (2x) \text{ to get } 4x \\
\hline
4x + 2 \\
4x + 2 \\
\hline
0
\end{array}
$$

Why can we not divide $3x^2 + x + 2$ by $2x + 1$, where both polynomials are over $\mathbb{Z}_6$?

1443

$\mathbb{Z}_6$ is not a division ring. We cannot eliminate the 2 in the divisor to get the required 3 in the first term of the dividend because 2 does not have a multiplicative inverse in $\mathbb{Z}_6$.

Polynomials have a ***division algorithm*** that is analogous to the division algorithm for the integers: Let R be a division ring. Suppose $a(x)$ and $d(x) \in R[x]$ with $d(x) \neq 0$. Then there exist unique polynomials $q(x)$ and $r(x)$ in $R[x]$ (called the ***quotient*** and ***remainder***, respectively) such that $a(x) = d(x)q(x) + r(x)$, and either $r(x)$ is the zero polynomial, or deg $r(x)$ < deg $d(x)$. We call $a(x)$ the ***dividend*** and $d(x)$ the ***divisor***.

Roots of a Polynomial

1444

Suppose F is a field. If $f(x) \in F[x]$, and $a \in F$, then $f(a)$ is the element of F obtained by replacing every occurrence of x in $f(x)$ with a. a is a ***root*** (or a ***zero***) of $f(x)$ if and only if $f(a) = 0$. We say that $f(x) \neq 0$ is a ***divisor*** of $g(x)$ or $f(x)$ ***divides*** $g(x)$ (denoted by $f(x) \mid g(x)$) if and only if there exists a polynomial $h(x)$ such that that $f(x)h(x) = g(x)$. Equivalently, we say that $f(x)$ is a ***factor*** of $g(x)$. You probably have seen a version of the following theorem in high school. It is called the ***remainder theorem***:

Suppose F is a field, $a \in F$, and $f(x) \in F[x]$. Then $f(a)$ is the remainder that results when $f(x)$ is divided by $x - a$.

Prove this theorem. *Hint*: Using the division algorithm, divide $f(x)$ by $x - a$. Then determine $f(a)$.

1445

By the division algorithm, $f(x) = (x - a)q(x) + r$, where r is a constant polynomial. Replacing x with a, we get $f(a) = (a - a)q(a) + r = 0q(a) + r = r$. ∎

Suppose $f(x) = x^2 + 2x + 1$ and $x - 1$ are polynomials over $\mathbb{Z}$. What is the remainder when $f(x)$ is divided by $x - 1$?

1446

$f(1) = 4$

The remainder theorem as stated above applies to polynomials over a field. But in the example above, we are applying it to the polynomials over $\mathbb{Z}$, which is not a field. As we saw earlier, we cannot always divide polynomials over $\mathbb{Z}$ (see frame **1439**). However, we can always divide by a polynomial of the form $x - a$, which is the type of division required by the remainder theorem.

Suppose $f(x) = x^2 + 2x + 1$ and $x + 2$ are polynomials over $\mathbb{Z}$. What is the remainder when $f(x)$ is divided by $x + 2$? *Hint*: To get the remainder on division by $x - a$, we replace x in the polynomial with what follows the *minus* sign in $x - a$. That is, we replace x with a. Note that $x + 2 = x - (-2)$. Thus, to determine the remainder on division by $x + 2 = x - (-2)$, we replace x in the polynomial with -2.

1447

$f(-2) = 1$

Another theorem from high school is the ***factor theorem***:

Suppose F is a field, $a \in F$, and $f(x) \in F[x]$. The element a is a root of $f(x) \Leftrightarrow x - a$ is a factor of $f(x)$.

Prove the "⇒" part of this biconditional: the element a is a root of $f(x) \Rightarrow x - a$ is a factor of $f(x)$. *Hint*: Use the remainder theorem.

1448

By the remainder theorem, $f(x) = (x - a)q(x) + f(a)$. If a is a root, then $f(a) = 0 \Rightarrow f(x) = (x - a)q(x) \Rightarrow x - a$ is a factor.

Now show the converse: $x - a$ is a factor of $f(x) \Rightarrow a$ is a root of $f(x)$.

1449

If $x - a$ is a factor of $f(x)$, then $f(x) = (x - a)q(x)$ for some polynomial $q(x)$. Replacing x with a, we get $f(a) = (a - a)q(a) = 0q(a) = 0$. Thus, a is a root. ∎

Is $x - 2$ a factor of $f(x) = 3x^3 - 2x - x + 18$ in $\mathbb{Z}[x]$? Is $x + 2$ a factor?

1450

$f(2) = 36 \neq 0 \Rightarrow x - 2$ is not a factor.
$f(-2) = 0 \Rightarrow x - (-2) = x + 2$ is a factor.

Suppose $x - 1$ is a factor of $f(x) = c_n x^n + c_{n-1} x^{n-1} + \cdots + c_2 x^2 + c_1 x + c_0$. What can you conclude about $c_n + c_{n-1} + \cdots + c_0$? *Hint*: What is $f(1)$?

1451

$f(1) = c_n + c_{n-1} + \cdots + c_0$. Because $x - 1$ is a factor, $f(1) = 0$. Thus, $c_n + c_{n-1} + \cdots + c_0 = 0$.

Using the factor theorem, we can prove the following theorem on the factorization of a polynomial with n distinct roots: Suppose $f(x)$ is a polynomial over a field F, deg $f(x) = n > 0$, and the leading term of $f(x)$ has the coefficient c_n. If $f(x)$ has exactly n distinct roots, $r_1, r_2, \ldots, r_n$, then $f(x) = c_n(x - r_1)(x - r_2)\ldots(x - r_n)$. To prove this, we apply the factor theorem n times. Dividing $f(x)$ by $x - r_1$, we get $f(x) = (x - r_1)q_1(x)$. r_2 must be a root of $q_1(x)$ because it is not a root of $x - r_1$. Thus, we can apply the factor theorem to $q_1(x)$. We get $q_1(x) = (x - r_2)q_2(x)$. Substituting in the equation for $f(x)$, we get $f(x) = (x - r_1)(x - r_2)q_2(x)$. Repeating in this fashion, once for each root, we get $f(x) = (x - r_1)(x - r_2)\ldots(x - r_n)q_n(x)$. By the division algorithm, the degrees in the sequence of polynomials $f(x), q_1(x), q_2(x), \ldots, q_n(x)$ must be strictly decreasing (by 1 for each quotient). Thus, the degree of $q_n(x)$ must be 0. In other words, $q_n(x)$ is some nonzero constant c. If we multiply out all the factors in $(x - r_1)(x - r_2)\ldots(x - r_n)c$, the leading term in the result is cx^n. We are given that the coefficient of the leading term of $f(x)$ is c_n. Thus, $c = c_n$, and $f(x) = c_n(x - r_1)(x - r_2)\ldots(x - r_n)$. ∎

$f(x) = 3x^2 + 3x - 6$ is a polynomial over $\mathbb{R}$. Confirm that 1 and -2 are its roots.

1452

$f(1) = (3)(1)^2 + 3(1) - 6 = 0$
$f(-2) = (3)(-2)^2 + 3(-2) - 6 = 12 - 6 - 6 = 0$

Using the theorem in the preceding frame, express $f(x)$ in factored form.

1453

$f(x) = 3(x - 1)(x - (-2)) = 3(x - 1)(x + 2)$

Using the factor theorem, we can also prove the following theorem on the maximum number of roots of $f(x)$:

A polynomial of positive degree n over a field F has at most n distinct roots in F.

Suppose $f(x)$ is a polynomial of positive degree n over a field F, and $f(x)$ has at least n roots, $r_1, r_2, \ldots, r_n$. We will show $f(x)$ can have no more than n roots. Repeatedly factoring as we did in the proof of the theorem in frame **1451**, we get $f(x) = (x - r_1)(x - r_2)\ldots(x - r_n)c$, where $c \in F$. Can c be equal to 0?

1454

No. If $c = 0$ then $f(x)$ is the zero polynomial. But $f(x)$ cannot be the zero polynomial because $f(x)$ is a polynomial of positive degree.

Suppose $b \in F$, and $b \neq r_i$ for $i = 1, 2, \ldots, n$. Substitute b for x in $f(x) = (x - r_1)(x - r_2)\ldots(x - r_n)c$?

1455

$f(b) = (b - r_1)(b - r_2)\ldots(b - r_n)c$

Are all the factors of $f(b)$ in the preceding frame nonzero?

1456

Yes. b does not equal any of the n roots. Thus, each factor $b - r_i$ is nonzero. c is also nonzero. We can then conclude that $f(b)$ is nonzero. On what basis can we make this conclusion? *Hint*: F is a field.

1457

F is a field. Thus, it is an integral domain. An integral domain has no divisors of zero.

The only elements of F that make $f(x)$ equal 0 are the given n roots. In other words, $f(x)$ can have no more than n roots. ∎

In our proof of the "at most n roots" theorem, we use the fact that a field has no divisors of zero. This observation naturally leads to the following question: Can a polynomial of degree n over a ring that *has* divisors of zero have more than n roots? Let's look at an example.

Determine the number of roots $x^2 + 3x + 2$ has over $\mathbb{Z}_6$.

1458

x	$x^2 + 3x + 2$
0	2
1	0
2	0
3	2
4	0
5	0

$x^2 + 3x + 2$ has degree 2 but has four roots.

Review Questions

1. Convert (1, 0, 0, 3, 0, 2) to powers-of-x form.
2. Convert $2x^2 + x^4 + 3$ to n-tuple representation.
3. Add $a(x) = 2x^2 + x^4 + 3$ and $b(x) = x^3 + 2x^2 + 1$. Give your answer in n-tuple representation. Assume the polynomials are over $\mathbb{Z}_4$.
4. Multiply the polynomials in review question 3. Give your answer in powers-of-x form.
5. Divide $x^3 + x^2 + x + 1$ by $x^2 + 1$. Assume the polynomials are over $\mathbb{Z}$.
6. Divide the polynomials in review question 5, but assume the polynomials are over $\mathbb{Z}_2$.
7. Divide $x^3 + 1$ by $3x^2 + 2$. Assume the polynomials are over $\mathbb{Q}$.
8. What is the remainder when $f(x) = x^3 + 1$ is divided by $x - 5$. Assume the polynomials are over $\mathbb{Z}$.
9. Confirm your answer for review question 8 by performing the division.
10. Show that $I[x]$ has the superclosure property in $R[x]$ if I is an ideal of R.

Answers to the Review Questions

1. $x^5 + 3x^2 + 2$
2. (1, 0, 2, 0, 3)
3. (1, 1, 0, 0, 0)
4. $x^7 + 2x^6 + 2x^5 + x^4 + 3x^3 + 3$
5. Quotient $= x + 1$, remainder $= 0$
6. Same answer as in review question 5.
7. Quotient $= \frac{1}{3}x$, remainder $= -\frac{2}{3}x + 1$.
8. By the remainder theorem, remainder $= f(5) = 126$.

9.
```
            x² +  5x  +   25
       _____
x−5 ) x³              +    1
      x³ −5x²
      _____
          5x²         +    1
          5x² − 25x
          _____
                 25x  +    1
                 25x  −  125
                 _____
                          126
```

10. Let $i(x) \in I[x]$ and $r(x) \in R[x]$. We have to show that $i(x)r(x)$ and $r(x)i(x)$ are in $I[x]$. Every coefficient in the product polynomials $i(x)r(x)$ and $r(x)i(x)$ consists of one or more terms each of which has a factor from I. Because I has the superclosure property, each of these terms, and, therefore, the coefficient itself is in I. Thus, $i(x)r(x), r(x)i(x) \in I[x]$.

Homework Questions

1. Suppose $p(x), q(x) \in \mathbb{Q}[x]$ with $p(x) = \frac{1}{2}x^3 + 1$ and $q(x) = \frac{2}{3}x + 2$. Divide $p(x)$ by $q(x)$.
2. Suppose F is a field. What are the units in $F[x]$? Is $F[x]$ a field? Justify your answer.
3. Determine all the roots of $x^3 - 6x^2 + 11x - 6 \in \mathbb{Z}[x]$.
4. Show that $I[x]$ is an ideal in $R[x]$ if I is an ideal in R. See review question 10.
5. Divide $3x^3 + 5x^2 + 7x + 10$ by $4x^2 + 2$. Assume the polynomials are over $\mathbb{Z}_{11}$.
6. How many polynomials with degree 2 are in $\mathbb{Z}_n[x]$?
7. Suppose φ maps each polynomial in $R[x]$ to its constant term. Show that φ is a ring homomorphism from $R[x]$ to R.
8. If a ring R is not commutative, we have to be careful about order when dividing polynomials over R. Why is the start of the quotient in frame **1436** $a^{-1}bx^{n-m}$ rather than $ba^{-1}x^{n-m}$? *Hint*: Consider the division algorithm.
9. Suppose $a(x), b(x), d(x) \in F[x]$ where F is a field, and $d(x)$ is a common divisor of $a(x)$ and $b(x)$. If $c \in F$, is $c(d(x))$ also a common divisor? Justify your answer.
10. Give a definition of a greatest common divisor for polynomials over a field analogous to the definition of the gcd for integers. For any pair of integers, the gcd is unique. Similarly, for any pair of polynomials, the gcd should be unique.

26 Polynomials Part 2

Principal Ideals of F[x]

1459

In this chapter, we will work strictly with polynomials over a field. The principal reason for narrowing our focus to polynomials over fields is that we can use the division algorithm on polynomials over a field. Recall from frame **1437** that the division of polynomials over a ring R requires that the nonzero elements of R have inverses, which is the case if R is a field.

Let's start by reviewing some of the results of our investigation of $F[x]$, the polynomials over a field F:

- **P1:** $F[x]$ is a commutative ring with unity (frames **1414** and **1415**).
- **P2:** $F[x]$ has no divisors of zero (frame **1415**).
- **P3:** $F[x]$ is an integral domain (frame **1416**).
- **P4:** $F[x]$ is *not* a field (frame **1417**).
- **P5:** $\deg p(x)q(x) = \deg p(x) + \deg q(x)$ (frame **1409**).
- **P6:** If $\deg p(x) \neq \deg q(x)$ then $\deg [p(x) + q(x)] = \max[\deg p(x), \deg q(x)]$ (frame **1411**).
- **P7:** If $\deg p(x) = \deg q(x)$ then $\deg [p(x) + q(x)] \leq \deg p(x)$ (frame **1411**).

What is the degree of $p(x) + q(x)$ where $p(x) = x^2 + 2$ and $q(x) = 2x^2 + 1$ are polynomials over $\mathbb{Z}_3$?

1460

$p(x) + q(x) = 0 \Rightarrow$ a degree is not defined.

Recall from frame **1296** that $\langle a \rangle$, the principal ideal generated by an element a in a commutative ring with unity is the smallest ideal that contains a. More precisely, $\langle a \rangle$ is the ideal that contains a and is a subset of every ideal that contains a. By definition, $\langle a \rangle = aR = Ra$, where R is a commutative ring with unity. $F[x]$ is a commutative ring with unity. Thus, if $p(x) \in F[x]$ then $\langle p(x) \rangle = p(x)F[x] = F[x]p(x)$. That is, $\langle p(x) \rangle$, the principal ideal generated by $p(x)$, is the set of all polynomials that are the product of $p(x)$ and some polynomial in $F[x]$.

We are about to show that the principal ideals of $F[x]$, where F is a field, have properties I1 through I6 that follow. However, before we do, try to justify each one yourself, if only informally. Then read on.

- **I1:** $0 \in \langle p(x) \rangle$
- **I2:** $p(x) \in \langle p(x) \rangle$
- **I3:** For $p(x) \neq 0$, all the nonzero polynomials in $\langle p(x) \rangle$ have degree greater than or equal to $\deg p(x)$.
- **I4:** $\langle g(x)h(x) \rangle \subseteq \langle g(x) \rangle$
- **I5:** For $g(x) \neq 0$ and $h(x) \neq 0$, $\langle g(x) \rangle = \langle h(x) \rangle$ if and only if $g(x) = h(x)c$ for some nonzero constant $c \in F$.
- **I6:** $p(x)$ is a nonzero constant if and only if $\langle p(x) \rangle = F[x]$.

Show I1: $0 \in \langle p(x) \rangle$.

1461

The additive identity is in every subring. $\langle p(x) \rangle$ is an ideal, which is a special type of subring. Thus, the additive identity is in $\langle p(x) \rangle$.

Show I2: $p(x) \in \langle p(x) \rangle$.

1462

$1 \in F[x] \Rightarrow p(x) = p(x)1 \in p(x)F(x) = \langle p(x) \rangle$.

Show I3: For nonzero $p(x)$, all the nonzero polynomials in $\langle p(x) \rangle$ have degree greater than or equal to deg $p(x)$.

1463

Every nonzero element of $\langle p(x) \rangle$ is equal to $p(x)f(x)$ for some nonzero $f(x) \in F[x]$. deg $p(x)f(x) = [\deg f(x) + \deg p(x)] \geq \deg p(x)$.

Does $\langle p(x) \rangle$ have a polynomial whose degree is deg $p(x)$.

1464

Yes, for example, $p(x)$

Show I4: $\langle g(x)h(x) \rangle \subseteq \langle g(x) \rangle$.

1465

Let $p(x) \in \langle g(x)h(x) \rangle \Rightarrow$ for some $f(x) \in F[x]$, $p(x) = [g(x)h(x)]f(x) = g(x)[h(x)f(x)] \Rightarrow p(x) \in \langle g(x) \rangle$.

Suppose $g(x)$ and $h(x)$ are nonzero polynomials. If $\langle g(x) \rangle = \langle h(x) \rangle$, what can we conclude about the degrees of $g(x)$ and $h(x)$? *Hint*: Because $g(x) \in \langle g(x) \rangle$ and $\langle g(x) \rangle = \langle h(x) \rangle$, $g(x) \in \langle h(x) \rangle$.

1466

Because $g(x)$ and $h(x)$ are nonzero and $g(x) \in \langle h(x) \rangle$, by I3 deg $g(x) \geq$ deg $h(x)$. Similarly, because $g(x)$ and $h(x)$ are nonzero and $h(x) \in \langle g(x) \rangle$, deg $h(x) \geq$ deg $g(x)$. Thus, deg $g(x) =$ deg $h(x)$.

However, we can say even more about $g(x)$ and $h(x)$, as indicated by I5: For any nonzero $g(x)$ and nonzero $h(x)$, $\langle g(x) \rangle = \langle h(x) \rangle \Leftrightarrow g(x) = h(x)c$ for some nonzero constant $c \in F$. Let's show this.

We know from the preceding frame that $g(x) \in \langle h(x) \rangle$. Thus, $g(x) = h(x)f(x)$ for some $f(x) \in F[x]$. We also know that deg $g(x) =$ deg $h(x)$. What can we then conclude about $f(x)$?

1467

deg $f(x) = 0$ (otherwise, deg $g(x)$ would be greater than deg $h(x)$) $\Rightarrow f(x)$ is some nonzero constant $c \in F \Rightarrow g(x) = h(x)c$.

We showed in the preceding frame the "$\Rightarrow$" part of I5. The converse is also true: If two polynomials differ by only a multiplicative factor that is a nonzero constant, then the ideals they generate are equal. We can restate the converse more succinctly: $\langle g(x) \rangle = \langle g(x)c \rangle$ where c is any nonzero constant. To prove this equality, we have to show that $\langle g(x) \rangle$ and $\langle g(x)c \rangle$ are subsets of each other. We have already shown that $\langle g(x)c \rangle \subseteq \langle g(x) \rangle$ (this is a special case of I4). Thus, we need to show only that $\langle g(x) \rangle \subseteq \langle g(x)c \rangle$ where c is a nonzero constant. Let's do this.

Let $p(x) \in \langle g(x) \rangle \Rightarrow p(x) = g(x)f(x)$ for some $f(x) \in F[x]$. We need to show that $g(x)f(x) \in \langle g(x)c \rangle$. To do this, we need to show that there is some $h(x) \in F[x]$ such that $g(x)f(x) = (g(x)c)h(x)$. What is the required $h(x)$?

1468

The required $h(x)$ is $c^{-1}f(x)$.

Suppose $g(x) = x^2 + x + 1$ and $h(x) = x^2 + 1$ are polynomials over $\mathbb{R}$. Is $\langle g(x) \rangle = \langle h(x) \rangle$?

1469

No. There is no constant polynomial c such that $g(x) = h(x)c$.

Ideal Generated by a Nonzero Constant Polynomial

1470

Show that $x^2 + 1 \in \langle 3 \rangle$ where $x^2 + 1$ and 3 are polynomials in $\mathbb{Z}_5[x]$. To do this, find a polynomial $f(x)$ such that $3f(x) = x^2 + 1$. *Hint*: $\mathbb{Z}_5$ is a field. Thus, 3 has a multiplicative inverse.

1471

The $f(x)$ we need is $3^{-1}(x^2 + 1)$. Then $3f(x) = (3)(3^{-1})(x^2 + 1) = 1(x^2 + 1) = x^2 + 1 \Rightarrow x^2 + 1 \in \langle 3 \rangle$.

What is the multiplicative inverse of 3 in $\mathbb{Z}_5$?

1472

2 because $2 \cdot_5 3 = 1$.

Determine the polynomial equal to $3^{-1}(x^2 + 1)$.

1473

$3^{-1}(x^2 + 1) = 2(x^2 + 1) = 2x^2 + 2$.

Let's check that $2x^2 + 2$ is the $f(x)$ we need: $3f(x) = 3(2x^2 + 2) = (3 \cdot_5 2)x^2 + (2 \cdot_5 3) = x^2 + 1$.

We can similarly show that for *any* polynomial $p(x) \in F[x]$ and *any* nonzero constant polynomial $c \in F[x]$, $p(x) \in \langle c \rangle$. Simply let $f(x) = c^{-1}p(x)$. Then $p(x) = cf(x) \Rightarrow p(x) \in \langle c \rangle$. F is a field so every nonzero constant c has an inverse c^{-1}.

What can we conclude about $\langle c \rangle$ and $F[x]$?

1474

Every polynomial is in $\langle c \rangle$. Thus, $\langle c \rangle = F[x]$.

This result is no surprise. In frame **1352**, we learned that if an ideal I in a ring R contains a unit (i.e., an element with a multiplicative inverse), then $I = R$. Every nonzero constant polynomial c in $F[x]$ has a multiplicative inverse (the polynomial c^{-1}), and, therefore, is a unit. Thus, every ideal of $F[x]$ that contains a nonzero constant polynomial is equal to $F[x]$.

Now let's consider the converse. Suppose $\langle p(x) \rangle = F[x]$. What can we conclude about $p(x)$? *Hint*: What must be true of $p(x)$ for $\langle p(x) \rangle$ to contain all the constant polynomials in $F[x]$?

1475

By I3 in frame **1460**, all the nonzero polynomials in $\langle p(x) \rangle$ have degrees greater than or equal to deg $p(x)$. Thus, for $\langle p(x) \rangle$ to contain any of the nonzero constant polynomials in $F[x]$, $p(x)$ itself must be a nonzero constant polynomial.

In the preceding several frames, we have shown I6 in frame **1460**: $p(x)$ is a nonzero constant if and only if $\langle p(x) \rangle = F[x]$.

F[x] is a Principal Ideal Domain

1476

Suppose F is a field, and I is a nontrivial ideal of $F[x]$. Let $g(x)$ be any one of the polynomials in in I of the lowest degree among the polynomials in I. How do I and $\langle g(x) \rangle$ compare? Is one a proper subset of the other? Are they equal?

By frame **1296**, $\langle g(x) \rangle$ is the smallest ideal that contains $g(x)$. That is, $\langle g(x) \rangle$ is a subset of any ideal that contains $g(x)$. Thus, if I is an ideal that contains $g(x)$, then $\langle g(x) \rangle \subseteq I$. No surprise here. But what is surprising is that $I \subseteq \langle g(x) \rangle$.

Let's show that $I \subseteq \langle g(x) \rangle$, where $g(x)$ is any one of the polynomials of lowest degree in I. Let $p(x) \in I$. We need to show that $p(x) \in \langle g(x) \rangle$. To do this, we use the division algorithm. Using the division algorithm, divide $p(x)$ by $g(x)$.

1477

$p(x) = g(x)q(x) + r(x)$ for some $q(x), r(x) \in F[x]$.

By the superclosure property of *I*, what can we conclude about $g(x)q(x)$? *Hint*: We are given that $g(x) \in I$.

1478

$g(x)q(x) \in I$ which implies that $-(g(x)q(x)) \in I$.

What then can we conclude about $r(x) = p(x) + [-(g(x)q(x))]$? *Hint*: Both $p(x)$ and $-(g(x)q(x))$ are in *I*.

1479

$r(x) \in I$ because $p(x), -(g(x)q(x)) \in I$.

By the division algorithm, either deg $r(x) <$ deg $g(x)$, or $r(x)$ is the zero polynomial.

Why must $r(x)$ be the zero polynomial? *Hint*: What is special about $g(x)$?

1480

$g(x)$ is a polynomial of lowest degree in $I \Rightarrow$ deg $r(x)$ is not less than deg $g(x)$ (otherwise $r(x)$ would be a polynomial in *I* with a degree lower than $g(x)$) $\Rightarrow r(x) = 0 \Rightarrow p(x) = g(x)q(x)$. Thus, $p(x) \in <g(x)> \Rightarrow I \subseteq <g(x)>$.

Key idea: Every ideal of $F[x]$ that has at least one nonzero polynomial is generated by *any* polynomial of lowest degree in that ideal.

Is $\{0\}$ generated by a single polynomial?

1481

Yes. $\{0\} = <0>$. That is, the ideal $\{0\}$ is generated by the zero polynomial.

Thus, *every* ideal of $F[x]$—every ideal with at least one nonzero polynomial, and the ideal that contains only 0—is generated by a single polynomial. In other words, every ideal of $F[x]$ is a principal ideal.

A ***principal ideal domain*** (abbreviated PID) is an integral domain in which every ideal is a principal ideal. $F[x]$ is an integral domain in which every ideal is principal. Thus, $F[x]$ is a PID.

Cosets of <*p*(*x*)>

1482

Suppose $p(x) \in F[x]$, where *F* is a field. $<p(x)>$ is an ideal of $F[x]$. Thus, if $f(x) \in F[x]$, then $f(x) + <p(x)>$ is one of the cosets of $<p(x)>$.

How do we denote the set of all the cosets of $<p(x)>$ in $F[x]$?

1483

$F[x]/<p(x)>$

Using the division algorithm, let's develop a simple characterization of the cosets in $F[x]/<p(x)>$.

Divide $f(x)$ by $p(x)$. Represent the quotient and remainder with $q(x)$ and $r(x)$, respectively.

1484

$f(x) = p(x)q(x) + r(x)$ where $r(x) = 0$ or deg $r(x) <$ deg $p(x)$.

Now replace $f(x)$ in $f(x) + <p(x)>$ using the preceding expression for $f(x)$.

1485

$f(x) + <p(x)> = [p(x)q(x) + r(x)] + <p(x)>$.

Simplify the right side of the equation above. *Hint*: $p(x)q(x) \in \langle p(x) \rangle$ so you can use the coset absorption rule (K4 in frame **968**).

1486

$\langle p(x) \rangle$ absorbs $q(x)p(x)$. Thus, $f(x) + \langle p(x) \rangle = [p(x)q(x) + r(x)] + \langle p(x) \rangle = r(x) + \langle p(x) \rangle$, where $r(x) = 0$ or deg $r(x)$ < deg $p(x)$.

Let's denote the remainder polynomial that results when $f(x)$ is divided by $p(x)$ with $f(x)$ mod $p(x)$. Using this notation, we get

$$f(x) + \langle p(x) \rangle = [f(x) \bmod p(x)] + \langle p(x) \rangle.$$

In $\mathbb{Z}_3[x]$, give the coset equal to $x^3 + 2x + 1 + \langle x^2 + 1 \rangle$ whose representative is $(x^3 + 2x + 1) \bmod (x^2 + 1)$.

1487

$$\begin{array}{r} x \\ x^2 + 1 \overline{)\, x^3 + 2x + 1\,} \\ \underline{x^3 + x} \\ x + 1 \quad \leftarrow \text{remainder} \end{array}$$

Thus, we can represent the coset $x^3 + 2x + 1 + \langle x^2 + 1 \rangle$ with $x + 1$. That is, $x^3 + 2x + 1 + \langle x^2 + 1 \rangle = x + 1 + \langle x^2 + 1 \rangle$.

In $\mathbb{Z}_3[x]$, give a coset equal to $x^2 + 1 + \langle x^2 + 1 \rangle$ whose representative is the remainder that is produced when $x^2 + 1$ is divided by $x^2 + 1$.

1488

$$\begin{array}{r} 1 \\ x^2 + 1 \overline{)\, x^2 + 1\,} \\ \underline{x^2 + 1} \\ 0 \quad \leftarrow \text{remainder} \end{array}$$

Thus, we can represent the coset $x^2 + 1 + \langle x^2 + 1 \rangle$ with 0. That is, $x^2 + 1 + \langle x^2 + 1 \rangle = 0 + \langle x^2 + 1 \rangle = \langle x^2 + 1 \rangle$. We can also arrive at the same result by observing that $x^2 + 1 \in \langle x^2 + 1 \rangle$. Thus, $\langle x^2 + 1 \rangle$ absorbs $x^2 + 1$.

Whenever we divide a nonzero polynomial by itself, we get a quotient of 1 and a remainder of 0. Thus, for all nonzero polynomials $p(x)$, $p(x) + \langle p(x) \rangle = [p(x) \bmod p(x)] + \langle p(x) \rangle = 0 + \langle p(x) \rangle = \langle p(x) \rangle$. We can also come to this conclusion by observing that $p(x) \in \langle p(x) \rangle$. Then by the absorption rule, $p(x) + \langle p(x) \rangle = \langle p(x) \rangle$.

An interesting question to ask at this point is this: Suppose $r(x) + \langle p(x) \rangle = r'(x) + \langle p(x) \rangle$, where $r(x)$ and $r'(x)$ are zero or have degrees less than deg $p(x)$. Are $r(x)$ and $r'(x)$ necessarily equal? In other words, does each coset have a *unique* representative among the polynomials in the set of polynomials consisting of the zero polynomial and the polynomials with degrees less than the degree of $p(x)$?

Before we answer this question, we need to make an observation about the cosets of an ideal. Suppose I is an ideal. If $i \in I$, then by the definition of cosets, $a + i \in a + I$. If $a + I = b + I$, then $a + i \in b + I$. Thus, there must be an $i' \in I$ such that $a + i = b + i'$. Moving i to the right side and b to the left side in the preceding equation, we get $a + (-b) = i' + (-i)$. $i' + (-i) \in I$ because I is closed under addition. Thus, $a + (-b) \in I$.

Prove the converse: If $a + (-b) \in I$ then $a + I = b + I$.

1489

$a + (-b) \in I \Rightarrow a + (-b) = i$ for some $i \in I \Rightarrow a = b + i$. Then $a + I = b + i + I$. By the absorption rule for cosets, $b + i + I = b + I$. Thus, $a + I = b + I$. ∎

Key idea: $a + I = b + I$ if and only if $a + (-b) \in I$.

Let's now get back to our question: Suppose $r(x) + \langle p(x) \rangle = r'(x) + \langle p(x) \rangle$, where $r(x)$ is 0 or has degree less than deg $p(x)$, and similarly for $r'(x)$. Is $r(x)$ necessarily equal to $r'(x)$?

By the preceding key idea, if $r(x) + \langle p(x) \rangle = r'(x) + \langle p(x) \rangle$, then $r(x) + (-r'(x)) \in \langle p(x) \rangle$. We know $r(x) + (-r'(x))$ cannot equal any nonzero polynomial in $\langle p(x) \rangle$. Why not? *Hint*: Consider the degree of $r(x) + (-r'(x))$.

1490

If $r(x) + (-r'(x))$ is nonzero, then by the division algorithm, its degree is less than the degree of $p(x)$. But by I3 in frame **1460**, all the nonzero polynomials in $<p(x)>$ have a degree greater than or equal to the degree of $p(x)$. Thus, $r(x) + (-r'(x))$ must equal 0, in which case $r(x) = r'(x)$. ∎

Thus, every coset of $<p(x)>$ has a *unique* representative among the polynomials in the set of polynomials consisting of the zero polynomial and the polynomials with degrees less than the degree of $p(x)$.

List all the cosets of $<x^2 + 1>$ in $\mathbb{Z}_2[x]$.

1491

All the polynomials in $\mathbb{Z}_2[x]$ with degree less than 2 are x, $x + 1$, and 1. Thus, in addition to the coset $0 + <x^2 + 1> = <x^2 + 1>$, we have $x + <x^2 + 1>$, $x + 1 + <x^2 + 1>$, and $1 + <x^2 + 1>$.

What is the number of cosets of $<x^4 + 1>$ in $\mathbb{Z}_2[x]$? *Hint*: The polynomials with degree less than 4 and the zero polynomial are of the form $ax^3 + bx^2 + cx + d$.

1492

For each coefficient, a, b, c, and d, there are two possible values: 0 or 1. Thus, there is a total of $2 \cdot 2 \cdot 2 \cdot 2 = 2^4 = 16$ polynomials. Each is a representative of a distinct coset.

What is the number of cosets of $<x^{100} + 1>$ in $\mathbb{Z}_2[x]$? *Hint*: A polynomial with degree less than 100 has 100 coefficients.

1493

2^{100}

What is the number of cosets of $<x^{100} + 1>$ in $\mathbb{Z}_3[x]$? *Hint*: There are 100 coefficients, each of which can have one of three values.

1494

3^{100}

Irreducible Polynomials

1495

An *irreducible polynomial* over a field is a non-constant polynomial that cannot be factored into two non-constant polynomials. An irreducible polynomial is analogous to a prime number in the integers. A *reducible polynomial* over a field is a non-constant polynomial that can be factored into two non-constant polynomials.

For example, $x^2 + 1$ over $\mathbb{R}$ is irreducible even though it can be factored. For example, $x^2 + 1 = (\frac{1}{2})(2x^2 + 2)$. But $\frac{1}{2}$ is a constant polynomial. Thus, this factorization does not qualify $x^2 + 1$ as a reducible polynomial. $x^2 - 1$, on the other hand, is reducible. It factors into $(x + 1)(x - 1)$. Both $x + 1$ and $x - 1$ are non-constant polynomials. Thus, $x^2 - 1$ is reducible.

If a polynomial $f(x)$ is reducible, what must be true about its degree? *Hint*: It must be a product of two non-constant polynomials.

1496

If $f(x)$ is reducible, it must factor into $g(x)h(x)$, where $g(x)$ and $h(x)$ are non-constant polynomials. Thus, $g(x)$ and $h(x)$ must have degrees greater than or equal to 1, in which case $f(x)$ must have degree greater than or equal to 2.

Is $x - 6$ reducible in $\mathbb{R}[x]$?

1497

No. By the preceding frame, a reducible polynomial must have a degree greater than or equal to 2. All polynomials of degree 1 are irreducible.

Characterize the following polynomials over $\mathbb{R}$ as irreducible, reducible, or neither: 3, 6, $x - 3$, $x^2 - 1$, $x^2 + 1$.

1498

3: Neither (the terms irreducible and reducible apply only to non-constant polynomials)
6: Neither
$x - 3$: Irreducible (all polynomials with degree 1 are irreducible)
$x^2 - 1$: Equals $(x + 1)(x - 1)$ so it is reducible
$x^2 + 1$: Irreducible

A polynomial that is irreducible over one field may be reducible over another field. For example, $x^2 + 1$ in $\mathbb{R}[x]$ is irreducible but not in $\mathbb{C}[x]$, where $\mathbb{C}$ is the field of complex numbers.

Show that $x^2 + 1$ in $\mathbb{C}[x]$ is reducible. *Hint*: $i^2 = -1$, where i is the imaginary number $\sqrt{-1}$.

1499

$(x + i)(x - i) = x^2 - xi + xi - i^2 = x^2 - i^2 = x^2 - (-1) = x^2 + 1$.

Suppose a polynomial $f(x)$ over a field F has a factor of degree 1. That is, $f(x) = (ax + b)g(x)$ for some $a \neq 0$, $b \in F$, and $g(x) \in F[x]$. Show that $f(x)$ has a root.

1500

If $x = -a^{-1}b$, then $ax + b = a(-a^{-1}b) + b = -b + b = 0 \Rightarrow -a^{-1}b$ is a root of $f(x)$.

The factor theorem can be helpful in determining if a polynomial is reducible. If a polynomial $f(x)$ has degree greater than 1 and has a root r, then by the factor theorem, it can be factored in $(x - r)q(x)$. Moreover, because $\deg f(x) = \deg [(x - r)q(x)] > 1$, and $\deg(x - r) = 1$, then $\deg q(x) \geq 1$. That is, $q(x)$ is a non-constant polynomial. Thus, $f(x)$ is reducible. Let's summarize:

Roots exists for a polynomial of degree > 1 $\Rightarrow$ the polynomial is reducible

Is $x^5 - 32$ reducible in $\mathbb{R}[x]$? *Hint*: Does it have any roots?

1501

Yes, because $x = 2$ is a root.

Roots for a polynomial of degree greater than 1 implies that the polynomial is reducible. What about the converse? Does reducibility imply roots exist?

Consider the polynomial $f(x) = x^4 + 2x^2 + 1$ in $\mathbb{R}[x]$. Does $f(x)$ have any roots?

1502

$x^4 + 2x^2 + 1 \geq 1$ because x^4 and x^2 are always non-negative. Thus, $f(x)$ has no roots.

Because $f(x)$ has no roots, none of its factors can be of degree 1 (otherwise it would have roots). However, the lack of roots *does not rule out* the possibility that $f(x) = x^4 + 2x^2 + 1$ has factors of degree greater than 1. In fact, $f(x) = x^4 + 2x^2 + 1$ can be factored into two factors of degree 2: $f(x) = x^4 + 2x^2 + 1 = (x^2 + 1)(x^2 + 1)$. Thus, $f(x)$ is reducible.
    As this example illustrates, reducibility does not imply roots exist. However, if $\deg f(x)$ is 2 or 3, then reducibility does, in fact, imply roots exist. Here is the proof: If $\deg f(x) = 2$, and $f(x)$ is reducible, then $f(x)$ is factorable into two polynomials of degree 1. A factor of degree 1 implies there is a root. The form of a degree 1 polynomial is $ax + b$. If $x = -a^{-1}b$, then $ax + b = a(-a^{-1}b) + b = -b + b = 0$. Thus, $-a^{-1}b$ is a root. If $\deg f(x) = 3$, and $f(x)$ is reducible, then $f(x)$ is factorable into either three polynomials of degree 1, or two polynomials, one of degree 1 and one of degree 2. In either case, at least one of the factors has degree 1. Thus, $f(x)$ has at least one root. ∎
    Let summarize our conclusion in the preceding frames about polynomials of degree 2 or 3:

Reducibility in a polynomial of degree 2 or 3 $\Rightarrow$ roots exist

Combining this result with the result in frame **1500**, we get that

a polynomial of degree 2 or 3 is reducible if and only if it has one or more roots.

We call this rule the ***reducibility test for degrees 2 and 3***. This rule is a biconditional. That is, it is a two-way implication. Negating both sides of the biconditional, we get the equivalent biconditional, a polynomial of degree 2 or 3 is irreducible if and only if it has no roots.

From frame **1500**, we also have the ***reducibility test for degrees > 1***:

If a polynomial with degree > 1 has any roots, then it is reducible.

and its equivalent contrapositive:

If a polynomial with degree > 1 is irreducible, then it does not have any roots.

For example, consider the polynomial $x^2 + 1$ over $\mathbb{R}$. $x^2 + 1 \geq 1$ for all $x \in \mathbb{R}$ because x^2 is always non-negative $\Rightarrow x^2 + 1$ has no roots $\Rightarrow$ it is irreducible (by the reducibility test for degrees 2 and 3).

Is $f(x) = x^3 - x + 1$ reducible in $\mathbb{Z}_3[x]$?

1503

$f(0) = 1$
$f(1) = 1$
$f(2) = 1$

$f(x)$ has no roots and has degree 2 or 3 $\Rightarrow$ not reducible.

Is $f(x) = x^3 + x + 1$ reducible in $\mathbb{Z}_3[x]$?

1504

$f(0) = 1$
$f(1) = 0$
$f(2) = 2$

$f(x)$ has a root (1 is a root) and has degree > 1 $\Rightarrow$ reducible.

Let's summarize:

- A polynomial of degree 2 or 3 is reducible if and only if it has one or more roots.
- A polynomial of degree 2 or 3 is irreducible if and only if it has no roots.
- If a polynomial of degree greater that 1 has any roots, then it is reducible.
- If a polynomial of degree greater than 1 is irreducible, then it does not have any roots.

Monic Polynomials

1505

Recall that a ***monic polynomial*** is a polynomial whose leading term has a coefficient of 1. There is a simple procedure for enumerating all the irreducible monic polynomials over $\mathbb{Z}_p$, where p is a prime number (making $<\mathbb{Z}_p, +_p, \cdot_p>$ a field). We can show that for any $n \geq 1$, if an irreducible polynomial of degree n exists then a monic irreducible polynomial of degree n exists (you prove this in homework question 9). Thus, by working strictly with monic polynomials, we do not run the risk of not finding an irreducible polynomial of degree n if one exists.

Is $f(x) = 2x^2 + x + 1$ in $\mathbb{Z}_3[x]$ an irreducible polynomial? *Hint*: Use the reducibility test for degrees 2 and 3 in frame **1501**.

1506

$f(0) = 1$
$f(1) = 1$
$f(2) = 2$

$f(x)$ has no roots and is of degree 2 or 3 $\Rightarrow f(x)$ is irreducible by the reducibility test for degrees 2 and 3.

Determine a monic polynomial of degree 2 in $\mathbb{Z}_3[x]$ from the irreducible polynomial $f(x) = 2x^2 + x + 1$. *Hint*: Multiply $f(x)$ by the multiplicative inverse of the constant in its leading term—that is, by 2^{-1}.

1507

Because $2 \cdot_3 2 = 1$, 2 is its own inverse. Let $m(x) = 2f(x) = x^2 + 2x + 2$.

Confirm that $m(x)$ is irreducible.

1508

$m(0) = 2$
$m(1) = 2$
$m(2) = 1$

$m(x)$ has no roots and is of degree 2 or 3 $\Rightarrow$ $m(x)$ is irreducible. We can also confirm that $m(x)$ is irreducible with a proof by contradiction: Suppose to the contrary $m(x)$ is reducible $\Rightarrow$ $m(x) = g(x)h(x)$ for some non-constant polynomials $g(x)$ and $h(x)$. Then $f(x) = 2^{-1}m(x) = [2^{-1}g(x)]h(x) \Rightarrow f(x)$ is reducible. But $f(x)$ is not reducible. Thus, $m(x)$ is irreducible. ∎

Suppose the leading term of an irreducible polynomial $f(x)$ of degree n over a field is $c_n x^n$. How would you get from $f(x)$ an irreducible monic polynomial of degree n?

1509

Multiply $f(x)$ by c_n^{-1}.

Enumerating Irreducible Monic Polynomials over $\mathbb{Z}_p$

1510

How do we enumerate all the irreducible monic polynomials over the $\mathbb{Z}_p$, where p is a prime? Our enumeration procedure takes advantage of two properties of polynomials over $\mathbb{Z}_p$:

1. For any degree n, there are only a finite number of monic polynomials of degree n over $\mathbb{Z}_p$. For example, consider the monic polynomials of degree 1 over $\mathbb{Z}_2$. Every such polynomial is of the form $x + b$, where $b \in \mathbb{Z}_2$. b has two possible values: 0 and 1 (the two elements in $\mathbb{Z}_2$). Thus, there are only two monic polynomials of degree 1: $x + 0 = x$ and $x + 1$.

2. If a monic polynomial is reducible, it can be factored into non-constant monic polynomials of lower degree (you prove this in homework question 10). If these factors are themselves reducible, then they too can be factored into non-constant monic polynomials. We can continue this process until all the factors are irreducible monic polynomials. Thus, every reducible monic polynomial over $\mathbb{Z}_p$ can be expressed as the product of two or more irreducible monic polynomials of lower degree.

Let's start enumerating all the irreducible monic polynomials over $\mathbb{Z}_2$ in degree order. That is, we will first enumerate all the irreducible monic polynomials of degree 1, then all irreducible monic polynomials of degree 2, and so on. To determine the irreducible monic polynomials of degree $n > 1$, we make use of the irreducible monic polynomials of degrees less than n.

What are all the irreducible monic polynomials of degree 1 over $\mathbb{Z}_2$? *Hint*: Recall that every polynomial of degree 1 is irreducible.

1511

$x, x + 1$

Using these polynomials as factors, we can construct several polynomials of degree 2. List all these polynomials. *Hint*: There are three.

1512

$xx = x^2$, $x(x+1) = x^2 + x$, $(x+1)(x+1) = x^2 + 1$

Each of these polynomials is the product of two polynomials of degree 1. Thus, they are all reducible. By item 2 in frame **1510**, every reducible monic polynomial can be factored into irreducible monic polynomials. The list above includes every monic polynomial of degree 2 over $\mathbb{Z}_2$ whose factors are irreducible monic polynomials. Thus, x^2, $x^2 + x$, and $x^2 + 1$ are the *only* reducible monic polynomials of degree 2 over $\mathbb{Z}_2$.

Every monic polynomial over $\mathbb{Z}_2$ of degree 2 has the form $x^2 + bx + c$. There are two possible values for b and for c: each can be either 0 or 1. Thus, there are four possible assignments to b and c, each of which corresponds to a distinct polynomial, as indicated in the following table:

b	c	$x^2 + bx + c$	
0	0	x^2	reducible
0	1	$x^2 + 1$	reducible
1	0	$x^2 + x$	reducible
1	1	$x^2 + x + 1$	

We determined above that x^2, $x^2 + 1$, and $x^2 + x$ are all the reducible monic polynomials of degree 2. Thus, the only remaining polynomial of degree 2, $x^2 + x + 1$, must be irreducible.

Now that we know all the irreducible monic polynomials of degrees 1 and 2, we can determine all the irreducible monic polynomials of degree 3.

List all the monic polynomials of degree 3 that can be constructed from the irreducible monic polynomials of degree 1 (x and $x + 1$). *Hint*: There are four.

1513

$xxx = x^3$, $xx(x+1) = x^3 + x^2$, $x(x+1)(x+1) = x^3 + x$, $(x+1)(x+1)(x+1) = x^3 + x^2 + x + 1$

List all the monic polynomials of degree 3 that can be constructed from one irreducible monic polynomial of degree 1 (x and $x + 1$) and one irreducible monic polynomial of degree 2 ($x^2 + x + 1$). *Hint*: There are two.

1514

$x(x^2 + x + 1) = x^3 + x^2 + x$
$(x + 1)(x^2 + x + 1) = x^3 + 1$

List all the monic polynomials of degree 3 over $\mathbb{Z}_2$ in table format. Indicate which are reducible. *Hint*: Every monic polynomial of degree 3 has the format $x^3 + bx^2 + cx + d$.

1515

b	c	d	$x^3 + bx^2 + cx + d$	
0	0	0	x^3	reducible
0	0	1	$x^3 + 1$	reducible
0	1	0	$x^3 + x$	reducible
0	1	1	$x^3 + x + 1$	
1	0	0	$x^3 + x^2$	reducible
1	0	1	$x^3 + x^2 + 1$	
1	1	0	$x^3 + x^2 + x$	reducible
1	1	1	$x^3 + x^2 + x + 1$	reducible

In the preceding two frames, we determined all the reducible monic polynomials of degree 3. They are labeled in the preceding table. Thus, the remaining polynomials in the table, $x^3 + x + 1$ and $x^3 + x^2 + 1$, must be irreducible.

We can continue this procedure indefinitely. On each iteration, we determine the irreducible monic polynomials for the next higher degree.

What is the total number of monic polynomials—both reducible and non-reducible—over $\mathbb{Z}_2$ of degree n? *Hint*: Every monic polynomial over $\mathbb{Z}_2$ of degree n is of the form $x^n + c_{n-1}x^{n-1} + \cdots + c_1x + c_0$.

1516

There are two possible values (0 and 1) for each c_i. There are n of these coefficients. Thus, there are 2^n distinct assignments to these coefficients, each corresponding to a distinct polynomial.

What is the total number of monic polynomials—both reducible and non-reducible—over $\mathbb{Z}_p$ of degree n, where p is a prime?

1517

Each polynomial has n coefficients. Each coefficient can assume any one of p values. Thus, there are p^n possible assignments, each corresponding to a distinct polynomial.

As n increases, p^n increases rapidly. Because n is an exponent, we say that p^n **increases exponentially with n**. p^n increases so rapidly as n increases that p^n (the number of monic polynomials of degree n) is so large that there are always monic polynomials of degree n left after we eliminate all the reducible ones (the proof of this is complex so we will omit it). Thus, for every $n \geq 1$, there is at least one irreducible monic polynomial of degree n.

Irreducible Polynomials Are Maximal Ideals

1518

Here is the reason irreducible polynomials are important: a principal ideal generated by an irreducible polynomial over a field are maximal. Maximal ideals are important because the cosets of a maximal ideal form a field. Recall from frame **1367** that a maximal ideal of a ring R is an ideal that is not a proper subset of any ideal except R itself. The following theorem establishes the connection between irreducible polynomials and maximal ideals: Suppose F is a field, and $p(x) \in F[x]$. Then

$$p(x) \text{ is irreducible} \Leftrightarrow \langle p(x) \rangle \text{ is maximal in } F[x].$$

To prove the "$\Rightarrow$" part of this biconditional, we have to show that if $p(x)$ is irreducible then $\langle p(x) \rangle$ is maximal. We do this by showing that if $p(x)$ is irreducible and $\langle p(x) \rangle \subset I \subseteq F[x]$, where I is an ideal, then $I = F[x]$, which implies that $\langle p(x) \rangle$ is maximal.

Suppose $p(x)$ is an irreducible polynomial over F, and I is an ideal such that $\langle p(x) \rangle \subset I \subseteq F[x]$. From frame **1481**, we know that $F[x]$ is a principal ideal domain. That is, every ideal can be generated by a single element in $F[x]$. Thus, $I = \langle g(x) \rangle$ for some $g(x) \in F[x]$.

Is $p(x) \in \langle g(x) \rangle$?

1519

Yes, because $p(x) \in \langle p(x) \rangle \subset I = \langle g(x) \rangle$.

Thus, $p(x) = g(x)f(x)$ for some $f(x) \in F[x]$. What can we conclude about $g(x)$ and $f(x)$, given that $p(x)$ is irreducible?

1520

$g(x)$ is a constant or $f(x)$ is a constant. Otherwise, $p(x)$ would be reducible.

$f(x)$ cannot be a constant. Why not? *Hint*: I5 in frame **1460**.

1521

$p(x) = g(x)f(x)$. Thus, if $f(x)$ is a constant, then by I5, $\langle p(x) \rangle = \langle g(x) \rangle$. But $\langle p(x) \rangle$ is a proper subset of $\langle g(x) \rangle$.

Because $f(x)$ cannot be constant, $g(x)$ must be a constant. Then by I6 in frame **1460**, $\langle g(x) \rangle = I = F[x]$. We have shown that if $p(x)$ is irreducible and $\langle p(x) \rangle \subset I \subseteq F[x]$, then $I = F[x]$. In other words, $\langle p(x) \rangle$ is not properly contained by any ideal except $F[x]$. Thus, $\langle p(x) \rangle$ is maximal.

Now let's prove the "$\Leftarrow$" part of the biconditional in frame **1518**. We have to show that $p(x)$ is irreducible if $\langle p(x) \rangle$ is maximal. We will do this by showing that if $p(x)$ is maximal and $p(x) = g(x)h(x)$, then $g(x)$ or $h(x)$ must be a constant (which implies $p(x)$ is irreducible). Suppose $\langle p(x) \rangle$ is maximal. Then $p(x)$ cannot be a constant. If, to the contrary, $p(x)$ is a constant, then by I6 in frame **1460**, $\langle p(x) \rangle = F[x]$. But then $\langle p(x) \rangle$ is not maximal (a maximal ideal of $F[x]$ by definition is a proper subset of $F[x]$). Suppose $p(x) = g(x)h(x)$. By I4 in frame **1460**, $\langle p(x) \rangle \subseteq \langle g(x) \rangle$. $\langle g(x) \rangle$ is a subset of $F[x]$. Thus, we have $\langle p(x) \rangle \subseteq \langle g(x) \rangle \subseteq F[x]$.

We are given that $\langle p(x) \rangle$ is maximal. So what must true of $\langle g(x) \rangle$?

1522

$\langle g(x) \rangle = \langle p(x) \rangle = \langle g(x)h(x) \rangle$ or $\langle g(x) \rangle) = F[x]$.

If $\langle g(x)\rangle = \langle p(x)\rangle = \langle g(x)h(x)\rangle$, then by I5 frame **1460**, $h(x)$ must be a constant polynomial. If, on the other hand, $\langle g(x)\rangle = F[x]$, then by I6 in frame **1460**, $g(x)$ must be a constant. There is no factorization of $p(x)$ into two non-constant polynomials. Thus, $p(x)$ is irreducible. ∎

Constructing Fields Using Irreducible Polynomials

1523

Irreducible polynomials are important because we can use them to construct finite fields. In fact, the construction technique we will present allows us to construct *any* finite field. The fields we will construct will be the cosets of the ideal $\langle f(x)\rangle$, where $f(x)$ is an irreducible monic polynomial.

Recall from our discussion on maximal ideals in chapter 24 that if I is an ideal of a commutative ring R with unity, then R/I is a field if and only I is a maximal ideal. Thus, if $f(x)$ is an irreducible polynomial, then $\langle f(x)\rangle$ is a maximal ideal of $F[x]$, and $F[x]/\langle f(x)\rangle$ is a field. Let's do an example. Let's construct a field of order $2^3 = 8$. To do this, we will use an irreducible monic polynomial of degree 3 over the field $\mathbb{Z}_2$. There are two such polynomials: $x^3 + x + 1$ and $x^3 + x^2 + 1$. We can use either one. Let's use $x^3 + x + 1$. Consider the cosets of $\langle x^3 + x + 1\rangle$ in $\mathbb{Z}_2[x]$. By frame **1486**, we know that each coset is of the form $ax^2 + bx + c + \langle x^3 + x + 1\rangle$.

How many polynomials over $\mathbb{Z}_2$ are of the form $ax^2 + bx + c$?

1524

There are two possible values for each of the three coefficients. Thus, there are $2 \cdot 2 \cdot 2 = 2^3$ distinct polynomials. Corresponding to each polynomial is a coset of $\langle x^3 + x + 1\rangle$.

List all the cosets of $\langle x^3 + x + 1\rangle$.

1525

a	b	c	$ax^2 + bx + c + \langle x^3 + x + 1\rangle$
0	0	0	$0 + \langle x^3 + x + 1\rangle$
0	0	1	$1 + \langle x^3 + x + 1\rangle$
0	1	0	$x + \langle x^3 + x + 1\rangle$
0	1	1	$x + 1 + \langle x^3 + x + 1\rangle$
1	0	0	$x^2 + \langle x^3 + x + 1\rangle$
1	0	1	$x^2 + 1 + \langle x^3 + x + 1\rangle$
1	1	0	$x^2 + x + \langle x^3 + x + 1\rangle$
1	1	1	$x^2 + x + 1 + \langle x^3 + x + 1\rangle$

These eight cosets are the elements of the field $\mathbb{Z}_2[x]/\langle x^3+x+1\rangle$.

Suppose we want to construct a field with p^n elements, where p is a prime. To do this, we use an irreducible polynomial $f(x)$ over $\mathbb{Z}_p$ of degree n. The coset representatives of $\langle f(x)\rangle$ have n coefficients, each of which can have any one of p values. Thus, there are p^n distinct cosets of $\langle f(x)\rangle$. These cosets are the elements of the field $\mathbb{Z}_p[x]/\langle f(x)\rangle$.

We will not prove this, but every finite field has the order p^n for some prime p and integer $n \geq 1$. Moreover, for any given order p^n, there is only one field up to isomorphism. Thus, using our construction technique, we can construct *any* finite field.

Polynomials in Place of Cosets as Field Elements

1526

The elements of the field $\mathbb{Z}_2[x]/\langle x^3 + x + 1\rangle$ are the cosets of the maximal ideal $\langle x^3 + x + 1\rangle$, represented by the polynomials over $\mathbb{Z}_2$ with degrees less than 3. To add two cosets, we simply add the representatives of the two cosets. For example, the sum of $(x + 1 + \langle x^3 + x + 1\rangle)$ and $(x^2 + \langle x^3 + x + 1\rangle)$ is $(x + 1) + (x^2) + \langle x^3 + x + 1\rangle = x^2 + x + 1 + \langle x^3 + x + 1\rangle$. Multiplication of cosets is more complex. To multiply two cosets, we multiply their representatives. However, the result can be a polynomial with a degree greater than or equal to 3. Thus, to determine the representative with degree less than 3, we have to divide by $x^3 + x + 1$. By frame **1486**, the remainder is the desired representative. For example, $(x + \langle x^3 + x + 1\rangle)(x^2 + \langle x^3 + x + 1\rangle) = x^3 + (x^3 + x + 1)$. To get the representative we want, we divide x^3 by $x^3 + x + 1$, and take the remainder:

$$\begin{array}{r} 1 \\ x^3 + x + 1 \overline{\smash{)}\, x^3 } \\ x^3 + x + 1 \\ \hline -x - 1 = x + 1 \text{ (because } -1 = 1 \text{ in } \mathbb{Z}_2\text{)} \end{array}$$

The remainder, $x + 1$, is the representative we want. Its degree is less than 3, and $x^3 + \langle x^3 + x + 1 \rangle = x + 1 + \langle x^3 + x + 1 \rangle$. In general, if $g(x)$ and $h(x)$ are representatives of two cosets in $\mathbb{Z}_2[x]/\langle x^3 + x + 1 \rangle$, the representative of the sum is $g(x) + h(x)$. The representative of the product is $[g(x)h(x) \bmod (x^3 + x + 1)]$ (recall that mod means divide and take the remainder).

There is, however, a way to determine the product of two cosets without dividing. Observe that $\langle x^3 + x + 1 \rangle$ absorbs $x^3 + x + 1$ because $x^3 + x + 1 \in \langle x^3 + x + 1 \rangle$. Thus, $0 + \langle x^3 + x + 1 \rangle = x^3 + x + 1 + \langle x^3 + x + 1 \rangle$. The coset $x^3 + x + 1 + \langle x^3 + x + 1 \rangle$ in turn equals the sum of the cosets $x^3 + \langle x^3 + x + 1 \rangle$ and $x + 1 + \langle x^3 + x + 1 \rangle$. Thus,

$$0 + \langle x^3 + x + 1 \rangle = (x^3 + \langle x^3 + x + 1 \rangle) + (x + 1 + \langle x^3 + x + 1 \rangle)$$

The inverse of $x + 1 + \langle x^3 + x + 1 \rangle$ is $-x - 1 + \langle x^3 + x + 1 \rangle$, which in $\mathbb{Z}_2$ is equal to $x + 1 + \langle x^3 + x + 1 \rangle$. This coset is its own inverse. What do we get if we add of $x + 1 + \langle x^3 + x + 1 \rangle$ to both sides of

$$0 + \langle x^3 + x + 1 \rangle = (x^3 + \langle x^3 + x + 1 \rangle) + (x + 1 + \langle x^3 + x + 1 \rangle)$$

1527

$x + 1 + \langle x^3 + x + 1 \rangle = x^3 + \langle x^3 + x + 1 \rangle$. Thus, the representative we want corresponding to the coset $x^3 + \langle x^3 + x + 1 \rangle$ is $x + 1$.

We were able to determine the coset representative with degree less than 3 that corresponds to the representative x^3 without dividing. Moreover, now that we have our representative for x^3, we can easily determine the representative with degree less than 3 that corresponds to x^4. Simply multiply both sides of

$$x^3 + \langle x^3 + x + 1 \rangle = x + 1 + \langle x^3 + x + 1 \rangle$$

by $x + \langle x^3 + x + 1 \rangle$. Do this.

1528

$xx^3 + \langle x^3 + x + 1 \rangle = x(x + 1) + \langle x^3 + x + 1 \rangle$ from which we get $x^4 + \langle x^3 + x + 1 \rangle = x^2 + x + \langle x^3 + x + 1 \rangle$.

Armed with the representatives with degree less than 3 corresponding to x^3 and x^4, we can easily determine the product of any two cosets. If the product of their representatives have any x^3 or x^4 terms, we simply substitute $x + 1$ and $x^2 + x$, respectively.

For example, let's determine $(x^2 + 1 + \langle x^3 + x + 1 \rangle)(x^2 + x + \langle x^3 + x + 1 \rangle)$. First multiply $x^2 + 1$ by $x^2 + x$.

1529

$x^4 + x^3 + x^2 + x$.

Now substitute $x + 1$ and $x^2 + x$ for x^3 and x^4, respectively.

1530

$x^4 + x^3 + x^2 + x = (x^2 + x) + (x + 1) + x^2 + x = (1 +_2 1)x^2 + (1 +_2 1 +_2 1)x + 1$ which in $\mathbb{Z}_2$ equals $x + 1$. Thus, $(x^2 + 1 + \langle x^3 + x + 1 \rangle)(x^2 + x + \langle x^3 + x + 1 \rangle) = x + 1 + \langle x^3 + x + 1 \rangle$.

Construct the Cayley table for $\mathbb{Z}_2[x]/\langle x^3 + x + 1 \rangle$ under coset multiplication. To save space in your table, represent each coset with its polynomial representative only. For example, for the coset $x + 1 + \langle x^3 + x + 1 \rangle$, enter only $x + 1$ into the Cayley table.

1531

·	0	1	x	$x+1$	x^2	x^2+1	x^2+x	x^2+x+1
0	0	0	0	0	0	0	0	0
1	0	1	x	$x+1$	x^2	x^2+1	x^2+x	x^2+x+1
x	0	x	x^2	x^2+x	$x+1$	1	x^2+x+1	x^2+x
$x+1$	0	$x+1$	x^2+x	x^2+1	x^2+x+1	x^2	1	x
x^2	0	x^2	$x+1$	x^2+x+1	x^2+x	x	x^2+1	1
x^2+1	0	x^2+1	1	x^2	x	x^2+x+1	$x+1$	x^2+x
x^2+x	0	x^2+x	x^2+x+1	1	x^2+1	$x+1$	x	x^2
x^2+x+1	0	x^2+x+1	x^2+x	x	1	x^2+x	x^2	$x+1$

The nonzero elements of a field form a commutative _____.

1532

group

Thus, the nonzero rows and columns of our Cayley table in the preceding frame represent a group of order 7. Up to isomorphism, there is only one group of order 7. What is this group?

1533

the cyclic group of order 7

Because 7 is prime, every nonzero element except 1 generates the group represented by our Cayley table. Show that x generates this group.

1534

$$
\begin{aligned}
x &= x \\
x^2 &= x^2 \\
x^3 &= x(x^2) = x+1 \\
x^4 &= x(x^3) = x^2+x \\
x^5 &= x(x^4) = x(x^2+x) = x^3+x^2 = (x+1)+x^2 = x^2+x+1 \\
x^6 &= x(x^5) = x(x^2+x+1) = x^3+x^2+x = (x+1)+x^2+x = (x+x)+x^2+1 = x^2+1 \\
x^7 &= x(x^6) = 1
\end{aligned}
$$

Construct the Cayley table for $\mathbb{Z}_2[x]/<x^3+x+1>$ under addition.

1535

+	0	1	x	$x+1$	x^2	x^2+1	x^2+x	x^2+x+1
0	0	1	x	$x+1$	x^2	x^2+1	x^2+x	x^2+x+1
1	1	0	$x+1$	x	x^2+1	x^2	x^2+x+1	x^2+x
x	x	$x+1$	0	1	x^2+x	x^2+x+1	x^2	x^2+1
$x+1$	$x+1$	x	1	0	x^2+x+1	x^2+x	x^2+1	x^2
x^2	x^2	x^2+1	x^2+x	x^2+x+1	0	1	x	$x+1$
x^2+1	x^2+1	x^2	x^2+x+1	x^2+x	1	0	$x+1$	x
x^2+x	x^2+x	x^2+x+1	x^2	x^2+1	x	$x+1$	0	1
x^2+x+1	x^2+x	x^2+1	x^2	$x+1$	x	1	0	

The entries in the preceding Cayley tables are cosets. However, we could just as well view the entries simply as polynomials. In other words, we can take our constructed field to consist of the polynomials themselves (0, 1, x, $x+1$, x^2, x^2+1, x^2+x and x^2+x+1 under polynomial addition and multiplication modulo x^3+x+1 rather than the cosets represented by these polynomials. The Cayley tables for the cosets and the Cayley tables for the polynomials are identical. Thus, the two fields—the field of cosets and the field of polynomials—are isomorphic.

In the preceding frames, we constructed a field of order 8 consisting of polynomials over $\mathbb{Z}_2$. The field we constructed is an extension of $\mathbb{Z}_2$. That is, $\mathbb{Z}_2$ is isomorphic to a subfield of our constructed field. Give a bijection from $\mathbb{Z}_2$ to a subfield of our field of polynomials that confirms that our constructed field is an extension of $\mathbb{Z}_2$.

1536

316   Chapter 26: Polynomials Part 2

Let φ be a function from $\mathbb{Z}_2$ to our constructed field of polynomials defined as follows:

$$\varphi(0) = \text{the polynomial } 0$$
$$\varphi(1) = \text{the polynomial } 1$$

This property of the constructed field is true in general. Whenever we construct a field based on an irreducible monic polynomial over the field $\mathbb{Z}_p$ as illustrated in the preceding frames, the constructed field will be an extension of $\mathbb{Z}_p$. Even more interesting, however, is that the irreducible polynomial used in the construction of the extension field will have a root in the extension field. Let's explore this important property of the extension field.

$x^3 + x + 1$ is the irreducible polynomial that we used to construct the field represented by the Cayley tables in frames **1531** and **1535**. Using these Cayley tables, let's determine the value of $x^3 + x + 1$. First, determine what x^3 equals by determining from the multiplication table the cube of x.

1537

$x^3 = xxx = x^2 x = x + 1$

Thus, $x^3 + x + 1 = $ _____ .

1538

$(x + 1) + x + 1 = (1 +_2 1)x + (1 +_2 1) = 0$. Thus, $x^3 + x + 1 = 0$.

The polynomial x is, in fact, a root of the polynomial $x^3 + x + 1$. The polynomial $x^3 + x + 1$ has no root in $\mathbb{Z}_2$, but it has a root in the constructed field—namely, x. This result is no surprise: When we replace the indeterminate in the irreducible polynomial used to construct the field, we get the same irreducible polynomial. Then to get the corresponding field element, we divide by the irreducible polynomial and take the remainder. Because we are dividing the irreducible polynomial by itself, this division always yields a remainder of 0. You may be puzzled at this point because we are using x in two different ways: We are using x as the indeterminate in the polynomial $x^3 + x + 1$. Thus, x and its powers are functioning simply as position markers for their coefficients. But x is also an element of the extension field we constructed. When we plug in the element x into the polynomial $x^3 + x + 1$ and evaluate it according to the Cayley tables above, we get 0. Thus, x, one of the elements of our constructed field, is a root of the polynomial $x^3 + x + 1$.

It may be easier to understand that $x^3 + x + 1$ has a root simply by rewriting it using a different indeterminate, such as t. We then get the polynomial $t^3 + t + 1$. This is the same polynomial—but with a different indeterminate. Now substitute the element x from the field represented by the Cayley tables in frames **1531** and **1535**. We get 0 as before. Thus, x is a root of $t^3 + t + 1$.

Let's do another example. Let's construct a field of order 4 using the polynomial $x^2 + x + 1$, which is irreducible over $\mathbb{Z}_2$. List the elements in the field $\mathbb{Z}_2[x]/\langle x^2 + x + 1\rangle$.

1539

$0 + \langle x^2 + x + 1\rangle, 1 + \langle x^2 + x + 1\rangle, x + \langle x^2 + x + 1\rangle, x + 1 + \langle x^2 + x + 1\rangle$.

Which of the above cosets equals $x^2 + \langle x^2 + x + 1\rangle$? Determine this by computing $x^2 \bmod (x^2 + x + 1)$.

1540

$$\begin{array}{r} 1 \\ x^2 + x + 1 \overline{\smash{\big)}\, x^2} \\ \underline{x^2 + x + 1} \\ -x - 1 = x + 1 \text{ in } \mathbb{Z}_2 \end{array}$$

Thus, $x^2 + (x^2 + x + 1) = x + 1 + (x^2 + x + 1)$.

Construct the Cayley table for $\mathbb{Z}_2[x]/\langle x^2 + x + 1\rangle$ under coset multiplication.

1541

·	0	1	x	$x+1$
0	0	0	0	0
1	0	1	x	$x+1$
x	0	x	$x+1$	1
$x+1$	0	$x+1$	1	x

Construct the Cayley table for the elements of $\mathbb{Z}_2[x]/(x^2+x+1)$ under coset addition.

1542

+	0	1	x	$x+1$
0	0	1	x	$x+1$
1	1	0	$x+1$	x
x	x	$x+1$	0	1
$x+1$	$x+1$	x	1	0

Describe the Cayley tables for the set of polynomials, $\{0, 1, x, x+1\}$, under polynomial addition and multiplication modulo $x^2 + x + 1$.

1543

They are identical to the Cayley tables for $\mathbb{Z}_2[x]/\langle x^2 + x + 1\rangle$. This set of polynomials is isomorphic to $\mathbb{Z}_2[x]/\langle x^2 + x + 1\rangle$. Thus, we have constructed two isomorphic fields of order 4: $\mathbb{Z}_2[x]/\langle x^2 + x + 1\rangle$ under coset addition and multiplication, and $\{0, 1, x, x+1\}$ under polynomial addition and multiplication modulo $x^2 + x + 1$.

Which elements among $\{0, 1, x, x+1\}$ is a root of the polynomial $x^2 + x + 1$?

1544

x

Confirm that x is a root using the Cayley tables in frames **1541** and **1542**. That is, using the multiplication table, determine x^2. Then, using the addition table, determine $x^2 + x + 1$. If the result is 0, then x is a root.

1545

From the Cayley table, we see that $xx = x + 1$. Thus, $x^2 + x + 1 = x + 1 + x + 1 = (1 +_2 1)x + (1 +_2 1) = 0$.

We have been constructing extension fields using irreducible polynomials over the field $\mathbb{Z}_p$, where p is some prime. However, we can construct extension fields using irreducible polynomials over any field. For example, $x^2 + 1$ is irreducible over the field $\mathbb{Q}$ (the rationals). Thus, $\mathbb{Q}[x]/\langle x^2 + 1\rangle$ is a field. Moreover, it is an extension of $\mathbb{Q}$ ($\mathbb{Q}$ is isomorphic to the set of cosets in $\mathbb{Q}[x]/\langle x^2 + 1\rangle$ whose representatives are constants).

Fundamental Theorem of Field Theory

1546

From the preceding section, we know that if $f(x)$ is an irreducible polynomial over a field F, we can construct an extension field in which $f(x)$ has a root. But we can make an even stronger statement:

If $f(x)$ is a non-constant polynomial over a field F, we can construct an extension field in which $f(x)$ has a root.

This statement is referred to as the ***fundamental theorem of field theory***. Let's prove this theorem.

Case 1: $f(x)$ has a root in the field F.
Then F is an extension of itself in which $f(x)$ has a root.
Case 2: $f(x)$ does not have a root in the field F.
We have already shown in the preceding section that if $f(x)$ is irreducible in F, then $F[x]/\langle f(x)\rangle$ is an extension field in which $f(x)$ has a root. So suppose $f(x)$ is reducible in F. Then $f(x)$ can be factored into irreducible polynomials. Let $g(x)$ be one of these irreducible factors of $f(x)$. Then by the preceding section, $g(x)$ has a root in the extension field $F[x]/\langle g(x)\rangle$. Because $g(x)$ is a factor of $f(x)$, the root of $g(x)$ is also a root of $f(x)$. ∎

Let's look at an example. $f(x) = x^4 + x^3 + x + 2$ has no roots in $\mathbb{Z}_3$. Confirm this.

1547

$f(0) = 2$
$f(1) = (1 + 1 + 1 + 2) \bmod 3 = 5 \bmod 3 = 2$.
$f(2) = (16 + 8 + 2 + 2) \bmod 3 = 28 \bmod 3 = 1$.

$f(x)$ is reducible. Factor it into two irreducible polynomials.

1548

$f(x) = (x^2 + 1)(x^2 + x + 2)$

What are the elements of the extension field $\mathbb{Z}_3[x]/\langle x^2 + 1\rangle$? *Hint*: There are three possibilities (0, 1, and 2) for each coefficient.

1549

$0, 1, 2, x, x + 1, x + 2, 2x, 2x + 1, 2x + 2$

What does x^2 equal in this field? Hint: $x^2 + 1 = 0$.

1550

$x^2 = -1 = 2$ in $\mathbb{Z}_3$

What does x^3 equal?

1551

$x^3 = x^2 x = 2x$

What does x^4 equal?

1552

$x^4 = x^2 x^2 = 2 \cdot_3 2 = 1$

Using these values for x^2, x^3, and x^4, determine the value of $f(x) = x^4 + x^3 + x + 2$ when the indeterminate x is replaced with the field element x.

1553

$x^4 + x^3 + x + 2 = 1 + 2x + x + 2 = (2 +_3 1)x + (1 +_3 2) = 0$.

Thus, the element x in $\mathbb{Z}_3[x]/\langle x^2 + 1\rangle$ is a root of $f(x)$.

Review Questions

1. Show that $x^3 + x + 2 \in \langle 5\rangle$, where $x^3 + x + 2$ and 5 are polynomials over $\mathbb{Z}_{11}$.
2. Show that $f(x) \in \langle c\rangle$, where $f(x)$ and c are polynomials over a field F, and c is a nonzero constant polynomial.
3. Can an ideal of $F[x]$ contain both $x^3 + x^2$ and x^3?
4. Can an ideal of $F[x]$ with no polynomials with degree less than 3 contain both $x^3 + x^2$ and x^3? *Hint*: See I5 in frame **1460**.
5. List all the cosets of $\langle x^2 + 1\rangle$ in $\mathbb{Z}_3[x]$.
6. Is the polynomial $x^3 + x^2 + 3$ over $\mathbb{Z}_5$ irreducible? Justify your answer.
7. Is the polynomial $x^3 + x + 1$ over $\mathbb{Z}_2$ irreducible? Justify your answer.
8. List the elements in the field of order 2 constructed from the irreducible polynomial $x + 1$ over $\mathbb{Z}_2$.
9. The field constructed in review question 8 under addition is isomorphic to which group? Its nonzero elements under multiplication are isomorphic to which group?
10. Is $\mathbb{Z}_p[x]/\langle 0\rangle$ a field, if p is prime?

Answers to the Review Questions

1. $5^{-1} = 9$. Let $f(x) = 9(x^3 + x + 2)$. Then $f(x) \in \mathbb{Z}_{11}[x]$, and $5f(x) = 5[9(x^3 + x + 2)] = x^3 + x + 2 \Rightarrow x^3 + x + 2 \in \langle 5 \rangle$.
2. Because c is a nonzero constant, c^{-1} exists. We then have $c^{-1}f(x) \in F[x]$, and $c[c^{-1}f(x)] = f(x) \Rightarrow f(x) \in \langle c \rangle$.
3. Yes. $F[x]$ is such an ideal.
4. No. There is no constant polynomial c such that $x^3 + x^2 = cx^3$. Thus, by I5 in frame **1460**, no such ideal exists.
5. $0 + \langle x^2 + 1 \rangle, 1 + \langle x^2 + 1 \rangle, 2 + \langle x^2 + 1 \rangle, x + \langle x^2 + 1 \rangle, 2x + \langle x^2 + 1 \rangle, x + 1 + \langle x^2 + 1 \rangle, x + 2 + \langle x^2 + 1 \rangle, 2x + 1 + \langle x^2 + 1 \rangle, 2x + 2 + \langle x^2 + 1 \rangle$
6. Yes. 1 is a root, and it is of degree 2 or 3.
7. No. It has no roots, and it is of degree 2 or 3.
8. $0 + \langle x + 1 \rangle, 1 + \langle x + 1 \rangle$
9. C_2, C_1
10. $\mathbb{Z}_p[x]$ is not a field and $\mathbb{Z}_p[x] \cong \mathbb{Z}_p[x]/\langle 0 \rangle$. Thus, $\mathbb{Z}_p[x]/\langle 0 \rangle$ is not a field.

Homework Questions

1. How many cosets of $\langle x^2 + 1 \rangle$ are in $\mathbb{Z}_{17}[x]$?
2. Determine $(7x^2 + 5x - 1)(3x^3 + 2x^2 + 6)$ in $\mathbb{Z}_{15}[x]$.
3. Determine all the irreducible monic polynomials of degree 4 over $\mathbb{Z}_2$.
4. Determine all the irreducible monic polynomials of degree 2 over $\mathbb{Z}_3$.
5. Show that $x^2 + 1$ is irreducible over $\mathbb{Z}_7$.
6. Show that if $d(x) \mid p(x)$ then $\langle p(x) \rangle \subseteq \langle d(x) \rangle$.
7. Show that $\mathbb{Q}[x]/\langle x^2 - 2x + 1 \rangle$ is not an integral domain.
8. Does a field exist whose order is 121? If so, specify it.
9. Show that if $f(x)$ is an irreducible polynomial of degree n over a field F, then there is an irreducible monic polynomial of degree n over F.
10. Prove that if a monic polynomial is reducible, it can be factored into non-constant monic polynomials of lower degree.

27 Vector Spaces

Definition of a Vector Spaces

1554

You may have already studied vector spaces in a course named linear algebra. If you did, you may have found vector spaces a very puzzling topic, and with good reason. A vector space is a complex algebraic structure. It involves not one but two sets and four operations. A ***vector space V over F*** consists of

- A commutative group $<V, +>$ whose elements are called ***vectors***.
- A field $<F, +, \cdot>$ whose elements are called ***scalars***.
- A function from $F \times V$ to V called ***scalar multiplication***. Scalar multiplication "connects" the field F with the group V.

To facilitate distinguishing vectors from scalars, we will write vectors in boldface.

Scalar multiplication is not the multiplication of two scalars. Rather, it is the multiplication of a scalar and a vector, the result of which is a vector. Note that the domain of scalar multiplication is $F \times V$. That is, it consists of ordered pairs in which the first component is from the field F, and the second component is a vector from V. Thus, if $a \in F$ and $v \in V$, then the scalar multiplication of a and v is written as $a\mathbf{v}$, never as $\mathbf{v}a$.

A vector space consists of two sets and four operations. Particularly confusing is the dual use of the + sign (for vector addition and for addition on the field F), and the dual use of juxtaposition (for scalar multiplication and for multiplication on the field F). Thus, when we see an expression that uses + or juxtaposition, we have to be careful to interpret the expression correctly. To do this, simply look at the operands—the operands determine the operation. For example, if the operands are juxtaposed, and one is from F and one is from V, then the operation is scalar multiplication, not multiplication on F. If, on the other hand, the two operands are both from F, then the operation is multiplication on F.

Suppose $a, b \in F$ and $\mathbf{u}, \mathbf{v} \in V$. In $a(\mathbf{u} + \mathbf{v})$, what operation is specified by the +. What operation is specified by the juxtaposition of a and $(\mathbf{u} + \mathbf{v})$?

1555

The operands surrounding the + are vectors. Thus, this + represents vector addition, the result of which is a vector. The juxtaposition is the product of a scalar (the element a) and the vector equal to $\mathbf{u} + \mathbf{v}$. Thus, this juxtaposition represents scalar multiplication.

In the expression $(ab)\mathbf{v}$, which operations do the two juxtapositions represent?

1556

In the juxtaposition ab, both a and b are elements of F. Thus, this juxtaposition represents multiplication on F. $\mathbf{v}$ is a vector. Thus, the juxtaposition of ab and $\mathbf{v}$ is scalar multiplication.

Which operations do the two juxtapositions in $a(b\mathbf{v})$ represent?

1557

The juxtaposition of b and $\mathbf{v}$ represents scalar multiplication. Because $b\mathbf{v}$ is a vector, the juxtaposition of a and $b\mathbf{v}$ also represents scalar multiplication.

Multiplication on a field F is associative. That is, $(ab)c = a(bc)$ for all $a, b, c \in F$. In $(ab)c$, ab is the product of two elements in F; in $a(bc)$, bc is also the product of two elements in F. The operation in ab is the same as the operation in bc. However, in

$(ab)v$ and $a(bv)$, the operation in ab is not the same as the operation in bv. In ab, the operation is multiplication on the field F; in bv, the operation is scalar multiplication.

What are the operations in the expression $(a + b)v$?

1558

a and b are elements of F. Thus, the $+$ operation is addition on F. Because $a + b$ is an element of F, the juxtaposition of $a + b$ and v represents scalar multiplication.

A set with an operation has to have certain properties to qualify as a group. Similarly, a commutative group V and a field F along with scalar multiplication has to have certain properties to qualify as a vector space. In particular, the following properties must hold for all $a, b \in F$ and $u, v \in V$:

V1: $a(u + v) = au + av$
V2: $(a + b)v = av + bv$
V3: $a(bv) = (ab)v$
V4: $1v = v$, where 1 is the multiplicative identity in F.

It is easy to remember the defining properties of a group (closed, associative, identity, and inverses) and the defining properties of a ring (commutative group under addition, semigroup under multiplication, distributive), but not so for vector spaces. This is not unexpected. A vector space has two sets and four operations. It may be helpful to remember the word AID, for associative, identity, and distributive. V3 is an associative property, V4 is an identity property, and V1 and V2 are distributive properties. Be sure to also remember that in scalar multiplication, the scalar is always the left operand, and the vector is always the right operand.

A vector space has three distinct identities. What are they?

1559

1. the additive identity in group V (called the ***zero vector*** and represented by **0**)
2. the additive identity in field F (represented by 0)
3. the multiplicative identity in field F (represented by 1)

Show that if V is a vector space over F, and $v \in V$, then $0v = \mathbf{0}$ (0 in regular typeface is the additive identity in F, and **0** in boldface type is the additive identity in V). *Hint*: Replace 0 with $0 + 0$. Then use V2 in the preceding frame.

1560

$0v = (0 + 0)v = 0v + 0v$. Recall that an idempotent in a group is the identity. $0v$ is idempotent in the group V. Thus, $0v$ equals the identity in V. That is, $0v = \mathbf{0}$.

Show that $(-1)v = -v$. That is, the scalar product of the additive inverse of 1 in F and a vector v equals the additive inverse of v in $<V, +>$. *Hint*: Start with $\mathbf{0} = 0v$, and replace 0 with $(-1 + 1)$.

1561

0
$= 0v$ preceding frame
$= (-1 + 1)v$ -1 is the additive inverse of 1.
$= (-1)v + 1v$ V2 in frame **1558**
$= (-1)v + v.$ V4 in frame **1558**

So $\mathbf{0} = (-1)v + v$. Adding the additive inverse of v to both sides, we get $-v = (-1)v$. ∎

In addition to the defining properties of a vector space given in frame **1558**, be sure to remember the results of the two preceding frames:

V5: $0v = \mathbf{0}$
V6: $(-1)v = -v$

Examples of Vector Spaces

1562

Consider the set $\mathbb{R}^2 = \mathbb{R} \times \mathbb{R}$. This is the set of ordered pairs of real numbers. Define addition on $\mathbb{R}^2$ as follows: $(a, b) + (c, d) = (a + c, b + d)$ for all $a, b, c, d \in \mathbb{R}$. Define scalar multiplication as follows: $c(a, b) = (ca, cb)$ for all $a, b, c \in \mathbb{R}$. Then $\mathbb{R}^2$ is a vector space over $\mathbb{R}$. Accordingly, we call the elements of $\mathbb{R}^2$ (i.e., the ordered pairs) vectors, and the elements of $\mathbb{R}$ scalars.

Let's show that $\mathbb{R}^2$ is a vector space. To do this, we show that

- $\mathbb{R}^2$ under the addition operation defined above is a commutative group, and
- scalar multiplication yields a vector, and
- the four required properties listed in frame **1558** hold.

Let's show that $\mathbb{R}^2$ under the operation addition defined above is a commutative group. Let $a, b, c, d, e, f \in \mathbb{R}$.

Closure: Because $\mathbb{R}$ is closed under addition, $a + c, b + d \in \mathbb{R}$. Thus, $(a, b) + (c + d) = (a + c, b + d) \in \mathbb{R}^2$.
Associativity:
$[(a, b) + (c, d)] + (e, f)$	
$= (a + c, b + d) + (e, f)$	vector addition
$= ([a + c] + e, [b + d] + f)$	vector addition
$= (a + [c + e], b + [d + f])$	$+$ is associative on $\mathbb{R}$.
$= (a, b) + (c + e, d + f)$	vector addition
$= (a, b) + [(c, e) + (d, f)]$	vector addition

Show that the identity, inverses, and commutative properties hold.

1563

Identity:
$(0, 0) + (a, b)$	
$= (0 + a, 0 + b)$	vector addition
$= (a, b)$	0 is the additive identity in $\mathbb{R}$.

Similarly, $(a, b) + (0, 0) = (a + 0, b + 0) = (a, b)$. Thus, $(0, 0)$ is the identity.
Inverses:
$(-a, -b) + (a, b)$	$a, b \in \mathbb{R} \Rightarrow -a, -b \in \mathbb{R} \Rightarrow (-a, -b) \in \mathbb{R}^2$
$= (-a + a, -b + b)$	vector addition
$= (0, 0)$	$-a, -b$ are the inverses of a, b.

Similarly, $(a, b) + (-a, -b) = (a + [-a], b + [-b]) = (0, 0)$. Thus, $(-a, -b)$ is the inverse of (a, b).
Commutativity:
$(a, b) + (c, d)$	
$= (a + c, b + d)$	vector addition
$= (c + a, d + b)$	$+$ is commutative on $\mathbb{R}$.
$= (c, d) + (a, b)$	vector addition ∎

Show that scalar multiplication as defined in frame **1562** yields a vector (i.e., an element in $\mathbb{R}^2$).

1564

$c(a, b) = (ca, cb)$. $ca, cb \in \mathbb{R}$ because $\mathbb{R}$ is closed under multiplication $\Rightarrow c(a, b) \in \mathbb{R}^2$ ∎

We have one more thing to do to establish that $\mathbb{R}^2$ is a vector space: We have to show that the four properties in frame **1558** hold. Let's show that V1, $a(\mathbf{u} + \mathbf{v}) = a\mathbf{u} + a\mathbf{v}$, holds. Let $\mathbf{u} = (c, d)$ and $\mathbf{v} = (e, f)$. Then

$a(\mathbf{u} + \mathbf{v})$	
$= a[(c, d) + (e, f)]$	substitution
$= a(c + e, d + f)$	vector addition
$= (a[c + e], a[d + f])$	scalar multiplication
$= (ac + ae, ad + af)$	multiplication is distributive on $\mathbb{R}$.
$= (ac, ad) + (ae, af)$	vector addition
$= a(c, d) + a(e, f)$	scalar multiplication
$= a\mathbf{u} + a\mathbf{v}$	substitution ∎

Show that V2, $(a + b)v = av + bv$, holds. Let $v = (c, d)$.

1565

$(a + b)v$
$= (a + b)(c, d)$ substitution
$= ([a + b]c, [a + b]d)$ scalar multiplication
$= (ac + bc, ad + bd)$ multiplication is distributive on $\mathbb{R}$.
$= (ac, ad) + (bc + bd)$ vector addition
$= a(c, d) + b(c, d)$ scalar multiplication
$= av + bv$ substitution ∎

Show that V3, $a(bv) = (ab)v$, holds. Let $v = (c, d)$.

1566

$a(bv)$
$= a(b(c, d))$ substitution
$= a(bc, bd)$ scalar multiplication
$= (a[bc], a[bd])$ scalar multiplication
$= ([ab]c, [ab]d)$ multiplication on $\mathbb{R}$ is associative
$= ab(c, d)$ scalar multiplication
$= (ab)v$ substitution ∎

Show that V4, $1v = v$, holds. Let $v = (c, d)$.

1567

$1v$
$= 1(c, d)$ Substitution
$= (1c, 1d)$ Scalar multiplication
$= (c, d)$ 1 is multiplicative identity in $\mathbb{R}$
$= v$ Substitution ∎

We have shown that $\mathbb{R}^2$ is a vector space over $\mathbb{R}$. We can similarly show that $\mathbb{R}^n$ for $n > 2$ are all vector spaces over $\mathbb{R}$ ($\mathbb{R}^n$ is the set of all n-tuples whose components are from $\mathbb{R}$). What about $\mathbb{R}^1 = \mathbb{R}$? For this case, what would be the scalars, what would be the vectors?

1568

$\mathbb{R}$ would be both the set of scalars and the set of vectors.

Give a quick argument that $\mathbb{R}$ is a vector space over $\mathbb{R}$.

1569

We already know that $\mathbb{R}$ is a commutative group. Scalar multiplication is the product of two elements of $\mathbb{R}$. Because $\mathbb{R}$ is closed under multiplication, scalar multiplication yields a vector (i.e., an element in $\mathbb{R}$). Finally, the four required properties in frame **1558** are properties of $\mathbb{R}$: Properties V1 and V2 are the distributive properties, V3 is the associative property of multiplication, and V4 is the multiplicative identity property.

Field over a Subfield

1570

A **subfield** K of a field F is a subset of F that is a field itself under the operations of F restricted to K.

Let K be a subfield of the field F. If $K \neq F$, which one of the following statements is true? Which is false?

1. K is a vector space over F.
2. F is a vector space over K.

Chapter 27: Vector Spaces

To determine which statement is true, let's start by checking if the first statement above is consistent with scalar multiplication. If it is not, then clearly K is not a vector space over F.

Suppose K is a vector space over F, and $K \neq F$. Let f be any element in F not in K (there has to be at least one such element because $K \neq F$ and $K \subseteq F$). Let $\boldsymbol{k}$ be any nonzero element in K (there has to at least one nonzero element in K because K is a field). Then $f\boldsymbol{k}$, the scalar multiplication of f and $\boldsymbol{k}$, must be a vector—that is, $f\boldsymbol{k}$ must be in K.

Show that if $f\boldsymbol{k} \in K$, then $f \in K$, thereby contradicting our assumption that $f \notin K$. *Hint*: If $f\boldsymbol{k} \in K$, then $f\boldsymbol{k} = \boldsymbol{k}'$ for some $\boldsymbol{k}' \in K$.

1571

Multiplying both sides of $f\boldsymbol{k} = \boldsymbol{k}'$ by $\boldsymbol{k}^{-1}$, we get $f\boldsymbol{k}\boldsymbol{k}^{-1} = \boldsymbol{k}'\boldsymbol{k}^{-1}$, which simplifies to $f = \boldsymbol{k}'\boldsymbol{k}^{-1} \in K$. But $f \notin K$.

How do we resolve this contradiction?

1572

We conclude that our initial assumption—that K is a vector space over F—is incorrect.

Now let's see if F is a vector space over K. Take any $k \in K$ and any $f \in F$. Because K is a subset of F, both k and f are elements of F. Thus, $kf \in F$. That is, scalar multiplication always yields a vector (i.e., an element of F) as it should. Moreover, we know that F is a commutative group under addition, and it has the properties listed in frame **1558** (V1 and V2 are the distributive properties, V3 is the associative property under multiplication, and V4 is the multiplicative identity property. Thus, F is a vector space over K.

Vector Space of Polynomials

1573

Let's consider one more example of a vector space. If F is a field, then $F[x]$, the polynomials over F is a vector space over F. $F[x]$ under polynomial addition is a commutative group. Scalar multiplication—the product of an element in F with a polynomial—yields a vector (i.e., a polynomial). Finally, the four properties listed in frame **1558** hold (we will not show this, but the properties follow easily from the properties of polynomials over a field)

Subspaces

1574

Suppose V is a vector space over F. A **subspace** S of V is a subset of V that is also a vector space over F under the operations of V restricted to S. To show a subset S of a vector space V is a subspace, we use the NCC subspace test. In the *NCC subspace test*, we show that S is a subspace of a vector space V by showing that

1. S is not empty, and
2. S is closed under vector addition, and
3. S is closed under scalar multiplication.

"NCC" stands for "nonempty", "closed under vector addition", and "closed under scalar multiplication." If a subset S of a vector space V satisfies these three properties, then S is a subspace of V. Let's confirm that if S satisfies the NCC test, then S is a subspace.

If properties 1 and 2 above hold, then by the CI subgroup test, S is a subgroup of V. Let $v \in S$. We know from frame **1561** that $(-1)v = -v$. Then if property 3 above holds, $-v \in S$. Thus, S has the inverses property under vector addition. S inherits commutativity under addition from V. It also inherits the four properties in frame **1558**. Thus, if S has the three properties specified by the NCC test, then it is a subspace of V.

For example, consider the subset $S[x]$ of polynomials in $\mathbb{Z}_2[x]$ consisting of all polynomials in $\mathbb{Z}_2[x]$ except those of degree greater than 5. The zero vector is in $S[x]$. Thus, $S[x]$ is nonempty. $S[x]$ is closed under vector addition and scalar multiplication (both produce polynomials with degrees less than or equal to 5). Thus, by the NCC test, $S[x]$ is a subspace of $\mathbb{Z}_2[x]$.

Linear Combinations of Vectors

1575

Let V be a vector space over a field F. Suppose $u, v \in V$, and $a, b \in F$. Then $au + bv$ is a **linear combination** of u and v. Let S be the set of all linear combinations of u and v. Using the NCC test, let's show that S is a subspace of V.

First show S is nonempty by showing that the zero vector is in S.

1576

$0u + 0v$
$= 0 + 0$ V5 in frame **1561**
$= 0$ 0 is the additive identity in V.

Next, show that S is closed under addition—that is, show that the sum of two linear combinations is itself a linear combination.

1577

$(au + bv) + (cu + dv)$
$= (au + cu) + (bv + dv)$ vector addition is associative and commutative.
$= (a + c)u + (b + d)v \in S$ V2 in frame **1558**

Finally, show that S is closed scalar multiplication.

1578

$c(au + bv)$
$= c(au) + c(bv)$ V1 in frame **1558**
$= (ca)u + (cb)v \in S$ V3 in frame **1558** ∎

Thus, S is a subspace of V. We say that S is the subspace **spanned** by u and v.

We showed above that the set of all linear combinations of two vectors in a vector space V is a subspace of V. It is easy to see that the steps we took to show this can be extended to the general case. That is, if $v_1, v_2, \ldots, v_n$ are vectors in a vector space V over F, then $\{a_1v_1 + a_2v_2 + \cdots + a_nv_n : a_1, a_2, \ldots, a_n \in F\}$ is a subspace of V. We call this subspace the **subspace spanned by** $v_1, v_2, \ldots, v_n$.

Linear Independence

1579

Let $v_1, v_2, \ldots, v_n$ be vectors in the vector space V over F. If there exists $a_1, a_2, \ldots, a_n$ in F not all zero such that $a_1v_1 + a_2v_2 + \cdots + a_nv_n = 0$, then we say that the vectors $v_1, v_2, \ldots, v_n$ are **dependent**.

A set of vectors is dependent if and only if at least one vector can be expressed as a linear combination of the others. Let's show this.

Suppose $a_1v_1 + a_2v_2 + \cdots + a_nv_n = 0$, and one of the coefficients is nonzero (i.e., $v_1, v_2, \ldots, v_n$ is a dependent set of vectors). Without loss of generality, assume $a_1 \neq 0$. Solve for v_1.

1580

Moving a_1v_1 to the right side, we get $a_2v_2 + \cdots + a_nv_n = -a_1v_1$. Multiplying through by the multiplicative inverse of $-a_1$, and switching sides, we get

v_1
$= ([-a_1]^{-1})(a_2v_2) + \cdots + ([-a_1]^{-1})(a_nv_n)$
$= ([-a_1]^{-1}a_2)v_2 + \cdots + ([-a_1]^{-1}a_n)v_n.$

Thus, v_1 a linear combination of $v_2, v_3, \ldots,$ and v_n.

Let's now consider the converse: If a vector in a set of vectors S can be expressed as a linear combination of the other vectors in S, then S is a dependent set of vectors. Suppose v_1 can be expressed as a linear combination of $v_2, v_3, \ldots, v_n$. That is, $v_1 = b_2 v_2 + \cdots + b_n v_n$. Moving v_1 to the right side, we get

$$\mathbf{0} = -v_1 + b_2 v_2 + \cdots + b_n v_n.$$

By V6 in frame **1561**, $-v_1 = (-1)v_1$. Substituting, we get $\mathbf{0} = -v_1 + b_2 v_2 + \cdots + b_n v_n = (-1)v_1 + b_2 v_2 + \cdots + b_n v_n$. The coefficient of v_1 is -1, a nonzero value. We have a linear combination of vectors equal to $\mathbf{0}$ in which not all the coefficients are 0. Thus, the set of vectors $\{v_1, v_2, \ldots, v_n\}$ is dependent. ∎

If a set of vectors is not dependent, we say that the set is ***independent***.

Show that $\{(1, 1), (2, 2)\}$ is a dependent set of vectors from $\mathbb{R}^2$.

1581

$-2(1, 1) + 1(2, 2) = (-2, -2) + (2, 2) = (0, 0)$. $(0, 0)$ is the zero vector in $\mathbb{R}^2$. Thus, the set is dependent.

Show that $\{(1, 0), (1, 1)\}$ is an independent set of vectors. *Hint*: Determine what the coefficients must be in a linear combination of these vectors that equals the zero vector. That is, solve for a and b in $a(1, 0) + b(1, 1) = (0, 0)$.

1582

Suppose $a(1, 0) + b(1, 1) = (0, 0)$. Then $(a, 0) + (b, b) = (0, 0)$. Adding the vectors $(a, 0)$ and (b, b), we get $(a + b, b) = (0, 0)$ $\Rightarrow a + b = 0$ and $b = 0 \Rightarrow a = 0$. The only linear combination of $(1, 0)$ and $(1, 1)$ that equals $(0, 0)$ has its two coefficients equal to 0. Thus, $\{(1, 0), (1, 1)\}$ is an independent set.

Show that $(1, 0)$ and $(0, 1)$ are independent.

1583

Suppose $a(1, 0) + b(0, 1) = (0, 0)$. Then $(a, 0) + (0, b) = (0, 0)$. Adding the two vectors, we get $(a, b) = (0, 0) \Rightarrow a = b = 0 \Rightarrow$ the vectors are independent.

Basis of a Vector Space

1584

The vectors $(1, 0)$ and $(0, 1)$ are not only independent, they also span $\mathbb{R}^2$. That is, every vector in $\mathbb{R}^2$ can be expressed as a linear combination of $(1, 0)$ and $(0, 1)$. It is easy to show this. Let (x, y) be an arbitrary vector in $\mathbb{R}^2$. We have to produce a linear combination of $(1, 0)$ and $(0, 1)$ that equals (x, y). Here is such a linear combination: $x(1, 0) + y(0, 1) = (x, 0) + (0, y) = (x, y)$.

$(1, 0)$ and $(1, 1)$ are independent. Show that they span $\mathbb{R}^2$.

1585

Let (x, y) be an arbitrary vector in $\mathbb{R}^2$. $(x - y)(1, 0) + y(1, 1) = (x - y, 0) + (y, y) = (x, y)$.

Let V be a vector space over a field F. A subset B of V is called a ***basis*** for V if the vectors in B are linearly independent and span V. For example, we showed above that $(1, 0)$ and $(0, 1)$ are linearly independent and span $\mathbb{R}^2$. Thus, $\{(1, 0), (0, 1)\}$ is a basis for $\mathbb{R}^2$. We also showed the same for $(1, 0)$ and $(1, 1)$. Thus, $\{(1, 0), (1, 1)\}$ is also a basis for $\mathbb{R}^2$. Clearly, the basis for a vector space is not unique. However, for any given vector space, the number of vectors in a basis is invariant. For example, every basis for $\mathbb{R}^2$ has exactly two vectors.

Give a basis for the subspace that consists of all the polynomials over $\mathbb{R}$ except for the polynomials of degree greater than 3.

1586

$\{1, x, x^2, x^3\}$

Suppose a set S of vectors is dependent. Then by frame **1580** at least one vector in S—let's call it v—can be expressed as a linear combination of the other vectors in S. If we discard v, the remaining set of vectors span the same subspace as the original set of vectors. Justify this assertion.

Any linear combination of vectors in S that includes v can be written without v by replacing v with its linear combination equivalent.

However, we cannot discard any vectors in an independent set and still span the same subspace. For this reason, a basis is often called a ***minimal spanning set***. It is minimal in the sense none of its vectors can be discarded without affecting the subspace spanned. For example, suppose the set of vectors, $v_1, v_2, \ldots, v_n$, is independent. The subspace spanned by this set of vectors includes v_1 (take the linear combination in which the coefficient of v_1 is 1 and all the remaining coefficients are 0). However, if we discard v_1, then the subspace spanned by the remaining vectors does not include v_1. If, to the contrary, it did include v_1, then v_1 could be expressed as a linear combination of the vectors $v_2, v_3, \ldots, v_n$. But that would mean the original set of vectors is dependent.

If a vector space has a basis with n vectors, then we say it has the ***dimension*** n. If a basis for a vector space is finite, we say that the vector space is ***finite dimensional***. Otherwise, it is ***infinite dimensional***.

Give a basis for the vector space consisting of all polynomials over $\mathbb{R}$. *Hint*: This vector space is infinite dimensional.

1588

$\{1, x, x^2, x^3, \ldots\}$

Review Questions

1. If e is a vector in the expression $a(b(c(cd + e))$, which variables represent vectors? If e is a scalar, which variables represent vectors?
2. Is the function $f: \mathbb{R} \to \mathbb{R}$ defined by $f(x) = x$ a subspace of $\mathbb{R}^2$? *Hint*: f defines a set of ordered pairs.
3. Is $\mathbb{Z}_4 \times \mathbb{Z}_4$ a vector space over $\mathbb{Z}_4$, where vection addition and scalar multiplication is defined as in the vector space $\mathbb{R}^2$. Justify your answer.
4. u and v are vectors in a vectors space V over $\mathbb{R}$. What is wrong with the following expresion $(2 + 3)uv$.
5. Describe the subspace of $\mathbb{R}[x]$ spanned by $\{1, x, x^2\}$.
6. Show that $S = \{(x, x) : x \in \mathbb{R}\}$ is a subspace of $\mathbb{R}^2$.
7. Does the set $\{(0, 1), \{1, 2)\}$ span $\mathbb{R}^2$? Justify your answer.
8. Does the set $\{1, x\}$ span $\mathbb{R}[x]$. Justify your answer.
9. Show that $\{(1, 1, 0), (0, 1, 1), (0, 1, 0)\}$ is a linearly independent set of vectors of $\mathbb{R}^3$.
10. Describe the subspace of $\mathbb{R}[x]$ spanned by $\{x^3\}$.

Answers to the Review Questions

1. If e is a vector, then then d and only d is also a vector. If e is a scalar then $a, b, c, d,$ and e are all scalars.
2. Yes. $f = \{(x, x) : x \in \mathbb{R}\} \Rightarrow f$ is not empty. Let $u, v \in f$. Then $u = (x, x)$ and $v = (y, y)$ for some $x, y \in \mathbb{R}$. $u + v = (x, x) + (y, y) = (x + y, x + y) \in f \Rightarrow$ closed under vector addition. $cu = c(x, x) = (cx, cx) \in f \Rightarrow$ closed under scalar multiplication. Thus, by the NCC subspace test, f is a subspace.
3. No. $\mathbb{Z}_4$ is not a field (2 has no multiplicative inverse).
4. The product uv is not defined.
5. $\{ax^2 + bx + c : a, b, c \in \mathbb{R}\}$
6. Same problem as review problem 2.
7. Yes. $a(0, 1) + b(1, 2) = (x, y) \Rightarrow (b, a + 2b) = (x, y) \Rightarrow b = x$ and $a = y - 2b = y - 2x$. Thus, for any (x, y), there is a linear combination of $(0, 1)$ and $(1, 2)$ equal to (x, y). For example, to get $(2, 3)$, we need $a = y - 2x = 3 - 2(2) = -1$ and $b = 2$. Let verify this: $-1(0, 1) + 2(1, 2) = (0, -1) + (2, 4) = (2, 3)$.
8. No. None of the linear combinations of 1 and x have powers of x greater than 1.
9. Suppose $a(1, 1, 0) + b(0, 1, 1) + c(0, 1, 0) = (0, 0, 0) \Rightarrow (a, a, 0) + (0, b, b) + (0, c, 0) = (0, 0, 0) \Rightarrow a = 0, a + b + c = 0,$ and $b = 0 \Rightarrow a = b = c = 0 \Rightarrow$ the set of vectors is independent.
10. $\{ax^3 : a \in \mathbb{R}\}$

Homework Questions

1. Is $S = \{(x, x+1) : x \in \mathbb{R}\}$ a subspace of $\mathbb{R}^2$? Justify your answer.
2. Is $S = \{(x, y, z) : x, y, z \in \mathbb{R} \text{ and } x + y + z = 0\}$ a subspace of $\mathbb{R}^3$. Justify your answer.
3. Is $\{t = (2, 4, 6), u = (3, 2, 1), v = (10, 12, 14)\}$ a linearly dependent set of vectors of $\mathbb{R}^3$? Justify your answer.
4. Is $\{ax + b : a, b \in \mathbb{R}\}$ a subspace of $\mathbb{R}[x]$? Justify your answer.
5. Suppose that $S = \{v_1, v_2, \ldots, v_n\}$ spans a vector space V. Show that some subset of S is the basis for V.
6. Show that any set of vectors that contains the **0** vector is not linearly independent.
7. Show $S = \{p(x) : p(1) = 0\}$ is a subspace of $\mathbb{R}[x]$.
8. Suppose S_1 and S_2 are subspaces of the vector space V such that $S_1 \cup S_2$ is a subspace of V. Show that $S_1 \subseteq S_2$ or $S_2 \subseteq S_1$.
9. Suppose the vectors u, v in a vector space V over $\mathbb{R}$ are linearly independent. Show that $\{u + v, u - v\}$ is also linearly independent.
10. Does $S = \{1, x, x^2, x^3\}$ span the same vector space as $T = \{1 + x, x + x^2, x^2 + x^3\}$? Justify your answer.

28 Partial Orders, Lattices, and Boolean Algebra

Partial Orders

1589

A *partial order*, like an equivalence relation, is a relation on a set that is reflexive and transitive. But, unlike an equivalence relation, a partial order is antisymmetric rather than symmetric. An *antisymmetric relation* is a relation in which no connection between two distinct elements is a "two-way street." That is, it cannot have any two elements connected this way:

We will use the symbol "≼" to represent a partial order unless it already has a well-established symbol.

Draw the arrow diagram for the relation on the set $P = \{a, b, c, d, e\}$ given by $\{(a, a), (b, b), (c, c), (d, d), (c, a), (c, b), (d, c), (d, a), (d, b)\}$.

1590

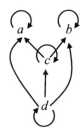

Does this arrow diagram represent a partial order?

1591

Yes. Each element has a self-loop, so the relation is reflexive. There are no two-way streets, so the relation is antisymmetric. Where there is a path from one element to another element, there is an arrow from the former pointing directly to the latter, so the relation is transitive. For example, there is a path from d to a via c. Transitivity requires that there be an arrow from d pointing directly to a, and there is.

Let's denote the partial order in the preceding frame with ≼. We can indicate that an ordered pair is in this relation in several ways. For example, to indicate that (c, b) is in the relation, we can write any of the following:

- $(c, b) \in$ ≼
- c ≼ b, or, equivalently, b ≽ c
- c is related to b.
- c is less than or equal to b.
- b is greater than or equal to c.

One of the characteristics of an arrow diagram for a partial order is that it has no cycles other that the self-loops on each element. Suppose a relation R has the ordered pairs (a, b), (b, c) and (c, a) in addition to all the pairs required by reflexivity and transitivity. This relation has a cycle that is not a self-loop: a to b, b to c, and c back to a. Here is its arrow diagram (we have omitted the arrows required by reflexivity and transitivity):

Show that this relation cannot be a partial order because it lacks the antisymmetry property.

1592

$b\,R\,c$ and $c\,R\,a \Rightarrow b\,R\,a$ (by transitivity). We also have that $a\,R\,b$. By antisymmetry, a must equal b. But a and b are distinct elements. Thus, the relation is not antisymmetric.

Because of the lack of cycles in a partial order, we can draw an arrow diagram for a partial order so all the arrows, except for the self-loops, are pointing in an upward direction. Thus, the arrowhead on each arrow will be higher in the diagram than the tail of the arrow. With this understanding, we do not need to show the arrowheads at all—the relative position of the endpoints of each arrow indicate where the arrowhead is. We also do not need to show the self-loops or the arrows implied by transitivity.

Using these simplifications, redraw the diagram in frame **1590**.

1593

Arrow diagrams of partial orders simplified in this way are called **Hasse diagrams**.

We call a set with a partial order a ***partially ordered set***. For example, the set P in frame **1589** is a partially ordered set.

Lattices

1594

Consider the following Hasse diagram of a partial order $\leqslant$ on the set $P = \{a, b, c, d, e, f\}$:

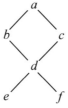

Let X be a subset of P. An element $p \in P$ is an **upper bound** of X if and only if for all $x \in X$, $x \leqslant p$.

What are the upper bounds of $X = \{e, f\}$ in the partially ordered set given by the Hasse diagram above? *Hint*: X has four upper bounds.

1595

a, b, c, and d

A *least upper bound* (abbreviated **lub**) of a subset X of a partially ordered set is an upper bound of X that is less or equal to *every* upper bound of X. For example, in the Hasse diagram in the preceding frame, d has the property that it is an upper bound of $\{e, f\}$, and is less than or equal every upper bound of X. Thus, d is a lub of $\{e, f\}$.

What are the least upper bounds of the following subsets of the partially ordered set in the preceding frame: $\{e\}$, $\{d, e\}$, $\{a, f\}$, and $\{b, c\}$.

1596

lub of $\{e\}$ = e, lub of $\{d, e\}$ = d, lub of $\{a, f\}$ = a, lub of $\{b, c\}$ = a.

Is it possible that a lub may not exist for a subset X of some partially ordered set?

1597

Yes. Consider

The only upper bound of $\{a\}$ is a; the only upper bound of $\{b\}$ is b. There is no element that is an upper bound of both a and b. Thus, there is no lub of $\{a, b\}$.

It is possible for subset of a partial order to have an upper bound but no lub. Give a Hasse diagram for which this is the case.

1598

$a \quad b$
$| \bowtie |$
$c \quad d$

The subset $\{c, d\}$ has two upper bounds: a and b. But neither a nor b is a lub. The element a is not a lub because it is not less or equal to b (by definition a lub of a subset X is less or equal to *every* upper bound of X). Similarly, b is not a lub because it is not less or equal to a.

Is it possible for a subset X of a partial order to have more than one lub? Let's investigate. Suppose u_1 and u_2 are two least upper bounds for a subset X. How do we know that $u_1 \leqslant u_2$?

1599

u_2 is an upper bound, and u_1 is a lub. Thus, by the definition of a lub, $u_1 \leqslant u_2$.

How do we know that $u_2 \leqslant u_1$?

1600

u_1 is an upper bound, and u_2 is a lub. Thus, by the definition of a lub, $u_2 \leqslant u_1$.

$u_1 \leqslant u_2$ and $u_2 \leqslant u_1$. What can we conclude?

1601

By the antisymmetry property of partial orders, $u_1 = u_2$. Thus, the lub of X, if it exists, is unique.

Using the definitions of upper bound and least upper bound as models, define a lower bound and a greatest lower bound.

1602

Let X be a subset of a partially order set p. An element $p \in P$ is a *lower bound* of X if and only if for all $x \in X, p \leqslant x$. A *greatest lower bound* (abbreviated **glb**) of a subset X is a lower bound of X that is greater than or equal to *every* lower bound of X.

In frame **1594**, what are the lower bonds of $\{b, c\}$ and the glb of $\{b, c\}$?

1603

d, e, and f are the lower bounds of $\{b, c\}$; d is the glb.

Like a lub, a glb is unique if a subset X of a partial order P has a glb. Incidentally, "lub" is pronounced as it is written; "glb" is pronounced "glub." We are now ready to define a lattice: A *lattice* is a partially ordered set in which every pair of elements has a lub and a glb. Which of the following partially ordered sets is a lattice:

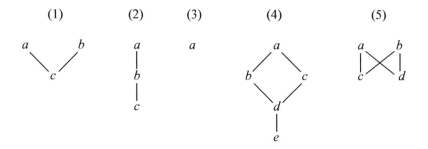

1604

Diagrams 2, 3, and 4 are lattices. In 1, a and b have no lub. In 5, a and b have neither a lub nor a glb; c and d also have neither a lub nor a glb.

What is the lub and glb of $\{a, b\}$ in diagram 2 in the preceding frame?

1605

lub of $\{a, b\} = a$, glb of $\{a, b\} = b$.

Let's define two binary operations—the join and the meet—on the elements of a lattice, the former based on the lub, and the latter based on the glb. Let $a, b \in L$, where L is a lattice. Then the *join* of a and b (denoted by $a + b$) is the least upper bound of a and b. The *meet* of a and b (denoted by $a \cdot b$ or ab) is the glb of a and b.

In diagram 4 in frame **1603**, what do $b + c$, bc, $b + e$, and be equal?

1606

$b + c = a$, $bc = d$, $b + e = b$, $be = e$.

Construct the Cayley tables for the join and meet operations for the following lattice:

1607

·	a	b	c	d
a	a	b	c	d
b	b	b	d	d
c	c	d	c	d
d	d	d	d	d

+	a	b	c	d
a	a	a	a	a
b	a	b	a	b
c	a	a	c	c
d	a	b	c	d

In a lattice with elements a and b, does $a + b = b + a$?

1608

Yes. $x + y$ is the least upper bound of the set $\{x, y\}$. $y + x$ is the least upper bound of $\{y, x\}$. $\{x, y\}$ and $\{y, x\}$ are the same set. Thus, the two upper bounds are equal.

In a lattice, with elements x and y, is $x + y$ an upper bound of x? Is xy a lower bound of x?

1609

By definition of the join operator +, the join $x + y$ is an upper bound of both x and y. By definition of the meet operator, the meet xy is a lower bound of both x and y.

In a lattice with elements a, b, and c, is $(a + b) + c$ is an upper bound of $\{a, b, c\}$?

1610

By the preceding frame, $(a + b) + c$ is an upper bound of $a + b$, and $a + b$ is an upper bound of a and b. Thus, by the transitivity property of partial orders, $(a + b) + c$ is an upper bound of both a and of b. By the preceding frame, $(a + b) + c$ is also an upper bound of c. Thus, $(a + b) + c$ is an upper bound of a, b, and c.

Show that $(a + b) + c$ is the least upper bound of $\{a, b, c\}$. *Hint*: Show that for any upper bound u of $\{a, b, c\}$, $(a + b) + c$ is less than or equal to u. Then by the definition of a lub, $(a + b) + c$ is the least upper bound of $\{a, b, c\}$.

1611

Suppose u is an upper bound of $\{a, b, c\}$. Then u is an upper bound of $\{a, b\}$. $a + b$ is the *least* upper bound of $\{a, b\}$. Thus, u is an upper bound of $a + b$. u is also an upper bound of c. $(a + b) + c$ is the *least* upper bound of $a + b$ and c. Thus, u is an upper bound of $(a + b) + c$. We have shown that any upper bound of $\{a, b, c\}$ is greater than or equal to $(a + b) + c$. Thus, $(a + b) + c$ is a least upper bound of $\{a, b, c\}$.

Using similar reasoning, we can show that $a + (b + c)$ is also a least upper bound of $\{a, b, c\}$. What can we then conclude about $(a + b) + c$ and $a + (b + c)$?

1612

Because the lub of $\{a, b, c\}$ is unique, $(a + b) + c = a + (b + c)$. In other words, the join operator + is associative.

In a lattice, does $a + a = a$? That is, is every element in a lattice idempotent?

1613

Yes. $a + a$ is the least upper bound of a. In a Hasse diagram, any upper bound of a not equal to a must be above a. Thus, a is the least upper bound of a. That is, $a + a = a$.

In a lattice, does $a + ab = a$?

1614

By the definition of a glb, ab is a lower bound of a. Thus, in the Hasse diagram for the lattice, ab is lower than a, and connected to a:

$$a$$
$$\vdots$$
$$ab$$

Thus, the least upper bound of a and ab is a. That is, $a + ab = a$. We call this property of lattices the **absorption law**. The a term in "$a + ab$" absorbs the ab term, yielding a as the result.

Suppose you look at a Hasse diagram in its normal orientation while you are standing on your head. If p is an upper bound of some set X, to you it will look like a lower bound of X. If p is the lub of X, to you it will look like the glb of X. If $a \leqslant b$, to you it will look like $a \geqslant b$. In your upside-down world, lower bounds are upper bounds, upper bounds are lower bounds, lub's are glb's, glb's are lub's, and $\leqslant$ is $\geqslant$.

Now suppose we have an assertion about a partially ordered set. If we look at the corresponding Hasse diagram upside down, we can make the same assertion, but we have to change upper bound to lower bound, lower bound to upper bond, lub to glb, glb to lub, and so on. This observation gives leads us to an important property of partially ordered sets call the ***principle of duality***:

> A statement about a partially order set remains true if we interchange "lub" and "glb" (or, equivalently, "+" and "·"), upper bound and lower bound, and "$\leqslant$" and "$\geqslant$".

A lattice is a special type of partial order. Thus, the principle of duality also applies to lattices. For example, we showed that in a lattice L, $a + (a \cdot b) = a$ for all $a, b \in L$. Then by the principle of duality, $a \cdot (a + b) = a$.

334 Chapter 28: Partial Orders, Lattices, and Boolean Algebra

Let's summarize the properties of lattice we have shown in the preceding frames. We also give the dual of each property, which by the principle of duality, is also true. Let L be a lattice with $a, b, c \in L$. Then

Commutativity:	$a + b = b + a$	$a \cdot b = b \cdot a$
Associativity:	$a + (b + c) = (a + b) + c$	$a \cdot (b \cdot c) = (a \cdot b) \cdot c$
Idempotency:	$a + a = a$	$a \cdot a = a$
Absorption:	$a + ab = a$	$a \cdot (a + b) = a$

A **largest element** in a partially ordered set is an element that is as large or larger than every element in the set. A **smallest element** is an element that is as small or smaller than every element in the set. We call a lattice with both a largest element and a smallest element a **bounded lattice**.

Show that a partially ordered set can have at most one largest element and at most one smallest largest element.

1615

Suppose to the contrary that a partially ordered set has two largest elements l_1 and l_2. Because l_2 is a largest element, l_1 is less than or equal to l_2. Because l_1 is a largest element, l_2 is less than or equal to l_1. Then by the antisymmetry property of partial orders, $l_1 = l_2$.

Give a Hasse diagram of a partially ordered set that does not have a largest element. Give a Hasse diagram of a partially ordered set that does not have a smallest element.

1616

In the diagram on the left, a is not larger than or equal to b. Similarly, b is not larger than or equal to a. Thus, neither a nor b can be the largest element. In the diagram on the right, neither e nor f is the smallest.

Using the diagrams above, give an informal argument that every finite lattice has a largest element and a smallest element.

1617

Every pair of elements in a lattice has a lub and a glb. Thus, the forking structures in the Hasse diagrams in the preceding frame (with no element greater than or equal to a and b, and with no element less than or equal to e and f) cannot appear in a Hasse diagram for a lattice.

When we are discussing lattices in general, we will use 1 to represent the largest element and 0 to represent the smallest element. When applying the principle of duality, in addition to the interchanges specified in frame **1614**, we interchange the largest element with the smallest element. For example, $a + 0 = a$. Then by the principle of duality, $a \cdot 1 = a$.

Suppose $0, 1, x \in L$, where L is a lattice, 0 is the smallest element, and 1 is the largest element. What does $x + 1, x + 0, x \cdot 1$ and $x \cdot 0$ equal?

1618

$x + 1 = 1, x + 0 = x, x \cdot 1 = x$, and $x \cdot 0 = 0$.

If in a lattice L, $a \cdot (b + c) = (a \cdot b) + (b \cdot c)$ for all $a, b, c \in L$, we say the lattice is **distributive**. In other words, a distributive lattice is a lattice in which the meet operation ($\cdot$) distributes over the join operation ($+$). By the principle of duality, in a distributive lattice, the join operator distributes over the meet operator. That is, $a + (b \cdot c) = (a + b) \cdot (a + c)$.

Show that the lattice corresponding to the following Hasse diagram is not distributive:

1619

$b \cdot (c + d) = b \cdot a = b$, but $(b \cdot c) + (b \cdot d) = e + e = e$. Thus, the lattice is not distributive. Distributivity is *not* a property of all lattices.

Suppose L is a lattice with a largest element 1 and a smallest element 0. If for every $a \in L$ there is an element $a^c \in L$ such that $a + a^c = 1$, and $a \cdot a^c = 0$, we say the lattice L is **complemented**. We call a^c a **complement** of a. For example, the element d in the diagram in the preceding frame is a complement of b because

- $b + d = a$, and a is the largest element, and
- $b \cdot d = e$, and e is the smallest element.

It follows from the definition of a complemented lattice that a complemented lattice is necessarily a bounded lattice (i.e., it has a largest element and a smallest element).

In the preceding frame, is c also the complement of b?

1620

Yes. Thus, a complement of an element is not necessarily unique.

What is the complement of a in the Hasse diagram in frame **1618**?

1621

e

In a complemented lattice, the complement of the largest element is the smallest element; the complement of the smallest element is the largest element.

Every element in the Hasse diagram in frame **1618** has a complement. Thus, it is a complemented lattice. Is the following lattice a complemented lattice:

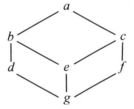

1622

No. b and f are complements, d and c are complements, a and g are complements, but e has no complement. This lattice is bounded but not complemented. A bounded lattice is *not* necessarily complemented.

In a complemented lattice, complements are *not* necessarily unique. For example, in the Hasse diagram in frame **1618**, b has both c and d as complements). However, in a complemented lattice that is also distributive, complements are unique. That is, each element has one and only one complement. Let's prove this. Our approach will be this: We assume that an element a has two complements: b and c. We then show that $b = c$. Thus, the element a cannot have more than one complement. Here is the proof: Suppose L is a bounded distributive lattice with largest element 1 and smallest element 0 in which an element a has complements b and c. Thus, $a + b = 1$, $a + c = 1$, $a \cdot b = 0$, and $a \cdot c = 0$.

b
$= b \cdot 1$ frame **1618**
$= b \cdot (a + c)$ c is the complement of a
$= b \cdot a + b \cdot c$ distributive law
$= a \cdot b + b \cdot c$ commutative law
$= 0 + b \cdot c$ b is the complement of a
$= b \cdot c$ frame **1618**

Similarly show that $c = b \cdot c$

1623

$c = c \cdot 1 = c \cdot (a + b) = c \cdot a + c \cdot b = 0 + c \cdot b = c \cdot b = b \cdot c$. We have shown that $b = b \cdot c$, and $c = b \cdot c$. Thus, $b = c$. ∎

Let summarize:

- A partially ordered set in which every pair of elements has a lub and a glb is a lattice.
- A lattice is not necessarily distributive.
- A lattice is not necessarily complemented.
- In a complemented lattice, the complement of an element is not necessarily unique.
- In a complemented distributive lattice, the complement of each element is unique.

All lattices with three elements have the same Hasse diagram, except for the names of the elements. Draw this diagram.

1624

```
a
|
b
|
c
```

Draw all the possible Hasse diagrams for a lattice with four elements.

1625

Boolean Algebra

1626

A Boolean algebra is a complex algebraic structure. However, we can define it easily as a special type of lattice: A **Boolean algebra** is a complemented distributive lattice. Using our knowledge of lattices, let's list the properties that a Boolean algebra must have. Because a Boolean algebra is a lattice, for all elements a, b,

Commutativity:	$a + b = b + a$	$a \cdot b = b \cdot a$
Associativity:	$a + (b + c) = (a + b) + c$	$a \cdot (b \cdot c) = (a \cdot b) \cdot c$
Idempotency:	$a + a = a$	$a \cdot a = a$
Absorption:	$a + ab = a$	$a \cdot (a + b) = a$

Because a Boolean algebra is complemented, it is necessarily bounded. Thus, it has a largest element 1 and a smallest element 0. It then follows that for each element a,

Identity:	$a + 0 = a$	$a \cdot 1 = a$
Domination:	$a + 1 = 1$	$a \cdot 0 = 0$

Because a Boolean algebra is a complemented lattice, for every element a there is an element a^c such that

Complements:	$a + a^c = 1$	$a \cdot a^c = 0$

0 and 1 are complements. That is,

Complements:	$0 + 1 = 1$	$0 \cdot 1 = 0$

Because a Boolean algebra is a distributive lattice, for all elements a, b, c,

Distributivity: $\quad a \cdot (b + c) = a \cdot b + a \cdot c \qquad a + (b \cdot c) = (a + b) \cdot (a + c)$

Here are two more properties of Boolean algebra (see homework problem 5):

DeMorgan's laws: $\quad (a + b)^c = a^c \cdot b^c \qquad\qquad (a \cdot b)^c = a^c + b^c$
Cancellation law: $\quad a + b = a + c \Rightarrow b = c \qquad a \cdot b = a \cdot c \Rightarrow b = c$

$\mathcal{P}(A)$, the set of all subsets of a set A, is partially ordered by $\subseteq$. Draw the Hasse diagram.

1627

From frame **1625**, we know this diagram represents a lattice. Its largest element is $A = \{a, b\}$; its smallest element is $\emptyset$. What are the complements of each of its elements?

1628

$\{a, b\}$ and $\emptyset$ are complements of each other; $\{a\}$ and $\{b\}$ are complements of each other.

Is the lattice in the preceding frame also distributive?

1629

Yes. Thus, it is a complemented, distributive algebra—it is a Boolean algebra.

Associated with a lattice are two binary operations—one based on the lub and one based on the glb. What are two binary operations for the lattice in frame **1627**?

1630

The lub of any two elements in $\mathcal{P}(A)$ is the union of those elements; the glb of any two elements is the intersection of those two elements. The complement of any element is the set complement with respect to the largest element $\{a, b\}$.

Let's list the features of this Boolean algebra:

Set of elements:	$\mathcal{P}(A) = \{\emptyset, \{a\}, \{b\}, \{a, b\}\}$
Partial order:	$\subseteq$
Largest element	$A = \{a, b\}$
Smallest element:	$\emptyset$
Binary operations:	$\cup$ (set union), which yields the lub of any two elements.
	$\cap$ (set intersection), which yields the glb of any two elements
Unary operation:	Set complementation (the complement with respect to the set $\{a, b\}$)

In this example, $\mathcal{P}(A)$ under union, intersection, and complementation is a Boolean algebra, where A has two elements. This result extends to the general case: $\mathcal{P}(A)$ is a Boolean algebra for any finite set A. Here is an interesting theorem on finite Boolean algebras:

If B is a finite Boolean algebra, then there is a finite set A such that B is isomorphic to $\mathcal{P}(A)$.

From this theorem, what can we conclude about the size of any finite Boolean algebra? *Hint*: $|\mathcal{P}(A)| = 2^{|A|}$.

1631

The size of any finite Boolean algebra is a power of 2.

Chapter 28: Partial Orders, Lattices, and Boolean Algebra

Review Questions

1. A set has six elements. Can it be a partially ordered set? Can it be a lattice? Can it be a Boolean algebra?
2. Is the set of all positive divisors of 12 a partially ordered set under the relation "divides"?
3. Is the set of all positive divisors of 12 a lattice under the relation "divides"?
4. Is the set of all positive divisors of 12 a Boolean algebra under the relation "divides"?
5. Is $\mathbb{Z}$ a partially ordered set under the relation "is equal to"?
6. Is $\mathbb{Z}^+$ a partially ordered set under the relation "is not equal to"?
7. Describe the Hasse diagram for $\mathbb{Z}$ under the relation "is equal to."
8. What is the largest element in the Boolean algebra $\mathcal{P}(\mathbb{Z})$?
9. Draw the Hasse diagram for $\mathcal{P}(A)$ under $\subseteq$, where $A = \{a, b, c\}$.
10. What does the Hasse diagram look like for a partial order in which for every pair of elements a and b, either $a \leqslant b$ or $a \geqslant b$.

Answers to the Review Questions

1. Yes, yes, no (size of every finite Boolean algebra is a power of 2).
2. Yes ("divides" for this set is reflexive, antisymmetric, and transitive).
3. Yes. lub is lcm; glb is gcd.
4. No. The number of positive divisors is not a power of 2.
5. Yes. The relation is reflexive, antisymmetric, and transitive.
6. No. It is not reflexive, not antisymmetric, and not transitive.
7. No two distinct integers are related. Thus, the Hasse diagram has no connecting lines.
8. $\mathbb{Z}$.
9.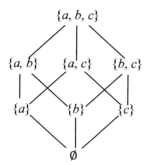
10. It is a single chain with no branches.

Homework Questions

1. Draw all the possible Hasse diagrams for a lattice with five elements.
2. Is $\mathbb{Z}^+$ a partially ordered set under "divides"?
3. Is $\mathbb{Z}$ a partially ordered set under the relation "divides"?
4. Is $\mathbb{Z}$ a partially ordered set under the relation "less than"?
5. Prove that DeMorgan's law, $(a + b)^c = a^c \cdot b^c$, holds in a Boolean algebra. *Hint*: Show that $(a + b) + a^c \cdot b^c = 1$.
6. Simplify this Boolean algebra expression $xyz + xy^c z$.
7. Simplify the Boolean algebra expression $x + xy$.
8. Simplify the Boolean algebra expression $x + x^c y$.
9. Simplify the Boolean algebra expression $(a + b) \cdot (a + b^c)$.
10. Simplify the Boolean algebra expression $x^c y^c z^c + xy^c z^c + x^c yz^c + xyz^c$.

Final Exam

Chapter 1

1. List the elements in the power set of $\{a, \{b, c\}\}$.
2. What is the power set of $\{a, \{a\}\}$.
3. Is the power set of a set A always bigger than the set A?
4. How many subsets does $\{1, a, 3.5\}$ have?

Chapter 2

5. Is "This sentence is false" a mathematical statement?
6. Negate $(\forall x)(\exists y)(x < y)$.
7. Restate using a quantifier: 2 is the smallest prime.
8. Restate using the existential quantifier "there exists": Not all math students play baseball.

Chapter 3

9. Determine the disjunction of conjunctions that corresponds to

p	q	r	?
F	F	F	T
F	F	T	T
F	T	F	F
F	T	T	T
T	F	F	F
T	F	T	F
T	T	F	T
T	T	T	F

10. Are the following statement and its converse true? If the sum x and y is even, then x and y are even.
11. Is $p \wedge p \equiv p$?
12. Is $p \vee (p \wedge q) \equiv p$?

Chapter 4

13. Prove by induction that $2^{2n} - 1$ is divisible by 3 for all $n \in \mathbb{Z}^+$.
14. Prove that if x is an integer, then x or $x + 1$ is even.
15. Prove that all positive integers are interesting. *Hint*: Us the well-ordering principle.

Chapter 5

16. Complement using DeMorgan's laws: $[(A \cup B \cup C) \cap (D \cup E)] \cup F$.
17. Determine the number of numbers from 1 to 1000 that are divisible by 3 or 7. Use the principle of inclusion-exclusion.

Chapter 6

18. Show that the function $f(x) = x^2 + x$ on $\mathbb{R}$ is not one-to-one. *Hint*: Find two values that map to the same value.
19. Show that $f(x) = 100x + 200$ on $\mathbb{R}$ is an onto function.
20. Specify an injection from $\mathbb{Z}^+$ to $T = \{x \in \mathbb{R} : 0 < x \leq 1\}$. What does the existence of such an injection imply about the cardinality of $\mathbb{Z}^+$ and T?

Chapter 7

21. A binary operation on S maps ordered pairs of elements of S. A ternary operation on S maps ordered triples. How many binary operations on S exist if S has n elements? How many ternary operations?
22. Does $\mathbb{Z}$ have an identity element under the operation $\oplus$ defined by $x \oplus y = x + y - 5$?

23. Suppose a set has a left identity l and a right identity r under the operation *. Does l necessarily equal r?
24. Is r in the preceding question an identity?

Chapter 8

25. Let x be the product of all the elements in a finite commutative group. Now consider xx. Can the individual elements in xx be rearranged so that each successive pair of elements contains elements that are inverses of each other? What does xx equal?
26. Let G be the set of nonnegative integers. Is $<G, \otimes>$ a group, where $a \otimes b = |a - b|$ for all $a, b \in G$?
27. Show that in a group if x has an inverse y and a right inverse r, then y and r are the same element.
28. Show that if x has and left inverse l and a right inverse r, then $l = r$.
29. Show that in a group, if $ab = e$, then $ba = e$.

Chapter 9

30. What is the order of the element with the highest order in S_{12}? *Hint*: (1 2 3 4 5 6 7) ∘ (8 9 10 11 12) has order 35, but there are elements of higher order.
31. How many elements in A_7 are 3-cycles?
32. Write the permutation $\begin{pmatrix} 1 & 2 & 3 & 4 & 5 & 6 & 7 & 8 & 9 & 10 \\ 6 & 5 & 7 & 9 & 2 & 1 & 4 & 8 & 10 & 3 \end{pmatrix}$ as a product of disjoint cycles and as a product of transpositions.
33. Is there an element in S_4 with order 3? With order 4? With any order greater than 4?
34. Express $(2\ 1\ 5) \circ (3\ 1\ 4\ 2)^{-1}$ as the product of disjoint cycles.

Chapter 10

35. How many trailing zeros are in 5! expressed as a binary (base 2) number?
36. How many trailing zeros are in 5! expressed as an octal (base 8) number?
37. Prove that if $\gcd(a, b) = 1$ and $c \mid b$, then $\gcd(a, c) = 1$.
38. Is 2 relatively prime to 2? Is 1 relatively prime to 1?
39. What two primes sum to 20? To 22? To 100?

Chapter 11

40. Let $R = \{(1, 2), (2, 2), (3, 4), (4, 3)\}$ be a relation on the set $S = \{1, 2, 3, 4, 5\}$. Specify the equivalence relation on S with the least number of elements that has all the elements of R.
41. What is the partition that corresponds to the equivalence relation in final exam question 40?
42. Is "has the same cardinality" an equivalence relation?

Chapter 12

43. Determine 6^{1789} mod 7.
44. Determine 10999 mod 9.
45. Prove that if $c \neq 0$ and $ac \equiv bc \pmod{mc}$, then $a \equiv b \pmod{m}$.

Chapter 13

46. Simplify the following product of elements from D_3: $(f)(fr)(fr^2)(f)(r^2)(fr)$.
47. We defined an f motion in D_3 as a flip around the altitude drawn through vertex 1. What elements of D_3 correspond to a flip around the altitude drawn through vertex 2? Through vertex 3? How do you know without taking powers that the order of these elements is 2?
48. Describe the group of symmetries of a rectangle whose length does not equal its width. Is it isomorphic to C_4 or the Klein four group?
49. How many symmetries does the small letter e have?

Chapter 14

50. Is $\{0, 3, 6\}$ under addition modulo 9 a subgroup of $<\mathbb{Z}_9, +_9>$?
51. Show that $H = \{g \in G : g^n = e\}$ is a subgroup of a commutative group G for any positive integer n.
52. Show that under addition the set of integers that yield a remainder of 0 on division by 5 is a subgroup of $<\mathbb{Z}, +>$.

Chapter 15

53. Show that $\langle \mathbb{Z}, + \rangle$ is isomorphic to $\langle 3\mathbb{Z}, + \rangle$.
54. Are all groups whose order is the product of the same two primes isomorphic to each other?

Chapter 16

55. What is the order of a^6 in the cyclic group of order 20 generated by a?
56. How many elements are in the subgroup $\langle 18 \rangle$ of $\langle \mathbb{Z}_{30}, + \rangle$?
57. How many subgroups does the cyclic group of order $p^i q^j$ have if p and q are distinct primes?
58. Find all the generators of $\mathbb{Z}_{30}$.

Chapter 17

59. Is it possible for a group to have multiple subgroups all with the same order? Justify your answer.
60. Show that if a nontrivial cyclic group is simple, then its order is prime.
61. Use a property of Cayley tables for groups to show that $|gG| = |G|$ for all $g \in G$, where G is a group.

Chapter 18

62. How many quotient groups does C_{15} have?
63. $H = \{(1), (1\ 2) \circ (3\ 4), (1\ 3) \circ (2\ 4), (1\ 4) \circ (2\ 3)\}$ is a subgroup of A_4. What are the left cosets of A_4/H? What is the common name for the subgroup H?

Chapter 19

64. Let φ be a homomorphism from G_1 to G_2. Show that $|\varphi(G_1)|$ divides $|G_1|$. *Hint*: $\varphi(G_1)$ is isomorphic to what group?
65. Define a homomorphism from S_n to $\mathbb{Z}_2$.
66. Show that $\varphi: \langle \mathbb{R}, + \rangle \to \langle \mathbb{R}^+, \cdot \rangle$ defined by $\varphi(x) = 10^x$ for all $x \in \mathbb{R}$ is an isomorphism.

Chapter 20

67. Find the kernel of the homomorphism $\varphi: \mathbb{Z}_8 \to \mathbb{Z}_8$ given that φ maps 1 to 6.
68. Is $C_4 \times C_5$ isomorphic to C_{20}? Justify your answer.
69. List all the commutative groups of order 28.
70. Are the groups $\mathbb{Z}_3 \times \mathbb{Z}_8 \times \mathbb{Z}_{10}$ and $\mathbb{Z}_{24} \times \mathbb{Z}_{10}$ isomorphic? Justify your answer.
71. What are the orders of the elements of $\mathbb{Z} \times \mathbb{Z}_5$?
72. For what values of n is Dic_{4n} commutative?
73. Find the multiplicative inverse of every nonzero element in $\mathbb{Z}_7$.
74. Draw the graph of the quaternion group based on the generating set $\{i, j\}$.
75. From the graph in final exam question 74, determine the graph of the quotient group $Q_8/\{1, -1\}$.

Chapter 21

76. Is the set of all the subsets of a nonempty set A a ring under union and intersection? Justify your answer.
77. Is the union of $\{0\}$ and the set of all odd integers a ring under addition and multiplication? Justify your answer.
78. Is $\{a + b\sqrt{2} : a, b \in \mathbb{Z}\}$ under addition and multiplication a ring? If so, is it a commutative ring? Is it a ring with unity?

Chapter 22

79. Is $\{x + iy : x, y \in \mathbb{Z}\}$ an ideal of the complex numbers? Justify your answer.
80. Let $\langle R, +, \cdot \rangle$ be a ring. Let $Z(R)$ be the center of the group $\langle R, + \rangle$ (see review question 6 in chapter 14). Show that $Z(R)$ is a subring of R.
81. Under what circumstances is $\langle m \rangle \cap \langle n \rangle$ the ideal $\langle mn \rangle$ in $\mathbb{Z}$?

Chapter 23

82. Specify a ring homomorphism from $\mathbb{Z}_4$ to $\mathbb{Z}_3$.
83. Specify a ring homomorphism from $\mathbb{Z}$ to $\mathbb{Z}_n$.

84. Does a ring isomorphism exist from $\mathbb{Z}$ to $\mathbb{Q}$?

Chapter 24

85. Does $\mathbb{Z}_{11}$ have any maximal ideals? Justify your answer.
86. Is $6\mathbb{Z}$ a prime ideal of $<\mathbb{Z}, +, \cdot>$? Justify your answer.
87. Let F be the field of quotients of an integral domain D. Show that every nonzero element in D is a unit in F.

Chapter 25

88. Is the set of all constant polynomials in the ring $\mathbb{Q}[x]$ an ideal? Justify your answer.
89. Determine the quotient and remainder when $p_1 = (3, 0, 2, 0, 0, 1, 4)$ is divided by $p_2 = (1, 0, 3, 2)$, where p_1 and p_2 are polynomials over $\mathbb{Z}_5$.
90. Determine the product of $p_1 = (3, 0, 2, 5)$ and $p_2 = (1, 0, 5)$, where p_1 and p_2 are polynomials over $\mathbb{Z}_{11}$.
91. Why can $p_1 = 3x^2$ not be divided by $p_2 = 2x$, where p_1 and p_2 are polynomials over $\mathbb{Z}$?

Chapter 26

92. Describe the steps in the construction of a finite field of order p^{10}, where p is a prime number.
93. Construct the Cayley table for the multiplication operation on the finite field constructed from the irreducible polynomial $x^2 + 1$ over $\mathbb{Z}_3$. Represent the elements of the field with polynomials in tuple form. For example, represent $x + 1$ with 11 (shorthand for (1, 1)).
94. Is $p(x) = x^3 + x^2 + x + 1$ reducible over $\mathbb{Z}_4$? Over $\mathbb{Z}_3$? Justify your answers.

Chapter 27

95. Suppose $\{u, v, x\}$ is an independent set of vectors over a field F. Show that if $au + bv + cx = a'u + b'v + c'x$ for $a, a', b, b', c, c' \in F$, then $a = a'$, $b = b'$, and $c = c'$.
96. Suppose V is a vector space of dimension 5 over $<\mathbb{Z}_3, +, \cdot>$. How many elements are in V?
97. Is the set of vectors $\{(-2, 2, 0), (1, 1, 1), (8, -4, 2)\}$ over $\mathbb{R}$ linearly dependent or independent? Justify your answer.

Chapter 28

98. Is the set of all humans a partially ordered set under the relation "has DNA identical to"?
99. Simplify the Boolean expression $x^c y^c z^c + x^c y^c z + x^c yz + x^c yz^c$.
100. If a Hasse diagram is a chain with no branches, is the represented partial order a lattice?

Answers to the Final Exam

Chapter 1

1. $\{\,\}, \{a\}, \{\{b, c\}\}, \{a, \{b, c\}\}$
2. $\{\{\,\}, \{a\}, \{\{a\}\}, \{a, \{a\}\}\}$
3. Yes, even for infinite sets.
4. 2^3

Chapter 2

5. No. If it is false, then it is true. So it cannot be false. If it is true, then it is false. So it cannot be true. Because it is neither true nor false, it is not a mathematical statement.
6. $(\exists x)(\forall y)(x \geq y)$
7. For all primes p, $2 \leq p$.
8. There exists a math student who does not play baseball.

Chapter 3

9. $\sim p \sim q \sim r \lor \sim p \sim q r \lor \sim p q r \lor p q \sim r$
10. The statement is false (counterexample: $3 + 5 = 8$); the converse is true.
11. Yes. $p \land p$ and p have identical truth tables.
12. Yes. If p is true, then $p \lor (p \land q)$ is true; if p is false, then $p \lor (p \land q)$ is false.

Chapter 4

13. $2^{2(1)} - 1 = 4 - 1 = 3$, which is divisible by 3. Thus, S_1 is true. Assume S_n is true $\Rightarrow 2^{2n} - 1 = 3i$ for some $i \in \mathbb{Z}^+ \Rightarrow 2^{2n} = 3i + 1$. Then $2^{2(n+1)} - 1 = 2^{2n+2} - 1 = 2^{2n}2^2 - 1 = 2^{2n}4 - 1 = (3i + 1)4 - 1 = 12i + 3$, which is divisible by 3. Thus, S_{n+1} is true. ∎
14. Assume x is not even $\Rightarrow x = 2k + 1$ for some $k \in \mathbb{Z} \Rightarrow x + 1 = 2x + 2 = 2(x + 1) \Rightarrow x + 1$ is even. ∎
15. Assume there are some uninteresting positive integers. Then, by the well-ordering principle, there is a smallest uninteresting positive number n. But that makes n interesting. This contradiction implies that our initial assumption—that there are some uninteresting positive numbers—is incorrect.

Chapter 5

16. $[(A^c \cap B^c \cap C^c) \cup (D^c \cap E^c)] \cap F^c$
17. Let S_3 and S_7 be the set of numbers divisible by 3 and 7, respectively. $|S_3 \cap S_7|$ is the number of numbers divisible by both 3 and 7 (it is equal to the number of numbers divisible by 21). Then $|S_3 \cup S_7| = |S_3| + |S_7| - |S_3 \cap S_7| = 333 + 142 - 47 = 428$.

Chapter 6

18. $f(0) = f(-1) \Rightarrow f$ is not one-to-one
19. Let $y \in \mathbb{R}$. If $x = (y - 200)/100$, then $x \in \mathbb{R}$ and $f(x) = y$. ∎
20. Map $d_n \ldots d_0$ to $0.d_0 \ldots d_n$. This injection implies that the cardinality of $\mathbb{Z}^+$ is less than or equal to the cardinality of T.

Chapter 7

21. n^3, n^4
22. Yes. $x \oplus 5 = (x + 5) - 5 = x$; $5 \oplus x = (5 + x) - 5 = x$.
23. $l = l * r$ (because r is a right identity) $= r$ (because l is a left identity).
24. Yes. By the preceding question $l = r$. Thus, r is both a left identity and a right identity, and therefore, an identity.

Chapter 8

25. Because the group is commutative, we can rearrange the order of the elements in xx. Each element in the group that is its own inverse can be paired with itself in xx because each element appears twice in xx. Each other element in the group can be paired with its inverse in x, so the pair is repeated in xx. For example, consider the cyclic rotation group $C_3 = \{e, r, r^2\}$. Then $xx = (err^2)(err^2) = (ee)(rr^2)(rr^2)$. After rearrangement, every pair in xx is the identity. Thus, xx equals the identity.
26. No; $\otimes$ is not associative: $(2 \otimes 3) \otimes 4 = |1 - 4| = 3$, but $2 \otimes (3 \otimes 4) = |2-1| = 1$.
27. $r = e * r = (y * x) * r = y * (x * r) = y * e = y$.
28. $l = l * e = l * (x * r) = (l * x) * r = e * r = r$.
29. Multiplying both sides of $ab = e$ by a^{-1}, we get $b = a^{-1}$. Multiplying both sides by a, we get $ba = a^{-1}a = e$.

Chapter 9

30. $(1\ 2\ 3) \circ (4\ 5\ 6\ 7\ 8) \circ (9\ 10\ 11\ 12)$ has order 60.
31. There are 35 distinct subsets of $\{1, 2, 3, 4, 5, 6, 7\}$ that have three elements. Corresponding to each subset are two permutations. For example, corresponding to $\{1, 2, 3\}$, there are the permutations $(1\ 2\ 3)$ and $(1\ 3\ 2)$. Thus, there are a total of $2 \cdot 35 = 70$ 3-cycles.
32. $(1\ 6) \circ (2\ 5) \circ (3\ 7\ 4\ 9\ 10)$, $(1\ 6) \circ (2\ 5) \circ (3\ 10) \circ (3\ 9) \circ (3\ 4) \circ (3\ 7)$
33. All permutations in S_4 are of one of the following forms: (1) has order 1, $(1\ 2)$ and $(1\ 2) \circ (3\ 4)$ have order 2, $(1\ 2\ 3)$ has order 3, $(1\ 2\ 3\ 4)$ has order 4. No other orders are possible.
34. $(2\ 1\ 5) \circ (3\ 1\ 4\ 2)^{-1} = (2\ 1\ 5) \circ (1\ 3\ 2\ 4) = (1\ 3) \circ (2\ 4\ 5)$

Chapter 10

35. Three, one for each 2 factor.
36. One, one for each triple of 2 factors.
37. $\gcd(a, b) = 1 \Rightarrow ax + by = 1$ for some $x, y \in \mathbb{Z}$; $c\,|\,b \Rightarrow b = kc$ for some $k \in \mathbb{Z}$; substituting we get $ax + (ky)c = 1 \Rightarrow \gcd(a, c) = 1$.
38. $\gcd(2, 2) = 2 \Rightarrow 2$ is not relatively prime to 2; $\gcd(1, 1) = 1 \Rightarrow 1$ is relatively prime to 1.
39. $20 = 7 + 13$, $22 = 5 + 17$, $100 = 47 + 53$. Goldbach's conjecture asserts that every even integer greater than 2 is the sum of two primes. This conjecture has never been proven or disproven.

Chapter 11

40. $\{(1, 1), (2, 2), (3, 3), (4, 4), (5, 5), (1, 2), (2, 1), (3, 4), (4, 3)\}$
41. $\{\{1, 2\}, \{3, 4\}, \{5\}\}$
42. Yes. This follows from the properties of a bijection. Specifically, the identity mapping is a bijection; the inverse of a bijection is a bijection; the composition of two bijections is a bijection. Reflexivity, symmetry, and transitivity, respectively, follow directly from these properties.

Chapter 12

43. $10^{999} \bmod 1 = 1^{999} \bmod 1 = 1 \bmod 9 = 1$
44. $6^{1789} \bmod 7 = (-1)^{1789} \bmod 7 = -1 \bmod 7 = 6$
45. $c \neq 0$ and $ac \equiv bc \pmod{mc} \Rightarrow ac = bc + mck$ for some $k \in \mathbb{Z}$. Canceling the c's, we get $a = b + mk \Rightarrow a \equiv b \pmod{m}$.

Chapter 13

46. $(f)(fr)(fr^2)(f)(r^2)(fr) = f$
47. fr^2, fr. Two successive flips of the same kind is the identity motion.
48. Flip around horizontal axis that bisects the rectangle, vertical axis that bisects the rectangle, 180° rotation around perpendicular axis through center, and the identity motion. Klein four group (because the flips and rotation all have order 2).
49. One: the identity motion.

Chapter 14

50. Yes, by the CF subgroup test—it is finite and closed: $3 +_9 3 = 6$, $3 +_9 6 = 6 +_9 3 = 0$, $6 +_9 6 = 3$.
51. *Closure*: Let $x, y \in H$. Then $x^n = y^n = e$. $e = ee = x^n y^n = (xy)^n$ (by commutativity) $\Rightarrow xy \in H$. *Inverses*: Let $x \in H$. Then $x^n = e$. $(x^{-1})^n = (x^n)^{-1} = e^{-1} = e \Rightarrow x^{-1} \in H$.
52. *Closure*: Suppose $x \bmod 5 = y \bmod 5 = 0 \Rightarrow x = 5i, y = 5j$ for some $i, j \in \mathbb{Z} \Rightarrow x + y = 5i + 5j = 5(i+j) \Rightarrow (x+y) \bmod 5 = 0$. *Inverses*: $-x = -5i = 5(-i) \Rightarrow (-x) \bmod 5 = 0$.

Chapter 15

53. Let $\varphi: \mathbb{Z} \to 3\mathbb{Z}$ be the function defined by $\varphi(x) = 3x$ for all $x \in \mathbb{Z}$. Let $y \in 3\mathbb{Z} \Rightarrow y = 3k$ for some $k \in \mathbb{Z} \Rightarrow \varphi(k) = 3k = y \Rightarrow \varphi$ is onto. Let $\varphi(x) = \varphi(y) \Rightarrow 3x = 3y \Rightarrow x = y \Rightarrow \varphi$ is one-to-one. $\varphi(x + y) = 3(x + y) = 3x + 3y = \varphi(x) + \varphi(y) \Rightarrow \varphi$ is operation preserving.
54. No. Counterexample: $|C_6| = |D_3| = 2 \cdot 3$, but C_6 and D_3 are not isomorphic.

Chapter 16

55. $20/\gcd(6, 20) = 20/2 = 10$
56. $30/\gcd(18, 30) = 30/6 = 5$
57. $(i + 1) \cdot (j + 1)$
58. Any element not relatively prime to 30: 7, 11, 13, 17, 19, 23, 29.

Chapter 17

59. Yes. V_4 has three subgroups of order 2.
60. Suppose G is a simple nontrivial cyclic group of order n. By frame **952**, if $k \mid n$, then there is a subgroup of H order k. Moreover, H is normal because G is commutative. If $1 < k < n$, H is neither the trivial subgroup nor G. But then G is not simple. Thus, there is no k such that $k \mid n$ and $1 < k < n \Rightarrow n$ is prime.
61. gH is the g row of the Cayley table for H. It is a permutation of the column labels (which are the element of H). Thus, $|gH| = |H|$.

Chapter 18

62. C_{15} is commutative $\Rightarrow$ all its subgroups are normal $\Rightarrow C_{15}$ has a quotient group for each subgroup: $\{e\}$, $\{e, r^5, r^{10}\}$, $\{e, r^3, r^6, r^9, r^{12}\}$ and C_{15}, where r is the generator of C_{15}.
63. H, $\{(1\ 2\ 3), (1\ 3\ 4), (2\ 4\ 3), (1\ 4\ 2)\}$, $\{(1\ 3\ 2), (2\ 3\ 4), (1\ 2\ 4), (1\ 4\ 3)\}$. H is the Klein four group.

Chapter 19

64. $|G_1/\ker \varphi| \cdot |\ker \varphi| = |G_1| \Rightarrow |G_1/\ker \varphi|$ divides $|G_1|$. We also know that $G_1/\ker \varphi \cong \varphi(G_1) \Rightarrow |\varphi(G_1)| = |G_1/\ker \varphi|$. Thus, $|\varphi(G_1)|$ divides $|G_1|$.
65. $\varphi(x) = 0$ if x is an even permutation; $\varphi(x) = 1$ if x is an odd permutation.
66. *Onto*: Let $y \in \mathbb{R}+$. Then $\log y \in \mathbb{R}$ and $\varphi(\log y) = 10^{\log y} = y$. *One-to-one*: $\varphi(x) = \varphi(x) \Rightarrow 10^x = 10^y \Rightarrow \log(10^x) = \log(10^y) \Rightarrow x = y$. *Operation preserving*: $\varphi(x + y) = 10^{x+y} = 10^x \cdot 10^y = \varphi(x) \cdot \varphi(y)$. Using φ, we can determine the sum of x and y by multiplying $\varphi(x)$ and $\varphi(y)$, and then mapping the product back to the desired sum using the inverse isomorphism φ^{-1}. For example, to compute $1 + 2$, using φ we map 1 and 2 to get 10 and 100. We multiple 10 and 100 to get 1000. We then map 1000 to 3 using the inverse isomorphism φ^{-1}. We, of course, would normally not want to do this: It transforms an easy problem (adding two numbers) to a harder problem (multiplying two numbers).

Chapter 20

67. $\varphi(0) = 0$
    $\varphi(1) = 6$
    $\varphi(2) = \varphi(1 +_8 1) = \varphi(1) +_8 \varphi(1) = 6 +_8 6 = 4$
    $\varphi(3) = \varphi(2 +_8 1) = \varphi(2) +_8 \varphi(1) = 4 +_8 6 = 2$
    $\varphi(4) = \varphi(3 +_8 1) = \varphi(3) +_8 \varphi(1) = 2 +_8 6 = 0$
    $\varphi(5) = \varphi(4 +_8 1) = \varphi(4) +_8 \varphi(1) = 0 +_8 6 = 6$
    $\varphi(6) = \varphi(5 +_8 1) = \varphi(5) +_8 \varphi(1) = 6 +_8 6 = 4$

$\varphi(7) = \varphi(6 +_8 1) = \varphi(6) +_8 \varphi(1) = 4 +_8 6 = 2$
ker $\varphi = \{0, 4\}$
68. Yes, because gcd(4, 5) = 1.
69. $C_2 \times C_2 \times C_7, C_4 \times C_7$
70. Yes. $\mathbb{Z}_3 \times \mathbb{Z}_8 \cong \mathbb{Z}_{24}$ because gcd(3, 8) = 1.
71. (0, 0) has order 1, (0, 1), (0, 2), (0, 3), and (0, 4) all have order 5. All other elements have infinite order.
72. In Dic_{4n}, $y^{-1}xy = x^{-1} \Rightarrow xy = yx^{-1}$. In Dic_{4n}, $x^{2n} = e \Rightarrow x = x^{-1}$ if and only if $n = 1$. Thus, only Dic_4 is commutative.
73. $1 \cdot_7 1 = 1, 2 \cdot_7 4 = 1, 3 \cdot_7 5 = 1, 6 \cdot_7 6 = 1.$ $1^{-1} = 1, 2^{-1} = 4, 3^{-1} = 5, 4^{-1} = 2, 5^{-1} = 3, 6^{-1} = 6$
74. 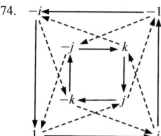 Key: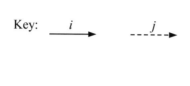

75. $Q_8/\{1, -1\}$ is V_4.

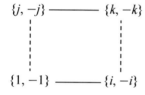 Key: $\underline{\{i, -i\}}$ $\underline{\{j, -j\}}$

Chapter 21

76. No. Under union and intersection, the inverses property does not hold.
77. No. It is lacking closure under addition.
78. It is a commutative ring with unity.

Chapter 22

79. No, it does not have the superclosure property: $(0.5 + 0i)(1 + 1i) = 0.5 + 0.5i \notin \{x + iy : x, y \in \mathbb{Z}\}$.
80. $Z(R)$ is a subgroup of $<R, +>$ by review question 6 in chapter 14. *Closure under multiplication*: Let $a, b \in Z(R)$ and $x \in R$. Then $abx = axb$ (because $b \in Z(R)$) $= xab$ (because $a \in Z(R)$).
81. gcd(m, n) = 1

Chapter 23

82. $\varphi(x) = 0$ for all $x \in \mathbb{Z}_4$
83. $\varphi(x) = x$ mod n
84. No. In $\mathbb{Q}$, every nonzero element has a multiplicative inverse, but in $\mathbb{Z}$, only 1 and -1 have multiplicative inverses.

Chapter 24

85. Yes. By Lagrange's theorem, the only proper subgroup of $\mathbb{Z}_{11}$ is $\{0\}$. Thus, $\{0\}$ is maximal.
86. No. It does not have the $oo = o$ property: $2, 3 \notin <6>$, but $2 \cdot 3 = 6 \in <6>$.
87. Let $\frac{a}{b}$ be an element of the field of quotients. $\frac{a}{b} \neq 0 \Rightarrow a \neq 0 \Rightarrow \frac{b}{a}$ is also in the field of quotients. $(\frac{a}{b})(\frac{b}{a}) = 1 \Rightarrow \frac{a}{b}$ is a unit.

Chapter 25

88. No. It does not have the superclosure property. Counterexample: if c is a constant, then cx is not a constant.

89.
```
              3 0 3 4
      1032)3 0 2 0 0 1 4
            3 0 4 1
            ─────
              3 4 0 1
              3 0 4 1
              ─────
                4 1 0 4
                4 0 2 3
                ─────
                  1 3 1
```

90.
```
                    1    0    5
              3     0    2    5
              ──────────────
                    5    0    3
              2     0   10
         0    0     0
    3    0    4
    ───────────────────────────
    3    0    6    5   10    3
```

91. There is no integer a such that $(2x)(ax) = 3x^2$.

Chapter 26

92. Find an irreducible monic polynomial $f(x)$ of degree 10 over $\mathbb{Z}_p$. Then $\mathbb{Z}_p[x]/f(x)$ is the desired field.

93.

·	00	01	02	10	11	12	20	21	22
00	00	00	00	00	00	00	00	00	00
01	00	01	02	10	11	12	10	21	22
02	00	02	01	20	22	21	10	12	22
10	00	10	20	02	12	22	01	11	21
11	00	11	22	12	20	01	21	02	10
12	00	12	21	22	01	10	11	20	02
20	00	20	10	01	21	11	02	22	12
21	00	21	12	11	02	20	22	10	01
22	00	22	11	21	10	02	12	01	20

Chapter 27

94. $\mathbb{Z}_4$ case: $p(1) = 0 \Rightarrow$ reducible. $\mathbb{Z}_3$ case: $p(0) = 1$, $p(1) = 1$, $p(2) = 0 \Rightarrow$ reducible.

95. $a\mathbf{u} + b\mathbf{v} + c\mathbf{x} = a'\mathbf{u} + b'\mathbf{v} + c'\mathbf{x} \Rightarrow (a - a')\mathbf{u} + (b - b')\mathbf{v} + (c - c')\mathbf{x} = 0$. Because $\{\mathbf{u}, \mathbf{v}, \mathbf{x}\}$ is an independent set, $a - a' = 0$, $b - b' = 0$, $c - c' = 0 \Rightarrow a = a'$, $b = b'$, $c = c'$.

96. Each vector is of the form $a\mathbf{v}_1 + b\mathbf{v}_2 + c\mathbf{v}_3 + d\mathbf{v}_4 + e\mathbf{v}_5$, where $a, b, c, d, e \in \mathbb{Z}_3$. There are three choices for each of these coefficients $\Rightarrow$ there are 3^5 vectors, all of which are distinct by final exam question 95.

97. $-3(-2, 2, 0) + 2(1, 1, 1) + (-1)(8, -4, 2) = (0, 0, 0) \Rightarrow$ linearly dependent.

Chapter 28

98. No. Identical twins have identical DNA. Thus, the relation is not antisymmetric.
99. x^c
100. Yes. In each pair, one element is the lub, and the other element is the glb.

Index (with page numbers)

A

absorption by coset 191
absorption law 49, 333
absurdity 28
additive notation 85
algorithm 109
antecedent 23
associative law 49
automorphism 177, 215

B

biconditional 24
binary operation
    additive inverse 78
    associative 75
    commutative 75
    definition 73
    identity 76
    inverse 77
    juxtaposition 74
    left identity 76
    multiplicative inverse 78
    operand 74
    operator 74
    right identity 76
block 129
Boolean algebra 336
bound variable 15

C

$C(g)$ 172
cancellation law 138
    for groups 87
    for integral domains 272
    for rings 247
Cantor, Georg 71
Cayley table 74, 79
Cayley's theorem 175
centralizer 172
common divisor 111
commutative law 47, 49
complement 9, 45, 335
complement law 49
composite number 113
composition 60
compound statement 18
congruence 133
    cancellation law 138
    classes 136
    modulus 133
    quotient group 140
    replacement rules 137
congruence modulo n 133

conjunction 18
consequent 23
contradiction 28
contrapositive 29
converse 28
coset 305
    absorption rule in groups 191
    absorption rule in rings 259
    addition 258, 263
    decomposition 190, 258
    generalized absorption rule 259
    Lagrange's theorem 192
    left 189
    multiplication 200, 258, 263
    properties 189
    representative 189
    right 189
cube 225
cyclic groups 178
    order of r^k 182
    order of subgroup 186
    subgroups 184

D

DeMorgan's law 49
dicyclic group 156
digraph 53
direct product 219
directed graph 53
disjoint 46
distributive law 49
dividend 107
divides 33
divisible by 33
division algorithm
    for integers 107
    for polynomials 298
division ring 296
divisor 33
divisor of zero 271
dodecahedron 225
domination law 49

E

ellipsis 3
equivalence 25
Euclid's lemma 114
Euclidean algorithm 112
even number 31
exactly one 36
exclusive or 19
existential quantifier 12
extension 293

F

factor 33, 298
factor theorem 298
field 274
    constructing 313
    definition 274
    fundamental theorem of field theory 317
    Galois field 274
    of quotients 281
flip-rotate commutation rule 158
follow-the-arrows 155
fourth roots of unity 145
free variable 15
FTGH 214
FTOA 115
function 55
    bijection 58
    bijection in table form 66
    codomain 55
    composition 60
    domain 55
    function on a set 59
    image 57
    injection 58
    inverse 62
    one-to-one correspondence 58
    one-to-one function 58
    onto function 57
    postfix notation 61
    preimage 57
    proving one-to-one 60
    proving onto 59
    representing with tables 65
    strictly increasing 71
    surjection 57
fundamental theorem of arithmetic 115
fundamental theorem of finite Abelian groups 222
fundamental theorem of group homomorphisms 210
fundamental theorem of ring homomorphisms 269

G

gcd 111
generating set 153, 168
glb 331
greatest common divisor 111
group
    A_4 230
    A_5 230
    Abelian 85
    alternating 103
    automorphism 177, 215
    C_1 151
    C_2 150
    C_3 149
    C_4 150
    C_5 150
    cancellation law 87
    center of 171, 205
    C_n 150
    commutative 85
    cyclic 178
    cyclic rotation group 149
    D_3 159
    definition 84
    dicyclic group 232
    dihedral group 156
    finite 85
    graph 153, 203
    homomorphism 206
    infinite 85
    isomorphism 144, 173, 206
    Klein four group 90, 92, 160, 176, 205, 340
    of integers modulo n 144
    order 85
    properties of Cayley tables 90
    quaternion 230
    quaternion isomorphsim 234
    quotient 258
    S_4 230
    S_5 230
    simple 216, 230
    S_n 94
    socks-shoes property 91
    solving equations 89
    symmetric 94, 153
    tetrahedral group 225
    uniqueness of identity 85
    uniqueness of inverse 85
    with order ≤ 12 234

H

homomorphism
    fundamental theorem 206, 268
    group 206
    identity mapping rule 207
    inverse mapping rule 207
    kernel 210, 267
    ring 267

I

icosahedron 225
ideal 260
    anti-superclosure 260
    generated by polynomial 304
    maximal 279, 312
    prime 276
    principal 261, 302
    SS ideal test 261
    superclosure 260
idempotent 92
identity law 49
implication 23
    antecedent 23
    consequent 23
    definitions that use 25
infix notation 54
inherit 92, 165
integral domain 254

characteristic of 273
definition 271
inverse 62
irrational number 3
isomorphism 145
    cyclic groups 181
    group 173, 206
    ring 245

K

kernel 210, 267

L

Lagrange's theorem 192
    consequences 196
lattice 330
    bounded 334
    complemented 335
    greatest lower bound 331
    join 332
    least upper bound 331
    meet 332
    upper bound 330
lcm 105
least common multiple 105, 120
left distributive law 237
logarithm 209
logical operator
    and 18
    biconditional 24
    exclusive or 19
    implication 23
    or 18
lub 331

M

mod 109, 133
modular arithmetic 142
    multiplication 143
multiple 33, 120
multiplicative notation 85

N

n-ary operation 74
natural number 3
negating
    compound statements 26
    predicates 10
    quantified statements 13
    statements 9
nilpotent 255
n-tuple 74, 286

O

octahedron 225
odd number 31
oo = o property 278

operation preserving 174
order
    of a group 104
    of an element 104, 168, 182, 241
    of element in direct product 220
    of r^k 182

P

partial order 329
    Hasse diagram 330
    largest element 334
    smallest element 334
partition 129
permutation 94
    cycle representation 95
    even 102
    odd 102
    standard cycle form 97
    transposition 100
pigeonhole principle 42
PIR 262
polynomial 286
    addition and multiplication 287
    adjoining x to a ring 293
    coefficient 291
    constant 286
    degree 286
    determining degree 289
    division 294
    divisor 298
    factor theorem 298
    indeterminate 291
    irreducible 307, 312
    leading term 286
    like terms 292
    monic 286, 309
    n-tuple representation 286
    powers-of-x representation 291
    product 287
    product diagram 287
    reducibililty tests 308
    reducible 307
    remainder theorem 298
    root 298
    term 291
    zero 286
power set 7
predicate 10
prime ideal 278
prime number 113
principal ideal domain 304
principal ideal ring 262
principle of inclusion-exclusion 50
proof 31
    biconditional 35
    by cases 41
    by contradiction 34
    by induction 39
    contrapositive 35
    corollary 114
    counterexample 76
    direct proof 31

disjunction...36
divisibility...109
lemma...114
one-to-one function...60
onto function...59
pigeonhole principle...42
two sets are equal...48
using the well-ordering principle...37
without loss of generality...42
proposition...9

Q

quantifier...11
    bound variable...15
    domain of quantification...11
    existential quantifier...12
    free variable...15
    multiple quantifiers...15
    universal quantifier...12
    universe...11
quasi-commutative law...199
quotient...107
quotient structure...131, 141

R

rational number...3
real number...3
reducibility test for degrees > 1...309
reducibility test for degrees 2 and 3...308
regular polygon...149
regular polyhedra...225
relation...52, 125, 231
    antisymmetric...131, 329
    arrow diagram...53
    defining set of...231
    directed graph...53
    equivalence relation...128
    implies...231
    infix notation...54
    inverse...54
    on A...53
    reflexive...125
    reflexive closure...126
    relational operator...54
    stronger...231
    symmetric...126
    symmetric closure...127
    transitive...127
    transitive closure...127
relatively prime...119
relatively prime divisor theorem...119
remainder...107
remainder operator...109, 133
remainder theorem...298
right distributive law...237
ring...237
    additive identity...237
    basic properties...238
    cancellation law...247

characteristic of...241, 273
commutative...237
definition...237
direct sum...249
division...296
divisors of zero...246
extension...284
field...254, 274
generalized distributive law...241
homomorphism...267
identity...237
identity-mapping rule...267
integral domain...254, 271
isomorphism...245
multiplicative identity...237
quotient...263
quotient ring...260
skew field...296
subring...256
terminology...237
trivial ring...244
unit...247, 274, 276
unity...237
with unity...237

S

self-loop...125
semigroup...78, 236
set...1
    cardinality...68
    closure...73
    complement...45
    countable...69
    countably infinite...69
    disjoint...46
    empty...2
    intersection...45
    pairwise disjoint...46
    partially ordered set...330
    power set...6
    proper subset...6
    proving two sets are equal...48
    set difference...46
    set-builder notation...3
    subset...5
    superset...5
    symmetric difference...50, 251
    uncountable...69
    union...45
    universal...4, 45
    Venn diagram...45
    well defined...1
skew field...296
smallest positive multiplier theorem...123
S_n...94
square root...113
statement...9
subfield...323
subgroup
    CF subgroup test...169

CI subgroup test 167
conjugate 172
cyclic 181
defintion 165
generated by a set 168
normal 194
of direct product 223
one-step subgroup test 170
proper 165
trivial 165
subring
definition 256
properties 257
SC test 256
switching-sides rule 89
symmetric difference 50
symmetry 147

T

tautology 28
ternary operation 74
tetrahedron 225
theorem 32
transitivity law 41
transposition 100
truth table 18
truth table to statement 19

U

$U(n)$ 119
unary operation 74
universal quantifier 12

V

vacuously true 126
variable 10
vector space
basis 326
definition 320
dimension 327
examples 322
finite dimensional 327
infinite dimensional 327
linear combination of vectors 325
linear independence 325
minimal spanning set 327
NCC subspace test 324
scalar 320
scalar multiplication 320
subspace 324
vector 320
Venn diagram 45

W

well defined 139, 200, 264
well-ordering principle 37, 38
word 231

Z

$Z(G)$ 171, 205
$\mathbb{Z}_n$ 144, 217

Made in the USA
Middletown, DE
11 November 2018